# Statistical Inference, Econometric Analysis and Matrix Algebra

Bernhard Schipp • Walter Krämer
Editors

# Statistical Inference, Econometric Analysis and Matrix Algebra

## Festschrift in Honour of Götz Trenkler

Springer

*Editors*

Prof. Dr. Bernhard Schipp
TU Dresden
Fakultät Wirtschaftswissenschaften
Professur für Quantitative Verfahren, insb.
Ökonometrie
Mommsenstr. 12
01062 Dresden
Germany
bernhard.schipp@tu-dresden.de

Prof. Dr. Walter Krämer
TU Dortmund
Fakultät Statistik
Lehrstuhl für Wirtschafts-
und Sozialstatistik
Vogelpothsweg 78
44221 Dortmund
Germany
walterk@statistik.tu-dortmund.de

ISBN: 978-3-7908-2577-0 e-ISBN: 978-3-7908-2121-5



*Cover design*: WMXDesign GmbH, Heidelberg

Printed on acid-free paper

9 8 7 6 5 4 3 2 1

springer.com

*Prof. Dr. Götz Trenkler*

# Preface

This Festschrift is dedicated to Götz Trenkler on the occasion of his 65th birthday.

As can be seen from the long list of contributions, Götz has had and still has an enormous range of interests, and colleagues to share these interests with. He is a leading expert in linear models with a particular focus on matrix algebra in its relation to statistics. He has published in almost all major statistics and matrix theory journals. His research activities also include other areas (like nonparametrics, statistics and sports, combination of forecasts and magic squares, just to mention a few).

Götz Trenkler was born in Dresden in 1943. After his school years in East Germany and West-Berlin, he obtained a Diploma in Mathematics from Free University of Berlin (1970), where he also discovered his interest in Mathematical Statistics. In 1973, he completed his Ph.D. with a thesis titled: *On a distance-generating function of probability measures*. He then moved on to the University of Hannover to become Lecturer and to write a habilitation-thesis (submitted 1979) on alternatives to the Ordinary Least Squares estimator in the Linear Regression Model, a topic that would become his predominant field of research in the years to come.

In 1983 Götz Trenkler was appointed Full Professor of Statistics and Econometrics at the Department of Statistics at the University of Dortmund, where he continues to teach and do research until today. He served as dean of the department from 1987 to 1990 and declined an offer from Dresden University of Technology in 1993. He has been visiting Professor at the University of California at Berkeley, USA, and the University of Tampere, Finland, and is a regular contributor to international conferences on matrix methods in statistics. Currently, he is the Coordinating Editor of *Statistical Papers*, Associate Editor of several other international journals and recently the twice-in-a-row recipient of the best-teacher-award of the department.

Among Götz Trenkler's extracurricular activities are tennis, chess and the compilation of a unique collection of *Aphorisms in Statistics*, samples of which can be found at the beginning of the chapters of this book. He certainly would do the scientific community a great service by having them published at some time.

The editors are grateful to all contributors, many of whom are not only scientific colleagues but also his personal friends.

We express our appreciation for editorial and LaTeX-assistance to Sabine Hegewald, and in particular to Matthias Deutscher, who managed to edit successfully almost 30 manuscripts that were characterized by a great variety of individual preferences in style and layout, and to Alice Blanck and Werner A. Müller from Springer Publishing for their support.

Dresden and Dortmund
July 2008

*Bernhard Schipp*
*Walter Krämer*

# Contents

# List of Contributors

**S. Ejaz Ahmed** Department of Mathematics and Statistics, University of Windsor, 401 Sunset Avenue, Windsor, ON, Canada N9B 3P4, seahmed@uwindsor.ca

**Oskar Maria Baksalary** Faculty of Physics, Adam Mickiewicz University, ul. Umultowska 85, 61-614 Poznań, Poland, baxx@amu.edu.pl

**Björn Bornkamp** Fakultät Statistik, Technische Universität Dortmund, 44221 Dortmund, Germany, bornkamp@statistik.tu-dortmund.de

**Herbert Büning** Freie Universität Berlin, 14195 Berlin, Germany, Herbert.Buening@fu-berlin.de

**Ka Lok Chu** Department of Mathematics, Dawson College, 3040 ouest, rue Sherbrooke, Westmount, QC, Canada H3Z 1A4, ka.chu@mail.mcgill.ca

**Richard William Farebrother** 11 Castle Road, Bayston Hill, Shrewsbury SY3 0NF, UK, R.W.Farebrother@Manchester.ac.uk

**Roland Fried** Fakultät Statistik, Technische Universität Dortmund, 44221 Dortmund, Germany, fried@statistik.tu-dortmund.de

**Arno Fritsch** Fakultät Statistik, Technische Universität Dortmund, 44221 Dortmund, Germany, arno.fritsch@statistik.tu-dortmund.de

**Ursula Gather** Fakultät Statistik, Technische Universität Dortmund, 44221 Dortmund, Germany, gather@statistik.tu-dortmund.de

**Jürgen Groß** Carl von Ossietzky Universität Oldenburg, Fakultät V, Institut für Mathematik, 26111 Oldenburg, Germany, j.gross@uni-oldenburg.de

**Christoph Hanck** Department Quantitative Economics, Universiteit Maastricht, 6211 LM Maastricht, The Netherlands, c.hanck@ke.unimaas.nl

**Joachim Hartung** Fakultät Statistik, Technische Universität Dortmund, 44221 Dortmund, Germany, hartung@statistik.tu-dortmund.de

**Wilfried Hazod** Fakultät für Mathematik, Technische Universität Dortmund, 44221 Dortmund, Germany, wilfried.hazod@mathematik.tu-dortmund.de

**Christian Heumann** Institut für Statistik, Universität München, 80799 München, Germany, christian.heumann@stat.uni-muenchen.de

**Rodrigo Herrera** Fakultät Wirtschaftswissenschaften, Technische Universität Dresden, 01062 Dresden, Germany, rherrera@gmx.net

**Ludwig A. Hothorn** Institut für Biostatistik, Leibniz Universität Hannover, D-30419 Hannover, Germany, hothorn@biostat.uni-hannover.de

**Jeffrey J. Hunter** Institute of Information & Mathematical Sciences, Massey University, Private Bag 102-904, North Shore Mail Centre, Auckland 0754, New Zealand, j.hunter@massey.ac.nz

**Katja Ickstadt** Fakultät Statistik, Technische Universität Dortmund, 44221 Dortmund, Germany, ickstadt@statistik.tu-dortmund.de

**George G. Judge** University of California-Berkeley, 207 Giannini Hall, Berkeley, CA 94720, judge@are.berkeley.edu

**Guido Knapp** Fakultät Statistik, Technische Universität Dortmund, 44221 Dortmund, Germany, knapp@statistik.tu-dortmund.de

**Walter Krämer** Fakultät Statistik, TU Dortmund, 44221 Dortmund, Germany, walterk@statistik.uni-dortmund.de

**Joachim Kunert** Fakultät Statistik, Technische Universität Dortmund, 44221 Dortmund, Germany, kunert@statistik.tu-dortmund.de

**Oliver Kuss** Institut für Medizinische Epidemiologie, Biometrie und Informatik, Martin-Luther-Universität Halle-Wittenberg, 06097 Halle (Saale), Germany, oliver.kuss@medizin.uni-halle.de

**Sandra Leissen** Fakultät Statistik, Technische Universität Dortmund, 44221 Dortmund, Germany, leissen@statistik.tu-dortmund.de

**Uwe Ligges** Fakultät Statistik, Technische Universität Dortmund, 44221 Dortmund, Germany, ligges@statistik.tu-dortmund.de

**Antti Liski** Tampere University of Technology, Tampere, Finland, Antti.Liski@tut.fi

**Erkki P. Liski** University of Tampere, Tampere, Finland, Erkki.Liski@uta.fi

**Shuangzhe Liu** Faculty of Information Sciences and Engineering, University of Canberra, Canberra ACT 2601, Australia, Shuangzhe.Liu@canberra.edu.au

**R.J. Martin** Wirksworth, DE4 4EB, UK, r.j.martin@sheffield.ac.uk

**Ron C. Mittelhammer** School of Economic Sciences, Washington State University, 101C Hulbert Hall, Pullman, WA 99164-6210, mittelha@wsu.edu

**Oliver Morell** Fakultät Statistik, Technische Universität Dortmund, 44221 Dortmund, Germany, morell@statistik.tu-dortmund.de

**Thomas Mühlenstädt** Fakultät Statistik, Technische Universität Dortmund, 44221 Dortmund, Germany, muehlens@statistik.tu-dortmund.de

**Markus Neuhäuser** Fachbereich Mathematik und Technik, RheinAhr-Campus Remagen, 53424 Remagen, Germany, neuhaeuser@rheinahrcampus.de

**Sévérien Nkurunziza** Department of Mathematics and Statistics, University of Windsor, 401 Sunset Avenue, Windsor, ON, Canada N9B 3P4, severien@uwindsor.ca

**Ingram Olkin** Department of Statistics, Sequoia Hall, 390 Serra Mall, Stanford University, Stanford, CA 94305-4065, USA, iolkin@stat.Stanford.EDU

**Tomasz Owsiany** Institute of Mathematics, University of Rzeszów, Al. Rejtana 16 A, 35-959 Rzeszów, Poland, towsiany@wp.pl

**D.S.G. Pollock** Department of Economic, University of Leicester, Leicester LE1 7RH, UK, d.s.g.pollock@le.ac.uk

**Simo Puntanen** Department of Mathematics and Statistics, University of Tampere, 33014 Tampere, Finland, Simo.Puntanen@uta.fi

**Jörg Rahnenführer** Fakultät Statistik, Technische Universität Dortmund, 44221 Dortmund, Germany, rahnenfuehrer@statistik.tu-dortmund.de

**Sabine Rothe** Fakultät Statistik, Technische Universität Dortmund, 44221 Dortmund, Germany, srothe@statistik.tu-dortmund.de

**Philip Sander** Universität Hamburg, Institut für Statistik und Ökonometrie, Von-Melle-Park 5, 20146 Hamburg, Germany, philip.sander@gmx.de

**Razaw al Sarraj** Department of Energy and Technology, Box 7032, 750 07 Uppsala, Sweden, Razaw.Al-Sarraj@etsm.slu.se

**Bernhard Schipp** Fakultät Wirtschaftswissenschaften, Technische Universität Dresden, 01062 Dresden, Germany, bernhard.schipp@tu-dresden.de

**Karsten Schmidt** Fakultät Wirtschaftswissenschaften, Fachhochschule Schmalkalden, 98574 Schmalkalden, Germany, kschmidt@fh-sm.de

**Shalabh** Department of Mathematics & Statistics, Indian Institute of Technology Kanpur, Kanpur 208016, India, shalab@iitk.ac.in, shalabh1@yahoo.com

**Peter Stahlecker** Universität Hamburg, Institut für Statistik und Ökonometrie, Von-Melle-Park 5, 20146 Hamburg, Germany, peter.stahlecker@uni-hamburg.de

**Czesław Stępniak**, Institute of Mathematics, University of Rzeszów, Al. Rejtana 16 A, 35-959 Rzeszów, Poland, cees@univ.rzeszow.pl

**George P.H. Styan** Department of Mathematics and Statistics, McGill University, 1005-805 ouest, rue Sherbrooke, Montréal QC, Canada H3A 2K6, styan@math.mcgill.ca

**Lauri Tarkkonen** Department of Mathematics and Statistics, PO Box 54, University of Helsinki, 00014 Helsinki, Finland, Lauri.Tarkkonen@helsinki.fi

**Helge Toutenburg** Institut für Statistik, Universität München, 80799 München, Germany, toutenb@stat.uni-muenchen.de

**Dietrich Trenkler** Fachbereich Wirtschaftswissenschaften, Universität Osnabrück, 49069 Osnabrück, Germany, Dietrich.Trenkler@Uni-Osnabrück.de

**Kimmo Vehkalahti** Department of Mathematics and Statistics, PO Box 68, University of Helsinki, 00014 Helsinki, Finland, Kimmo.Vehkalahti@helsinki.fi

**Dietrich von Rosen** Department of Energy and Technology, Box 7032, 750 07 Uppsala, Sweden, dietrich.von.rosen@et.slu.se

**Claus Weihs** Fakultät Statistik, Technische Universität Dortmund, 44221 Dortmund, Germany, weihs@statistik.tu-dortmund.de

**Hans Joachim Werner** Wirtschaftswissenschaftlicher Fachbereich, Statistische Abteilung, Universität Bonn, 53113 Bonn, Germany, hjw.de@uni-bonn.de

# Part I
# Nonparametric Inference

*It is proven that the celebration of birthdays is healthy. Statistics show that those people who celebrate the most birthdays become the oldest.*

S. den Hartog

# Adaptive Tests for the c-Sample Location Problem

Herbert Büning

**Abstract** This paper deals with the concept of adaptive tests and with an application to the $c$-sample location problem. Parametric tests like the ANOVA F-tests are based on the assumption of normality of the data which is often violated in practice. In general, the practising statistician has no clear idea of the underlying distribution of his data. Thus, an adaptive test should be applied which takes into account the given data set. We use the concept of Hogg [21], i.e. to classify, at first, the unknown distribution function with respect to two measures, one for skewness and one for tailweight, and then, at the second stage, to select an appropriate test for that classified type of distribution. It will be shown that under certain conditions such a two-staged adaptive test maintains the level. Meanwhile, there are a lot of proposals for adaptive tests in the literature in various statistical hypotheses settings. It turns out that all these adaptive tests are very efficient over a broad class of distributions, symmetric and asymmetric ones.

## 1 Introduction

In the parametric case of testing hypotheses the efficiency of a test statistic strongly depends on the assumption of the underlying distribution of the data, e.g. if we assume normality then optimal tests are available for the one- two- and c-sample location or scale problem such as t-tests, F-tests and Chi-square-tests. In the nonparametric case the distribution of the test statistic is not based on a special distribution of the data like the normal, only the assumption of continuity of the distribution is needed in general. It is well known, however, that the efficiency of nonparametric tests depends on the underlying distribution, too, e.g. the Kruskal–Wallis test in the c-sample location problem has high power for symmetric and medium- up to long-tailed distributions in comparison to its parametric and nonparametric competitors whereas the Kruskal–Wallis test can be poor for asymmetric distributions.

Herbert Büning
Freie Universität Berlin, D-14195 Berlin, Germany
Herbert.Buening@fu-berlin.de

But for the practising statistician it is more the rule rather than the exception that he has no clear idea of the underlying distribution of his data. Consequently, he should apply an adaptive test which takes into account the given data set.

At present, we register a lot of papers on adaptive tests in the literature, concerning one-, two- and c-sample location or scale problems with two-sided and one-sided ordered alternatives as well as umbrella alternatives.

Most of these adaptive tests are based on the concept of Hogg [21], that is, to classify, at first, the type of the underlying distribution with respect to some measures like tailweight and skewness and then to select an appropriate rank test for the classified type of distribution. It can be shown that this two-staged test procedure is distribution-free, i.e. it maintains the level over the class of all continuous distribution functions.

In our paper Hogg's concept of adaptive tests is presented and demonstrated by a real data set. Adaptive tests are generally not the best ones for a special distribution but mostly second best whereas the parametric competitors are poor in many cases. That is just the philosophy of an adaptive test to select the best one for a given data set. It works in the sense of "safety first" principle. For clarity of exposition we confine our attention to the $c$-sample location problem. A power comparison by means of Monte Carlo simulation shows that the adaptive test is very efficient over a broad class of distributions in contrary to its parametric and nonparametric competitors.

## 2 Model, Hypotheses and Data Example

We consider the following c-sample location model:

Let $X_{i1},\dots,X_{in_i}$, $i=1,\dots,c$, be independent random variables with $X_{ij} \sim F_X(x-\theta_i)$, $j=1,\dots,n_i$, $\theta_i \in \mathbb{R}$,
where the distribution function $F_X$ is assumed to be continuous. We wish to test

$$\mathrm{H}_0 : \theta_1 = \cdots = \theta_c.$$

As alternative hypotheses we consider

the two-sided alternative $\mathrm{H}_1^{(1)} : \theta_r \neq \theta_s$ for at least one pair $(r,s),\ r \neq s$,
the ordered alternative $\mathrm{H}_1^{(2)} : \theta_s \leq \cdots \leq \theta_c$ with at least one strict inequality,
the umbrella alternative $\mathrm{H}_1^{(3)} : \theta_1 \leq \cdots \leq \theta_{l-1} \leq \theta_l \geq \theta_{l+1} \geq \cdots \geq \theta_c$
with at least one strict inequality for peak $l$, $2 \leq l \leq c-1$.

Now, let us present a data example for $\mathrm{H}_1^{(1)}$, the example is given by Chatfield ([13] p. 101).

*Example 1.* A study was carried out at a major London hospital to compare the effects of different types of anaesthetic used in major operations. Eighty patients undergoing a variety of operations were randomly assigned to one of the four anaesthetics and a variety of observations were taken on each patient before and after the operation. This exercise concentrates on just one of the response variables, namely the time, in minutes, from the reversal of the anaesthetic until the patient opened his or her eyes.

The data are shown in Table 1.

Figure 1 shows the boxplots of the data.

Obviously, we cannot assume normality for that kind of data, the underlying distributions might be skewed to the right. Thus, what is an appropriate test for testing $H_0$? An answer will be given at the end of Sect. 3.3.

Data examples for testing $H_0$ against the alternatives $H_1^{(2)}$ and $H_1^{(3)}$ can be found in Hand et al. ([18], p. 212) and Simpson and Margolin [35], respectively.

**Table 1** Time in minutes, from reversal of anaesthetic until the eyes open for each of 20 patients treated by one of four anaesthetics A,B,C or D

| | | | | | | | | | | | | | | | | | | | | |
|---|---|---|---|---|---|---|---|---|---|---|---|---|---|---|---|---|---|---|---|---|
| A | 3 | 2 | 1 | 4 | 3 | 2 | 10 | 12 | 12 | 3 | 19 | 1 | 4 | 5 | 1 | 1 | 7 | 5 | 1 | 12 |
| B | 6 | 4 | 1 | 1 | 6 | 2 | 1 | 10 | 1 | 1 | 1 | 2 | 10 | 2 | 2 | 2 | 2 | 1 | 3 | 7 |
| C | 3 | 5 | 2 | 4 | 2 | 1 | 6 | 13 | 1 | 1 | 1 | 4 | 1 | 1 | 1 | 8 | 1 | 2 | 4 | 0 |
| D | 4 | 8 | 2 | 3 | 2 | 3 | 6 | 2 | 3 | 4 | 8 | 5 | 10 | 2 | 0 | 10 | 2 | 3 | 9 | 1 |

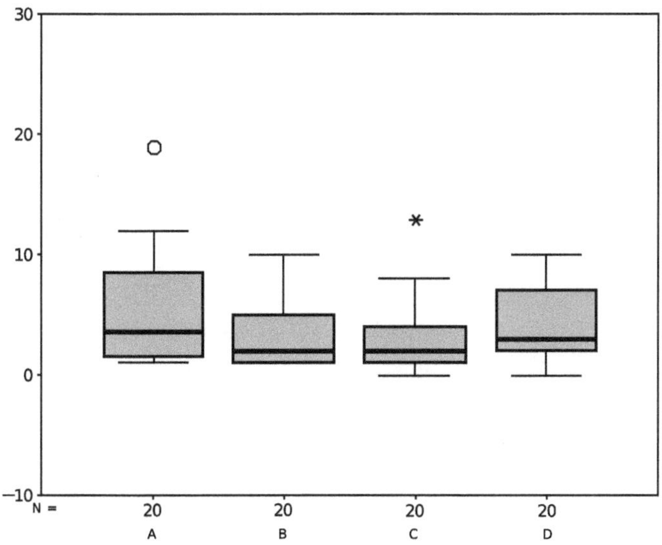


**Fig. 1** Boxplots of the data of Example 1

# 3 Tests for Two-sided Alternatives

## 3.1 Parametric F-test

Let $X_{i1},\dots,X_{in_i}$, $i=1,\dots,c$, be independent and normally distributed random variables, i.e.

$$X_{ij} \sim N(\mu_i,\sigma_i^2),\ j=1,\dots,n_i \text{ with } \sigma_1^2=\cdots=\sigma_c^2=\sigma^2.$$

We wish to test

$$\mathrm{H}_0:\ \mu_1=\cdots=\mu_c \quad \text{versus} \quad \mathrm{H}_1:\mu_r\neq\mu_s \text{ for at least one pair } (r,s),\ r\neq s.$$

Then the likelihood ratio $F$-test is based on the statistic

$$F=\frac{(N-c)\sum\limits_{i=1}^{c} n_i(\bar{X}_i-\bar{X})^2}{(c-1)\sum\limits_{i=1}^{c}\sum\limits_{j=1}^{n_i}(X_{ij}-\bar{X}_i)^2},\ \text{where } N=\sum_{i=1}^{c} n_i,\ \overline{X}_i=\frac{1}{n_i}\sum_{j=1}^{n_i}X_{ij} \text{ and } \overline{X}=\frac{1}{N}\sum_{i=1}^{c} n_i\overline{X}_i.$$

Under $\mathrm{H}_0$, the statistic $F$ has an F-distribution with $c-1$ and $N-c$ degrees of freedom. If we assume non-normal distributions with at least finite second moments it can be shown that, under $\mathrm{H}_0$, F has asymptotically a chi-square distribution with $c-1$ degrees of freedom, see, e.g. Tiku et al. [37].

## 3.2 Rank Tests

Let $X_{(1)},\dots,X_{(N)}$ be the combined ordered sample of $X_{11},\dots,X_{1n_1},\dots,X_{c1},\dots,X_{cn_c}$, $N=\sum_{i=1}^{c} n_i$.

We define indicator variables $V_{ik}$ by

$$V_{ik}=\begin{cases}1 & \text{if } X_{(k)} \text{ belongs to the ith sample}\\ 0 & \text{otherwise.}\end{cases}$$

Furthermore, we have real valued scores $a(k),\ k=1,\dots,N$, with mean $\bar{a}=\frac{1}{N}\sum\limits_{k=1}^{N} a(k)$.

Now, we define for each sample a statistic $A_i$ in the following way

$$A_i=\frac{1}{n_i}\sum_{k=1}^{N} a(k)V_{ik},\ 1\le i\le c.$$

$A_i$ is the average of the scores for the ith sample. Then the linear rank statistic $L_N$ is given by

$$L_N = \frac{(N-1)\sum_{i=1}^{c} n_i(A_i - \bar{a})^2}{\sum_{k=1}^{N} (a(k) - \bar{a})^2}.$$

Under $H_0$, $L_N$ is distribution-free and has asymptotically a chi-square distribution with $c-1$ degrees of freedom, that means, $H_0$ has to be rejected in favour of $H_1^{(1)}$ if $L_N \geq \chi^2_{1-\alpha}(c-1)$.

In the following, some examples of rank tests are given; for references, see, e.g. Gastwirth [14], Randles and Wolfe [32], Büning [3, 5], Gibbons and Chakraborti [15] as well as Büning and Trenkler [12]. In parenthesis that type of distribution is indicated for which the test has high power.

*Example 2 (Gastwirth test G (short tails)).*

$$a_G(k) = \begin{cases} k - \frac{N+1}{4} & \text{if } k \leq \frac{N+1}{4} \\ 0 & \text{if } \frac{N+1}{4} < k < \frac{3(N+1)}{4} \\ k - \frac{3(N+1)}{4} & \text{if } k \geq \frac{3(N+1)}{4}. \end{cases}$$

*Example 3 (Kruskal–Wallis test KW (medium tails)).*

$$a_{KW}(k) = k.$$

As an efficient test for long tails Büning [5] proposed the so called $LT$-test with scores chosen analogously to Huber‘s $\Psi$-function referring to $M$-estimates, see Huber [25].

*Example 4 (LT-test (long tails)).*

$$a_{LT}(k) = \begin{cases} -\left(\left[\frac{N}{4}\right]+1\right) & \text{if } k < \left[\frac{N}{4}\right]+1 \\ k - \frac{N+1}{2} & \text{if } \left[\frac{N}{4}\right]+1 \leq k \leq \left[\frac{3(N+1)}{4}\right] \\ \left[\frac{N}{4}\right]+1 & \text{if } k > \left[\frac{3(N+1)}{4}\right]. \end{cases}$$

[x] denotes the greatest integer less than or equal to x.

*Example 5 (Hogg–Fisher–Randles test HFR (right-skewed)).*

$$a_{HFR}(k) = \begin{cases} k - \frac{N+1}{2} & \text{if } k \leq \frac{N+1}{2} \\ 0 & \text{if } k > \frac{N+1}{2}. \end{cases}$$

For left-skewed distributions interchange the terms $k-(N+1)/2$ and 0 in the above definition.

All these four rank tests are included in our simulation study in Sect. 4. They are "bricks" of the adaptive tests proposed in the next section.

For the case of ordered alternatives $H_1^{(2)}$ and umbrella alternatives $H_1^{(3)}$ the most familiar rank tests are the tests of Jonckheere [27] and Mack and Wolfe [28], respectively. They are based on pairwise two-sample Wilcoxon statistics computed on the ith sample vs. the combined data in the first $i-1$ samples, $2 \leq i \leq c$. It is well known that both tests have high power for symmetric and medium-tailed distributions. Büning [6], Büning and Kössler [9] modifies these tests by using two-sample statistics of Gastwirth and Hogg–Fisher–Randles rather than the Wilcoxon statistic. These so called Jonckheere-type- and Mack–Wolfe-type tests are very efficient for short-tailed and asymmetric distributions.

## 3.3 Adaptive Tests

Husková [26] and Hájek et al. [16] distinguishes between two different concepts of adaptive procedures, nonrestrictive and restrictive ones. In the case of nonrestrictive procedures the optimal scores $a_{\text{opt}}(k)$ for the locally most powerful rank test, which depend on the (unknown) underlying distribution function F and its density f, are estimated directly from the data. This approach is applied, e.g. by Behnen and Neuhaus [1] in many testing situations. We will apply the adaptive procedure of Hogg [21] which belongs to the class of restrictive procedures, i.e. a "reasonable" family of distributions and a corresponding class of "suitable" tests are chosen. The adaptive test of Hogg is a two-staged one. At the first stage, the unknown distribution function is classified with respect to some measures like tailweight and skewness. At the second stage, an appropriate test for that classified type of distribution is selected and then carried out. Hogg [22] states: *"So adapting the test to the data provides a new dimension to nonparametric tests which usually improves the power of the overall test."*

This two-staged adaptive test maintains the level $\alpha$ for all continuous distribution functions as shown by the following

**Lemma 1.** *(1) Let $\mathscr{F}$ denote the class of distribution functions under consideration. Suppose that each of m tests based on the statistics $T_1,\ldots,T_m$ is* distribution-free *over the class $\mathscr{F}$; i.e. $P_{H_0}(T_h \in C_h|F) = \alpha$ for each $F \in \mathscr{F}$, $h = 1,\ldots,m$.*

*(2) Let S be some statistic that is* independent *of $T_1,\ldots,T_m$ under $\mathbf{H}_0$ for each $F \in \mathscr{F}$. Suppose we use S to decide which test $T_h$ to conduct. (S is called a selector statistic.). Specially, let $U_S$ denote the set of all values of S with the following decomposition:*

$$U_S = D_1 \cup D_2 \cup \cdots \cup D_m,\ D_h \cap D_k = \varnothing \text{ for } h \neq k,$$

*so that $S \in D_h$ corresponds to the decision to use the test $T_h$. The overall testing procedure is then defined by:*

*If $S \in D_h$ then reject* $H_0$ *if $T_h \in C_h$.*

*This two-staged adaptive test is, under* $H_0$, *distribution-free over the class $\mathscr{F}$, i.e. it maintains the level $\alpha$ for each $F \in \mathscr{F}$.*

*Proof.*
$$P_{\mathrm{H}_0}(\text{reject } \mathrm{H}_0|\, F) = P_{\mathrm{H}_0}\left(\bigcup_{h=1}^{m} (S \in D_h \wedge T_h \in C_h|\, F)\right)$$
$$= \sum_{h=1}^{m} P_{\mathrm{H}_0}(S \in D_h \wedge T_h \in C_h|F)$$
$$= \sum_{h=1}^{m} P_{\mathrm{H}_0}(S \in D_h|F) \cdot P_{\mathrm{H}_0}(T_h \in C_h|F)$$
$$= \alpha \cdot \sum_{h=1}^{m} P_{\mathrm{H}_0}(S \in D_h|F) = \alpha \cdot 1 = \alpha. \quad \square$$

Let us apply this Lemma on our special problem:

1. $\mathscr{F}$ is the class of all continuous distribution functions $F$ and $T_1,\ldots,T_m$ are linear rank statistics. Then $T_h$ is distribution-free over $\mathscr{F}$, $h = 1,\ldots,m$.

2. $S$ is a function of the order statistics of the combined sample. Under $\mathbf{H}_0$, the order statistics are the complete sufficient statistics for the common, but unknown $F$, and therefore independent of every statistic whose distribution is free of $F$ (theorem of Basu, see, e.g. Roussas [33], p. 215). Thus, under $\mathbf{H}_0$, $S$ is independent of the linear rank statistics $T_1,\ldots,T_m$.

As a selector statistic $S$ we choose $S = (\hat{M}_S, \hat{M}_T)$, where $\hat{M}_S$ and $\hat{M}_T$ are measures of skewness and tailweight, respectively, defined by

$\hat{M}_S = \dfrac{\hat{x}_{0.975} - \hat{x}_{0.5}}{\hat{x}_{0.5} - \hat{x}_{0.025}}$ and

$\hat{M}_T = \dfrac{\hat{x}_{0.975} - \hat{x}_{0.025}}{\hat{x}_{0.875} - \hat{x}_{0.125}}$ with the $p$-quantile $\hat{x}_p$ given by

$$\hat{x}_p = \begin{cases} X_{(1)} & \text{if } p \le 0.5/N \\ (1-\lambda)X_{(j)} + \lambda X_{(j+1)} & \text{if } 0.5/N < p \le 1 - 0.5/N \\ X_{(N)} & \text{if } p > 1 - 0.5/N \end{cases}$$

where $X_{(1)},\ldots,X_{(N)}$ again are the order statistics of the combined $c$ samples and $j = [np + 0.5]$, $\lambda = np + 0.5 - j$. Obviously, $\hat{M}_S < 1$, if $F$ is skewed to the left, $\hat{M}_S = 1$, if $F$ is symmetric and $\hat{M}_S > 1$, if $F$ is skewed to the right. $\hat{M}_T \ge 1$, the longer the tails the greater $\hat{M}_T$. The measures $\hat{M}_S$ and $\hat{M}_T$ are location and scale invariant.

In Table 2 values of the corresponding theoretical measures $M_S$ and $M_T$ are presented for some selected distributions where CN1, CN2 and CN3 are contaminated normal distributions:

$CN_1 = 0.95N(0,1) + 0.05N(0,3^2)$, $CN_2 = 0.9N(0,1) + 0.1N(0,5^2)$, both symmetric, and $CN_3 = 0.5N(1,4) + 0.5N(-1,1)$, a distribution skewed to the right.

We see, the exponential distribution is extremely right-skewed and the Cauchy has very long tails. Now, two questions arise:

*First, what is an appropriate number m of categories $D_1,\ldots,D_m$?*

Such a number $m$ may be three, four or five, in most proposals four categories are preferred, three for symmetric distributions (short, medium, long tails) and one for distributions skewed to the right. A fifth category can be defined for left-skewed distributions.

**Table 2** Theoretical values of $M_S$ and $M_6$

| Distributions | $M_S$ | $M_T$ |
|---|---|---|
| Uniform | 1.000 | 1.267 |
| Normal | 1.000 | 1.704 |
| CN1 | 1.000 | 1.814 |
| Logistic | 1.000 | 1.883 |
| Double exp. | 1.000 | 2.161 |
| CN2 | 1.000 | 2.606 |
| Cauchy | 1.000 | 5.263 |
| CN3 | 1.765 | 1.691 |
| Exponential | 4.486 | 1.883 |

*Second, how do we fix the bounds of the categories?*

The bounds depend on the theoretical values of $M_S$ and $M_T$ (see Table 2) in order to consider different strength of skewness and tailweight. Simulations by trial and error may improve the bounds in the adaptive scheme. To our experience, however, the very special choice of the bounds is not the crucial point, it is much more important to include efficient rank tests in the adaptive scheme, an efficient rank test not only for the corresponding category but also in the neighbourhood of that category because of possible misclassifications, see Table 3.

Now, for our special $c$-sample location problem we propose the following four categories which are based on $S$:

$$D_1 = \{S|0 \le \hat{M}_S \le 2;\ 1 \le \hat{M}_T \le 1.5\}$$
$$D_2 = \{S|0 \le \hat{M}_S \le 2;\ 1.5 < \hat{M}_T \le 2\}$$
$$D_3 = \{S|\hat{M}_S \ge 0;\ \hat{M}_T > 2\}$$
$$D_4 = \{S|\hat{M}_S > 2;\ 1 \le \hat{M}_T \le 2\}.$$

This means, the distribution is classified as symmetric with short- or medium tails, if $S$ falls in the category $D_1$ or $D_2$, respectively, as long-tailed if $S$ belongs to $D_3$ and as skewed to the right with short- or medium tails if $S$ falls in $D_4$.

We now propose the following adaptive test $A$:

$$A = \begin{cases} G & \text{if } S \in D_1 \\ KW & \text{if } S \in D_2 \\ LT & \text{if } S \in D_3 \\ HFR & \text{if } S \in D_4. \end{cases}$$

Figure 2 shows the adaptive scheme of test A.

The adaptive test above is based on the measures $\hat{M}_S$ and $\hat{M}_T$ calculated from the combined ordered sample $X_{(1)},\ldots,X_{(N)}$ in order to guarantee that the resulting test is distribution-free in the sense of the Lemma. Another way is to calculate the measures $\hat{M}_S$ and $\hat{M}_T$ from each of the $c$ samples separately and then to consider the weighted sum of these measures, that is

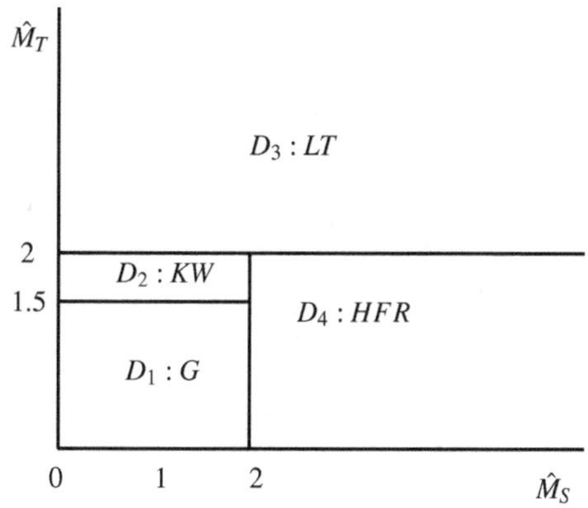


**Fig. 2** Adaptive scheme

$$\bar{M}_S = \frac{n_1\hat{M}_{S1} + \cdots + n_c\hat{M}_{Sc}}{N} \text{ and } \bar{M}_T = \frac{n_1\hat{M}_{T1} + \cdots + n_c\hat{M}_{Tc}}{N},$$

where $\hat{M}_{Si}$ and $\hat{M}_{Ti}$ are the measures for skewness and tailweight of the ith sample, $i = 1, \ldots, c$.

The adaptive test based on the measures from the *combined* sample is denoted by *AC* and that based on the measures from the *single* samples by *AS*. The adaptive test *AC* is distribution-free, the measures $\hat{M}_S$ and $\hat{M}_T$, however, are affected by the amount of the shift under $H_1$, whereas the adaptive test *AS* is *not* distribution-free, but $\bar{M}_S$ and $\bar{M}_T$ are not affected by the shift.

Table 3 shows the classification performance of $(\hat{M}_S, \hat{M}_T)$ and $(\bar{M}_S, \bar{M}_T)$ for the case of $c = 4$, $n_1 = n_2 = n_3 = n_4 = 20$. The data were generated by simulation (10,000 replications) from the uniform (Uni), normal (Norm), logistic (Log), double exponential (Dexp), Cauchy (Cau), the contaminated normal CN3 and the exponential (Exp) distribution.

The amount of shift is determined by the parameters $\theta_i = k_i\sigma_F$, $i = 1, \ldots, 4$, where $\sigma_F$ is the standard deviation of the underlying distribution function $F$. For the Cauchy we choose $\sigma_{\text{Cau}} = F^{-1}_{\text{Cau}}(0.8413) = 1.8373$ because of $\Phi(1) = 0.8413$ where $\Phi$ is the standard normal distribution function.

Let us consider, as an example, the *AC*-test with data from the uniform distribution and $k_i = 0$, $i = 1, \ldots, 4$. Then in 9,911 of 10,000 cases these data were (correctly) classified as symmetric and short-tailed ($D_1$), in 72 cases as symmetric and medium-tailed ($D_2$), in 0 cases as long-tailed ($D_3$) and in 17 cases as skewed to the right ($D_4$). Under the null hypothesis the classification schemes based on $(\hat{M}_S, \hat{M}_T)$ and $(\bar{M}_S, \bar{M}_T)$ are quite effective for all distributions considered.

In contrary to the *AS*-test the *AC*-test – based on the classification performance of $(\hat{M}_S, \hat{M}_T)$ – is strongly affected by the amount of shift for the uniform, double exponential and the two distributions skewed to the right, CN3 and Exp. As

**Table 3** Skewness and tailweight classification of the adaptive tests *AC* and *AS*, $c = 4$, $n_1 = n_2 = n_3 = n_4 = 20$

| $k_1, k_2, k_3, k_4$ | Uni | Norm | Log | Dexp | Cau | CN3 | Exp |
|---|---|---|---|---|---|---|---|
| **0,0,0,0** | | | | | | | |
| $D_1$ | | | | | | | |
| AC | 9,911 | 1,171 | 247 | 47 | 0 | 1041 | 5 |
| AS | 9,786 | 1,484 | 410 | 66 | 1 | 1083 | 4 |
| $D_2$ | | | | | | | |
| AC | 72 | 8,131 | 6,538 | 3,082 | 13 | 5,711 | 4 |
| AS | 151 | 7,901 | 6,704 | 3,544 | 14 | 4,977 | 4 |
| $D_3$ | | | | | | | |
| AC | 0 | 690 | 3,203 | 6,855 | 9,987 | 822 | 3,409 |
| AS | 0 | 577 | 2,816 | 6,313 | 9,983 | 901 | 3,319 |
| $D_4$ | | | | | | | |
| AC | 17 | 8 | 12 | 16 | 0 | 2,426 | 6,582 |
| AS | 63 | 38 | 70 | 77 | 2 | 3,039 | 6,673 |
| **0,0.2,0.4,0.6** | | | | | | | |
| $D_1$ | | | | | | | |
| AC | 9,407 | 1,101 | 308 | 52 | 0 | 1,088 | 70 |
| AS | 9,799 | 1,416 | 423 | 83 | 0 | 1,092 | 13 |
| $D_2$ | | | | | | | |
| AC | 588 | 8,154 | 6,807 | 3,571 | 9 | 6,393 | 166 |
| AS | 140 | 7,915 | 6,726 | 3,585 | 15 | 4,952 | 5 |
| $D_3$ | | | | | | | |
| AC | 0 | 735 | 2,826 | 6,366 | 9,990 | 801 | 3,852 |
| AS | 0 | 634 | 2,797 | 6,268 | 9,984 | 900 | 3,350 |
| $D_4$ | | | | | | | |
| AC | 5 | 10 | 17 | 11 | 1 | 1,718 | 5,912 |
| AC | 61 | 35 | 54 | 64 | 1 | 3,056 | 6,632 |
| **0,0.4,0.8,1.2** | | | | | | | |
| $D_1$ | | | | | | | |
| AC | 6,128 | 1,191 | 450 | 95 | 0 | 1,161 | 315 |
| AS | 9,765 | 1,442 | 376 | 60 | 0 | 1,072 | 9 |
| $D_2$ | | | | | | | |
| AC | 3,868 | 8,171 | 7,285 | 4,830 | 20 | 7,366 | 2,123 |
| AS | 181 | 7,922 | 6,656 | 3,505 | 8 | 5,034 | 3 |
| $D_3$ | | | | | | | |
| AC | 1 | 634 | 2,248 | 5,058 | 9,980 | 774 | 3,132 |
| AS | 0 | 598 | 2,898 | 6,371 | 9,989 | 949 | 3,322 |
| $D_4$ | | | | | | | |
| AC | 3 | 4 | 17 | 17 | 0 | 699 | 4,430 |
| AS | 54 | 38 | 70 | 64 | 3 | 2,945 | 6,666 |

the differences of $\theta_1, \ldots, \theta_4$ increase, all these four distributions tend to be classified more as having medium tails. But for large differences of the location parameters each of the tests in the adaptive scheme should reveal these differences. For the normal and the Cauchy distribution the classification performance of $(\hat{M}_S, \hat{M}_T)$ is hardly affected by the shift. Similar results hold for $c = 3$ samples and other sizes.

Now, let us analyze the data Example 1 from Sect. 2. What is an appropriate test for these data? First, we calculate the measures $\hat{M}_S$ and $\hat{M}_T$ of the combined ordered sample $X_{(1)},\ldots,X_{(N)}$ of the four samples in order to guarantee that the resulting adaptive test *AC* maintains the level. For the data we get $\hat{M}_S = 3.80$ and $\hat{M}_T = 1.41$, i.e. the distribution of the data is extremely skewed to the right, see Table 2. The selector statistic $S = (3.80, 1.41)$ belongs to $D_4$ and we have to apply the *HFR*-test. Because of $HFR = 5.636 < \chi^2_{0.95}(3) = 7.815$, $\mathbf{H}_0$ is not rejected at level $\alpha = 5\,\%$. It should be noted that the adaptive test *AC* is only asymptotically distribution-free because an asymptotical critical value of *HFR* is used.

If we calculate the measures $\bar{M}_S$ and $\bar{M}_T$ from each of the four samples separately, we get $\bar{M}_S = 5.51$ and $\bar{M}_T = 1.79$. Thus, we have to apply the *HFR*-test, too, and we get the same test decision. But notice, the adaptive test *AS* based on the selector statistic $S = (5.51, 1.79)$ is not distribution-free.

In the same sense as described above adaptive tests may be constructed for ordered alternatives $\mathbf{H}_1^{(2)}$ and umbrella alternatives $\mathbf{H}_1^{(3)}$ by including Jonckheere-type or Mack–Wolfe-type tests in the adaptive scheme, see Büning [6] and Büning and Kössler [9].

## 4 Power Study

We investigate via Monte Carlo methods (10,000 replications) the power of all the tests from Sect. 3. The selected distributions are the same as in Table 2 where each of them has mean or median (Cauchy) equal to zero. Here, we again consider only the case of $c = 4$ samples with equal sizes $n_i = 20,\ i = 1,\ldots,4$. The location parameters $\theta_i$ are defined by $\theta_i = k_i\sigma_F$ as in Sect. 3.3. The nominal level of the tests is $\alpha = 5\,\%$. Table 4 presents the power values.

We can state:

The F-test maintains the level $\alpha$ quite well for all distributions considered with the exception of the Cauchy for which finite moments do not exist. In this sense, the approximation of the distribution of $F$ by the chi-square distribution does not work, see Sect. 3.1. Thus, for the Cauchy a power comparison of the $F$-test with the other tests becomes meaningless.

For each of the distributions (with exception of the normal) there is a linear rank test which has higher power than the $F$-test, e.g. the Gastwirth test for the uniform, the Kruskal–Wallis test for CN1 and the logistic, the *LT*-test for the double exponential and CN2 and the Hogg–Fisher–Randles test for both distributions skewed to the right, CN3 and Exp.

The adaptive tests, *AC* and *AS*, are the best ones over this broad class of distributions. The *AS*-test has (slightly) higher power than the *AC*-test, but since in all cases the actual level of the *AS*-test starts higher than the level of the *AC*-test, it is difficult to assess the higher power values of the *AS*-test in comparison to the *AC*-test. Except for the normal distribution the *AC*-test is more powerful than the $F$-test for all symmetric and asymmetric distributions.

**Table 4** Power of some tests (in percent) under selected distributions $\alpha = 5\%$, $c = 4$, $(n_1, n_2, n_3, n_4) = (20, 20, 20, 20)$

| Tests | $k_1, k_2, k_3, k_4$ | Uni | Norm | CN1 | Log | Dexp | CN2 | Cau | CN3 | Exp |
|---|---|---|---|---|---|---|---|---|---|---|
| *F* | 0, 0, 0, 0 | 4.8 | 4.9 | 4.9 | 4.8 | 4.8 | 4.3 | 1.8 | 5.2 | 4.3 |
| | 0, 0.2, 0.4, 0.6 | 3.5 | 33.7 | 36.4 | 35.4 | 34.7 | 40.3 | | 34.7 | 36.9 |
| | 0, 0.3, 0.6, 0.9 | 68.5 | 68.8 | 69.3 | 68.9 | 69.2 | 71.3 | | 69.3 | 69.9 |
| *G* | 0, 0, 0, 0 | 4.5 | 4.6 | 4.9 | 4.6 | 4.9 | 4.8 | 5.1 | 4.7 | 4.2 |
| | 0, 0.2, 0.4, 0.6 | 50.4 | 27.6 | 33.4 | 28.2 | 25.0 | 51.9 | 12.1 | 32.7 | 70.7 |
| | 0, 0.3, 0.6, 0.9 | 85.1 | 59.5 | 65.2 | 57.2 | 52.6 | 84.1 | 21.0 | 65.6 | 90.7 |
| *KW* | 0, 0, 0, 0 | 4.6 | 4.7 | 4.7 | 4.4 | 4.9 | 5.0 | 5.0 | 4.8 | 4.3 |
| | 0, 0.2, 0.4, 0.6 | 29.3 | 31.7 | 39.3 | 36.7 | 45.4 | 70.5 | 31.2 | 37.6 | 64.2 |
| | 0, 0.3, 0.6, 0.9 | 61.5 | 65.8 | 75.2 | 71.4 | 81.0 | 96.7 | 60.8 | 72.3 | 92.1 |
| *LT* | 0, 0, 0, 0 | 4.6 | 4.9 | 4.9 | 4.8 | 4.7 | 5.1 | 5.3 | 4.8 | 4.5 |
| | 0, 0.2, 0.4, 0.6 | 18.7 | 28.9 | 36.0 | 35.1 | 49.2 | 70.0 | 39.9 | 43.2 | 53.4 |
| | 0, 0.3, 0.6, 0.9 | 40.8 | 60.2 | 71.4 | 69.2 | 84.2 | 97.0 | 72.7 | 66.8 | 87.7 |
| *HFR* | 0, 0, 0, 0 | 4.5 | 4.7 | 4.9 | 4.9 | 4.6 | 5.0 | 5.1 | 4.7 | 4.8 |
| | 0, 0.2, 0.4, 0.6 | 23.3 | 24.8 | 31.3 | 29.6 | 36.5 | 59.0 | 26.8 | 43.5 | 86.1 |
| | 0, 0.3, 0.6, 0.9 | 50.0 | 54.2 | 63.9 | 60.0 | 70.6 | 90.0 | 50.7 | 78.6 | 99.2 |
| *AC* | 0, 0, 0, 0 | 4.5 | 4.8 | 4.8 | 4.4 | 4.7 | 5.1 | 5.3 | 4.7 | 4.7 |
| | 0, 0.2, 0.4, 0.6 | 49.1 | 30.8 | 37.9 | 35.9 | 47.6 | 70.5 | 39.9 | 37.6 | 72.9 |
| | 0, 0.3, 0.6, 0.9 | 78.9 | 64.1 | 73.6 | 70.1 | 82.5 | 97.0 | 72.7 | 71.3 | 94.1 |
| *AS* | 0, 0, 0, 0 | 5.1 | 5.3 | 5.2 | 4.9 | 5.0 | 5.2 | 5.4 | 6.0 | 4.8 |
| | 0, 0.2, 0.4, 0.6 | 50.5 | 32.5 | 38.7 | 37.1 | 48.7 | 70.4 | 39.9 | 41.8 | 75.8 |
| | 0, 0.3, 0.6, 0.9 | 84.9 | 66.0 | 74.1 | 71.2 | 83.2 | 97.1 | 72.7 | 75.9 | 96.2 |

The adaptive test *AC* is not the best one for a special distribution but mostly second or third best. That is just the philosophy of an adaptive test, to select the best one for a given data set.

## 5 Outlook

In our paper we studied an adaptive $c$-sample location test which behaves well over a broad class of distributions, symmetric ones with different tailweight and right-skewed distributions with different strength of skewness. Further adaptive tests for the two- and $c$-sample location problem can be found in Hogg et al. [23], Ruberg [34], Hill et al. [20], Hothorn and Liese [24], Büning [4, 5, 6], Beier and Büning [2], Sun [36], O'Gorman [30], Büning and Kössler [9], Büning and Rietz [10] and Neuhäuser et al. [29]. For an adaptive two-sample scale test, see Hall and Padmanabhan [17] and Büning [8] and for an adaptive two-sample location-scale test of Lepage-type, see Büning and Thadewald [11]. An adaptive test for the general two sample problem based on Kolmogorov–Smirnov- and Cramér- von Mises-type tests has been proposed by Büning [7]. A very comprehensive survey of adaptive procedures is given by O'Gorman [31].

In our proposal for an adaptive test in Sect. 3.3 we restrict our attention to two measures for skewness and tailweight, $\hat{M}_S$ and $\hat{M}_T$. Other measures for skewness and tailweight are discussed in the literature, see, e.g. the measures $\hat{Q}_1$ and $\hat{Q}_2$ of Hogg [21]. Of course, we may add other types of measures in order to classify the unknown distribution function possibly more correctly, e.g. we can include an additional measure for peakedness, see Büning [3] and Hogg [21]. In this case we have a three dimensional selector statistic $S$ defining our adaptive scheme. To our experience, there is, however, no remarkable gain in power of the adaptive test by adding the peakedness measure, see Handl [19]. Thus, we propose to use only two measures, one for skewness and one for tailweight.

As a result of all our studies on adaptive tests we can state without any doubt, that adaptive testing is an important tool for any practising statistician and it would be a profitable task to add adaptive procedures to statistical software packages.

## References

[1] Behnen, K., Neuhaus, G.: Rank tests with estimated scores and their application. Teubner, Stuttgart (1989)

[2] Beier, F., Büning, H.: An adaptive test against ordered alternatives. Comput. Stat. Data Anal. **25**, 441–452 (1997)

[3] Büning, H.: Robuste und adaptive tests. De Gruyter, Berlin (1991)

[4] Büning, H.: Robust and adaptive tests for the two-sample location problem. OR Spektrum **16**, 33–39 (1994)

[5] Büning, H.: Adaptive tests for the c-sample location problem - the case of two-sided alternatives. Commun. Stat. Theor. Meth. **25**, 1569–1582 (1996)

[6] Büning, H.: Adaptive Jonckheere-type tests for ordered alternatives. J. Appl. Stat. **26**, 541–551 (1999)

[7] Büning, H.: An adaptive distribution-free test for the general two-sample problem. Comput. Stat. **17**, 297–313 (2002)

[8] Büning, H.: An adaptive test for the two-sample scale problem. Diskussionsbeiträge des Fachbereichs Wirtschaftswissenschaft der Freien Universität Berlin, Nr. 2003/10 (2003)

[9] Büning, H., Kössler, W.: Adaptive tests for umbrella alternatives. Biom. J. **40**, 573–587 (1998)

[10] Büning, H., Rietz, M.: Adaptive bootstrap tests and its competitors in the c-sample location problem. J. Stat. Comput. Sim. **73**, 361–375 (2003)

[11] Büning, H., Thadewald, T.: An adaptive two-sample location-scale test of Lepage-type for symmetric distributions. J. Stat. Comput. Sim. **65**, 287–310 (2000)

[12] Büning, H., Trenkler, G.: Nichtparametrische statistische Methoden. De Gruyter, Berlin (1994)

[13] Chatfield, C.: Problem-solving – A statistician's guide. Chapman & Hall, London (1988)

[14] Gastwirth, J.L.: Percentile modifications of two-sample rank tests. J. Am. Stat. Assoc. **60**, 1127–1140 (1965)
[15] Gibbons, J.D., Chakraborti, S.: Nonparametric statistical inference, 3rd ed. Dekker, New York (1992)
[16] Hájek, J., Sidák, Z.S., Sen, P.K.: Theory of rank tests. Academic, New York (1999)
[17] Hall, P., Padmanabhan, A.R.: Adaptive inference for the two-sample scale problem. Technometrics **39**, 412–422 (1997)
[18] Hand, D.J., Daly, F., Lunn, A.D., McConway, K.J., Ostrowski, E.: A handbook of small data sets. Chapman & Hall, London (1994)
[19] Handl, A.: Maßzahlen zur Klassifizierung von Verteilungen bei der Konstruktion adaptiver verteilungsfreier Tests im unverbundenen Zweistichproben-Problem. Unpublished Dissertation, Freie Universität Berlin (1986)
[20] Hill, N.J., Padmanabham, A.R., Puri, M.L.: Adaptive nonparametric procedures and applications. Appl. Stat. **37**, 205–218 (1988)
[21] Hogg, R.V.: Adaptive robust procedures. A partial review and some suggestions for future applications and theory. J. Am. Stat. Assoc. **69**, 909–927 (1974)
[22] Hogg, R.V.: A new dimension to nonparametric tests. Commun. Stat. Theor. Meth. **5**, 1313–1325 (1976)
[23] Hogg, R.V., Fisher, D.M., Randles, R.H.: A two-sample adaptive distribution-free test. J. Am. Stat. Assoc. **70**, 656–661 (1975)
[24] Hothorn, L., Liese, F.: Adaptive Umbrellatests - Simulationsuntersuchungen. Rostocker Mathematisches Kolloquium **45**, 57–74 (1991)
[25] Huber, P.J.: Robust estimation of a location parameter. Ann. Math. Stat. **35**, 73–101 (1964)
[26] Husková, M.: Partial review of adaptive procedures. In: Sequential Methods in Statistics **16**, Banach Center Publications, Warschau (1985)
[27] Jonckheere, A.R.: A distribution-free k-sample test against ordered alternatives. Biometrika **41**, 133–145 (1954)
[28] Mack, G.A., Wolfe, D.A.: K-sample rank tests for umbrella alternatives. J. Am. Stat. Assoc. **76**, 175–181 (1981)
[29] Neuhäuser, M., Büning, H., Hothorn, L.: Maximum test versus adaptive tests for the two-sample location problem. J. Appl. Stat. **31**, 215–227 (2004)
[30] O'Gorman, T.W.: A comparison of an adaptive two-sample test to the t-test, rank-sum, and log-rank tests. Commun. Stat. Simul. Comp. **26**, 1393-1411 (1997)
[31] O'Gorman, T.W.: Applied adaptive statistical methods – tests of significance and confidence intervals. ASA-SIAM Ser. Stat. Appl. Prob., Philadelphia (2004)
[32] Randles, R.H., Wolfe, D.A.: Introduction to the theory of nonparametric statistics. Wiley, New York (1979)
[33] Roussas, G.G.: A first course in mathematical statistics. Addison-Wesley, Reading, MA (1973)

[34] Ruberg, S.J.: A continuously adaptive nonparametric two-sample test. Commun. Stat. Theor. Meth. **15**, 2899–2920 (1986)

[35] Simpson, D.G., Margolin, B.H.: Recursive nonparametric testing for dose-response relationships subject to downturns at high doses. Biometrika **73**, 589–596 (1986)

[36] Sun, S.: A class of adaptive distribution-free procedures. J. Stat. Plan. Infer. **59**, 191–211 (1997)

[37] Tiku, M.L., Tan, W.Y., Balakrishnan, N.: Robust inference. Dekker, New York (1986)

# On Nonparametric Tests for Trend Detection in Seasonal Time Series

Oliver Morell and Roland Fried

**Abstract** We investigate nonparametric tests for identifying monotone trends in time series as they need weaker assumptions than parametric tests and are more flexible concerning the structure of the trend function. As seasonal effects can falsify the test results, modifications have been suggested which can handle also seasonal data. Diersen and Trenkler [5] propose a test procedure based on records and Hirsch et. al [8] develop a test based on Kendall's test for correlation. The same ideas can be applied to other nonparametric procedures for trend detection. All these procedures assume the observations to be independent. This assumption is often not fulfilled in time series analysis. We use the mentioned test procedures to analyse the time series of the temperature and the rainfall observed in Potsdam (Germany) from 1893 to 2008. As opposed to the rainfall time series, the temperature data show positive autocorrelation. Thus it is also of interest, how the several test procedures behave in case of autocorrelated data.

## 1 Introduction

One interest in time series analysis is to detect monotonic trends in the data. Several parametric and nonparametric procedures for trend detection based on significance tests have been suggested. Parametric methods rely on strong assumptions for the distribution of the data, which are difficult to check in practice and possibly not fulfilled. Furthermore a parametric form of the trend has to be specified, where only some unknown parameters need to be estimated. Nonparametric test procedures are more flexible as they afford only rather general assumptions about the distribution. Also the trend often only needs to be monotonic without further specifications.

---

Oliver Morell (✉)
Fakultät Statistik, Technische Universität Dortmund, D-44221 Dortmund, Germany
morell@statistik.tu-dortmund.de

First ideas for nonparametric test procedures based on signs (see e.g. [3] or [13]), ranks (see e.g. [4] or [12]) and records [7] have been developed early. However, all these approaches need the assumption of i.i.d. random variables under the null hypothesis. For time series with seasonal behavior this assumption is not valid. One way to handle this problem is to estimate and subtract the seasonality. Another approach is to use tests which are robust against seasonal effects. Hirsch et. al. [8] develop a test procedure based on Kendall's test of correlation [10]. Diersen and Trenkler [5] propose several tests based on records. They show that splitting the time series increases the power of the record tests, especially when seasonal effects occur. The procedures of Hirsch et. al. and Diersen and Trenkler use the independence of all observations to calculate a statistic separately for each period and sum them to get a test statistic for a test against randomness. The same ideas can be used for the above mentioned tests based on signs or ranks.

We apply the procedures to two climate time series from a gauging station in Potsdam, Germany: mean temperature and total rainfall. Such climate time series often show seasonality with a period of one year. Section 2 introduces the test problem of the hypothesis of randomness against a monotonic trend as well as test procedures which can also be used for seasonal data, namely some tests based on records for the splitted time series [5] and the seasonal Kendall–Test [8]. We also modify other nonparametric test statistics to consider seasonality. The mentioned sign– and rank–tests are transformed to new seasonal nonparametric tests. In Sect. 3 we compare the power of the several test procedures against different types of monotone trends and in the case of autocorrelation. In Sect. 4 the two climate time series are analysed. In particular, the test procedures are used to check the hypothesis of randomness. Section 5 summarizes the results.

## 2 Nonparametric Tests of the Hypothesis of Randomness

A common assumption of statistical analysis is the hypothesis of randomness. It means that some observed values $x_1,\dots,x_n$ are a realisation of independent and identically distributed (i.i.d.) continuous random variables (rv) $X_1,\dots,X_n$, all with the same cumulative distribution function (cdf) $F$. There are several test procedures which can be used to test the hypothesis of randomness $\mathrm{H}_0$ against the alternative $\mathrm{H}_1$ of a monotonic trend. However, in time series analysis the observed values $x_1,\dots,x_n$ are a realisation of a stochastic process and can be autocorrelated, implying a lack of independence of $X_1,\dots,X_n$. Additionally, many time series show seasonal effects and so $X_1,\dots,X_n$ are not identically distributed, even if there is no monotonic trend. We modify the hypothesis of randomness for seasonal data to handle at least the second problem:

Firstly, if there is a cycle of $k$ periods, the random sample $\mathbf{X} = (X_1,\dots,X_n)$ is splitted into $k$ parts

$$\mathbf{X} = (\mathbf{X}_1,\mathbf{X}_2,\dots,\mathbf{X}_k) \text{ with } \mathbf{X}_j = (X_{1,j},X_{2,j},\dots,X_{n_j,j}) \text{ and } X_{i,j} = X_{k(i-1)+j} \quad (1)$$

for $j = 1,\dots,k$ and $i = 1,\dots,n_j$. $\mathbf{X}_j$ thus includes all $n_j$ observations of season $j$. Under the null hypothesis $H_0$ of no trend the continuous rv $X_1,\dots,X_n$ are still considered to be independent but only for each $j$ the rv's $X_{1,j},\dots,X_{n_j,j}$ are identically distributed with common cdf $F_j$. Under the alternative $H_1$ of a monotonic trend there are values $0 = a_{1,j} \le a_{2,j} \le \dots \le a_{n_j,j}$ with $a_{i,j} < a_{i+1,j}$ for at least one $i \in \{1,\dots,n_j - 1\}$ and $j \in \{1,\dots,k\}$ such that $F_{i,j}(x) = F_j(x - a_{i,j})$ in case of an increasing and $F_{i,j}(x) = F_j(x + a_{i,j})$ in case of a decreasing trend. Under $H_0$ the hypothesis of randomness within each period is fulfilled. In the following we denote the test problem of the hypothesis of randomness for seasonal data against a monotone trend alternative with $\mathscr{H}_R$ and introduce test procedures for $\mathscr{H}_R$.

## 2.1 Tests Based on Record Statistics

Foster and Stuart [7] introduce a nonparametric test procedure for $\mathscr{H}_R$ based on the number of upper and lower records in the sequence $X_1,\dots,X_n$ and the reversed sequence $X_n,\dots,X_1$. A test procedure for $\mathscr{H}_R$ based on this approach which is robust against seasonality is introduced by Diersen and Trenkler [5]. A first application of their procedure is given in [6].

Using (1) we define upper and lower record statistics $U^o_{i,j}$, $L^o_{i,j}$, $U^r_{i,j}$ and $L^r_{i,j}$ of the original and the reversed sequence for all periods $j = 1,\dots,k$ at $i = 2,\dots,n_j$ as

$$U^o_{i,j} = \begin{cases} 1\ , \text{ if} & X_{i,j} > \max\{X_{1,j}, X_{2,j},\dots,X_{i-1,j}\} \\ 0 \ \text{ otherwise} & \end{cases} \tag{2}$$

$$L^o_{i,j} = \begin{cases} 1\ , \text{ if} & X_{i,j} < \min\{X_{1,j}, X_{2,j},\dots,X_{i-1,j}\} \\ 0 \ \text{ otherwise} & \end{cases} \tag{3}$$

$$U^r_{n_j-i+1,j} = \begin{cases} 1\ , \text{ if} & X_{n_j-i+1,j} > \max\{X_{n_j-i+2,j}, X_{n_j-i+3,j},\dots,X_{n_j,j}\} \\ 0 \ \text{ otherwise} & \end{cases} \tag{4}$$

$$L^r_{n_j-i+1,j} = \begin{cases} 1\ , \text{ if} & X_{n_j-i+1,j} < \min\{X_{n_j-i+2,j}, X_{n_j-i+3,j},\dots,X_{n_j,j}\} \\ 0 \ \text{ otherwise} & \end{cases} \tag{5}$$

with

$$U^o_{1,j} = L^o_{1,j} = U^r_{n_j,j} = L^r_{n_j,j} = 1 \tag{6}$$

as the first value of a sequence is always an upper and a lower record.

Under $H_0$ for a larger $i$ the probability of a record will get smaller. Therefore Diersen and Trenkler [5] recommend to use linear weights $w_i = i - 1$ for a record at the $i$–th position of the original or reversed sequence. The sum of the weighted records of the original sequence

$$U^o = \sum_{j=1}^{k} \sum_{i=1}^{n_j} w_i U^o_{i,j} \text{ and } L^o = \sum_{j=1}^{k} \sum_{i=1}^{n_j} w_i L^o_{i,j}\ , \tag{7}$$

and the sum of the records of the reversed series

$$U^r = \sum_{j=1}^{k}\sum_{i=1}^{n_j} w_i U^r_{n_j-i+1,j} \text{ and } L^r = \sum_{j=1}^{k}\sum_{i=1}^{n_j} w_i L^r_{n_j-i+1,j} \tag{8}$$

can be used as test statistics for $\mathscr{H}_R$. They are sums of independent rv and all have the same distribution under $H_0$. The expectations and variances are given by

$$\mathrm{E}(U^o) = \sum_{j=1}^{k}\sum_{i=1}^{n_j} \frac{w_i}{i} \text{ and } \mathrm{Var}(U^o) = \sum_{j=1}^{k}\sum_{i=1}^{n_j} w_i^2 \frac{i-1}{i^2} \tag{9}$$

and especially

$$\mathrm{E}(U^o) = k\sum_{i=1}^{n_1} \frac{i-1}{i} \text{ and } \mathrm{Var}(U^o) = k\sum_{i=1}^{n_1} \frac{(i-1)^3}{i^2} \tag{10}$$

if linear weights $w_i = i-1$ are used and all periods $j$ have the same number of observations $n_1$.

If an upward trend exists, $U^o$ and $L^r$ become large while $L^o$ and $U^r$ become small. The opposite is true, if a downward trend exists. These informations can be used to combine the sums in (8) and (9) and to use the statistics

$$T_1 = U^o - L^o,\ T_2 = U^o - U^r,\ T_3 = U^o + L^r,\ T_4 = U^o - U^r + L^r - L^o \tag{11}$$

for $\mathscr{H}_R$. Under $H_0$ the distributions of $T_1$, $T_2$ and $T_3$ will not change, if $\widetilde{T}_1 = L^r - U^r$, $\widetilde{T}_2 = L^r - L^o$ and $\widetilde{T}_3 = U^r + L^o$, respectively, are taken instead of the sums given in (11). From these statistics, only

$$T_1 = U^o - L^o = \sum_{j=1}^{k}\sum_{i=1}^{n_j} w_i \left(U^o_{i,j} - L^o_{i,j}\right) \tag{12}$$

can be expressed as a sum of independent rv, because here records from the same sequence are combined. We have under $H_0$

$$\mathrm{E}(T_1) = 0 \text{ and } \mathrm{Var}(T_1) = 2\sum_{j=1}^{k}\sum_{i=1}^{n_j} \frac{w_i^2}{i}\,. \tag{13}$$

In contrast to $T_1$, in $T_2$, $T_3$ and $T_4$ we use records from the original sequence as well as from the reversed sequence. So the summands here are not independent. We get the expectations

$$\mathrm{E}(T_2) = \mathrm{E}(T_4) = 0 \text{ and } \mathrm{E}(T_3) = 2\sum_{j=1}^{k}\sum_{i=1}^{n_j} \frac{w_i}{i}\,. \tag{14}$$

while the variances of $T_2$, $T_3$ and $T_4$ become unwieldly expressions and are given in [6] for the case $n_1 = \ldots = n_k$.

Diersen and Trenkler [6] recommend a splitting with large $k$ and small $n_j$, $j = 1,\dots,k$. The first reason for this are the asymptotic properties of the statistics in (11). With $X_1,\dots,X_n$ assumed to be independent and $n_1 = \dots = n_k$, the statistics $T_1$, $T_2$, $T_3$ and $T_4$ are the sum of $k$ i.i.d. rv. So for $k \to \infty$ all four test statistics are asymptotically normal distributed. These asymptotics are not fulfilled, if the statistics in (11) are only weighted but not splitted. Diersen and Trenkler [5] showed for this case that the asymptotic distribution is not a normal one. The second reason is that compared to the best parametric test in the normal linear regression model and the (non seasonal) Kendall–Test the asymptotic relative efficiency vanishes for fixed $k$ and increasing $n_j$. So it is also an interesting question if the efficiency of other nonparametric tests can be increased, if the time series is splitted with a large $k$ and a small number $n_j$ of observations in each period $j$.

## 2.2 The Seasonal Kendall-test

Mann [12] introduced a test for $\mathscr{H}_R$ based on Kendall's test for independence of two random variables in a bivariate distribution [10]. It was modified by Hirsch et al. [8] to robustify the test statistic against seasonal effects. Taking the splitted series in (1), they use the test statistic

$$S = \sum_{j=1}^{k} S_j \quad \text{with} \quad S_j = \sum_{i=1}^{n_j-1} \sum_{i'=i+1}^{n_j} \operatorname{sgn}(X_{i',j} - X_{i,j}) \tag{15}$$

for $\mathscr{H}_R$. So in $S_j$ the number of pairs $(X_{i,j}, X_{i',j})$ with $X_{i,j} < X_{i',j}$ is subtracted from the number of pairs $(X_{i,j}, X_{i',j})$ with $X_{i,j} > X_{i',j}$, $i < i'$, for period $j$. If there is a positive (negative) monotonic trend in period $j$, the statistic $S_j$ is expected to be large (small) while it will probably realise a value near 0 if there is no monotonic trend. If the same positive (negative) monotonic behavior can be observed for all periods, the statistic $S$ will also become large (small). $S$ will also take a value close to 0, if no monotonic trend exists.

The exact distribution of $S$ under $H_0$ is symmetric with

$$E(S) = \sum_{j=1}^{k} E(S_j) = 0 \tag{16}$$

and if there are no identical values (ties) in the observations of any period $j$, the variance is given by

$$\operatorname{Var}(S) = \sum_{j=1}^{k} \operatorname{Var}(S_j) = \sum_{j=1}^{k} \frac{n_j(n_j-1)(2n_j+5)}{18} \tag{17}$$

as $S_1,\dots,S_k$ are independent. A pair of observations is called a tie of extend $\delta$, if $\delta$ observations of $x_1,\dots,x_n$ have the same value. If $X_1,\dots,X_n$ are continuous rv, the probability of a tie is zero, but for rounded values, ties can be observed. Let $n_{\delta,j}$ be the number of ties within $\mathbf{X}_j$ with extend $\delta$. Then the variance of $S$ becomes smaller:

$$\mathrm{Var}(S) = \sum_{j=1}^{k} \frac{\left(n_j(n_j-1)(2n_j+5) - \sum_{\delta=1}^{n_j} n_{\delta,j}\delta(\delta-1)(2\delta+5)\right)}{18} \tag{18}$$

As every $S_j$ is asymptotically normally distributed for $n_j \to \infty$, the statistic $S$ as a finite sum of independent asymptotically normally distributed rv is asymptotically normal, too, if $n_j$ converges to infinity for each $j$. The exact distribution of $S$ under $\mathrm{H}_0$ (neglecting ties) can be determined by enumerating all permutations of $X_{1,j},\ldots,X_{n_j,j}$ for each $j$ and calculating the values of $S_j$ for every permutation of each $j$. The individual values and their frequencies can be easily calculated with Chap. 5 of [11]. According to the frequencies of the single values for each $S_j$, the distribution of $S$ can be obtained by reconsidering every possible combination of the values and multiplying the corresponding frequencies. However, for large $n$ calculating the exact distribution of $S$ is time consuming, so the normal approximation should be used whenever possible. Hirsch et al. [8] state that already for $k = 12$ and $n_j = 3$ the normal approximation of $S_j$ works well. They also claim that their test is robust against seasonality and departures from normality, but not robust against dependence. Hirsch and Slack [9] develop a test for $\mathscr{H}_R$, which performs better than $S$ if the data are autocorrelated. This test uses estimates of the covariances between two seasons based on Spearman's rank correlation coefficient. The estimated covariances are used to correct the variance of $S$ in the normal approximation.

## 2.3 Some Rank Statistics for $\mathscr{H}_R$

Aiyar et al. [1] compare the asymptotic relative efficiencies of many nonparametric tests for the hypothesis of randomness against trend alternatives. They consider mostly linear and nonlinear rank statistics, which we will use in the following for $\mathscr{H}_R$:

Taken the splitted series from (1) let $R(X_{1,j}),\ldots,R(X_{n_j,j})$ be the ranks of the continuous rv $X_{1,j},\ldots,X_{n_j,j}$, for $j \in \{1,\ldots,k\}$. Then two linear rank test statistics based on Spearman's rank correlation coefficient are given by

$$R_1 = \sum_{j=1}^{k} \widetilde{R}_{1,j} \text{ with } \widetilde{R}_{1,j} = \sum_{i=1}^{n_j} \left(i - \frac{n_j+1}{2}\right)\left(R(X_{i,j}) - \frac{n_j+1}{2}\right) \text{ and} \tag{19}$$

$$R_2 = \sum_{j=1}^{k} \widetilde{R}_{2,j} \text{ with } \widetilde{R}_{2,j} = \sum_{i=1}^{n_j} \left(i - \frac{n_j+1}{2}\right) \mathrm{sign}\left(R(X_{i,j}) - \frac{n_j+1}{2}\right).$$

Both statistics are symmetric and have an expected value of 0. Their variances are given by

$$\mathrm{Var}(R_1) = \sum_{j=1}^{k} \mathrm{Var}(\widetilde{R}_{1,j}) = \sum_{j=1}^{k} \frac{n_j^2(n_j+1)^2(n_j-1)}{144} \quad \text{and}$$

$$\mathrm{Var}(R_2) = \sum_{j=1}^{k} \mathrm{Var}(\widetilde{R}_{2,j}) \text{ with } \mathrm{Var}(\widetilde{R}_{2,j}) = \begin{cases} \sum_{j=1}^{k} \frac{n_j^2(n_j+1)}{12} & , n_j \text{ even} \\ \sum_{j=1}^{k} \frac{n_j(n_j-1)(n_j+1)}{12} & , n_j \text{ odd} . \end{cases} \tag{20}$$

Instead of considering all rv like in (19), the $(1-2\gamma)$ truncated sample can be taken for all periods, with $\gamma \in (0, 0.5)$. Like [1] we define

$$c_{i,j} = \begin{cases} -1 , & 0 < i \leq \lfloor \gamma n_j \rfloor \\ 0 , & \lfloor \gamma n_j \rfloor < i \leq n_j - \lfloor \gamma n_j \rfloor \\ +1 , & n_j - \lfloor \gamma n_j \rfloor < i \leq n_j \end{cases} \tag{21}$$

so that the two statistics

$$R_3 = \sum_{j=1}^{k} \widetilde{R}_{3,j} \text{ with}$$

$$\widetilde{R}_{3,j} = \sum_{i=1}^{n_j} c_{i,j} \left( R(X_{i,j}) - \frac{n_j+1}{2} \right) = \sum_{j=1}^{k} \left( \sum_{i=n_j-\lfloor \gamma n_j \rfloor+1}^{n_j} R(X_{i,j}) - \sum_{i=1}^{\lfloor \gamma n_j \rfloor} R(X_{i,j}) \right) \text{ and}$$

$$R_4 = \sum_{j=1}^{k} \widetilde{R}_{4,j} \text{ with}$$

$$\begin{aligned} \widetilde{R}_{4,j} &= \sum_{i=1}^{n_j} c_{i,j} \operatorname{sign}\left( R(X_{i,j}) - \frac{n_j+1}{2} \right) \\ &= \sum_{i=n_j-\lfloor \gamma n_j \rfloor+1}^{n_j} \operatorname{sign}\left( R(X_{i,j}) - \frac{n_j+1}{2} \right) - \sum_{i=1}^{\lfloor \gamma n_j \rfloor} \operatorname{sign}\left( R(X_{i,j}) - \frac{n_j+1}{2} \right) \end{aligned} \tag{22}$$

compare the sum of the most recent $\lfloor \gamma n_j \rfloor$ ranks (signs) with the sum of the first $\lfloor \gamma n_j \rfloor$ ranks (signs). Again the expectation of $R_3$ and $R_4$ is 0. Under the null hypothesis, the variances are given by

$$\mathrm{Var}(R_3) = \sum_{j=1}^{k} \frac{n_j(n_j+1)\lfloor \gamma n_j \rfloor}{6} \quad \text{and}$$

$$\mathrm{Var}(R_4) = \sum_{j=1}^{k} \mathrm{Var}(\widetilde{R}_{4,j}) \quad \text{with } \mathrm{Var}(\widetilde{R}_{4,j}) = \begin{cases} 2\frac{n_j}{n_j-1} \lfloor \gamma n_j \rfloor & , n_j \text{ even} \\ 2\lfloor \gamma n_j \rfloor & , n_j \text{ odd} . \end{cases} \tag{23}$$

Again the above variances are only valid if all observations have different values. If ties occur, one possibility, which leads to a loss of power but keeps the variances from (20) and (23) under the null hypothesis is to give random ranks to tied observations. Alternatives like average ranks, which reduce the loss of power compared to random ranks, are not considered here.

In addition to this, [1] also consider nonlinear rank statistics. In analogy to them we define for each period $j$

$$I_{i,i',j} = \begin{cases} 1 \text{ , if } X_{i,j} < X_{i',j} \\ 0 \text{ , otherwise} \end{cases} , \tag{24}$$

$i,i' \in \{1,\ldots,n\}, i \neq i'$. Under the null hypothesis of randomness, we have

$$\mathrm{E}(I_{i,i',j}) = \frac{1}{2} \text{ and } \mathrm{Var}(I_{i,i',j}) = \frac{1}{4} . \tag{25}$$

Based on the sign difference test [13] we define for $\mathscr{H}_R$

$$N_1 = \sum_{j=1}^{k} \widetilde{N}_{1,j} \text{ with } \widetilde{N}_{1,j} = \sum_{i=2}^{n_j} I_{i-1,i,j} \tag{26}$$

which counts the number of pairs for each period $j$, where the consecutive observation has a larger value and then sums these pairs over all periods. For each $j$ we have $n_j - 1$ differences. Under $\mathrm{H}_0$ and from (25) we get

$$\mathrm{E}(N_1) = \sum_{j=1}^{k} \frac{1}{2}(n_j - 1) \text{ and } \mathrm{Var}(N_1) = \sum_{j=1}^{k} \frac{1}{12}(n_j + 1) . \tag{27}$$

For each $j$ the distribution of $\sum_{i=2}^{n_j} I_{i-1,i,j}$ converges to a normal distribution [13]. Therefore the distribution of $N_1$ converges to a normal distribution, too.

Another test for $\mathscr{H}_R$ based on Cox and Stuart [3] is given by

$$N_2 = \sum_{j=1}^{k} \widetilde{N}_{2,j} \text{ with } \widetilde{N}_{2,j} = \sum_{i=1}^{\lfloor n_j/2 \rfloor} (n_j - 2i + 1) I_{i,n_j-i+1,j} . \tag{28}$$

Cox and Stuart [3] show that $N_2$ leads to the best weighted sign test with respect to the efficiency of a sign test of $\mathscr{H}_R$. The linear rank test statistics $R_1$ and $R_2$ and the procedure $S$ of Kendall compare all pairs of observations, while in (28) each observation is taken only for one comparison. Using (25) we get under $\mathrm{H}_0$

$$\mathrm{E}(N_2) = \sum_{j=1}^{k} \mathrm{E}(\widetilde{N}_{2,j}) \quad \text{with} \quad \mathrm{E}(\widetilde{N}_{2,j}) = \begin{cases} \frac{n_j^2}{8} & , n_j \text{ even} \\ \frac{n_j^2 - 1}{8} & , n_j \text{ odd} \end{cases}$$

$$\text{and } \mathrm{Var}(N_2) = \sum_{j=1}^{k} \frac{1}{24} n_j (n_j^2 - 1) . \tag{29}$$

Cox and Stuart [3] also introduce a best unweighted sign test, which can be formulated for $\mathscr{H}_R$ as follows

$$N_3 = \sum_{j=1}^{k} \widetilde{N}_{3,j} \text{ with } \widetilde{N}_{3,j} = \sum_{i=1}^{\nu_j} I_{i,n_j-\nu_j+i,j} \,. \tag{30}$$

The value $\nu_j \leq \frac{1}{2}n_j$ is taken to compare observations further apart. We get

$$\mathrm{E}(N_3) = \sum_{j=1}^{k} \frac{\nu_j}{2} \text{ and } \mathrm{Var}(N_3) = \sum_{j=1}^{k} \frac{\nu_j}{4} \tag{31}$$

under $\mathrm{H}_0$. Cox and Stuart [3] recommend $\nu_j = \frac{1}{3}n_j$.

Again a splitting with small $n_1 = \ldots = n_k$ and large $k$ leeds asymptotically to a normal distribution for all introduced test statistics, as $k$ i.i.d. rv are added.

## 3 Comparison of the Nonparametric Tests for $\mathscr{H}_R$

Now we compare the different tests presented in Sect. 2 for different sample sizes and splitting factors and for various alternatives. We consider the time series model

$$X_{i,j} = a_{i,j} + E_{i,j} \; j = 1,\ldots,k, \; i = 1,\ldots,n_j, \tag{32}$$

where $E_{1,1},\ldots,E_{n_k,k}$ are Gaussian white noise with expected value 0 and constant variance $\sigma_E^2 = 1$. $X_{i,j}$ is the $i$–th observation for season $j$. For simplicity we fix the number of seasons to $k = 4$ and assume that each season has the same sample size $n_1$. Furthermore, the slopes are given by $a_{1,j} \leq \ldots \leq a_{n_1,j}$. We are interested in particular in three different kinds of monotone trends, with the same trend structure in each season. This means that for each $j$ we have the same slopes. With $a_{i,j} = i\theta$ we achieve a linear trend, where the parameter $\theta$ controls the slope of the straight line. We also consider a concave case with $a_{i,j} = \theta\sqrt{n_1 i}$, and a convex case with $a_{i,j} = \theta i^2/n_1$, so that all trends increase to $\theta n_1$. We consider sample sizes $n \in \{12, 24, 32, 48, 64, 96, 120\}$ and splittings into $\widetilde{k} \in \{1, 4, 8, 12, 16, 24, 32\}$ groups whenever $\widetilde{n}_1 = n/\widetilde{k}$ is an integer. We do not consider splittings with $\widetilde{n}_1 = 2$ as here $R_3$ and $R_4$ for $\gamma = \frac{1}{3}$ as well as $N_3$ with $\nu_1 = \ldots = \nu_{\widetilde{k}} = \frac{1}{3}$ are not defined. The other test statistics are equivalent in this case, as they all consider an unweighted ranking of two observations in each splitting. With $\widetilde{k} = 1$ the unsplitted case is also taken into account. In case of seasonal effects the power of all tests will probably be reduced if $\widetilde{k} = 1$ is chosen. We compare the power of the tests of Sect. 2 for all reasonable combinations of $\widetilde{k}$ and $n$ from above and take 1000 random samples from (32) for each combination. We use the asymptotic versions of the tests at a significance level of $\alpha = 0.05$. The percentage cases of rejections of $\mathrm{H}_0$ estimate the power of the several test procedures. Here we only consider the case of an upward trend, i.e. $\theta > 0$. We consider the linear, the convex and the concave case from above and calculate the power of all tests for $\theta \in \{0.01, 0.02, \ldots, 0.49, 0.50\}$. To achieve monotone power functions, we use the R–function `isotone` from the R–package `EbayesThresh` for monotone least squares regression to smooth the simulated power curves ([14, 15]).

Firstly we compare the weighted record statistics. For $n \geq 64$ all power functions take values close to 1, independently of the splitting factor $\widetilde{k}$, if a linear trend with $\theta > 0.1$ exists. In the concave case only $U^o$ and $T_2$ with $\widetilde{k} = 1$ perform worse for $n = 64$. An explanation for this is the strength of the slope. A positive concave trend increases less towards the end of the time series. Hence there will be fewer records at the end of the time series and $U^o$ will perform worse than $L^r$. As our version of $T_2$ also uses $U^o$ we receive similar results for this test statistic. In the convex case similar results can be obtained for $L^r$ as a convex upward trend of the original sequence means a concave downward trend of the negative reversed series. The power functions of the record tests for $\widetilde{k} = 1$ and $\widetilde{k} = 4$ can be seen in Fig. 1 for the linear, the concave and the convex case. Looking also at other sample sizes $n$ in the linear case (see Fig. 2), we find that $T_3$ performs best among the record tests in most of the cases. Generally, the power of the record tests gets larger in the above situations, if a larger $\widetilde{k}$ is chosen. Only $T_3$ performs better for a medium value of $\widetilde{k}$, e.g. $\widetilde{k} = 4$ for $n = 32$ or $\widetilde{k} = 12$ for $n = 96$. The previous findings are confirmed in the case of a convex or concave trend.

In Fig. 3 the power functions of the rank tests are shown, when different $\widetilde{k}$ for a fixed $n = 64$ are used. We show the concave case here, because the differences are qualitatively the same, but slightly bigger than for the linear or the convex trend. The seasonal Kendall–Test $S$ and Spearman–Test $R_1$ perform best, when a small $\widetilde{k}$ is used. Conclusions about an optimal splitting for the other rank tests are hard to state. If $\widetilde{k}$ is large compared to $n$, the power of the tests is reduced for most of the situations. However, generally we observe for all these tests (except $N_1$) good results, if $\widetilde{k} = 4$ is chosen. $N_1$ performs worse than the other tests in most situations even though it is the only test statistic with an increasing power in case of a larger splitting factor $\widetilde{k}$. From the rank tests $S$ and $R_1$ achieve the largest power in most situations. Comparing the best rank tests $S$ and $R_1$ with $\widetilde{k} = 4$ and the best record tests $T_3$ and $T_4$ with a large splitting factor $\widetilde{k} = 4$, $S$ and $R_1$ have a larger power in every situation.

Next we consider a situation with autocorrelated data. Here the hypothesis of randomness is not fulfilled, but no monotone trend exists. It is interesting which test procedures are sensitive to autocorrelation in the sense that they reject $H_0$ even though there is no monotone trend. We consider an autoregressive process of first order (AR(1))

$$E_t = \rho E_{t-1} + \varepsilon_t \ , \ t = 1, \ldots, n \ , \tag{33}$$

with autocorrelation coefficient $\rho$, i.e. we assume the sequence $E_1, \ldots, E_n$ to be autocorrelated with correlation $\rho$ and hence the autocorrelation within $E_{1,j}, \ldots, E_{n_1,j}$ with $E_{i,j} = E_{k(i-1)+j}$ is smaller than $\rho$. The innovations $\varepsilon_{1,j}, \ldots, \varepsilon_{n_1,j}$ are i.i.d. normally distributed random variables with expectation 0 and variance $\sigma_\varepsilon^2$, where

$$\sigma_\varepsilon^2 = (1-\rho^2)\sigma_E^2 = (1-\rho^2) \tag{34}$$

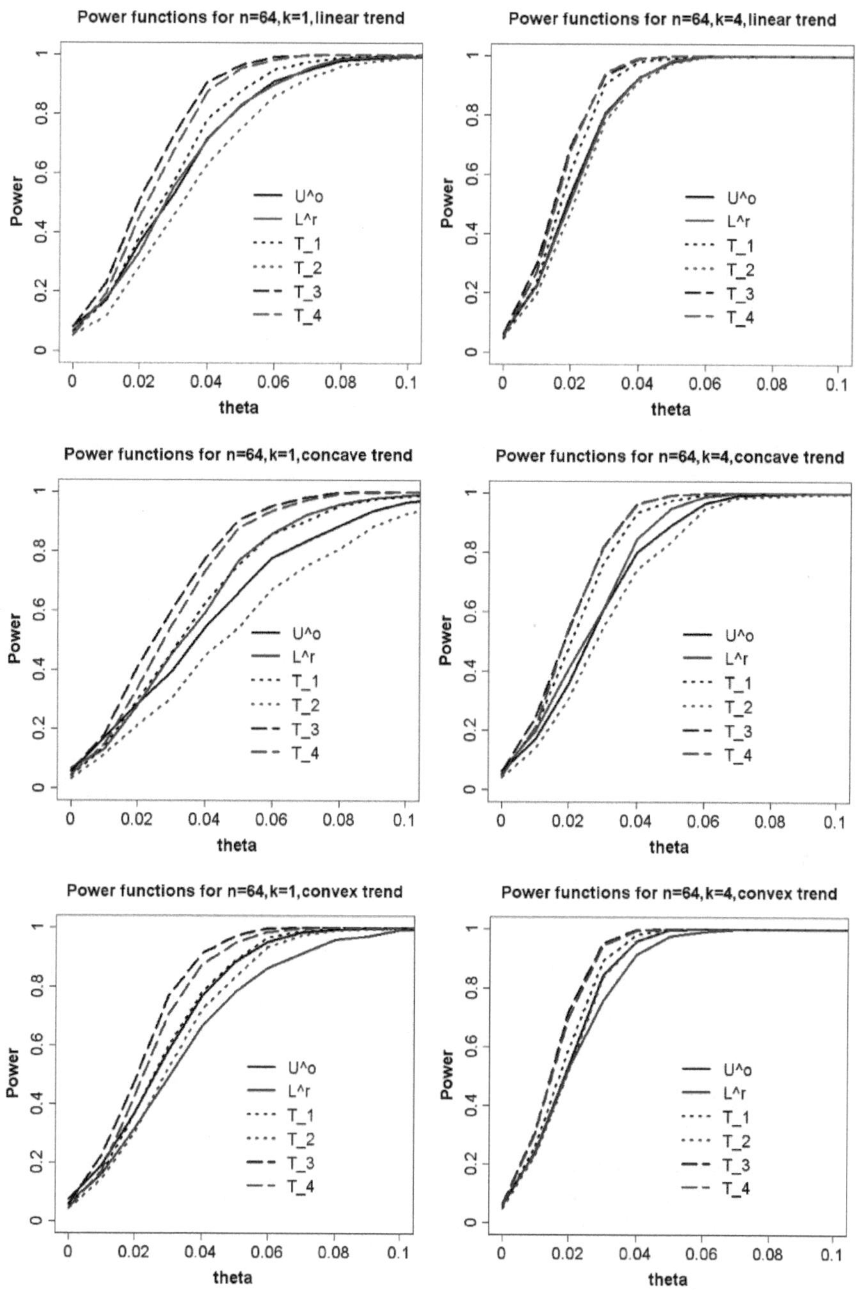


**Fig. 1** Power functions of the record tests for $n = 64$, small $\theta$ and $\widetilde{k} = 1$ (*left*) and $\widetilde{k} = 4$ (*right*)

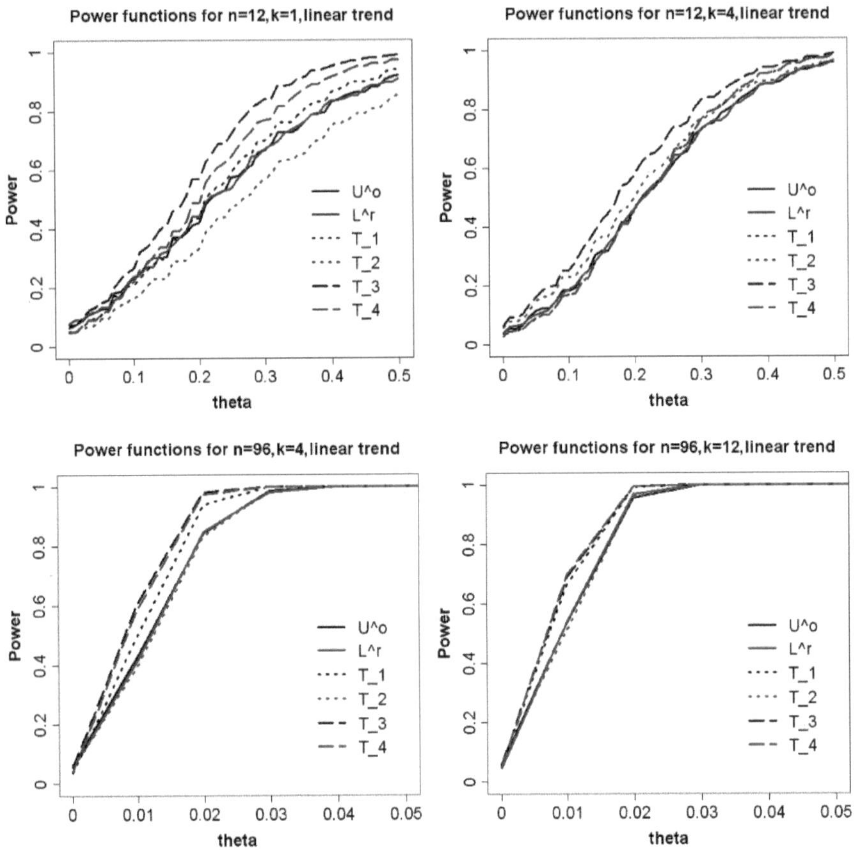


**Fig. 2** Power functions of the record tests for $n = 12$ (*top*) and $n = 96$ (*bottom*) for different $\widetilde{k}$

as we want to keep $\sigma_E^2$ equal to 1 again. We vary $\rho$ in $\{0.025, 0.05, \ldots, 0.875, 0.9\}$. The resulting detection rates of the record tests can be seen in Fig. 4 for $n = 96$ and different values of $\widetilde{k}$. $T_3$ is more sensitive to positive autocorrelation than $T_1$, $T_2$ and $T_4$ if a small $\widetilde{k}$ is used, but this difference vanishes for a large $\widetilde{k}$. The better performance of $T_1$, $T_2$ and $T_4$ for small $\widetilde{k}$ can be explained by the fact that they subtract statistics which become large in case of monotonically decreasing sequences from statistics which become large in case of monotonically increasing sequences. Positive autocorrelations cause both patterns to occur so that the effects cancel out.

For the rank tests we get the following findings: $N_2$ becomes robust against autocorrelations $\rho \leq 0.6$ for larger sample sizes $n \geq 48$, if we choose $\widetilde{k}$ so that we have three observations in each split. We observe for the pairs $n = 48$, $\widetilde{k} = 16$ and $n = 96$, $\widetilde{k} = 32$ for most of the values of $\rho$ a power of less than $\alpha = 0.05$. If we choose a splitting factor leading to $n_1 > 3$ this robustness is lost (see Fig. 5). $N_1$ behaves the most insensitive against autocorrelation for a large $\widetilde{k}$, but $N_1$ was also the test with the smallest power if a trend exists. For the other tests we have for

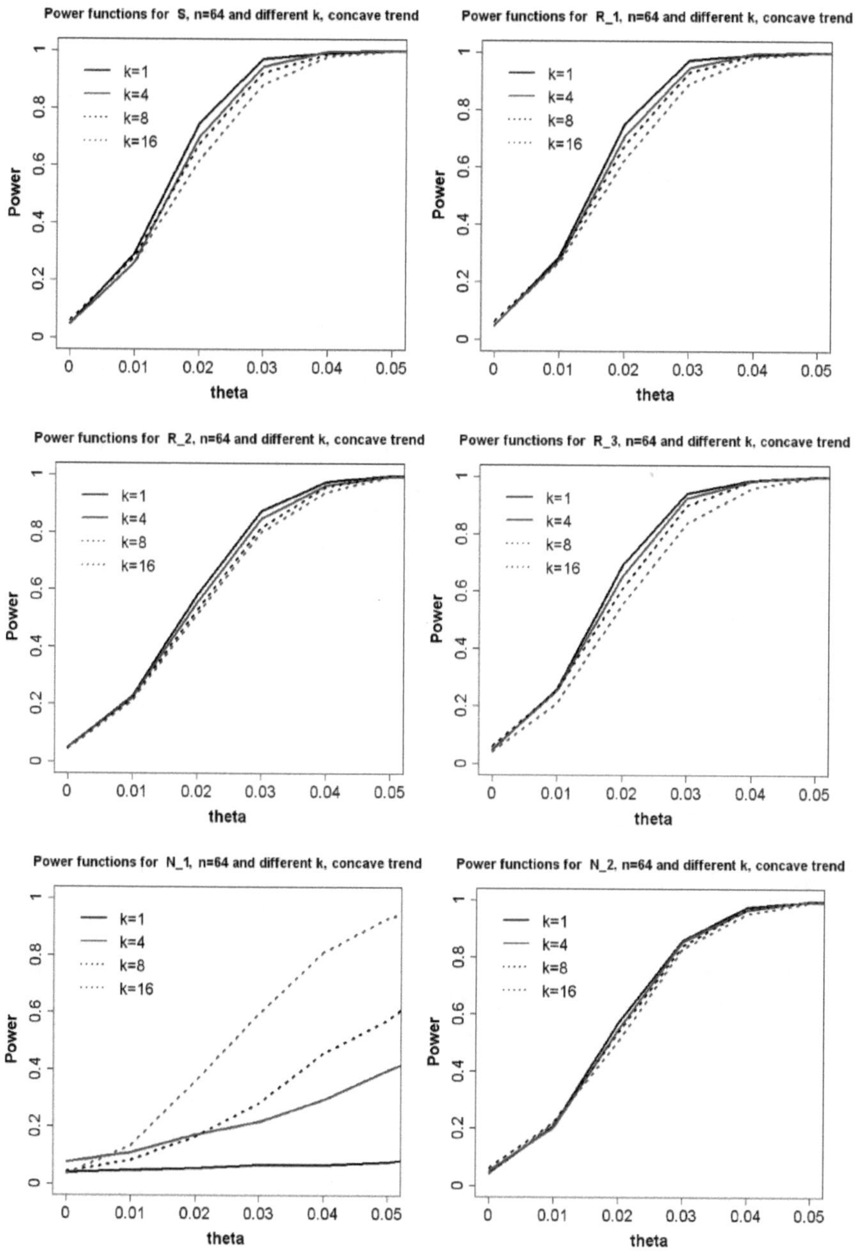


**Fig. 3** Power functions of the rank tests for different $\widetilde{k}$ with $n = 64$ and a concave trend

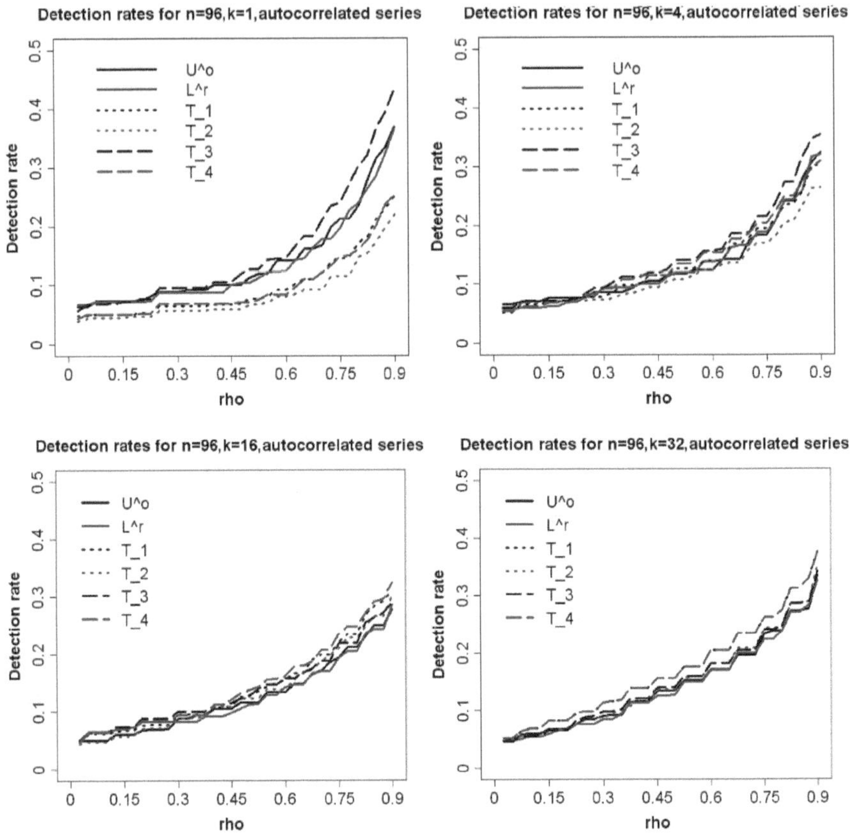


**Fig. 4** Detection rates of the record tests for $n = 96$ and different $\tilde{k}$ with autocorrelation

a fixed $n$ a higher detection rate, when a smaller splitting factor $\tilde{k}$ is used. If we compare the record tests with the rank tests, we find that $T_3$ reacts less sensitive to autocorrelation than the rank tests in most situations.

## 4 Analysis of the Climate Time Series from Potsdam

Now the methods from Sect. 2 are applied to some real time series data. The two series analysed here consist of the monthly observations of the mean air temperature and the total rainfall in Potsdam between January 1893 and April 2008. There are no missing values. The secular station in Potsdam is the only meteorological station in Germany for which daily data have been collected during a period of over 100 years without missings. The measures are homogeneous, what is due to the facts that the

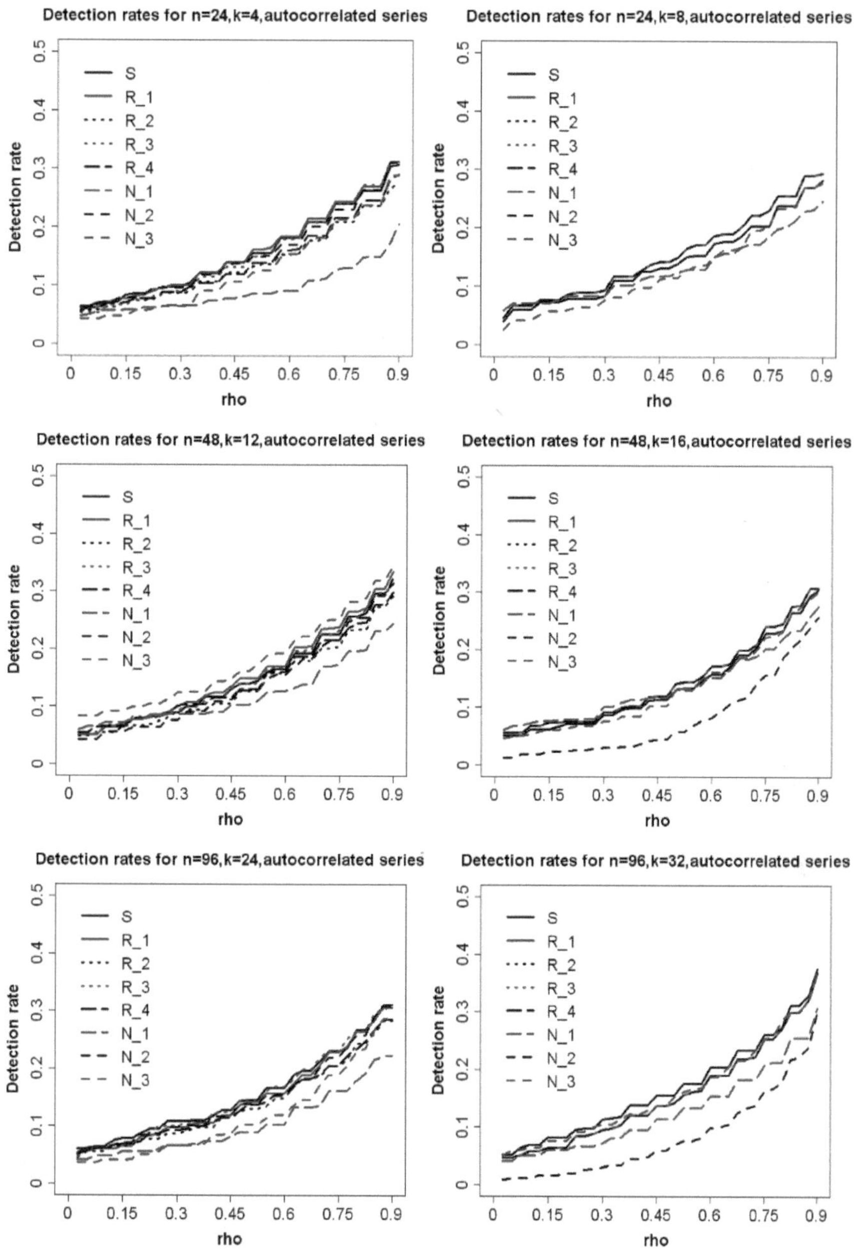


**Fig. 5** Detection rates of the rank tests with $n_1 = 4$ (*top*) and $n_1 = 3$ (*bottom*) observations in each splitting with autocorrelation

station has never changed its position, the measuring field stayed identical and the sort of methods, prescriptions and instruments, which are used for the measuring, have been kept.

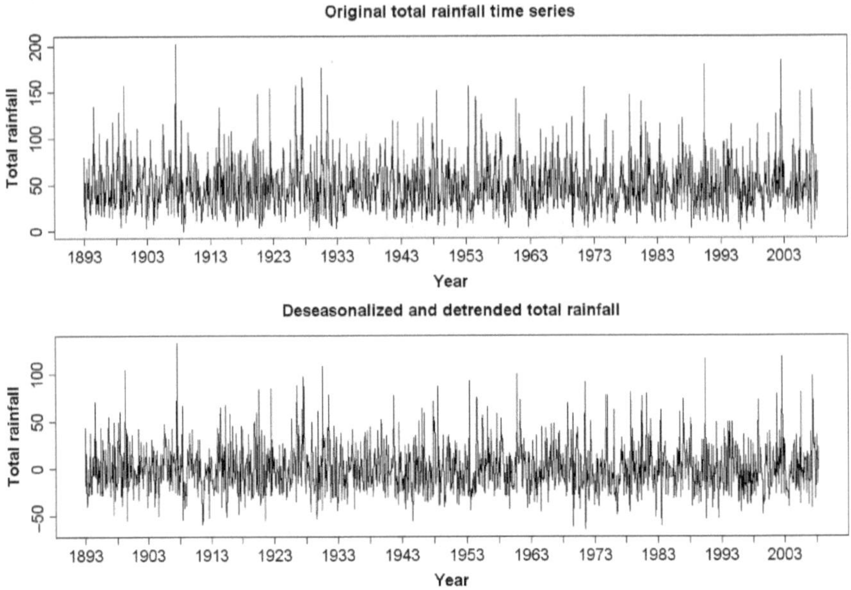


**Fig. 6** Original (*top*) and detrended and deseasonalized (*bottom*) total rainfall time series

Before the methods from Sect. 2 can be applied, we have to check if the assumptions are fulfilled. Independence of the observations can be checked with the autocorrelation function (ACF) and the partial autocorrelation function (PACF). Before this we detrend the time series by subtracting a linear trend. We also deseasonalize the time series by estimating and subtracting a seasonal effect for each month. The original and the detrended deseasonalized time series can be found in Fig. 6 for the total rainfall and in Fig. 7 for the mean temperature. The autocorrelation functions of the detrended and deseasonalized time series show positive autocorrelations at small time lags in case of the temperature and no correlation in case of the rainfall (see Fig. 8). In the former case, a first order autoregressive model with a moderately large AR(1) coefficient gives a possible description of the correlations. We use the test statistics from Sect. 2 to test the hypothesis of randomness against the alternative of an upward trend in both time series.

We consider all test statistics except $L^o$ and $U^r$ as these tests are only useful to detect a downward trend. As we have in both time series monthly observations for more than 115 years, we choose the splitting factor $\widetilde{k}$ as multiples of 12, more precisely $\widetilde{k} \in \{12, 24, 60, 120, 240, 360\}$. This guarantees that even $R_3$, $R_4$ (with $\gamma = \frac{1}{3}$) and $N_3$ (with $\nu_j = \frac{1}{3}n_j$) can be computed for each split. For every test procedure we use the asymptotic critical values, which seems to be reasonable for the above $\widetilde{k}$. The resulting p–values can be seen in Table 1 for the total rainfall time series and in Table 2 for the mean temperature.

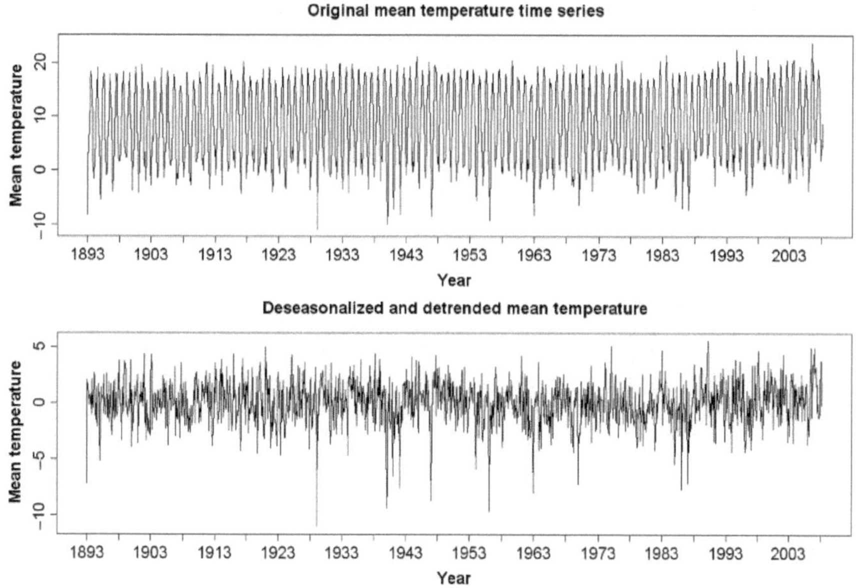


**Fig. 7** Original (*top*) and detrended and deseasonalized (*bottom*) mean temperature time series

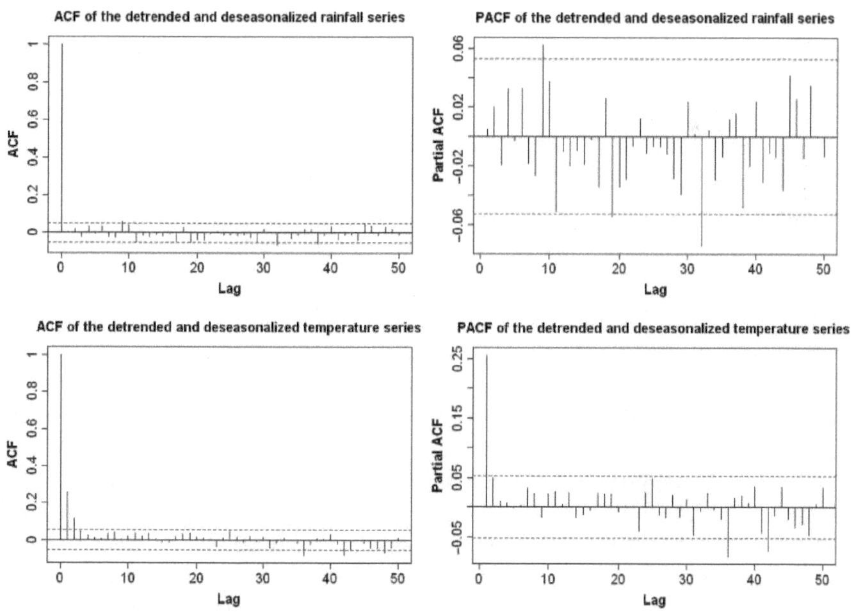


**Fig. 8** Autocorrelation (*left*) and partial autocorrelation function (*right*) of the detrended and deseasonalized rainfall (*top*) and temperature time series (*bottom*)

**Table 1** p–values for the total rainfall time series (in percent)

| $\widetilde{k}$ | 12 | 24 | 60 | 120 | 240 | 360 |
|---|---|---|---|---|---|---|
| $U^o$ | 6.4 | 40.9 | 11.7 | 18.3 | 11.1 | 6.1 |
| $L^r$ | 9.3 | 21.3 | 32.4 | 26.8 | 38.7 | 7.9 |
| $T_1$ | 4.2 | 34.9 | 14.2 | 7.9 | 14.8 | 9.8 |
| $T_2$ | 4.3 | 31.8 | 3.3 | 11.9 | 12.8 | 7.4 |
| $T_3$ | 2.3 | 23.7 | 12.9 | 15.7 | 17.8 | 4.6 |
| $T_4$ | 1.9 | 22.5 | 6.0 | 7.8 | 17.6 | 7.5 |
| $S$ | 17.2 | 12.8 | 28.1 | 25.6 | 24.1 | 9.1 |
| $R_1$ | 19.4 | 15.7 | 33.2 | 39.2 | 37.5 | 13.0 |
| $R_2$ | 26.7 | 19.2 | 36.3 | 42.2 | 33.1 | 26.5 |
| $R_3$ | 44.0 | 38.6 | 57.0 | 58.9 | 45.5 | 11.1 |
| $R_4$ | 48.7 | 44.8 | 63.4 | 61.8 | 41.2 | 20.5 |
| $N_1$ | 8.2 | 35.6 | 32.4 | 18.6 | 5.1 | 5.8 |
| $N_2$ | 4.6 | 5.1 | 58.4 | 61.7 | 49.1 | 20.0 |
| $N_3$ | 61.1 | 61.1 | 46.1 | 46.1 | 46.1 | 14.6 |

**Table 2** p–values for the mean temperature time series (in percent)

| $\widetilde{k}$ | 12 | 24 | 60 | 120 | 240 | 360 |
|---|---|---|---|---|---|---|
| $U^o$ | 0.00 | 0.00 | 0.00 | 0.00 | 0.00 | 0.00 |
| $L^r$ | 0.00 | 0.03 | 0.01 | 0.00 | 0.00 | 0.00 |
| $T_1$ | 0.00 | 0.00 | 0.00 | 0.00 | 0.00 | 0.00 |
| $T_2$ | 0.00 | 0.00 | 0.00 | 0.00 | 0.00 | 0.00 |
| $T_3$ | 0.00 | 0.00 | 0.00 | 0.00 | 0.00 | 0.00 |
| $T_4$ | 0.00 | 0.00 | 0.00 | 0.00 | 0.00 | 0.00 |
| $S$ | 0.00 | 0.00 | 0.00 | 0.00 | 0.00 | 0.00 |
| $R_1$ | 0.00 | 0.00 | 0.00 | 0.00 | 0.00 | 0.00 |
| $R_2$ | 0.00 | 0.00 | 0.00 | 0.00 | 0.00 | 0.00 |
| $R_3$ | 0.00 | 0.00 | 0.00 | 0.00 | 0.00 | 0.00 |
| $R_4$ | 0.00 | 0.00 | 0.00 | 0.00 | 0.00 | 0.00 |
| $N_1$ | 97.42 | 13.40 | 5.04 | 21.07 | 0.05 | 0.06 |
| $N_2$ | 0.00 | 0.00 | 0.00 | 0.00 | 0.00 | 0.00 |
| $N_3$ | 0.00 | 0.00 | 0.00 | 0.00 | 0.00 | 0.00 |

For the total rainfall time series the record tests $T_1$, $T_2$, $T_3$ and $T_4$ with $\widetilde{k} = 12$ detect a monotone trend at a significance level of $\alpha = 0.05$. From the rank tests only $N_2$ finds a monotone trend at this $\alpha$. Using a larger splitting factor we only find a monotone trend with $T_2$ for $\widetilde{k} = 60$. Of course we need to keep in mind that we perform multiple testing and thus expect about four significant test statistics among the more than 80 tests performed here even if there is no trend at all.

All tests except $N_1$ detect a monotone trend in the temperature time series for all splittings $\widetilde{k}$. The statistic $N_1$ only detects a monotone trend, if $\widetilde{k}$ is large. But as all tests need the assumption of independence, the results of Table 2 can not be interpreted as p–values of unbiased tests. This is why we deseasonalize the temperature time series and fit an AR(1)–Model to the deseasonalized series by maximum likelihood. If the data generating mechanism is an AR(1) process with uncorrelated innovations, then the residuals of the fitted AR(1) model are asymptotically uncorrelated. The residuals are even asymptotically independent, if the innovations are i.i.d. The residuals are asymptotically normal, if the innovations are normally distributed (see Section 5.3 of [2]). Looking at the plot of the scaled residual time series in Fig. 9 and its ACF in Fig. 10, we do not find significant autocorrelations between the residuals. However, the residuals do not seem to be identically normally distributed, as we can find some outliers in the residual plot. Table 3 shows the p–values of the record and rank tests for the residuals. We find mostly larger p–values than in Table 2, but again all tests except $N_1$ detect a positive monotone trend at $\alpha = 0.05$, what confirms the previous findings.

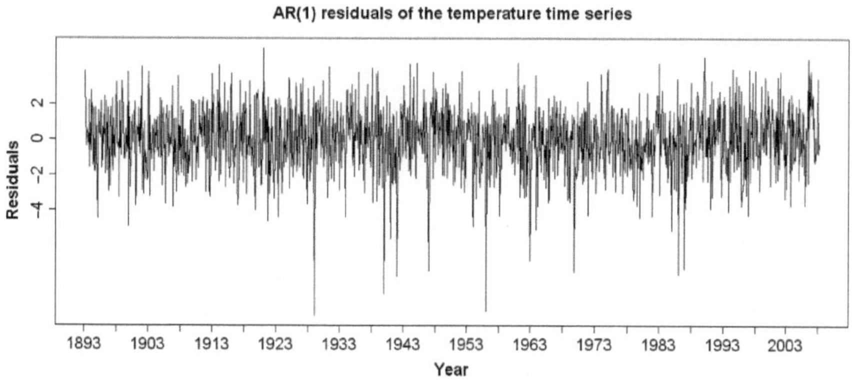


**Fig. 9** Residuals of the temperature time series obtained from fitting an AR(1) model to the deseasonalized temperature time series

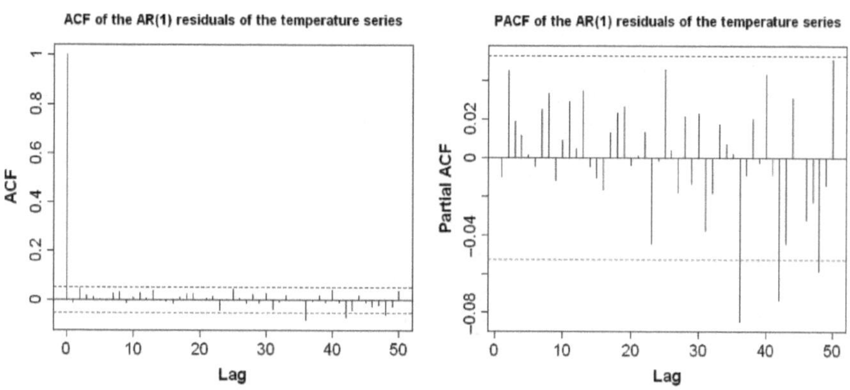


**Fig. 10** ACF (*left*) and PACF (*right*) of the AR(1) residuals of the deseasonalized temperature series

**Table 3** p–values for the residual temperature time series (in percent)

| $\tilde{k}$ | 12 | 24 | 60 | 120 | 240 | 360 |
|---|---|---|---|---|---|---|
| $U^o$ | 0.30 | 0.19 | 0.07 | 0.07 | 0.00 | 0.15 |
| $L^r$ | 2.77 | 0.41 | 0.13 | 0.93 | 0.24 | 0.09 |
| $T_1$ | 0.01 | 0.01 | 0.00 | 0.00 | 0.00 | 0.01 |
| $T_2$ | 0.44 | 0.07 | 0.05 | 0.08 | 0.00 | 0.12 |
| $T_3$ | 0.05 | 0.01 | 0.00 | 0.01 | 0.00 | 0.03 |
| $T_4$ | 0.02 | 0.00 | 0.00 | 0.01 | 0.00 | 0.01 |
| $S$ | 0.00 | 0.00 | 0.00 | 0.00 | 0.00 | 0.01 |
| $R_1$ | 0.00 | 0.00 | 0.00 | 0.00 | 0.00 | 0.00 |
| $R_2$ | 0.00 | 0.00 | 0.00 | 0.00 | 0.00 | 0.02 |
| $R_3$ | 0.00 | 0.00 | 0.00 | 0.00 | 0.00 | 0.00 |
| $R_4$ | 0.00 | 0.00 | 0.00 | 0.00 | 0.00 | 0.01 |
| $N_1$ | 93.10 | 23.01 | 11.80 | 53.56 | 0.10 | 1.91 |
| $N_2$ | 0.00 | 0.00 | 0.00 | 0.00 | 0.01 | 0.01 |
| $N_3$ | 0.01 | 0.03 | 0.00 | 0.00 | 0.00 | 0.00 |

## 5 Conclusions

We have considered nonparametric tests for trend detection in time series. We have not found large differences between the power of the different tests. All tests based on records or ranks react sensitive to autocorrelations. Our results confirm findings by Diersen and Trenkler that $T_3$ can be recommended among the record tests because of its good power and its simplicity. Robustness of $T_3$ against autocorrelation can be achieved for the price of a somewhat reduced power by choosing a large splitting factor $\tilde{k}$. However, even higher power can be achieved by applying a nonparametric rank test like the seasonal Kendall–Test $S$ or the Spearman–Test $R_1$ with a small $\tilde{k}$, even though for the price of a higher sensitivity against positive autocorrelation. The power of all rank tests except $N_1$ gets smaller, if a larger splitting factor is used. For $N_1$ a larger splitting factor enlarges the power, but $N_1$ is not recommended to use, as even with a large splitting factor it is less powerful than the other tests. From the rank tests the test $N_2$ seems robust against autocorrelations below 0.6, if only a few observations are taken in each block. Another possibility to reduce the sensitivity to autocorrelation is to fit a low order AR model and consider the AR residuals. We have found a significant trend in the time series of the monthly mean temperature in Potsdam both when using the original data and the AR(1) residuals. Since in the plot of the scaled residuals for this series we find some outliers, another interesting question for further research is the robustness of the several tests against atypical observations.

## References

[1] Aiyar, R.J., Guillier, C.L., Albers, W.: Asymptotic relative efficiencies of rank tests for trend alternatives. J. Am. Stat. Assoc. **74**, 227–231 (1979)
[2] Brockwell, P.J., Davis, R.A.: Introduction to time series and forecasting. Springer, New York (2002)
[3] Cox, D.R., Stuart, A.: Some quick sign tests for trend in location and dispersion. Biometrika **42**, 80–95 (1955)
[4] Daniels, H.E.: Rank correlation and population models. J. Roy. Stat. Soc. B **12**, 171–181 (1950)
[5] Diersen, J., Trenkler, G.: Records tests for trend in location. Statistics **28**, 1–12 (1996)
[6] Diersen, J., Trenkler, G.: Weighted record tests for splitted series of observations, In: Kunert, J., Trenkler, G. (eds.) Mathematical Statistics with Applications in Biometry, Festschrift in Honour of Prof. Dr. Siegfried Schach, pp. 163–178. Eul, Lohmar (2001)
[7] Foster, F.G., Stuart, A.: Distribution-free tests in time-series based on the breaking of records. J. Roy. Stat. Soc. B **16**, 1–22 (1954)
[8] Hirsch, R.M., Slack, J.R., Smith, R.A.: Techniques of trend analysis for monthly water quality data. Water Resour. Res. **18**, 107–121 (1982)
[9] Hirsch, R.M., Slack. J.R.: A nonparametric trend test for seasonal data with serial dependence. Water Resour. Res. **20**, 727–732 (1984)
[10] Kendall, M.G.: A new measure of rank correlation. Biometrika **30**, 81–93 (1938)
[11] Kendall, M.G., Gibbons, J.D.: Rank correlation methods. Arnold, London (1990)
[12] Mann, H.B.: Non-parametric tests against trend. Econometrica **13**, 245–259 (1945)
[13] Moore, G.H., Wallis, W.A.: Time series significance tests based on signs of differences. J. Am. Stat. Assoc. **38**, 153–164 (1943)
[14] R Development Core Team: R: A language and environment for statistical computing, R Foundation for Statistical Computing. Vienna, Austria, ISBN 3-900051-07-0, URL: http://www.R-project.org (2008)
[15] Silverman, B.W.: EbayesThresh: Empirical Bayes thresholding and related methods. R package version 1.3.0., URL: http://www.bernardsilverman.com (2005)

# Nonparametric Trend Tests for Right-Censored Survival Times

Sandra Leissen, Uwe Ligges, Markus Neuhäuser, and Ludwig A. Hothorn

**Abstract** In clinical dose finding studies or preclinical carcinogenesis experiments survival times may arise in groups associated with ordered doses. Here interest may focus on detecting dose dependent trends in the underlying survival functions of the groups. So if a test is to be applied we are faced with an *ordered alternative* in the test problem, and therefore a *trend test* may be preferable. Several trend tests for survival data have already been introduced in the literature, e.g., the logrank test for trend, the one by Gehan [4] and Mantel [12], the one by Magel and Degges [11], and the modified ordered logrank test by Liu et al. [10], where the latter is shown to be a special case of the logrank test for trend. Due to their similarity to *single contrast tests* it is suspected that these tests are more powerful for certain trends than for others. The idea arises whether *multiple contrast tests* can lead to a better overall power and a more symmetric power over the alternative space. So based on the tests mentioned above two new multiple contrast tests are constructed. In order to compare the *conventional* with the *new tests* a simulation study was carried out. The study shows that the new tests preserve the nominal level satisfactory from a certain sample size but fail to conform the expectations in the power improvements.

## 1 Introduction

In clinical dose-finding studies we might look at the observed survival times of patients allocated to $k$ dose groups $x_{ij}, i = 1,\dots,k, j = 1,\dots,n_i$, where $n_i$ is the sample size of the $i$-th group. A common starting point for every patient is the beginning of the treatment with the $k$-th dose. The endpoint can be varying, eg., the release of a symptom or death due to a certain disease or condition, so that literally a survival time is observed. We want to concentrate on the latter case which we might

Sandra Leissen
Fakultät Statistik, Technische Universität Dortmund, D-44221 Dortmund, Germany
leissen@statistik.tu-dortmund.de

encounter in oncology or cardiology studies. The starting point is usually known here for every patient. However, due to losses-to-follow-up's or deaths for other reasons, the exact time of the occurrence of the endpoint may not be available for every patient. So we have to deal here (partly) with right-censored survival times.

In order to put this into a more formal framework, consider $T_{ij}, i = 1, \ldots, k, j = 1, \ldots, n_i$, to be the independent random variables of the survival times to occur with outcomes $t_{11}, \ldots, t_{kn_k}$. Further let $C_{ij}, i = 1, \ldots, k, j = 1, \ldots, n_i$, be independent random variables with outcomes $c_{11}, \ldots, c_{kn_k}$, where $C_{ij}$ reflects the censoring time corresponding to the survival time $T_{ij}$. For each of the $k$ groups it is assumed that the survival times $T_{i1}, \ldots, T_{in_k}$ of the group populations and the corresponding censoring times $C_{i1}, \ldots, C_{in_k}$ follow the same distribution respectively. Further it is supposed that the $(T_{ij}, C_{ij})$ are pairwise independent. It is not possible to observe both $t_{ij}$ and $c_{ij}$ for a study object $(i, j)$, but $x_{ij}$, the outcome of $X_{ij} = \min(T_{ij}, C_{ij})$. Additionally, the *status* of every survival time is known. If $t_{ij} \leq c_{ij}$ the observation is uncensored. If $t_{ij} > c_{ij}$ the observation is right-censored and we mark the censored survival time with $x_{ij}^*$.

The main question in such a dose-finding study could be whether the survival time increases with the dose. So we look at the survival functions of the $k$ groups given by

$$S_i(t) := P(T_{ij} \geq t) = 1 - F_i(t),\ i = 1, \ldots, k,$$

where $F_i(t)$ denotes the distribution function of the survival times of the patients in group $i$. Now we expect the order of the doses to be transferred to the corresponding survival functions of the groups. So if a test is to be used our test problem reads

$$\mathrm{H}_0 : S_1 = \ldots = S_k \quad \text{vs.} \quad \mathrm{H}_< : S_1 \leq \ldots \leq S_k \text{ with } S_1 < S_k. \tag{1}$$

Another example for such a situation arises in preclinical carcinogenesis experiments. Here the starting point might be again the begin of a treatment with a certain dose. The endpoint is often chosen to be the occurrence of the first tumour. The question of interest here is whether the risk of a tumour increases with the dose. The according test problem can be formulated as

$$\mathrm{H}_0 : S_1 = \ldots = S_k \quad \text{vs.} \quad \mathrm{H}_> : S_1 \geq \ldots \geq S_k \text{ with } S_1 > S_k. \tag{2}$$

Both test problems show a trend in the survival functions in the alternatives. Therefore, the test to be applied should be sensible for such *ordered alternatives*. Various so-called *trend tests* have already been proposed in the literature, eg., the logrank test for trend (cf. Collett [3]), the one by Gehan [4] and Mantel [12], the one by Magel and Degges [11], and the modified ordered logrank test by Liu et al. [10].

The alternative space of the test problem (1) and (2) is big for $k \geq 3$ and increases with $k$. Therefore, it is further desirable that a trend test shows a symmetric distribution of its power over the alternative space. It is suspected that the trend tests mentioned above are not satisfying in this regard. This suspicion is due to the structural similarity of the tests to *single contrast tests* as they are known for the classical $k$-sample problem with comparable ordered test problems to (1) and (2) (cf. section 3). For the latter setting Mukerjee et al. [13] describe *multiple contrast tests*. The idea arises to transfer this methodology to the setting of right-censored survival times with the test problems (1) and (2). From this we expect to gain *new trend tests* with a higher overall power and, more importantly, a more symmetric power over the alternative space. So a new trend test based on the logrank test for trend as well as a new one based on the trend test by Gehan [4] and Mantel [12] are built using the construction principle of multiple contrast tests. The modified logrank test is not further regarded since this test is a special case of the logrank test for trend (which is shown in the next section). A comprehensive simulation study including many different scenarios of, e.g., distributions, sample sizes, degrees of censoring and ties, is carried out to compare the *conventional* with the new tests.

The conventional tests are introduced in the next section. In section 3 the setting of single and multiple contrast tests is established and the two new trend tests are introduced. The experimental design and the results of the simulation study are presented in section 4. Note that only test problem (1) is considered in the following. For dealing with test problem (2) the order of the groups can be simply reversed, so that test problem (1) results again. Besides, for every test presented here, a corresponding version for test problem (2) as well as for the test problem

$$\mathrm{H}_0 : S_1 = \ldots = S_k \quad \text{vs.} \quad \mathrm{H}_{<,>} : \mathrm{H}_< \text{ or } \mathrm{H}_> ,$$

if possible, is given in Leissen [9].

## 2 Conventional Trend Tests

### *2.1 The logrank test for trend*

Consider $k$ groups of survival data with distinct event times $s_j, j = 1, \ldots, m$. For each event time let $n_{1j}, \ldots, n_{kj}$ be the number of objects at risk and $d_{1j}, \ldots, d_{kj}$ be the number of events respectively. Further let $n_{.j} := \sum_{i=1}^k n_{ij}$ and $d_{.j} := \sum_{i=1}^k d_{ij}$. For fixed $d_{.j}, n_{1j}, \ldots, n_{kj}$ the $d_{1j}, \ldots, d_{kj}$ build a random vector following under $\mathrm{H}_0$ a multivariate hypergeometric distribution with density

$$f(d_{1j}, \ldots, d_{kj}) = \frac{\binom{n_{1j}}{d_{1j}} \cdot \ldots \cdot \binom{n_{kj}}{d_{kj}}}{\binom{n_{.j}}{d_{.j}}}$$

and expected values

$$\mathrm{E}(d_{ij}) = n_{ij} \frac{d_{.j}}{n_{.j}} =: e_{ij}$$

in the marginal distributions. For each group consider the sum of the observed and expected number of events over the single event times

$$L_i = \sum_{j=1}^{m} (d_{ij} - e_{ij}),\ i = 1, \ldots, k, \tag{3}$$

and the weighted sum of the $L_i$ over the groups

$$LT_{(w)} = \sum_{i=1}^{k} w_i L_i = \sum_{i=1}^{k} w_i (d_{i.} - e_{i.}). \tag{4}$$

If $H_0$ is true, it holds that $E(LT_{(w)}) = 0$ and that

$$\mathrm{Var}(LT_{(w)}) = \sum_{j=1}^{m} \frac{d_{.j}(n_{.j} - d_{.j})}{(n_{.j} - 1)} \left( \sum_{i=1}^{k} w_i^2 \frac{n_{ij}}{n_{.j}} - \left( \sum_{i=1}^{k} w_i \frac{n_{ij}}{n_{.j}} \right)^2 \right) \tag{5}$$

(cf. Leissen [9]), so that the statistic

$$LRT_{(w)} = LT_{(w)} \Big/ \sqrt{\mathrm{Var}(LT_{(w)})} \tag{6}$$

follows a standard normal distribution asymptotically. The weights $w_i$ can be chosen arbitrarily (in particular they do not have to sum up to 1) although they should be chosen sensibly. Collett [3] indicates that linear weights are often chosen to express a linear trend among the groups. A relatively high value of a $L_i$ indicates that the survival function $S_i$ of group $i$ is stochastically smaller than those of the other groups since more events occur than expected under $H_0$. So with weights $(k, \ldots, 1)$ a sensible test with size $\alpha$ is constructed if $H_0$ is rejected in favour of $H_<$ if $LRT_{(w)} > u_{(1-\alpha)}$, where $u_{(1-\alpha)}$ denotes the $(1-\alpha)$-quantile of the standard normal distribution. For a more detailed description see Collett [3] and Leissen [9].

## 2.2 The modified ordered logrank test

Liu et al. [10] propose the so-called *modified ordered logrank test* with the test statistic

$$LIT = \frac{\sum_{r=1}^{k-1} L_{(r)}}{\sqrt{\mathrm{Var}\left(\sum_{r=1}^{k-1} L_{(r)}\right)}}, \tag{7}$$

where

$$L_{(r)} = \sum_{j=1}^{m} (d_{(1\ldots r)j} - e_{(1\ldots r)j}),\ r = 1, \ldots, k-1, \tag{8}$$

with $d_{(1\ldots r)j} = d_{1j} + \ldots + d_{rj}$ and $e_{(1\ldots r)j} = e_{1j} + \ldots + e_{rj}$, so that $L_{(r)}$ corresponds to the logrank statistic $L_i$ according to (3) for the combined sample $1, \ldots, r$. The variance in the nominator of the test statistic is given by Liu et al. [10] for the special case of no ties in the data. A general derivation can be found in Leissen [9].

It can be shown that the statistic *LIT* is a special case of the statistic $LRT_{(w)}$ given in (6). For the proof it is needed that

$$e_{..} = \sum_{j=1}^{m} \sum_{i=1}^{k} e_{ij} = \sum_{j=1}^{m} \frac{d_{.j}}{n_{.j}} \sum_{i=1}^{k} n_{ij} = \sum_{j=1}^{m} d_{.j} = d_{..} \,. \tag{9}$$

Now consider that the numerator of *LIT* can be rewritten as

$$\begin{aligned} \sum_{r=1}^{k-1} L_{(r)} &= \sum_{j=1}^{m} (d_{1j} - e_{1j}) + \ldots + \sum_{j=1}^{m} (d_{(1\ldots(k-1))j} - e_{(1\ldots(k-1))j}) \\ &= (d_{1.} - e_{1.}) + \ldots + \left((d_{1.} - e_{1.}) + \ldots + (d_{(k-1).} - e_{(k-1).})\right) \\ &= (k-1)(d_{1.} - e_{1.}) + \ldots + (d_{(k-1).} - e_{(k-1).}) \,. \end{aligned} \tag{10}$$

If weights $w_1, \ldots, w_k$ of equal distance $c \in \mathbb{R}$ are chosen, so that $w_1 - w_2 = c$, $w_1 - w_3 = 2c$, ..., $w_1 - w_k = (k-1)c$, then the numerator of $LRT_{(w)}$ can be written as:

$$\begin{aligned} LT_{(w)} &= w_1(d_{1.} - e_{1.}) + \ldots + w_k(d_{k.} - e_{k.}) \\ &= w_k\left((d_{1.} - e_{1.}) + \ldots + (d_{k.} - e_{k.})\right) + \\ &\quad (w_1 - w_k)(d_{1.} - e_{1.}) + \ldots + (w_{(k-1)} - w_k)(d_{(k-1).} - e_{(k-1).}) \\ &= w_k(d_{..} - e_{..}) + \ldots \\ &\overset{(9)}{=} (w_1 - w_k)(d_{1.} - e_{1.}) + \ldots + (w_{(k-1)} - w_k)(d_{(k-1).} - e_{(k-1).}) \\ &= c\left((k-1)(d_{1.} - e_{1.}) + \ldots + (d_{(k-1).} - e_{(k-1).})\right) \,. \end{aligned} \tag{11}$$

In comparing (10) with (11) we see that for the given weights above the numerator of the test statistic of the logrank test for trend is $c$ times bigger than the numerator of the test statistic of the modified ordered logrank test. If $c = 1$ the numerators are identical and thus the whole test statistics are identical as well. But also for all decreasing weights with equal distances (such as $w_1 = 4$, $w_2 = 3$, $w_3 = 2$, $w_4 = 1$ or $w_1 = 6$, $w_2 = 2$, $w_3 = -2$, $w_4 = -6$ for four groups, i.e. $c$ is positive) equal test statistics result. This is due to the normalisation in the denominator.

## 2.3 *The trend test by Gehan and Mantel*

In the U-statistic by Mann and Whitney (cf. Büning and Trenkler [2]) it is counted how many observations of one sample $(x_{i'1}, \ldots, x_{i'n_{i'}})$ are greater than those of another sample $(x_{i1}, \ldots, x_{in_i})$:

$$U'_{ii'} = \sum_{j=1}^{n_i} \sum_{j'=1}^{n_{i'}} W'_{jj'} \quad \text{with} \quad W'_{jj'} = \begin{cases} 1, & x_{ij} < x_{i'j'} \\ 0, & x_{ij} > x_{i'j'} \, . \end{cases} \tag{12}$$

In order to be able to apply this test to right-censored (and tied) survival times, Gehan [4] proposes the following modification of the rank statistic $W'_{jj'}$:

$$W_{jj'} = \begin{cases} 1, & x_{ij} < x_{i'j'} \text{ or } x_{ij} \leq x^*_{i'j'} \\ 0, & x_{ij} = x_{i'j'},\ (x^*_{ij}, x^*_{i'j'}),\ x^*_{ij} < x_{i'j'} \text{ or } x_{ij} > x^*_{i'j'} \\ -1, & x_{ij} > x_{i'j'} \text{ or } x^*_{ij} \geq x_{i'j'} \, . \end{cases} \tag{13}$$

Obviously $W_{jj'}$ allows for the comparison of the values of a censored and another (un-)censored observation. The resulting

$$U_{ii'} = \sum_{j=1}^{n_i} \sum_{j'=1}^{n_{i'}} W_{jj'} \tag{14}$$

is amongst others known as *Gehan's Generalized Wilcoxon-Statistic*. Let $n_{(ii')} := n_i + n_{i'}$. Mantel [12] determines the permutational distribution of $U_{ii'}$ and derives that under $\mathrm{H}_0 : S_i = S_{i'}$ it holds that

$$\mathrm{E}(U_{ii'}) = 0, \quad \mathrm{Var}(U_{ii'}) = \frac{n_i n_{i'} \sum_{j=1}^{n_{(ii')}} \left( \sum_{j'=1}^{n_{(ii')}} W_{jj'} \right)^2}{n_{(ii')} (n_{(ii')} - 1)}, \tag{15}$$

and that $U_{ii'}$ is asymptotically normally distributed.

Moreover, Terpstra [18] and Jonckheere [8] developed independently of each other an extension of the Mann-Whitney-Test for ordered alternatives. It is especially designed to detect trends in the distribution functions of $k$ ($k \geq 3$) groups. Their test statistic reads

$$JT = \sum_{i=1}^{k-1} \sum_{i'=i+1}^{k} U'_{ii'} \, .$$

In order to construct a trend test for the test problems (1) and (2), Gehan [4] uses his "trick" again and replaces $U'_{ii'}$ by $U_{ii'}$ in $JT$:

$$\begin{aligned} G &= \sum_{i=1}^{k-1} \sum_{i'=i+1}^{k} U_{ii'} \\ &= U_{12} + \ldots + U_{1k} + U_{23} \ldots + U_{2k} + \ldots + U_{k-1k} \\ &= U_{12} + U_{13} + U_{23} + \ldots + U_{1k} + \ldots + U_{k-1k} \\ &= \sum_{i=2}^{k} U_{(1 \ldots i-1)i} \, , \end{aligned}$$

where $U_{(1\ldots i-1)i} = U_{1i} + \ldots + U_{i-1i}$ denotes the U-statistic according to (14) for the combined samples of groups $1,\ldots,i-1$ and the sample of group $i$. Besides Terpstra [18] shows that $U'_{12},\ldots,U'_{(1\ldots k-1)k}$ are independent under the null hypothesis of equality of the distribution functions of the $k$ groups. It is expected that this also holds for $U_{12},\ldots,U_{(1\ldots k-1)k}$. Hence a meaningful statistic for testing of trend in survival functions of $k$ groups is constructed according to Gehan [4] and Mantel [12] with

$$GMT = \frac{G}{\sqrt{\sum_{i=2}^{k} \mathrm{Var}(U_{(1\ldots i-1)i})}}, \tag{16}$$

where $\mathrm{Var}(U_{(1\ldots i-1)i})$ is given through (15) with

$$\mathrm{Var}(U_{(1\ldots i-1)i}) = \frac{n_{(1\ldots i-1)} n_i \sum_{j=1}^{n_{(1\ldots i)}} \left(\sum_{j'=1}^{n_{(1\ldots i)}} W_{jj'}\right)^2}{n_{(1\ldots i)}(n_{(1\ldots i)}-1)}, \; n_{(1\ldots i)} := \sum_{i'=1}^{i} n_{i'}. \tag{17}$$

Gehan [4] assumes that $GMT$ follows a standard normal distribution asymptotically. Due to the results of the simulation study in section 4 it appears that this holds. As relatively big/small values of $GMT$ speak for a positive/negative trend in the survival functions of the $k$ groups, an appropriate test with level $\alpha$ for the test problem (1) is constructed if $H_0$ is rejected if $GMT > u_{1-\alpha}$. This test will be called *Gehan-Mantel-Test*.

## 2.4 The trend test by Magel and Degges

Let $X_{ij}$ i.i.d. with continuous cdf $F(X-\theta_i)$, $i=1,\ldots,k$, $j=1,\ldots,n_i$. Hettmansperger and Norton [5] look at the $k$-sample test problem with location shifts in the alternative

$$\mathrm{H}_0^{HN}: \; \theta_1 = \ldots = \theta_k \quad \text{vs.} \quad \mathrm{H}_1^{HN}: \; \theta_i = \theta_0 + \theta c_i \quad (\theta > 0, \, \theta_0 \in \mathbb{R})$$

with arbitrary but fixed $c_i$. Amongst others they propose the test statistic

$$HNT = \frac{HN}{\sqrt{\mathrm{Var}(HN)}} \quad \text{with} \quad HN = \sum_{i=1}^{k-1} \sum_{i'=i+1}^{k} \frac{(g_{i'} - g_i)}{n_i n_{i'}} U'_{ii'}, \tag{18}$$

where $U'_{ii'}$ is defined as in (12) and $g_i = \lambda_i(c_i - \bar{c}_w)$ with $\lambda_i = \frac{n_i}{n}$, $n = \sum_{i=1}^{k} n_i$, and $\bar{c}_w = \sum_{i=1}^{k} \lambda_i c_i$. If a trend is suspected in the data, but cannot be specified further, the authors recommend the usage of linear constants $c_i$, e.g., $c_i = 1,\ldots,c_i = k$ in the case of an increasing alternative. Although the test is introduced for a test problem with location shifts in the alternative, it may also be applied for test problems with general trends in the alternative, as in terms of (1) and (2).

In order to make this test usable for right-censored survival times, Magel and Degges [11] follow the idea of Gehan [4] and replace $U'_{ii'}$ by $U_{ii'}$ in $HN_p$, where the latter is given by (14). This results in the test statistic

$$MDT_{(c)} = \frac{MD_{(c)}}{\sqrt{\mathrm{Var}(MD_{(c)})}} \quad \text{with} \quad MD_{(c)} = \sum_{i=1}^{k-1} \sum_{i'=i+1}^{k} \frac{(g_{i'} - g_i)}{n_i n_{i'}} U_{ii'} \,. \tag{19}$$

Magel and Degges [11] show that $MDT_{(c)}$ is asymptotically standard normally distributed under $H_0$. Further they derive the formula of $\mathrm{Var}(MD_{(c)})$ for a data situation without ties. As this formula is very lengthy it is left out here. The simulation results in section 4 will show that the test tends to be conservative when there are ties in the data. This may be unsatisfactory in real applications, but here it entails that this test does not have to be excluded from the power comparison of all trend tests in scenarios with ties. In order to reach a sensible test for test problem (1) with asymptotic size $\alpha$, increasing weights $c_i$, e.g., (1, ..., k), must be chosen in $MD_{(c)}$, and $H_0$ must be rejected if $MDT_{(c)} > u_{1-\alpha}$. We will call this test *Magel-Degges-Test.*

## 3 Multiple Contrast Tests

An ordered alternative as in (1) or (2) contains several *partial alternatives*, whereas the number of these grows with the number of groups. If there are three groups the alternative in (1) includes the partial alternatives:

$$\text{(T.3.1)}\ S_1 = S_2 < S_3, \quad \text{(T.3.2)}\ S_1 < S_2 = S_3, \quad \text{(T.3.3)}\ S_1 < S_2 < S_3 \,.$$

In the case of four groups the number of partial alternatives increases to seven:

$$\begin{aligned}
&\text{(T.4.1)}\ S_1 = S_2 = S_3 < S_4, \\
&\text{(T.4.2)}\ S_1 = S_2 < S_3 = S_4, \\
&\text{(T.4.3)}\ S_1 < S_2 = S_3 = S_4, \\
&\text{(T.4.4)}\ S_1 = S_2 < S_3 < S_4, \\
&\text{(T.4.5)}\ S_1 < S_2 = S_3 < S_4, \\
&\text{(T.4.6)}\ S_1 < S_2 < S_3 = S_4, \\
&\text{(T.4.7)}\ S_1 < S_2 < S_3 < S_4 \,.
\end{aligned}$$

For every number of groups $k \geq 3$ the number of partial alternatives $\nu_k$ for (1) and (2) can be determined by $\nu_k = 2^{(k-1)} - 1$. As it is not known a-priori which of the partial alternatives is on hand, it is desirable that a trend test shows a symmetric power over the alternative space. It is suspected that the trend tests introduced in the last section rather offer an asymmetric power over the alternative space.

Therefor, consider that the $k$-sample problem introduced in section 1 and (1) can be understood as a nonparametric counterpart, based on survival times, for the classical parametric $k$-sample problem $X_{ij} \sim \mathrm{N}(\mu_i, \sigma^2), i = 1, \ldots, k, j = 1, \ldots, n_i$, with the ordered one-sided test problem

$$\mathrm{H}_0^\mu : \mu_1 = \ldots = \mu_k \quad \text{vs.} \quad \mathrm{H}_1^\mu : \mu_1 \leq \ldots \leq \mu_k \text{ with } \mu_1 < \mu_k .$$

For testing such a problem, a *single contrast test*

$$EKT_{(a)} = \frac{\sum_{i=1}^k a_i \bar{X}_i}{\sqrt{\widehat{\mathrm{Var}}\left(\sum_{i=1}^k a_i \bar{X}_i\right)}} \quad \text{with} \quad \sum_{i=1}^k a_i = 0 \tag{20}$$

can be used, where $\bar{X}_i = \sum_{j=1}^{n_i} X_{ij}$ denotes the sample mean of group $i$ and $a = (a_1, \ldots, a_k)$ is a contrast vector. Since this vector imposes an order and weighting of the groups, it appears to be feasible that the choice of the contrast is crucial for the distribution of the power of a test over the alternative space. In fact the power of a single contrast test depends strongly on the pattern of the true means $\mu_1, \ldots, \mu_k$. The extent of this dependence is shown, e.g., by Hothorn et al. [7] and Neuhäuser and Hothorn [15].

The idea arises whether more symmetry over the alternative space can be reached by combining several contrast vectors in one test statistic, as it is done in *multiple contrast tests*. These are described by Mukerjee et al. [13] for the situation specified above. There the test statistic of a multiple contrast test is defined as

$$MKT_{(b)} = \max(EKT_{(b_1)}, \ldots, EKT_{(b_q)}) . \tag{21}$$

The contrasts for the $q, q \geq 2$, statistics $EKT_{(b_1)}, \ldots,$ $EKT_{(b_q)}$ should be chosen in such a way that they cover the alternative space as good as possible. Bretz and Hothorn [1] present different tuple of contrast vectors and indicate that *step contrasts*

$$\tilde{b}_h = (\tilde{b}_{h1}, \ldots, \tilde{b}_{hk}),\ h = 1, \ldots, k-1 \quad \text{with}$$

$$\tilde{b}_{hi} = \begin{cases} -(k-h), & i = 1, \ldots, h \\ h, & i = h+1, \ldots, k \end{cases} \tag{22}$$

have proven to be a good choice in similar problems in the literature. The multiple contrast test with these $(k-1)$ contrasts for $k$ groups goes back to Hirotsu [6]. Their justification for the choice of the step contrasts is that they display the "edges" of the subspace restricted by $\mathrm{H}_1^\mu$ in the $(k-1)$ dimensional space. From their usage it follows that the first $h$ groups and the last $k-h$ groups are combined respectively and that the combined groups are set in contrast.

## 3.1 The logrank maximum test

If in $LT_{(w)}$, compare (4), weights are chosen which add up to 0, the logrank test for trend also represents a single contrast test. Here the deviation of observed and expected events of the groups are set into contrast. An analogous multiple contrast test to (21) with step contrasts for test problem (1) can be constructed on the basis of the statistic

$$LTM_{(b')} = \max(LT_{(b'_1)}, \ldots, LT_{(b'_{k-1})}), \tag{23}$$

where $b'_h = -\tilde{b}_h, h = 1, \ldots, k-1$. These inversed step contrasts are used in order to consider decreasing weights again, so that again relatively big values of $LT_{(w)}$ speak against $H_0$ (compare section 2.1).

The determination of the distribution of such a *maximum statistic* is very elaborate and therefore not accomplished here. Actually, the distribution of a statistic does not have to be known when a test is to carry out. It suffices that arbitrary quantiles of the distribution are determinable under $H_0$. Let $m_\alpha$ be the $\alpha$-quantile of the distribution of a maximum statistic of arbitrary statistics $T_i, i = 1, \ldots, k, k \in \mathbb{Z}^+$. Since it holds that

$$P\big(\max(T_1, \ldots, T_k) < m_\alpha\big) = P(T_1 < m_\alpha, \ldots, T_k < m_\alpha),$$

the quantiles of the distribution of a maximum statistic correspond to the equicoordinate quantiles of the mutual distribution of the single elements of the maximum statistic. Thus it is possible to use appropriate equicoordinate quantiles of the element-wise mutual distribution of a maximum statistic as critical values for the test decision of the resulting *maximum test*.

The statistics $LT_{(b'_1)}, \ldots, LT_{(b'_{k-1})}$ are all asymptotically standard normally distributed under $H_0$ (compare section 2.1). From this the multivariate standard normal distribution does not follow to be the asymptotic mutual distribution. But we shall assume so, since simulations (compare section 4.2) show that the test based on $LTM_{(b')}$ approaches the defined size $\alpha = 0.05$ for increasing sample size when the equicoordinate 0.95-quantile of the multivariate standard normal distribution is used. In order to build a test statistic, the vector

$$LTV_{(b')} = \left(LT_{(b'_1)} \ \cdots \ LT_{(b'_{k-1})}\right)^t$$

has to be standardised. Under $H_0$ the expectation vector is the null vector. The variance of $LT_{(b'_h)}, h = 1, \ldots, k-1$, and the covariance of $LT_{(b'_h)}$ and $LT_{(b'_{h'})}$ with $h = 1, \ldots, k-2$ and $h' = 2, \ldots, k-1, h \neq h'$ are derived in Leissen [9]. The variance is given by (5), whereas $LT_{(w)}$ must be replaced by $LT_{(b'_h)}$. The covariance reads

$$\mathrm{Cov}(LT_{(b'_h)}, LT_{(b'_{h'})}) = \sum_{j=1}^{m} \frac{d_{.j}(n_{.j} - d_{.j})}{n_{.j}^2(n_{.j} - 1)} \left( \sum_{i=1}^{k} \left( b'_{hi} n_{ij} \left( b'_{h'i} n_{.j} - \sum_{i=1}^{k} b'_{h'i} n_{ij} \right) \right) \right).$$

Finally the test statistic for a multiple contrast test based on the statistics $LT_{(w)}$ with step contrasts $b'_h$ is given by:

$$LRM_{(b')} = \max(LTV^t_{(b')} \cdot \Sigma^{-\frac{1}{2}}_{LTV_{(b')}}), \tag{24}$$

where $\Sigma^{-\frac{1}{2}}_{LTV_{(b')}}$ is the inverse square root of the covariance matrix of $LTV_{(b')}$. Let $z_{1-\alpha}$ be the equicoordinate $(1-\alpha)$-quantile of the multivariate standard normal distribution. A sensible test for the test problem (1) is given, if $H_0$ is rejected for $LRM_{(b')} > z_{1-\alpha}$.

## 3.2 *The Gehan-Mantel maximum test*

In fact the Gehan-Mantel-Test (compare section 2.3) does not represent a single contrast test as introduced in (20). So a multiple contrast test as in (21) based on the Gehan-Mantel-Test cannot be constructed. But consider that in the U-statistics $U_{(1\dots i-1)i}$, $i = 2,\dots,k$, the groups $1,\dots,i-1$ are combined and set in contrast to group $i$. Hence the sum $G$ reflects the contrasts

$$b^*_h = (b^*_{h1},\dots,b^*_{hk}),\ h = 1,\dots,k-1 \quad \text{with}$$

$$b^*_{hi} = \begin{cases} -1, & i = 1,\dots,h \\ h, & i = h+1 \\ 0, & i = h+2,\dots,k. \end{cases}$$

As sum of different contrasts, $G$ renders a certain contrast as well. Therefore, a maximum test with contrasts in the respective elements based on the Gehan-Mantel-Test is constructed. The contrasts $b^*_h$, $h = 1,\dots,k-1$, correspond to *Helmert contrasts of different dimensionalities*. They are also given by Bretz and Hothorn [1] as a possible choice for building a multiple contrast test. So simply the maximum of the single $U_{(1\dots i-1)i}$, $i = 2,\dots,k$, is taken to build a new maximum statistic. Under $H_0$ each $U_{(1\dots i-1)i}$ follows a normal distribution with zero mean asymptotically. Further the $U_{(1\dots i-1)i}$ are independent and their variance is given by (17). Altogether

$$GMM = \max\left(\frac{U_{12}}{\sqrt{\text{Var}(U_{12})}},\dots,\frac{U_{(1\dots k-1)k}}{\sqrt{\text{Var}(U_{(1\dots k-1)k})}}\right) \tag{25}$$

will be used as test statistic of another maximum test for the test problem (1). Due to the same reasons as stated in the last section, the statistic $GMM$ will be compared with the equicoordinate $(1-\alpha)$-quantile of the multivariate standard normal distribution. Again the simulations results of section 4.2 will show that then the test maintains the nominal level $\alpha = 0.05$ asymptotically.

A maximum test statistic for the test problem (1) based on the test statistic of the Magel-Degges-Test which is constructed analogously to *LRM* and *GMM* is also proposed in Leissen [9]. As the simulation results showed that this test does not preserve the defined nominal level of $\alpha = 0.05$, this test was discarded.

From now on we will also refer to all introduced tests by using the names of their corresponding test statistics.

# 4 Simulation Study

## 4.1 Layout of the simulation study

A simulation study was carried out to investigate the power of the five presented tests (namely the *LRT*, *LRM*, *GMT*, *GMM*, and *MDT*) over the alternative space and to compare the maximum tests with their sum based counterparts. Only data with a positive trend in the survival functions, as imposed by the alternative in (1), was simulated. In the *LRT* the recommended weights $(k,\ldots,1)$ were used (in the decreasing version, so that relatively big values of the statistic speak for a positive trend). Analogously, the weights $(1,\ldots,k)$ were chosen in the *MDT*. The test statistics of all the other tests are already constructed in such a way that relatively big values speak for the alternative in (1). Incidentally, due to symmetry properties of the test statistics and the underlying asymptotic normality, the power estimations are also valid for the test problem (2).

In real applications data sets with survival times vary regarding their degree of censoring and ties. The distribution of the observations depends on the endpoint of interest. The sample size of a study is determined depending on the research problem and costs. Last but not least the difference between survival functions can be of various kinds. It is plausible that the power of the tests varies for different scenarios of the data, and more importantly, that the power ratio of the tests to each other might change from one scenario to another. Therefore, the power of the tests was simulated for many different scenarios by choosing the following factors and settings:

1. *Distribution*
   By nature survival times follow a continuous distribution on the positive axis whereas further the distribution is often right-skewed by experience. For these reasons the Weibull, the Exponential, the Gamma and the Lognormal distribution are often used to model survival times. But situations have also encountered in which the survival times are approximately normally distributed. Therefore, a representative of each distribution mentioned was selected as setting: W(0.5, 1), Exp(1), Ga(3, 1), N(3, 1), LogN(0, 1). The parameters were chosen so that the distributions look as different as possible.
2. *Number of groups* (equivalently to number of partial alternatives)
   As the number of partial alternatives increases rapidly with the number of groups

(compare section 3), scenarios with three and four groups only were considered for the sake of clearness.

3. *Group sizes*
In order to preserve the here defined nominal level of $\alpha = 0.05$ satisfactory, it was decided to choose 20 observations per group as the minimum sample size (compare section 4.2). The group size of 50 observations each was also considered. Further an unbalanced design was also to be investigated. In order to reflect the situation of many clinical studies in which the control group contains more observations than the others, the designs (50, 20, 20) for three groups and (50, 20, 20, 20) for four groups were chosen.
4. *Shifts of the survival functions*
For generating the differences between the survival functions according to the alternative in (1) the following location and scale shifts were used:

$$S^k_{lo,li}(t) = S(t - \theta^k_{li}) \quad \text{and} \quad S^k_{sc,li}(t) = S\big(t\exp(-\theta^k_{li})\big) \quad \text{with} \quad \theta^k_{li} = a\frac{\psi^k_{li}}{\sqrt{n}}.$$

The survival function $S(t)$ corresponds to one of the initial distributions W(0.5, 1), Exp(1), Ga(3, 1), N(3, 1) or LogN(0, 1) given above. The sample size $n$ of the whole scenario serves as standardisation. The values $\psi^k_{li}$ depend on the group $i = 1,\ldots,k$, on the partial alternative $l = 1,\ldots,\nu_k$ (compare section 3), and on the number of groups $k = 3,4$. They can be taken from the matrices

$$\Psi^3 = (\psi^3_{li}) = \begin{pmatrix} 0 & 0 & 1 \\ 0 & 1 & 1 \\ 0 & \frac{1}{2} & 1 \end{pmatrix} \quad \text{and} \quad \Psi^4 = (\psi^4_{li}) = \begin{pmatrix} 0 & 0 & 0 & 1 \\ 0 & 0 & 1 & 1 \\ 0 & 1 & 1 & 1 \\ 0 & 0 & \frac{1}{2} & 1 \\ 0 & \frac{1}{2} & \frac{1}{2} & 1 \\ 0 & \frac{1}{2} & 1 & 1 \\ 0 & \frac{1}{3} & \frac{2}{3} & 1 \end{pmatrix}.$$

These matrices are sensible since each row expresses the course of a partial alternative. Note that the survival function of the first group always corresponds to one of the initial distributions. In every row of the two matrices a step between the survival functions with a total value of 1 is expressed. With the constant $a$ it is possible to influence this total value of the steps, so that it reflects the *intensity* of the shifts. In the simulations values of $a = 1,\ldots,9,11,13,15$ were considered.
5. *Degree of censoring*
The random variable which reflects the censoring time is assumed to follow a uniform distribution on the interval $[0,b]$ (cp. Magel and Degges [11]). If censoring underlies a scenario, a censoring time is drawn for every simulated survival time. If the censoring time is bigger than the survival time, nothing is changed, if it is smaller, the survival time is reduced to this censoring time and is regarded as a censored survival time. The right limit of the interval $[0,b]$ is determined for every scenario such that a desired percentage of censoring in the data is approximately achieved. As setting a *weak* (20%), a *medium* (30%), and a *strong* (50%) degree of censoring was considered.

**Table 1** Combinations of the factors *group sizes*, *degree of censoring*, and *degree of ties* in the simulation study

| combination | group sizes | degree of censoring | degree of ties |
|---|---|---|---|
| 1 | 20 each | null | null |
| 2 | 20 each | medium | null |
| 3 | 20 each | null | medium |
| 4 | 50 each | null | null |
| 5 | unbalanced | null | null |
| 6 | unbalanced | medium | medium |
| unbalanced = 1st group 50, thereafter 20 | | | |

6. *Degree of ties*
   Ties are generated in the data by rounding of the drawn random numbers. A *weak*, *medium*, and *strong* degree of ties indicates the rounding to two, one, and zero decimal places.

The execution of a full factorial design with all these factors and their settings would have been too computationally intensive. Therefore, only the number of groups with all partial alternatives, the distributions, and the kind of shifts with the given intensities were full factorially combined. All the resulting combinations were carried out once in conjunction with the six fixed combinations which are displayed in Table 1. For further explanations on the choice of the factors and settings as well as on the generation of the survival times due to the given constraints of the scenarios, we refer to Leissen [9].

Each scenario was simulated with 10000 runs. Then for each generated data set the test statistics of the five trend test were computed and compared with the (equicoordinate) 0.95-quantile of the (multivariate) standard normal distribution (as described in sections 2 and 3). Finally the power of every test for every scenario was estimated by dividing the resulting number of rejections by the number of runs.

Another part of the simulation study was to check if the trend tests (especially the maximum tests as the asymptotic normality is not proved here) preserve the defined level $\alpha = 0.05$. Therefore, the type I error rates of the tests were simulated for various scenarios.

All computations as well as the graphics in this paper were generated with R (R Development Core Team [17]). Most parts of the code for the reproducible simulation study (with detailed explanations) can be found in Leissen [9]. In the course of the simulation study 7200 simulations were carried out. Due to space restrictions only a fraction of the results can be presented. For a more comprehensive presentation we refer to Leissen [9].

## 4.2 Investigation of the compliance with the nominal level

In Table 2 the simulated type I error rates of the five trend tests under investigation are listed up for some different scenarios. When there are no censored and tied observations the *LRT*, *GMT*, *GMM*, and *MDT* approximate satisfactory the nominal

**Table 2** Simulated type I error rates for different scenarios with four groups

| | LRT | LRM | GMT | GMM | MDT | LRT | LRM | GMT | GMM | MDT |
|---|---|---|---|---|---|---|---|---|---|---|
| | no censoring, no ties group sizes: each 20 | | | | | no censoring, no ties group sizes: each 50 | | | | |
| W(0.5, 1) | .053 | .057 | .050 | .044 | .049 | .057 | .059 | .055 | .053 | .055 |
| Exp(1) | .054 | .068 | .051 | .052 | .050 | .050 | .054 | .045 | .047 | .045 |
| Ga(3, 1) | .053 | .065 | .053 | .052 | .053 | .052 | .055 | .050 | .044 | .050 |
| N(3, 1) | .050 | .060 | .050 | .049 | .049 | .050 | .054 | .047 | .047 | .047 |
| LogN(0, 1) | .051 | .063 | .047 | .050 | .048 | .050 | .060 | .050 | .050 | .050 |
| | strong degree of censoring, no ties group sizes: each 20 | | | | | no censoring, no ties group sizes: 50/20/20/20 | | | | |
| W(0.5, 1) | .048 | .053 | .044 | .042 | .032 | .045 | .055 | .046 | .047 | .047 |
| Exp(1) | .050 | .056 | .049 | .042 | .046 | .048 | .053 | .051 | .045 | .050 |
| Ga(3, 1) | .052 | .060 | .045 | .043 | .044 | .050 | .056 | .052 | .053 | .053 |
| N(3, 1) | .053 | .062 | .046 | .040 | .041 | .048 | .053 | .050 | .049 | .050 |
| LogN(0, 1) | .055 | .059 | .050 | .048 | .046 | .047 | .052 | .046 | .048 | .048 |
| | no censoring, strong degree of ties group sizes: each 20 | | | | | strong degree of censoring and ties group sizes: 50/20/20/20 | | | | |
| W(0.5, 1) | .055 | .064 | .054 | .054 | .038 | .048 | .047 | .050 | .049 | .058 |
| Exp(1) | .052 | .055 | .054 | .056 | .038 | .049 | .051 | .049 | .049 | .062 |
| Ga(3, 1) | .054 | .056 | .051 | .050 | .048 | .047 | .050 | .044 | .037 | .044 |
| N(3, 1) | .053 | .050 | .052 | .047 | .047 | .046 | .050 | .044 | .041 | .033 |
| LogN(0, 1) | .054 | .057 | .049 | .051 | .042 | .049 | .047 | .050 | .043 | .059 |

level of $\alpha = 0.05$ already for 20 observations in each group. Only the *LRM* still shows up to be quite anticonservative in this situation. The anticonservatism decreases when 50 observations per group or the unbalanced design underlies a scenario. The other tests are even conservative for some distributions and group sizes. When the data contains censored observations the size of the tests decreases all in all. So the *LRM* is less anticonservative for 20 observations in each group and even preserves the nominal level in the unbalanced design. Overall the *GMT*, *GMM*, and the *MDT* even appear quite conservative in these situations, although there are some exceptions for the *MDT*. Tied observations seem also to lower the type I error rate of the *LRM* and the *MDT* (compare section 2.4). For the other tests not much changes in this situation. The results of the simulations for three groups are similar to these of four groups (see Leissen [9]). The main difference is that the convergence is even stronger in the 3-sample case, particulary for the *LRM*.

## 4.3 Comparison of power

Figures 1–4 show the course of the power of each of the five trend tests under comparison over the partial alternatives $(T.4.1) - (T.4.7)$ in the 4-sample case for different scenarios. The lines that connect the points of the simulated power values are only an optical aid. In every figure each of the six plots correspond to one of the six constellations of group sizes, degree of censoring, and degree of ties given by Table 1.

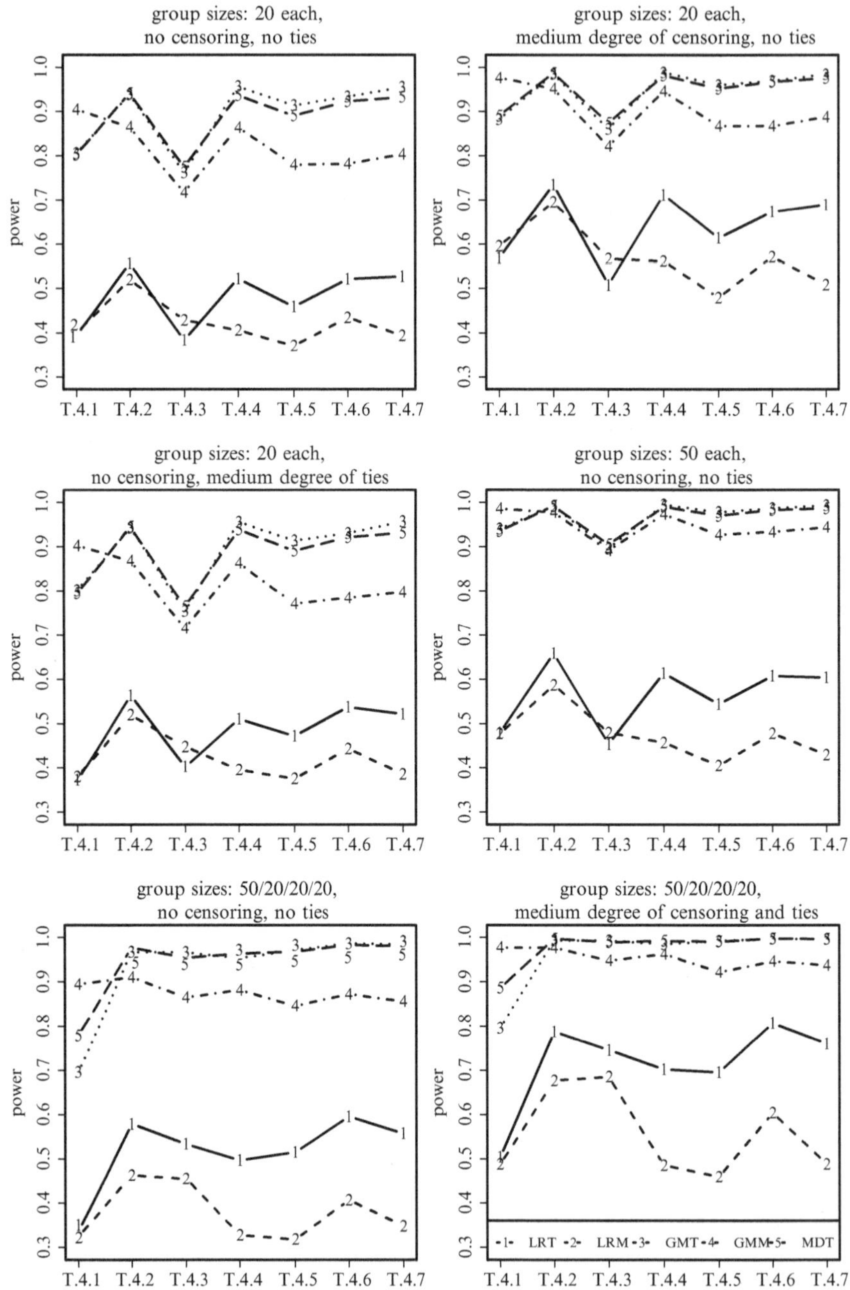


**Fig. 1** Course of estimated power for all tests under comparison over the 7 partial alternatives of 4 groups for all constellations after table 1, based on W(0.5, 1)-distributed observations with location shifts of intensity $a = 8$

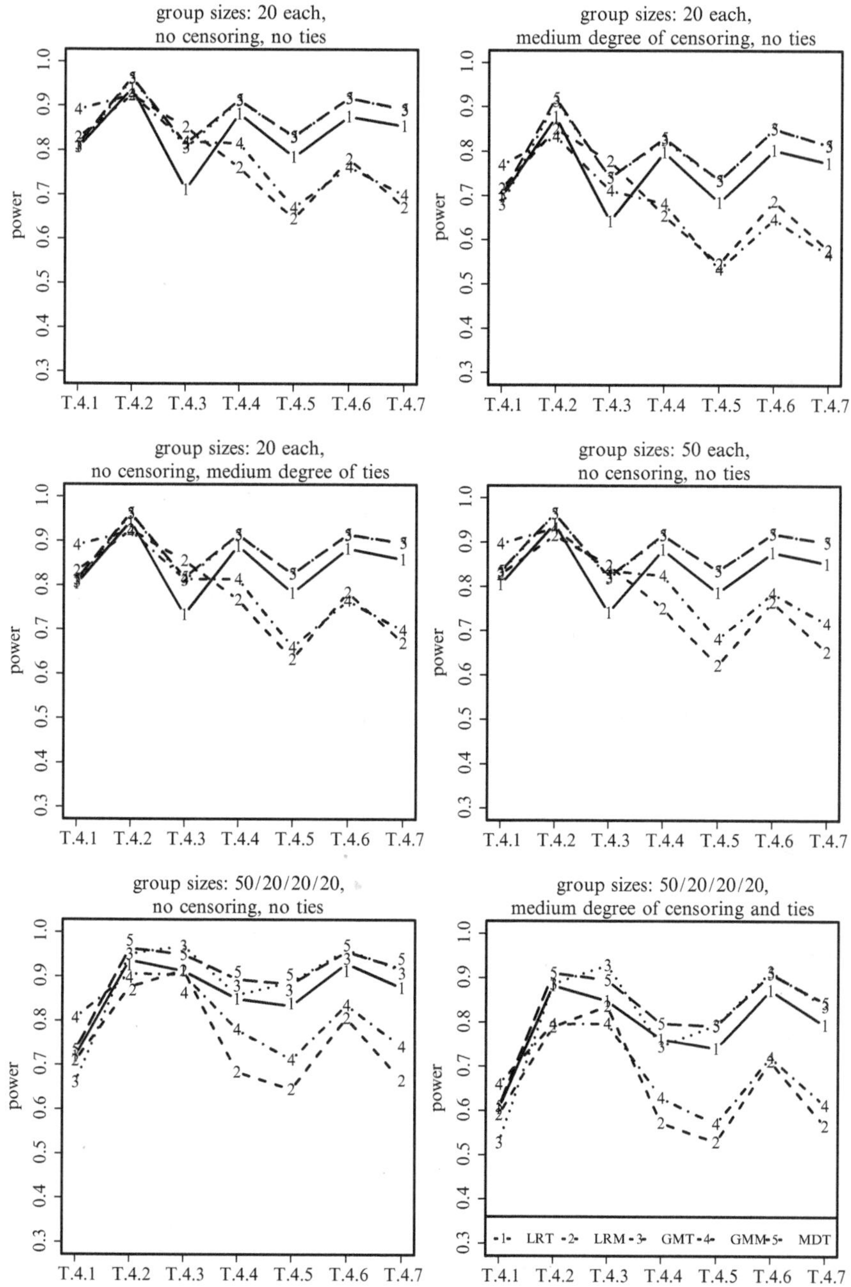


**Fig. 2** Course of estimated power for all tests under comparison over the 7 partial alternatives of 4 groups for all constellations after table 1, based on N(3, 1)-distributed observations with location shifts of intensity $a = 8$

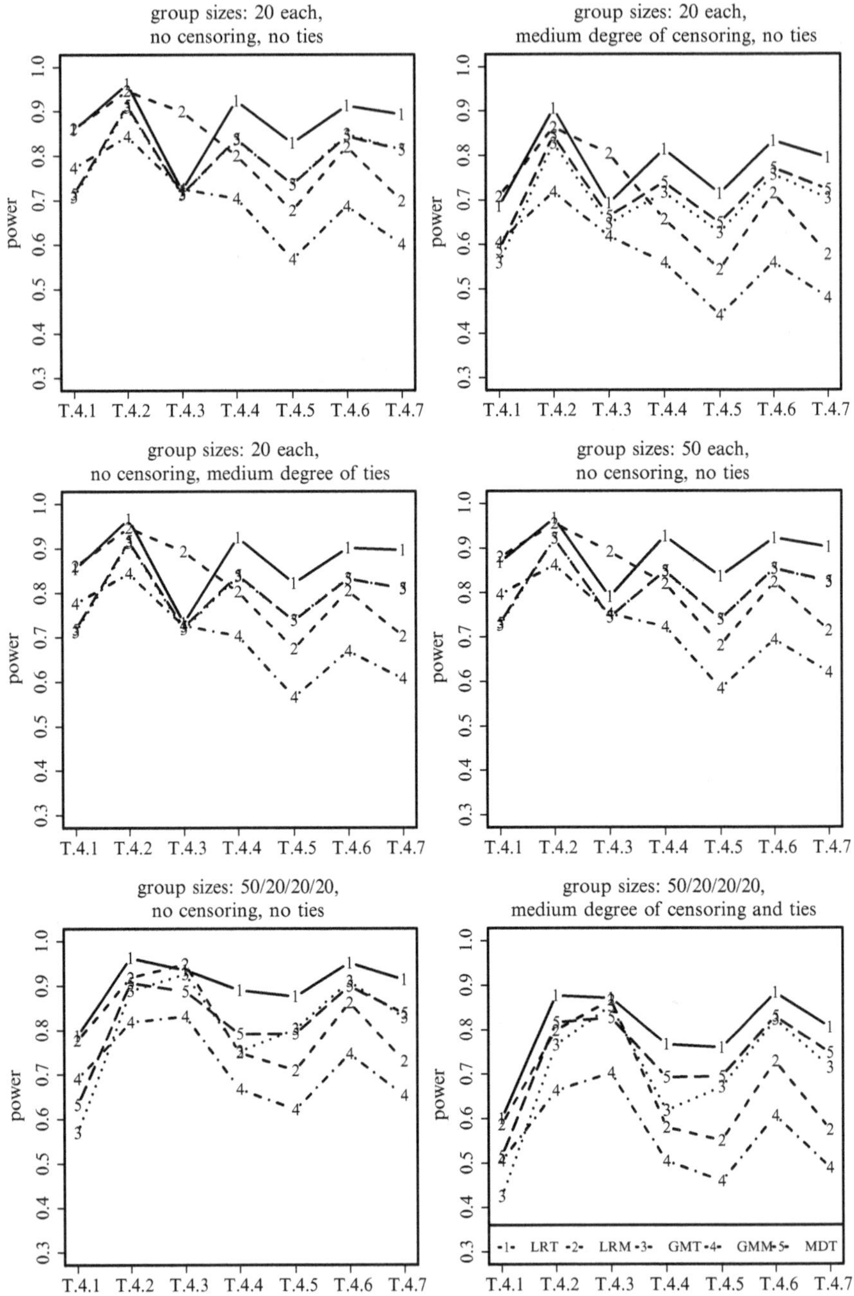


**Fig. 3** Course of estimated power for all tests under comparison over the 7 partial alternatives of 4 groups for all constellations after table 1, based on Exp(1)-distributed observations with scale shifts of intensity $a = 8$

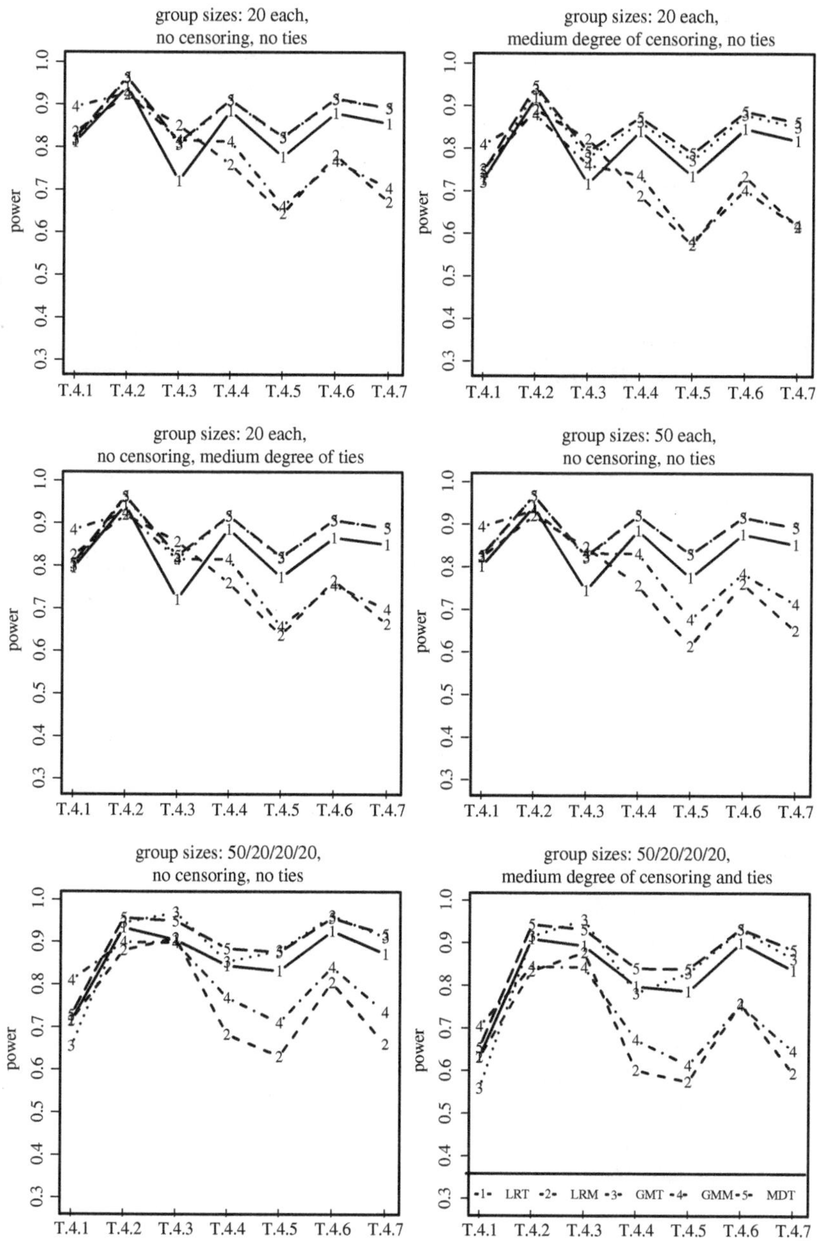


**Fig. 4** Course of estimated power for all tests under comparison over the 7 partial alternatives of 4 groups for all constellations after table 1, based on LogN(0, 1)-distributed observations with scale shifts of intensity $a = 8$

The kind of shifts and distribution of the data varies over the four figures. The intensity parameter $a$ is chosen to be 8 in every plot. The displayed scenarios are chosen exemplary. For a presentation of the results of the remaining scenarios we refer to Leissen [9]. In the 3-sample case the ranking of the power estimations is similar to that of the 4-sample case, but the extent of the differences between the power of the tests is usually less pronounced in the case of three groups.

In Fig. 1 it can be seen that in the scenarios with location shifts in W(0.5, 1)-distributed data, the power estimations of the *GMT* and *MDT* are almost always approximately identical. These two tests are overall the best tests here. The *GMM* shows a similar course of the power over the partial alternatives, but except for (T.4.1) the estimations are always a bit lower than the corresponding estimations for the *GMT* and *MDT*. Both the logrank tests take course a lot below the others, whereas the conventional one shows up better than the multiple contrast test except for (T.4.1) and (T.4.3) in the balanced designs.

In the scenarios with location shifts in N(3, 1)-distributed data (Fig. 2), the tests show a similar but more distinct course over the partial alternatives. The main difference to Fig. 1 is that the logrank tests approach the *GMT* and the *MDT* while the *GMM* departs from them. The two maximum tests are the worst prevalently. The *GMT* and the *MDT* are still the best while they are beaten by the *GMM* in (T.4.1) and by the *LRM* in (T.4.3) in the balanced designs only.

Looking at Figs. 3 and 4 for scale shifts in Exp(1)- and LogN(0,1)-distributed data one can see that again the course of power of the tests over the partial alternatives is similar to those in Figs. 1 and 2, but that the "level" of the courses is somehow shifted again. In Fig. 3 the *LRT* performs overall best, the *GMM* worst. The power estimations for the *LRM* are comparably high for the first three partial alternatives and the highest for (T.4.3). Figure 4 resembles Fig. 2 a lot.

## 5 Conclusions

Multiple contrast tests are useful in a variety of settings, as shown for normally distributed data, non-normal data (Neuhäuser et al. [16]) and dichotomous data (Neuhäuser and Hothorn [14]). However, for right-censored survival data this study could not demonstrate a superior power of multiple contrast tests in general, but only in specific situations. According to our simulations *GMT* and *MDT*, the trend tests by Gehan and Mantel as well as by Magel and Degges, can be recommended.

## References

[1] Bretz, F., Hothorn, L.A.: Statistical Analysis of Monotone or Non-Monotone Dose-Response Data from *In Vitro* Toxological Assays. Alternatives to Laboratory Animals **31**, 81–96 (2003)

[2] Büning, H., Trenkler, G.: Nichtparametrische Statistische Methoden (2nd ed.). De Gruyter, Berlin (1994)

[3] Collett, D.: Modelling Survival Data in Medical Research (2nd ed.). Chapman & Hall/CRC, Boca Raton (2003)

[4] Gehan, E.A.: A Generalized Wilcoxon Test for Comparing Arbitrarily Singly-Censored Samples. Biometrika **52**, 203–223 (1965)

[5] Hettmansperger, T.P., Norton, R.M.: Tests for Patterned Alternatives in $k$-Sample Problems. J. Am. Stat. Assoc. **82**, 292–299 (1987)

[6] Hirotsu, C.: The Cumulative Chi-Squares Method and a Studentised Maximal Contrast Method for Testing an Ordered Alternative in a One-Way Analysis of Variance Model. Rep. Stat.: Application Research **26**, 12–21 (1979)

[7] Hothorn, L.A., Neuhäuser, M., Koch, H.-F.: Analysis of Randomized Dose Finding Studies: Closure Test Modifications Based on Multiple Contrast Tests. Biometrical J. **39**, 467–479 (1997)

[8] Jonckheere, A.R.: A Distribution-Free $k$-Sample Test Against Ordered Alternatives. Biometrika **41**, 133–145 (1954)

[9] Leissen, S.: Nichtparametrische Trendtests zur Analyse von rechtszensierten Überlebenszeiten. Diplomarbeit, Fachbereich Statistik, Universität Dortmund, Dortmund (2007)

[10] Liu, P.-Y., Tsai, W.Y., Wolf, M.: Design and Analysis for Survival Data under Order Restrictions with a Modified Logrank Test. Stat. Med. **17**, 1469–1479 (1998)

[11] Magel, R.C., Degges, R.: Tests for Ordered Alternatives with Right Censored Data. Biometrical J. **40**, 495–518 (1998)

[12] Mantel, N.: Ranking Procedures for Abitrarily Restricted Observation. Biometrics **23**, 65–78 (1967)

[13] Mukerjee, H., Robertson, T., Wright, F.T.: Comparison of Several Treatments with a Control using Multiple Contrasts. J. Am. Stat. Assoc. **82**, 902–910 (1987)

[14] Neuhäuser, M., Hothorn, L.A.: Trend Tests for Dichotomous Endpoints with Application to Carcinogenicity Studies. Drug Inf. J. **31**, 463–469 (1997)

[15] Neuhäuser, M., Hothorn, L.A.: An Analogue of Jonckheere's Trend Test for Parametric and Dichotomous Data. Biometrical J. **40**, 11–19 (1998)

[16] Neuhäuser, M., Seidel, D., Hothorn, L.A., Urfer, W.: Robust Trend Tests with Application to Toxicology. Environ. and Ecol. Stat. **7**, 43–56 (2000)

[17] R Development Core Team: R: A Language and Environment for Statistical Computing. R Foundation for Statistical Computing, Vienna, Austria (2008), http://www.R-project.org. ISBN 3-900051-07-0

[18] Terpstra, T.J.: The Asymptotic Normality and Consistency of Kendall's Test Against Trend, when Ties are Present in One Ranking. Indigationes Mathematicae **14**, 327–333 (1952)

# Penalty Specialists Among Goalkeepers: A Nonparametric Bayesian Analysis of 44 Years of German Bundesliga

Björn Bornkamp, Arno Fritsch, Oliver Kuss, and Katja Ickstadt

**Abstract** Penalty saving abilities are of major importance for a goalkeeper in modern football. However, statistical investigations of the performance of individual goalkeepers in penalties, leading to a ranking or a clustering of the keepers, are rare in the scientific literature. In this paper we will perform such an analysis based on all penalties in the German Bundesliga from 1963 to 2007. A challenge when analyzing such a data set is the fact that the counts of penalties for the different goalkeepers are highly imbalanced, leading to the question on how to compare goalkeepers who were involved in a disparate number of penalties. We will approach this issue by using Bayesian hierarchical random effects models. These models shrink the individual goalkeepers estimates towards an overall estimate with the degree of shrinkage depending on the amount of information that is available for each goalkeeper. The underlying random effects distribution will be modelled nonparametrically based on the Dirichlet process. Proceeding this way relaxes the assumptions underlying parametric random effect models and additionally allows to find clusters among the goalkeepers.

## 1 Introduction

In modern football, penalty shots are of vital importance. The world cup finals in 1990, 1994, and 2006, for example, were all decided by penalties. Nevertheless, scientific investigations of penalty conversions or savings are rare. Shooting techniques and tactics, ball speed, anticipation of the keeper, stress management of the shooter, or empirical investigation of penalty myths have been the objects of investigation ([8, 12, 16, 15, 13, 21, 9]). However, we are not aware of studies which try to find rankings or clusters of successful penalty scorers or savers.

Björn Bornkamp
Fakultät Statistik, Technische Universität Dortmund, D-44221 Dortmund, Germany
bornkamp@statistik.tu-dortmund.de

This is astonishing as the perception of especially skilled goalkeepers seems to be commonplace. For example, the English Wikipedia page for ‘Penalty kick’ (http://en.wikipedia.org/wiki/Penalty_kick, accessed 15/04/2008) shows a list of goalkeepers (Carlo Cudicini, Peter Schmeichel, Pepe Reina, Oliver Kahn, Ricardo, Francesco Toldo, Brad Friedel, Artur Boruc, Jens Lehmann, Edwin van der Sar and Mark Schwarzer) who are ‘noted for their penalty-saving capabilities’, but there is no quantitative evidence to support the claim of existence of a group of ‘penalty specialists’. The German Wikipedia page on the penalty (http://de.wikipedia.org/wiki/Elfmeter, accessed 15/04/2008) asserts that there are some goalkeepers being able to save more penalties than the average goalkeeper and gives a ranking of the German goalkeepers with the largest number of saved penalties. It is interesting from a statistical viewpoint that this ranking contains only the absolute number of saved penalties, not accounting for the number of potentially savable penalties for the respective goalkeeper.

In this paper we approach the problem of ranking and clustering goalkeepers for their penalty-saving capabilities in a statistically more valid way. Our data set includes all 3,768 penalties from August 1963 to May 2007 from the German Bundesliga. Data were collected from three different sources. All penalties from August 1963 to May 1993 were taken from [7]. Penalties from August 1993 to February 2005 were made available by IMP AG, München (www.impire.de), a German company that collects data for a commercial football data base. The remaining penalties were found by a systematic internet search, their completeness was checked via the aggregated data published by the “kicker” (the leading German football magazine) in its annual review of the Bundesliga season. As we are focusing on the goalkeeper’s ability to save penalties, we removed all penalties that missed the goal or hit goal-post or crossbar. This resulted in 261 deletions with 3,507 penalties remaining for final analysis. Out of these 3,507 penalties 714 were saved by the goalkeeper corresponding to a rate of 20.4%. The following additional information was available for each penalty: goalkeeper, goalkeeper’s team, scorer, scorer’s team, experience of goalkeeper and scorer (in terms of penalties), home advantage, day and year of season, and, of course, successful conversion or saving of the penalty. In total 273 goalkeepers were involved in the 3,507 penalties, many of them having been faced only with a small number of penalties (94 were involved in three or less penalties, see also Fig. 1 (i)). Figure 1 (ii) shows the relative frequencies of saved penalties for all goalkeepers. The modes of the density at 0 and 1 are due to the goalkeepers that were involved in very few penalties and saved none or all. It is intuitively clear that a goalkeeper who was involved in only one single penalty during his career and saved this, should not be considered the best penalty saver despite his 100% saving rate. Consequently, the relative frequency of saved penalties is a bad estimator of the “true” ability of the goalkeeper, motivating the use of more sophisticated statistical procedures.

That is, we are faced with two main statistical challenges:

(i) How to derive a sound statistical model, which will produce more “reasonable” estimates for the goalkeepers effect than simple relative frequencies?

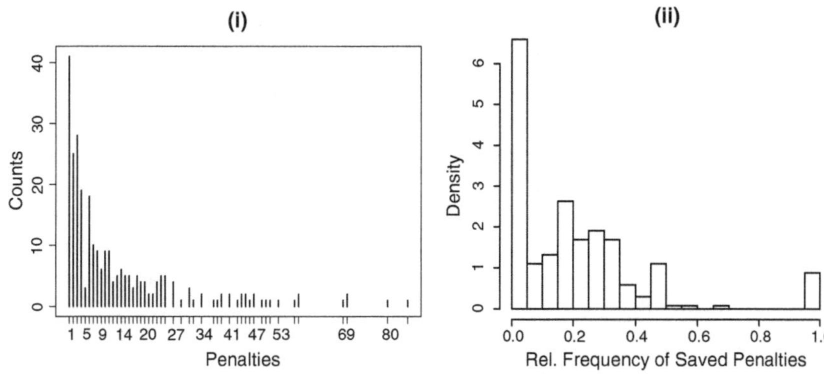


**Fig. 1** (i) Counts of penalties per goalkeeper and (ii) histogram of relative frequencies of saved penalties per goalkeeper

(ii) How to investigate whether the population of goalkeepers can be grouped into clusters containing, for example, a group of 'penalty specialists' and a group of 'penalty losers'?

In Section 2 we will introduce the statistical methods, which will allow us to approach (i) and (ii), while Section 3 is devoted to the analysis of the data. Final conclusions will be drawn in Section 4.

# 2 Statistical Methodology: Hierarchical Models and Bayesian Nonparametrics

In the first two parts of this section we will describe the statistical methodology, while the third part deals with the actual model and priors we will use for the analysis in Section 3. The material in this section is mainly based on [4], who provides a recent review of nonparametric modeling of random effects distributions in Bayesian hierarchical models, and [14], who also illustrate how to implement a related model in BUGS.

## *2.1 Hierarchical Models*

An appropriate tool to approach problem (i) from a statistical perspective is the hierarchical model. In its most simple form it can be described as follows: Suppose we observe a normally distributed random variable $Y_i$ once for each of $n$ subjects. The model for the data would then be

$$Y_i = \mu_i + \varepsilon_i, \quad i = 1, \ldots, n,$$

where $\varepsilon_i \overset{iid}{\sim} \mathcal{N}(0,\sigma^2)$ and $\mu_i$ some unknown *constant* effect of the $i$th subject. In a classical model, the maximum likelihood estimate for each of the subject effects $\mu_i$ would equal $y_i$. Typically, this will lead to highly variable estimates for the subject's effect as there are as many parameters as observations. However, if it is known (or at least reasonable to assume) that the subjects belong to the same population, a different approach is more appropriate. In this case one would model the subject effects $\mu_i$ as realizations from an unknown population (i.e., random effects) distribution $P$, rather than as unrelated constants. Consequently, in this model all realizations $y_i$ are used in estimating the random effects distribution $P$, which in turn leads to estimates of the individual effects $\mu_i$. However, these would be shrunken towards each other. That is, hierarchical models allow for sharing information across subjects, rather than treating subjects as completely unrelated. A Bayesian analysis with an implementation via Gibbs and general Markov chain Monte Carlo (MCMC) sampling is particularly suited for the analysis of more complex hierarchical models (while the standard frequentist approaches become infeasible). Such a Bayesian approach is taken in this article.

## *2.2 The Dirichlet Process*

Reformulating question (ii) statistically we would like to investigate, whether the random effects distribution $P$ of the goalkeeper effects is multimodal. Figure 1 (ii) suggests that this might be the case, even when ignoring the modes at 0 and 1. For this reason we base the analysis in this article on Bayesian nonparametric methodologies, as they allow to model a multimodal random effects distribution. Specifically, we will model the random effects distribution $P$ as a (location) mixture of normal distributions and assume a nonparametric prior for the mixing distribution. The motivation for using mixtures of normal distributions stems from the fact that any distribution on the real line can be approximated arbitrarily well by a mixture of normals ([2]). We hence model the density of the random effects distribution $P$ as $\int N(x|\theta,\sigma^2)Q(d\theta)$, where $N(.|\theta,\sigma^2)$ is a normal density with mean $\theta$ and variance $\sigma^2$ and $Q(d\theta)$ is a discrete mixing distribution. The main issue in this kind of Bayesian analysis is which prior to assume for the unknown discrete mixing distribution $Q(d\theta)$. A flexible and convenient solution is to use the Dirichlet process, dating back to [5]. The Dirichlet process is a random discrete probability measure, i.e., a stochastic process that realizes discrete probability measures. It is characterized by two parameters: A base probability measure $F_0$ and a positive real number $\alpha$. A random probability measure $Q$ follows a Dirichlet process prior if the distribution of $(Q(S_1),\ldots,Q(S_k))'$ for a partition $S_1,\ldots,S_k$ of the underlying sample space (in our case $\mathbb{R}$) has a Dirichlet distribution with parameter $(\alpha F_0(S_1),\ldots,\alpha F_0(S_k))'$. Hence $F_0$ is the underlying prior mean distribution (i.e., $\mathbb{E}(Q(S_i)) = F_0(S_i)$), while $\alpha$ is a precision parameter (for $\alpha \to \infty$ the realizations will be more and more similar to $F_0$). The main reason for the popularity of the Dirichlet process for Bayesian nonparametric applications is the fact that it has an important conjugacy property, which allows for an efficient exact implementation in many cases (see [5] for details). Another reason for the popularity of Dirichlet process priors is the constructive *stick-breaking* representation of the Dirichlet process given by [17]. Sethuraman showed that $Q$ has a Dirichlet process prior with parameters $\alpha$ and $F_0$ iff

$$Q(d\theta) = \sum_{h=1}^{\infty} \pi_h \delta_{\theta_h}(d\theta), \quad \text{with } \theta_h \overset{iid}{\sim} F_0, \tag{1}$$

where $\delta_\theta$ denotes the probability measure degenerated at $\theta$ and $\pi_h = V_h \prod_{l<h}(1 - V_h)$ with $V_h \overset{iid}{\sim} \text{Beta}(1,\alpha)$. The terminology *stick-breaking* is used, because starting with a probability stick of length one, $V_1$ is the proportion of the stick broken off and allocated to $\theta_1$, $V_2$ is the proportion of the remaining $1 - V_1$ stick length allocated to $\theta_2$, and so on (see also [6] for details on the general class of stick-breaking priors). From this stick-breaking representation it becomes obvious that the precision parameter $\alpha$ also determines the clustering properties of the Dirichlet process. For small $\alpha$, most probability mass will be distributed on the first realizations of $F_0$ leading to a clustering of observations. On the other hand for $\alpha \to \infty$ there will be many clusters and a specific realization of $Q$ will be more similar to $F_0$. For a review of Bayesian clustering procedures, including those based on the Dirichlet process see, for example, [10]. For a random sample of size $n$ from a probability distribution realized by a Dirichlet process [1] has shown that the prior density of the number of distinct values (clusters/components) $k$ in $n$ realizations is

$$p(k|\alpha,n) = c_n(k) n! \alpha^k \frac{\Gamma(\alpha)}{\Gamma(\alpha+n)}, \tag{2}$$

where $c_n(k) = \frac{\left[{n \atop k}\right]}{\sum_{j=0}^{n} \left[{n \atop j}\right]}$, and $\left[{n \atop j}\right]$ denotes a Stirling number of the first kind (to approximate Stirling numbers for large $n$ methods introduced by [20] can be used). The expected number of clusters $k$ in $n$ realizations is given by

$$E(k|\alpha,n) = \sum_{i=1}^{n} \frac{\alpha}{\alpha+i-1}. \tag{3}$$

Both formulas play an important role for the prior elicitation of the parameter $\alpha$.

The stick-breaking representation of the Dirichlet process is also useful because it directly leads to good finite dimensional approximations for the Dirichlet process by truncation of the sum (1). A finite dimensional approximation to the Dirichlet process is given by

$$Q(d\theta) = \sum_{h=1}^{N} \pi_h \delta_{\theta_h}(d\theta), \quad \text{with } \theta_h \overset{iid}{\sim} F_0$$

where $\pi_h = V_h \prod_{l<h}(1 - V_h)$, $V_h \overset{iid}{\sim} \text{Beta}(1,\alpha), h = 1,\ldots,N-1$, and $\pi_N = 1 - \sum_{h=1}^{N-1} \pi_h$. $N$ is a truncation parameter, which is chosen large enough to obtain a good

approximation. For small values of $\alpha$ a relatively small $N$ is sufficient to approximate the underlying Dirichlet process well. We refer to [14] for a detailed discussion of this aspect. In the following we will abbreviate the truncated prior distribution induced for the weights as $Seth_N(\alpha)$, i.e., $\boldsymbol{\pi} \sim Seth_N(\alpha)$.

## 2.3 Model and Prior Specification

We will model the $j$th observed penalty of the $i$th goalkeeper as a realization of a Bernoulli random variable with probability $\rho_{ij}$ that the goalkeeper saves the penalty. This probability $\rho_{ij}$ is hence modeled as a function of the $i$th goalkeeper and some additional covariates $\boldsymbol{x}_{ij}$. That is, we assume the model

$$\text{logit}(\rho_{ij}) = \gamma_i + \boldsymbol{\beta}'\boldsymbol{x}_{ij}, \quad i = 1, \ldots, 273, \ j = 1, \ldots, n_i,$$

where $\gamma_i$ is the random effect of the $i$th goalkeeper, $n_i$ is the number of penalties the $i$th goalkeeper was involved in, and $\boldsymbol{\beta}$ is the vector of regression coefficients for the covariates.

The $\gamma_i$ are modeled as iid realizations of a random effect distribution $P$, which in turn is modeled as a location mixture of normal distributions

$$\int N(x|\theta, \sigma^2) Q(d\theta) = \sum \pi_h N(x|\theta_h, \sigma^2),$$

and the Dirichlet process will be used as a prior for $Q(d\theta)$. The parameter $\alpha$ of the Dirichlet process will be chosen equal to $1/3$. Using (3) it can be seen that this leads to a prior mean of $\approx 2.91$ clusters/components for a sample of size 273. Calculation of (2) shows (see also Fig. 2), that the prior density for the number of components has peaks at 2 and 3 and then decreases rapidly, leaving virtually no probability mass for $k > 8$, which seems reasonable for our penalty data. As the expected number of components is relatively small it is sufficient to select the truncation parameter $N$ equal to 20. As the base measure $F_0$ of the Dirichlet process we will use a normal distribution with parameters 0 and variance 3.289. $F_0$ is chosen such that it is approximately equal to a uniform distribution on the probability scale. For the precision $\sigma^{-2}$ of the normal densities in the mixture we will use an exponential prior distribution with mean 16. The prior distribution for $\boldsymbol{\beta}$, the coefficients of the covariates, are chosen as vague uniform distributions. A concise summary of the model and its different hierarchies is given in Table 1.

To assess the merit of a nonparametric model of the random effects distribution via the proposed Dirichlet process model, we compare it to two less flexible models via the deviance information criterion (DIC) [18]. The DIC is similar to AIC or BIC but more suitable for hierarchical models. Defining $\boldsymbol{\rho}$ as the vector containing the probabilities $\rho_{ij}$ the deviance is in our case given by

**Table 1** Hierarchical model used for analysis

| Level | Parameters |
|---|---|
| I | $Y_{ij} \sim \text{Bernoulli}(\rho_{ij}),\ i = 1,\ldots,273\ j = 1,\ldots,n_i$ |
| II | $\text{logit}(\rho_{ij}) = \gamma_i + \boldsymbol{\beta}'\boldsymbol{x}_{ij}$ |
| III | $\gamma_i \overset{iid}{\sim} \sum_{h=1}^{20} \pi_h f(x,\theta_h,\sigma^2),\ i = 1,\ldots,273$ <br> $\boldsymbol{\beta} \sim U([-10,10]^p)$ |
| IV | $\sigma^{-2} \sim Exp(1/16), \boldsymbol{\pi} \sim Seth_{20}(\alpha = 1/3)$ <br> $\theta_h \overset{iid}{\sim} \mathcal{N}(0,3.289),\ h = 1,\ldots,20$ |

$$D(\boldsymbol{\rho}|y) = -2\sum_{i=1}^{273}\sum_{j=1}^{n_i} y_{ij}\log(\rho_{ij}) + (1-y_{ij})\log(1-\rho_{ij}).$$

The DIC is then defined as $\overline{D(\boldsymbol{\rho}|y)} + p_D$, where $\overline{D(\boldsymbol{\rho}|y)}$ is the average deviance over the MCMC draws measuring the model fit and $p_D = \overline{D(\boldsymbol{\rho}|y)} - D(\bar{\boldsymbol{\rho}}|y)$ is an estimate of the "effective" number of parameters penalizing the model complexity ($\bar{\boldsymbol{\rho}}$ is the average of $\boldsymbol{\rho}$ over the MCMC iterations). For more details on the DIC we refer to [18]. The first model that we will use for comparison, is a model that does not allow for individual goalkeeper effects at all, leading to $\text{logit}(\rho_{ij}) = \mu_0 + \boldsymbol{\beta}'\boldsymbol{x}_{ij}$, with a fixed common intercept $\mu_0$. Hence, by comparing this model with the Dirichlet process model in terms of the DIC we will be able to quantify the improvement of modeling individual goalkeeper effects. The second model we use for a comparison is a parametric normal random effects model, which can be obtained by setting $\gamma_i \overset{iid}{\sim} \mathcal{N}(\mu_0, \sigma_0^2)$ in level III of Table 1, and using suitable vague hyper-priors for $\mu_0$ and $\sigma_0^2$ (here we use $\mu_0 \sim \mathcal{N}(0,3.289)$ and $\sigma_0^2 \sim U([0,3])$). By comparing the Dirichlet process model with this parametric model we will be able to quantify the improvement of a nonparametric modeling of the random effects distribution. Subsequently the two restricted models will be referred to as 'Intercept' and 'Normal', our proposed model will be termed the 'Dirichlet' model.

## 2.4 Choice of Covariates

The main aim of this analysis is to model the goalkeepers effect on the probability of saving a penalty kick, but the effect of the scorer should also be taken into account. The logarithm of the number of taken penalties provides a good fit in an univariate logistic regression and is chosen to represent the penalty takers effect. For better interpretability the logarithm of base 2 is chosen. As home field advantage has an effect in many sports, the home field advantage of the goalkeeper is included as a binary covariate. To see whether there is a general time trend in the probability of saving a penalty, year is included as a covariate. "Year" here refers to a football season, which starts at the end of summer. A year effect could be due to improved techniques for saving or taking a penalty. In addition the day of the season is included as a covariate to account for possible time trends within a season. For model fitting all covariates are scaled to lie between 0 and 1.

### 2.5 Computation

The models described in Section 2.3 are fit to the data using the OpenBUGS software version 2.10. Further analysis is done in R 2.6.2 using the interface provided by the R2WinBUGS package [19].

For each model the MCMC sampler is run with two independent chains with a burn-in of 50,000 iterations followed by 100,000 iterations of which every 20th is kept. Trace plots of parameters did not indicate problems with convergence of the chains and the results of the independent chains are similar. The results presented are based on the pooled draws of the independent chains, leading to a total number of 10,000 draws for each model.

## 3 Results

First the overall fit of the models is compared with the DIC criterion. Table 2 shows the DIC and its components for the three models considered. Both the Normal and the Dirichlet model improve on the model with only an intercept, indicating some gain with the inclusion of a random effects distribution. The improvement is not very large, indicating that the probability of saving a penalty does not vary too much between goalkeepers. As it is more flexible, the Dirichlet model has a lower average deviance than the Normal model but also a larger number of effective parameters leading to a DIC that is only slightly lower.

To answer the question whether there are distinct clusters of goalkeepers with differing abilities we compare the posterior distribution of the number of distinct components $p(k|y, \alpha, n)$ to the prior computed via (2). Barplots of both distributions are shown in Fig. 2 (i). One can see that the posterior puts less mass on a higher number of components than the prior, with one single component having the highest posterior probability. The posterior mean is 1.98 components compared to the prior mean 2.91. Observing the posterior expectation of the random effects distributions

**Table 2** Average deviance, effective number of parameters and DIC for the different models

| Model | $\overline{D(\rho|y)}$ | $p_D$ | DIC |
|---|---|---|---|
| Intercept | 3,453.8 | 5.0 | 3,458.8 |
| Normal | 3,422.5 | 31.1 | 3,453.6 |
| Dirichlet | 3,414.8 | 36.8 | 3,451.6 |

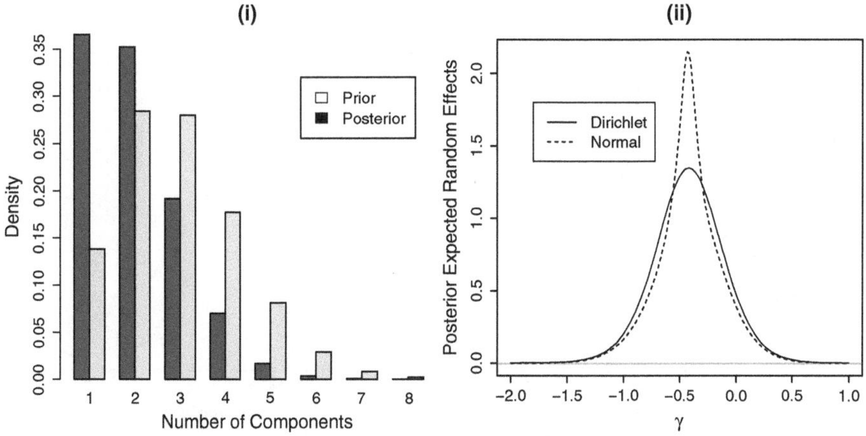


**Fig. 2** (i) Distribution of the number of distinct components $k$ for the Dirichlet model. (ii) Posterior expected random effects distribution $\gamma$ for the Normal and Dirichlet model

shown in Fig. 2 (ii), there is no sign of multimodality. Thus there is not much support for different clusters in the data. In the Dirichlet model even for parameter draws with several distinct components, the resulting distribution tended to be unimodal (a mixture of normal distribution does not have to be multimodal). However, the more flexible Dirichlet model leads to a distribution with heavier tails than the one resulting from the Normal model.

Next we take a look at the estimates for the goalkeepers' probabilities to save a penalty that can be derived from the models. For this we consider

$$E\left(\frac{\exp(\gamma_i + \boldsymbol{\beta}'\boldsymbol{x}_{med})}{1+\exp(\gamma_i + \boldsymbol{\beta}'\boldsymbol{x}_{med})}\bigg| y\right), \; i = 1,\ldots,273, \tag{4}$$

the posterior expectation of the goalkeepers' probabilities to save a penalty kick when the covariates take their respective median values $\boldsymbol{x}_{med}$. The median values stand for a scorer with 10 taken penalties, the season 1983/84 and the 17th day of the season. The binary variable home field advantage is set to 0, representing no home field advantage for the goalkeeper. Figure 3 shows the posterior mean probabilities of the goalkeepers (from (4)) for all goalkeepers smoothed by a kernel density estimate. Comparing Fig. 3 (i) to the distribution of the relative frequencies in Fig. 1 (ii) it can be seen that the probabilities are considerably shrunken towards each other. The range of estimates is only about 0.1. Figure 3 (ii) shows a close-up look at the distribution in (i), and as for the random effects distribution it can be seen that the estimates of the Normal and Dirichlet model differ mainly in the tails, with the Dirichlet model leading to more pronounced tails.

Regarding the question of identifying the best and worst keepers, the tails of the distribution are of importance. As the Dirichlet model is more flexible in the tails it is used to determine a ranking of the keepers. In performing the ranking (see Table 3)

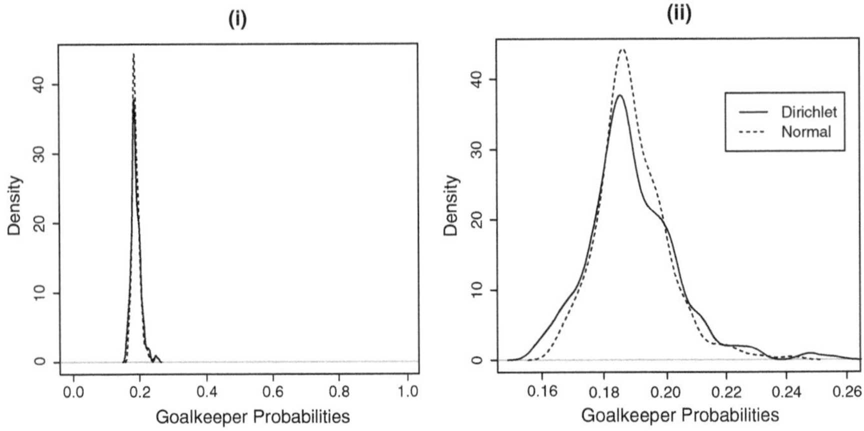


**Fig. 3** Posterior expected probabilities of saving a penalty for the Normal and Dirichlet model; (i) on the range [0,1] and (ii) on the range [0.15, 0.26]

we rely on the recommendations of [11] who argue that ranking should be based on the posterior expectation of the rank rather than the posterior expected effect. This explains the fact that in some cases a goalkeeper with a higher rank nevertheless has a higher posterior expected probability of saving a penalty.

Several interesting observations arise from the ranking in Table 3. Goalkeepers estimated saving probabilities are not really different, with the best keeper having 25.5% and the worst keeper having 16.0%, yielding only a 10%-points difference. Moreover, the credible intervals for the saving probabilities are seen to be pretty large, credible intervals for the best and the worst keeper overlap considerably. As such, saving capabilities are rather similar across goalkeepers, reflecting the fact that no explicit clusters of goalkeepers could be found in our analysis. It is nevertheless surprising, that the two German goalkeepers who are thought to be penalty specialists (Oliver Kahn and Jens Lehmann) rank relatively low, indicating that both of them perform rather badly in penalty saving. This is probably due to the perception of the German expertise in penalty shoot-outs in recent tournaments, with Kahn and Lehmann playing prominent roles on these occasions. The degree of shrinking from the Dirichlet model is quite impressive. To demonstrate this, we consider Michael Melka and Gerhard Teupel as two representatives of the goalkeepers who were faced with only one single penalty during their career in the German Bundesliga. Michael Melka who saved this single penalty (thus having an observed 100% saving rate), has an estimated saving probability of only 20.2%. Gerhard Teupel, not saving this single penalty (resulting in an observed 0% saving rate) estimated saving probability is 18.6%, not very different from Melka's probability. Another peculiarity might be the fact that 3 goalkeepers of Bayern München (Manfred Müller, Walter Junghans, and Sepp Maier, having played 588 games or more that 17 seasons for the team altogether) are among the worst 5 penalty savers. This is in strict contrast to the fact that Bayern München is the most successful team in the German Bundesliga. It is

**Table 3** Ranking of goalkeepers based on the average rank

| Goalkeeper | Rank | $P(\text{Saving}\|y)$ with 95% CI | % Saved | # Saved | # Penalties |
|---|---|---|---|---|---|
| Kargus, Rudolf | 1 | 0.255 [0.183, 0.354] | 0.329 | 23 | 70 |
| Enke, Robert | 2 | 0.248 [0.162, 0.418] | 0.500 | 9 | 18 |
| Pfaff, Jean–Marie | 3 | 0.247 [0.155, 0.483] | 0.545 | 6 | 11 |
| Köpke, Andreas | 4 | 0.228 [0.159, 0.324] | 0.317 | 13 | 41 |
| Radenkovic, Petar | 5 | 0.229 [0.158, 0.331] | 0.353 | 12 | 34 |
| ⋮ | ⋮ | ⋮ | ⋮ | ⋮ | ⋮ |
| Melka, Michael | 54 | 0.202 [0.121, 0.317] | 1.000 | 1 | 1 |
| ⋮ | ⋮ | ⋮ | ⋮ | ⋮ | ⋮ |
| Teupel, Gerhard | 154 | 0.186 [0.107, 0.285] | 0.000 | 0 | 1 |
| ⋮ | ⋮ | ⋮ | ⋮ | ⋮ | ⋮ |
| Kahn, Oliver | 224 | 0.178 [0.120, 0.245] | 0.172 | 10 | 58 |
| ⋮ | ⋮ | ⋮ | ⋮ | ⋮ | ⋮ |
| Lehmann, Jens | 228 | 0.178 [0.115, 0.248] | 0.176 | 6 | 34 |
| ⋮ | ⋮ | ⋮ | ⋮ | ⋮ | ⋮ |
| Schmadtke, Jörg | 269 | 0.162 [0.098, 0.227] | 0.098 | 4 | 41 |
| Müller, Manfred | 270 | 0.160 [0.082, 0.232] | 0.042 | 1 | 24 |
| Junghans, Walter | 271 | 0.158 [0.083, 0.230] | 0.042 | 1 | 24 |
| Rynio, Jürgen | 272 | 0.159 [0.088, 0.226] | 0.074 | 2 | 27 |
| Maier, Sepp | 273 | 0.160 [0.104, 0.218] | 0.130 | 9 | 69 |

**Table 4** Estimated odds ratios with 95% credible intervals in the Dirichlet model. For the penalty taker odds ratio is for a scorer with twice the number of penalties. The odds ratio for year compares the last to the first year, which is also the case for day of the season

| Covariate | OR with 95% CI |
|---|---|
| Scorer | 0.754 [0.711, 0.798] |
| Home Field Advantage | 0.956 [0.789, 1.145] |
| Year | 0.894 [0.637, 1.222] |
| Day of Season | 0.894 [0.674, 1.166] |

also astonishing that Sepp Maier ranks the worst. After all, he was the goalkeeper of the German team winning the 1974 world cup, and is still the German goalkeeper with the most international matches ($N = 95$) up to now.

Finally, we consider the effects of the covariates. Since a logistic regression model is fitted, $\exp(\beta_k)$ can be interpreted as the change in the odds of the event, if the $k$th covariate is risen by 1. Table 4 shows the estimated odds ratios for the Dirichlet model. As the credible interval for the odds ratio of the scorer effect does not contain 1 there is strong evidence that a scorer that has taken more penalties reduces the goalkeeper's probability of saving the penalty. This is a reasonable result, since players that are known to be good penalty takers are probably chosen more often to take a penalty kick. As the scorer effect is given on the log2 scale, we can interpret the odds ratio as follows: Faced with a scorer that scored twice as many penalties, the goalkeeper's odds of saving is multiplied by 0.754. For all the other covariates, 1 is clearly inside the credible interval. This implies that there is no evidence for a home field advantage for the goalkeeper. Additionally, evidence can neither be found for an overall time trend or a time trend within seasons. These conclusions are also obtained for the other two models.

## 4 Final Remarks and Outlook

In this article we analyzed the penalty saving abilities of goalkeepers in the first 44 years of the German Bundesliga. As is typical for such a data set, many goalkeepers were involved only in a few penalties. This poses the question on how to derive reasonable estimates for those keepers and how to compare keepers with a highly disparate number of penalties. We approached this issue by using Bayesian hierarchical models, i.e., the goalkeepers are modeled as realizations from a common random effects distribution $P$. This naturally allows for borrowing strength and hence shrinkage between the goalkeepers individual effect estimates. A major impetus for studying the data was to investigate whether there are certain groups of goalkeepers, such as 'penalty specialists' and 'penalty losers'. This motivated the use of Bayesian nonparametric approaches to model the random effects, as these techniques allow for modelling multimodal random effects distributions.

In the analyses we conducted in Section 3 we did not find any hint for multimodality. On the contrary, a-posteriori there was evidence that the number of components/clusters in the normal mixture model is even smaller than assumed a-priori (see Fig. 2 (i)). We also produced a ranking of the goalkeepers based on the average rank encountered during the MCMC runs. One observation is central: there is no strong evidence in the data that the different goalkeepers are highly different, for example, the credibility intervals for the goalkeeper ranking first (Rudolf Kargus) and last (Sepp Maier) overlap considerably.

From an application viewpoint it is somewhat surprising to see well-known goalkeepers like Sepp Maier ranking so low. This is a direct consequence of the shrinkage effect of the random effects model: As can be seen in Table 3, only goalkeepers who were involved in many penalties can rank at the top or the bottom of the list, while the goalkeepers with fewer penalties are all in the middle of the ranking. This is reasonable from a statistical point of view, as we can only make statistically accurate estimates for keepers with many penalties, while those with few penalties are shrunken towards the overall mean. This shrinkage effect should be kept in mind, when interpreting the ranking of goalkeepers from an application viewpoint. As can be seen in the tails of the random effects distribution and the estimated individual effects (Figs. 2 (ii) and 3 (ii)) the Dirichlet model already allows for a more realistic and flexible type of shrinkage than the normal model. There are however opportunities to model the random effects distribution even more flexible. The Dirichlet process may be replaced by another stochastic process, *e.g.*, a (normalized) $\alpha$-stable process or the used normal kernel may be replaced by a $t$-density. Both approaches would lead to even heavier tails of the random effects distribution and thus to a model representing less shrinkage.

In our analysis only the covariate we used as a substitute for the scorer effect seems to have an important effect. This motivates a further study, where the penalty scorer effect is also modeled by a random effects distribution instead of a simple fixed covariate. This might lead to a more realistic model and would allow for a ranking of the scorers as well. For the Dirichlet model a complication arises, however, if a second random effect is to be included. Then it is necessary to center the random effects distributions to have mean zero. Simply setting the mean of the base probability measure $F_0$ to zero is not sufficient to achieve zero mean of the random effects distribution, and more sophisticated procedures need to be applied such as the centered Dirichlet process [3], which we plan to do in future research.

**Acknowledgement** We are grateful to IMP AG, München, for providing parts of the underlying data set and Holger Rahlfs and Jörn Wendland of IMP AG for their kind cooperation. Mareike Kunze (IMEBI Halle) was very helpful in data entry, processing, and management. The work of Björn Bornkamp is supported by the Research Training Group "Statistical modelling" of the German Research Foundation (DFG).

## References

[1] Antoniak, C.E.: Mixtures of Dirichlet processes with applications to Bayesian nonparametric problems. Ann. Stat. **2**, 1152–1174 (1974)

[2] Diaconis, P., Ylvisaker, D.: Quantifying prior opinion. In: Bernardo, J., DeGroot, M., Lindley, D., Smith, A. (eds.) Bayesian Statistics 2, pp. 133–156. Elsevier, Amsterdam (1985)

[3] Dunson, D.B., Yang, M., Baird, D.: Semiparametric Bayes hierarchical models with mean and variance constraints. Technical Report 2007-08, Department of Statistical Science, Duke University, Durham, NC, USA (2007)

[4] Dunson, D.B.: Nonparametric Bayes applications to biostatistics. Technical Report 2008-06, Department of Statistical Science, Duke University, Durham (2008)

[5] Ferguson, T.S.: A Bayesian analysis of some nonparametric problems. Ann. Stat. **1**, 209–230 (1973)

[6] Ishwaran, H., James, L.F.: Gibbs sampling methods for stick-breaking priors. J. Am. Stat. Assoc. **96**, 161–173 (2001)

[7] Kropp, M., Trapp, A.: 35 Jahre Bundesliga-Elfmeter 1963–1999. Agon Statistics 36, 2nd edn. Agon-Sportverlag, Kassel (1999)

[8] Kuhn, W.: Penaltykick strategies for shooters and goalkeepers. In: Reilly, K.D.T., Lees, A., Murphy, W. (eds.) Science and football, pp. 489–492. E & FN Spon, London (1988)

[9] Kuss, O., Kluttig, A., Stoll, O.: The fouled player should not take the penalty himself: An empirical investigation of an old German football myth. J. Sport. Sci. **25**, 963–967 (2007)
[10] Lau, J.W., Green, P.J.: Bayesian model-based clustering procedures. J. Comput. Graph. Stat. **16**, 526–558 (2007)
[11] Lin, R., Louis, T.A., Paddock, S.M., Ridgeway, G.: Loss function based ranking in two-stage hierarchical models. Bayesian Anal. **1**, 915–946 (2006)
[12] Loy, R.: Handlungsstrategien von Torhütern und Schützen in der Strafstoßsituation des Fußballsports. In: Bäumler, G., Bauer, G. (eds.) Sportwissenschaft rund um den Fußball, pp. 67–78. Schriften der Deutschen Vereinigung für Sportwissenschaft 96, Sankt Augustin (1988)
[13] Morya, E., Ranvaud, R., Pinheiro, W.: Dynamics of visual feedback in a laboratory simulation of a penalty kick. J. Sport. Sci. **21**, 87–95 (2003)
[14] Ohlssen, D.I., Sharples, L.D., Spiegelhalter, D.J.: Flexible random-effects models using Bayesian semi-parametric models: Applications to institutional comparisons. Stat. Med. **26**, 2088–2112 (2007)
[15] Savelsbergh, G., van der Kamp, J., Williams, A., Ward, P.: Anticipation and visual search behaviour in expert soccer goalkeepers. Ergonomics **48**, 1686–1697 (2005)
[16] Savelsbergh, G., Williams, A., van der Kamp, J., Ward, P.: Visual search, anticipation and expertise in soccer goalkeepers. J. Sport Sci. **20**, 279–287 (2002)
[17] Sethuraman, J.: A constructive definition of Dirichlet priors. Stat. Sin. **4**, 639–650 (1994)
[18] Spiegelhalter, D.J., Best, N.G., Carlin, B.P., van der Linde, A.: Bayesian measures of model complexity and fit. J. Roy. Stat. Soc. B **64**, 583–639 (2002)
[19] Sturtz, S., Ligges, U., Gelman, A.: R2WinBUGS: A package for running WinBUGS from R. J. Stat. Softw. **12**, 1–16 (2005)
[20] Temme, N.M.: Asymptotic estimates of Stirling numbers. Stud. Appl. Math. **89**, 233–243 (1993)
[21] Van der Kamp, J.: A field simulation study of the effectiveness of penalty kick strategies in soccer: Late alterations of kick direction increase errors and reduce accuracy. J. Sport. Sci. **24**, 467–477 (2006)

# Permutation Tests for Validating Computer Experiments

Thomas Mühlenstädt and Ursula Gather

**Abstract** Deterministic computer experiments are of increasing importance in many scientific and engineering fields. In this paper we focus on assessing the adequacy of computer experiments, i.e. we test if a computer experiment is predicting a corresponding real world phenomenon. A permutation test is presented which can be adapted to different situations in order to achieve good power.

## 1 Introduction

Computer experiments are of great relevance especially in engineering. A broad variety of methods for analyzing, designing and predicting data from computer experiments has been proposed in the literature, see for example [1] for an overview. The advantages of computer experiments are obvious, they are often much faster, cheaper and generally easier to work with compared to the respective real world experiments on certain phenomena. However, validating a computer experiment remains an important problem. Here, permutation tests are a valuable tool as they are distribution free. Additionally, they can achieve the same asymptotic power as some corresponding uniformly best unbiased test, see [5] Chap. 15.2. Hence, we suggest a permutation test for the null hypothesis that a computer experiment is a correct predictor for a corresponding real world experiment.

Our article is organized as follows: In Sect. 2 we briefly give the theoretical background for permutation tests. In Sect. 3 permutation tests for validating computer experiments are proposed and in Sect. 4 an example illustrates the behavior of the suggested tests. A summary concludes our paper.

---

Thomas Mühlenstädt
Fakultät Statistik, Technische Universität Dortmund, D-44221 Dortmund, Germany
muehlens@statistik.tu-dortmund.de

## 2 Permutation Tests

In this section we review some properties of permutation tests. Throughout this paper we use the notation of [5]. Let $T(Y)$ be a real valued statistic for testing a hypothesis $H_0$ and $Y$ be a real valued random vector with observation $y \in \mathbf{Y}$, where $\mathbf{Y}$ is the sample space of $Y$. Let $\mathbf{G}$ be a finite group of transformations mapping $\mathbf{Y}$ onto $\mathbf{Y}$ with cardinality $M$, i.e. $\#\mathbf{G} = M$. For an observation $y$ of $Y$, let

$$T^{(1)}(y) \leq T^{(2)}(y) \leq \cdots \leq T^{(M)}(y)$$

be the ordered values $T(gy)$ of the test statistic for all $g$ in $\mathbf{G}$. We write

$$q := M - \lfloor M\alpha \rfloor,\ \alpha \in (0,1),$$

with $\lfloor . \rfloor$ being the floor function. Now, let $M^+(y) := \#\{T^{(j)}(y) | j = 1\ldots M, T^{(j)}(y) > T^{(q)}(y)\}$ and $M^0(y) := \#\{T^{(j)}(y) | j = 1\ldots M, T^{(j)}(y) = T^{(q)}(y)\}$. Define the randomization constant $a(y)$ as

$$a(y) := \frac{M\alpha - M^+(y)}{M^0(y)}.$$

A test $\Phi(y)$ is then defined as

$$\Phi(y) = \begin{cases} 1, & \text{if } T(y) > T^{(q)}(y) \\ a(y), & \text{if } T(y) = T^{(q)}(y) \\ 0, & \text{if } T(y) < T^{(q)}(y). \end{cases} \tag{1}$$

By construction every $y \in \mathbf{Y}$ fulfills:

$$\sum_{g \in \mathbf{G}} \Phi(gy) = M^+(y) + a(y)M^0(y) = M\alpha.$$

This test is a level $\alpha$ test due to the following theorem, which is an extension of Theorem 15.2.1 in [5].

**Theorem 1.** *Let $Y$ have distribution $P \in \mathcal{P}$. Consider the null hypothesis $\mathcal{P}_0 \subset \mathcal{P}$ and let $\mathbf{G}$ be a finite group of transformations mapping $\mathbf{Y}$ onto $\mathbf{Y}$. If for every $P \in \mathcal{P}_0$ the test statistic $T$ is invariant under transformations $g \in \mathbf{G}$, i.e. $T(Y)$ and $T(gY)$ have the same distribution for all $g \in \mathbf{G}$, then*

$$E_P[\Phi(T(Y))] = \alpha, \textit{ for all } P \in \mathcal{P}_0$$

*with $\Phi(T(Y))$ being the permutation test described above.*

*Proof.* By construction we have

$$\sum_{g \in \mathbf{G}} \Phi(T(gy)) = M^+(y) + a(y)M^0(y) = M\alpha \text{ for all } y \in \mathbf{Y}.$$

Hence

$$M\alpha = E_P[\sum_g \Phi(T(gY))] = \sum_g E_P[\Phi(T(gY))].$$

Under the null hypothesis we have $E_P[\Phi(T(gY))] = E_P[\Phi(T(Y))]$, such that

$$M\alpha = \sum_g E_P[\Phi(T(Y))] = ME_P[\Phi(T(Y))] \Leftrightarrow \alpha = E_P[\Phi(T(Y))].$$

Lehmann and Romano [5] require the assumption that the distribution of $Y$ is invariant under the null hypothesis. However, $\Phi(Y)$ is also a level $\alpha$ test if just the test statistic $T$ is invariant under transformations $g \in \mathbf{G}$. In the next section we omit the randomization constant $a(y)$ and reject the null hypothesis if $T(y) > T^{(q)}(y)$. It is easy to check that $\Phi(Y)$ then remains a level $\alpha$ test.

## 3 Permutation Tests for Computer Experiments

In the following we use an idea of Good [2] to propose local permutation tests for computer experiments as the test described in [2] is inconsistent for some alternatives. Here we interpret a computer experiment as an unknown function $f$ depending on $x \in \mathbb{R}^d$. Let further $Y_1 := Y(x_1), \dots, Y_n := Y(x_n)$ be random variables for which a regression model is assumed: $E(Y(x_i)) = g(x_i)$. We test $H_0 : g(x) = f(x)$ against $H_1 : \exists\, x \in \mathbb{R}^d : f(x) \neq g(x)$.

Now assume that $Y_1, \dots, Y_n$ are mutually independent and that $Y(x)$ possesses a continuous and symmetric distribution for each fixed $x \in \mathbb{R}^d$. Define a random variable

$$Z(x) := \begin{cases} 0, \text{ if } Y(x) - f(x) > 0; \\ 1, \text{ elsewhere.} \end{cases}$$

Good [2], p. 126 uses $T(x_1, \dots, x_n) := \sum_{i=1}^n Z(x_i)$ as test statistic and all possible combinations of sign changes of $Y(x_i) - f(x_i), i = 1, \dots, n$ as transformation group for testing the above hypothesis. Under the above null hypothesis we have $P(Y(x_i) - f(x_i) > 0) = P(Y(x_i) - f(x_i) < 0) = 0.5$ for all $i = 1, \dots, n$. Although the original differences are not necessarily identically distributed, the auxiliary variables $Z(x_i)$ are identically distributed under the null hypothesis. Hence, any test statistic based on these variables yields a level $\alpha$ test. Good's test statistic is binomially distributed $\mathbb{B}(n, 0.5)$ under the null hypothesis. However, this test is inconsistent for certain alternatives. As an example consider $f(x) = \beta x$ and $g(x) = \gamma x$ with $\beta \neq \gamma \in \mathbb{R}$ and $x \in [-1, 1]$. For data points $x_1, \dots, x_n$ equally distributed on $[-1, 1]$, we have for half of the data points $f(x_i) \leq g(x_i)$ and for the other half $f(x_i) \geq g(x_i)$. Although the alternative is true, the test statistic will presumably attain only medium sized values.

To avoid this inconsistency, we suggest to apply Good's test locally. This means we define a subset $s_i$ for every data point $(x_i, y_i)$ containing $k+1$ points ($(x_i, y_i)$ and $k$ additional points) and calculate the test statistic $T_i^k := T(s_i)$ for this subset of the data. Under $H_0$, the $T_i^k$ are identically distributed with $E(T_i^k) = \frac{k+1}{2}$ and $var(T_i^k) = \frac{(k+1)}{4}$. Now, with $D_i^k := \left(T_i^k - \frac{k+1}{2}\right)^2$, we get the test statistic

$$T^k := \sum_{i=1}^{n} D_i^k,$$

which is small under the null hypothesis. This yields a level $\alpha$ test due to the theorem in Sect. 2 if again all possible combinations of sign changes of $Y(x_i) - f(x_i), i = 1,\ldots,n$ are used as transformation group.

Now, an important question is how to define the subsets $s_i$. In the following we discuss two possibilities: Firstly, the points can be grouped according to their $k$ nearest neighbors (*knn*) w.r.t. the input values $x_i$. *knn*-rules are widely used in classification, see for example [3]. The $k$ nearest neighbors of a point $x_i$ are defined to be the $k$ points $x_i^{(1)},\ldots,x_i^{(k)}$ which have the $k$ smallest Euclidean distances from $x_i$ among all $n$ points $x_1,\ldots,x_n$. If the null hypothesis is rejected, the *knn* subsets with highest values of $D_i^k$ provide information on the local fit of the computer experiment. Note that the $k$ nearest neighbors are calculated from the standardized input variables. Otherwise, the $k$ nearest neighbors are defined by input variables with large ranges. We refer to this test as $x - knn$ test.

Often computer experiments come with high dimensional input space combined with the restriction that only a limited number of runs are possible. Then it is still possible to define $k$ nearest neighbors. But, due to the curse of dimensionality, this is not necessarily a good choice, see [3] Chap. 2.5. Therefore, we will group the observations according to their $y$ values. If the unknown function shows some kind of monotonic behavior, similar predictor values will likely have some common characteristics. Hence, grouping them might result in a high power for the permutation test. This version of the test is called $y - knn$ test. Again, if the null hypothesis is rejected, those $D_i^k$ with high values suggest that the fit of the computer experiment for $y$ values near to $y_i$ is poor.

Generally, the subsets $s_i$ can be defined in many different ways. The power of the test against certain kinds of alternatives can be controlled by the way subsets are chosen. If there is some a-priori knowledge about the function $g(x)$ it can be incorporated when defining appropriate subsets.

## 4 A Simulation Study

We consider a 4-dimensional example taken from [4] which is used to illustrate a sequential design algorithm for finding a robust control parameter configuration during simulations. The function given below is chosen to demonstrate this sequential algorithm:

$$g(x) := g(x_1,x_2,x_3,x_4) := \frac{1}{30}\psi(x_1,x_2)\psi(x_3,x_4) + (x_1 - \pi)^2$$

$$\text{with } \psi(x_1,x_2) = \left(x_2 - \frac{5.1}{4\pi^2}x_1^2 + \frac{5}{\pi}x_1 - 6\right)^2 + 10\left(1 - \frac{1}{8\pi}\right)\cos(x_1) + 10.$$

The function $\psi(x_1,x_2)$ is defined on $[-5,10] \times [0,15]$. The following regression model is assumed:

$$Y(x_1,x_2,x_3,x_4) = g(x_1,x_2,x_3,x_4) + \varepsilon(x_2)$$

with $\varepsilon(x_2) \sim \mathbb{N}(0,(3x_2+10)^2)$. Different predictors are used, once the true function and twice perturbed predictors:

$$\begin{aligned}
f_1(x) &= g(x) && \text{null hypothesis } H_0\\
f_2(x) &= g(x)(1+0.007x_3x_4^{1/8}) && \text{alternative } H_1\\
f_3(x) &= g(x)(1+0.007(x_2-6)^3) && \text{alternative } H_2
\end{aligned}$$

Here, two settings of the computer experiment are considered to check if the power of the tests depends on the design. In particular we are interested in experiments with a simplex based latin hypercube design [6] and experiments with random latin hypercubes [7]. For both settings all three functions $f_1, f_2, f_3$ are applied to simulate data resulting in six different combinations, see Table 1. For each combination, 1,000 data sets with sample size $n = 40$ are simulated and all three tests (Good's, $x-knn$, $y-knn$) are performed. For the tests $x-knn$ and $y-knn$, a random sample of 1,000 permutations is chosen as the complete group of transformations is too large (here $\#G = 2^{40}$). For both tests, $x-knn$ and $y-knn$, we have chosen $k = 8$.

The simulation results are summarized in Table 1. For $\alpha = 0.1, 0.05, 0.01$, the table shows the percentage of the tests rejected at the corresponding levels of

**Table 1** Simulation results

| | | $\alpha$ | Good | $x-knn$ | $y-knn$ |
|---|---|---|---|---|---|
| $H_0$ | Simplex | 0.1 | 0.079 | 0.109 | 0.078 |
| | | 0.05 | 0.030 | 0.045 | 0.038 |
| | | 0.01 | 0.002 | 0.008 | 0.004 |
| | LHD | 0.1 | 0.082 | 0.093 | 0.099 |
| | | 0.05 | 0.040 | 0.046 | 0.044 |
| | | 0.01 | 0.002 | 0.006 | 0.008 |
| $H_1$ | Simplex | 0.1 | 0.523 | 0.716 | 0.379 |
| | | 0.05 | 0.374 | 0.598 | 0.247 |
| | | 0.01 | 0.090 | 0.297 | 0.074 |
| | LHD | 0.1 | 0.426 | 0.633 | 0.284 |
| | | 0.05 | 0.275 | 0.478 | 0.186 |
| | | 0.01 | 0.065 | 0.184 | 0.066 |
| $H_2$ | Simplex | 0.1 | 0.274 | 0.682 | 0.682 |
| | | 0.05 | 0.270 | 0.708 | 0.679 |
| | | 0.01 | 0.113 | 0.397 | 0.332 |
| | LHD | 0.1 | 0.197 | 0.760 | 0.418 |
| | | 0.05 | 0.068 | 0.530 | 0.257 |
| | | 0.01 | 0.004 | 0.137 | 0.061 |

significance. The percentage of rejections under the null hypothesis is close to the corresponding level of significance for all considered tests. For both alternatives, $H_1$ and $H_2$, all three tests deliver comparable results for random latin hypercubes and for simplex based latin hypercubes. Thus, the power of the tests does not seem to depend on the design. For alternative $H_1$, the $x - knn$ test possesses the highest power in the simulation while Good's test performs slightly better than the $y - knn$ test. Again, for alternative $H_2$, the $x - knn$ test shows highest power. But here the test $y - knn$ delivers better results than Good's test. Hence, alternative $H_2$ is an alternative for which Good's test is almost inconsistent.

## 5 Summary

The presented permutation tests are an attractive tool for validating computer experiments as they are not difficult to apply and do not depend on strong assumptions. They provide a considerable improvement over the test described in [2]. Depending on the context, different ways of forming subsets can be used in order to incorporate prior knowledge about the behavior of the simulation and the real experiments. For the simulated example the $x - knn$ version has shown to be very efficient.

**Acknowledgement** Financial support of the DFG (research training group "Statistical Modelling") is gratefully acknowledged.

## References

[1] Fang, K.-T., Li, R., Sudjianto, A.: Design and Modeling for Computer Experiments. Chapman & Hall/CRC, New York (2006)
[2] Good, P.: Permutation Tests – A Practical Guide to Resampling Methods for Testing Hypotheses. Springer, New York (2000)
[3] Hastie, T., Tibshirani, R., Friedman, J.: The Elements of Statistical Learning. Springer, New York (2001)
[4] Lehman, J.S., Santner, T.J., Notz, W.I.: Designing Computer Experiments to Determine Robust Control Variables. Statistica Sinica **14**, 571–590 (2004)
[5] Lehmann, E.L., Romano, J.P.: Testing Statistical Hypotheses. Springer, New York (2005)
[6] Mühlenstädt, T.: Simplex based space filling designs. Technical Report, Dortmund University of Technology, Faculty of Statistics, Dortmund (2008)
[7] Santner, T.J., Williams, B.J., Notz, W.I.: The Design and Analysis of Computer Experiments, Springer, New York (2003)

# Part II
# Parametric Inference

*Statistics is the only profession which demands the right to make mistakes about five percent of the time.*

Anonymous

# Exact and Generalized Confidence Intervals in the Common Mean Problem

Joachim Hartung and Guido Knapp

**Abstract** Several exact confidence intervals for the common mean of independent normal populations have been proposed in the literature. Not all of these intervals always produce genuine intervals. In this paper, we consider three types of always genuine exact confidence intervals and compare these intervals with two known generalized confidence intervals for the common mean and a newly proposed one. Besides simulation results, two real data examples are presented illustrating the performance of the various procedures.

## 1 Introduction

Inference on the common mean problem has a long history in statistics. Graybill and Deal [4] pioneered the research on common mean estimation and since then, a lot of further research has been done on this problem, especially from a decision theoretic point of view, see Chap. 5 in [8] for a comprehensive presentation of these results.

The focus of this paper is on confidence intervals for the common mean. Large sample confidence intervals can be easily constructed around the Graybill-Deal estimator with estimated standard errors proposed by Meier [13] or Sinha [14].

Fairweather [3] was the first who proposed an exact confidence interval on the common mean which is based on a linear combination of $t$-test statistics. Also using $t$-test statistics, Cohen and Sackrowitz [1] developed two further exact confidence intervals on the common mean. Jordan and Krishnamoorthy [10] suggested using a linear combination of $F$-test statistics for constructing an exact confidence interval. Yu et al. [17], derived exact confidence intervals using $P$-values of $F$-test statistics. Using well-known methods of combining $P$-values, see [9], they constructed

Joachim Hartung
Fakultät Statistik, Technische Universität Dortmund, D-44221 Dortmund, Germany
hartung@statistik.uni-dortmund.de

the exact confidence intervals by inverting the acceptance region of a family of level-$\alpha$-tests. Recently, Hartung and Knapp [6] used $P$-values of $t$-test statistics and introduced two broad classes of exact confidence intervals for the common mean using weighted inverse normal and generalized inverse $\chi^2$-methods for combining $P$-values. Beside Fairweather's interval, the intervals proposed by Hartung and Knapp always yield genuine intervals. All the other exact confidence intervals do not necessarily provide genuine intervals.

Based on the concept of generalized confidence intervals introduced by Weerahandi [16], Krishnamoorthy and Lu [11] as well as Lin and Lee [12] proposed generalized pivotal quantities that can be used for calculating generalized confidence intervals on the common mean. In this paper, we will introduce a further generalized pivotal quantity.

The outline of this paper is as follows: In Sect. 2, we introduce the common mean problem. Section 3 contains the description of the exact confidence intervals where we restrict the presentation on the three types of intervals mentioned above which always yield genuine intervals. In Sect. 4, we describe the concept of generalized confidence intervals and present three generalized pivotal quantities for the common mean. In the common mean problem, experiments are designed to provide duplicate information about a parameter. In Sect. 5, we use two real data examples of such experiments and apply the intervals on both the data sets. In a simulation study, whose results are displayed in Sect. 6, we investigate the intervals with respect to their actual confidence levels, especially the generalized confidence intervals, and with respect to their expected lengths. At the end, some final remarks are given, especially with respect to the fields of applications.

## 2 Common Mean Problem

Let us consider $k$ independent normal populations where the $i$th population follows a normal distribution with mean $\mu \in \mathbb{R}$ and variance $\sigma_i^2 > 0$, $i = 1, \ldots, k$. Let $\bar{Y}_i$ denote the sample mean in the $i$th population, $S_i^2$ the sample variance, and $n_i$ the sample size, $i = 1, \ldots, k$. Then, we have

$$\bar{Y}_i \sim N\left(\mu\,,\, \frac{\sigma_i^2}{n_i}\right) \quad \text{and} \quad \frac{(n_i - 1)\, S_i^2}{\sigma_i^2} \sim \chi^2_{n_i - 1},\ i = 1, \ldots, k, \tag{1}$$

and the statistics are all mutually independent. Note that $(\bar{Y}_i, S_i^2, i = 1, \ldots, k)$ is minimal sufficient for $(\mu, \sigma_1^2, \ldots, \sigma_k^2)$ even though it is not complete.

If the population variances $\sigma_1^2, \ldots, \sigma_k^2$ are completely known, the maximum likelihood estimator of $\mu$ is given by

$$\hat{\mu} = \frac{\sum_{i=1}^{k} n_i\, \bar{Y}_i / \sigma_i^2}{\sum_{j=1}^{k} n_j / \sigma_j^2}. \tag{2}$$

The above estimator is also the minimum variance unbiased estimator under normality as well as the best linear unbiased estimator without normality for estimating $\mu$. The variance of $\hat{\mu}$ is given by

$$\mathrm{Var}(\hat{\mu}) = \left(\sum_{i=1}^{k} n_i/\sigma_i^2\right)^{-1}. \tag{3}$$

An estimator of the common mean given in a closed form can be obtained by replacing $\sigma_i^2$ by $S_i^2$ in (2). This yields the well-known Graybill-Deal [4] estimator given as

$$\hat{\mu}_{\mathrm{GD}} = \frac{\sum_{i=1}^{k} n_i \bar{Y}_i/S_i^2}{\sum_{j=1}^{k} n_j/S_j^2}. \tag{4}$$

Clearly, $\hat{\mu}_{\mathrm{GD}}$ is an unbiased estimator of the common mean $\mu$.

For calculating the variance of $\hat{\mu}_{\mathrm{GD}}$, a standard conditional argument first yields

$$\begin{aligned}\mathrm{Var}(\hat{\mu}_{\mathrm{GD}}) &= \mathrm{E}\left[\mathrm{Var}(\hat{\mu}_{\mathrm{GD}}|S_1,\ldots,S_k)\right] + \mathrm{Var}\left[\mathrm{E}(\hat{\mu}_{\mathrm{GD}}|S_1,\ldots,S_k)\right] \\ &= \mathrm{E}\left[\left(\sum_{i=1}^{k}\frac{n_i\,\sigma_i^2}{S_i^4}\right)\Big/\left(\sum_{i=1}^{k}\frac{n_i}{S_i^2}\right)^2\right].\end{aligned} \tag{5}$$

Meier [13] derived a first order approximation of the variance of $\hat{\mu}_{\mathrm{GD}}$ as

$$\mathrm{Var}(\hat{\mu}_{\mathrm{GD}}) = \frac{1}{\sum_{i=1}^{k} n_i/\sigma_i^2}\left[1 + 2\sum_{i=1}^{k}\frac{1}{n_i-1}\,c_i\,(1-c_i) + O\left(\sum_{i=1}^{k}\frac{1}{(n_i-1)^2}\right)\right] \tag{6}$$

with

$$c_i = \frac{n_i\,/\,\sigma_i^2}{\sum_{j=1}^{k} n_j\,/\,\sigma_j^2},\ i = 1,\ldots,k.$$

For further statistical inference on the common mean, an estimator of the variance of $\hat{\mu}_{\mathrm{GD}}$ should be available. Sinha [14] derived an unbiased estimator of the variance of $\hat{\mu}_{\mathrm{GD}}$ that is a convergent series. A first order approximation of this estimator is

$$\widehat{\mathrm{Var}}_{(1)}(\hat{\mu}_{\mathrm{GD}}) = \tag{7}$$

$$\frac{1}{\sum_{i=1}^{k} n_i/S_i^2}\left[1 + \sum_{i=1}^{k}\frac{4}{n_i+1}\left(\frac{n_i\,/\,S_i^2}{\sum_{j=1}^{k} n_j\,/\,S_j^2} - \frac{n_i^2\,/\,S_i^4}{\left(\sum_{j=1}^{k} n_j\,/\,S_j^2\right)^2}\right)\right].$$

This estimator is comparable to the approximate estimator

$$\widehat{\mathrm{Var}}_{(2)}\left(\hat{\mu}_{\mathrm{GD}}\right) = \tag{8}$$

$$\frac{1}{\sum_{i=1}^{k} n_i/S_i^2}\left[1+\sum_{i=1}^{k}\frac{4}{n_i-1}\left(\frac{n_i\,/\,S_i^2}{\sum_{j=1}^{k} n_j\,/\,S_j^2}-\frac{n_i^2\,/\,S_i^4}{\left(\sum_{j=1}^{k} n_j\,/\,S_j^2\right)^2}\right)\right].$$

due to Meier [13].

Using the Graybill–Deal estimator (4) for the common mean and an appropriate variance estimator of it, for instance (7) or (8), large sample $100(1-\alpha)\%$ confidence intervals for $\mu$ can be constructed as

$$\hat{\mu}_{\mathrm{GD}} \pm \sqrt{\widehat{\mathrm{Var}}\left(\hat{\mu}_{\mathrm{GD}}\right)}\, z_{1-\alpha/2}$$

with $z_{1-\alpha/2}$ the $(1-\alpha/2)$-quantile of the standard normal distribution.

## 3 Exact Confidence Intervals

Since

$$t_i = \frac{\sqrt{n_i}\,(\bar{Y}_i-\mu)}{S_i} \sim t_{n_i-1} \tag{9}$$

are test statistics for testing hypotheses about $\mu$ based on the $i$th sample, suitable linear combinations of these test statistics or other functions thereof can be used as a pivotal quantity to construct exact confidence intervals for $\mu$.

Fairweather [3] suggested using a weighted linear combination of the $t_i$'s, namely

$$W_t = \sum_{i=1}^{k} u_i\, t_i, \quad u_i = \frac{\left[\mathrm{Var}(t_i)\right]^{-1}}{\sum_{j=1}^{k}\left[\mathrm{Var}(t_j)\right]^{-1}},\; i=1,\ldots,k. \tag{10}$$

Let $b_{1-\alpha/2}$ denote the quantile of the distribution of $W_t$ satisfying the equation

$$1-\alpha = \mathrm{P}\left(|W_t| \le b_{1-\alpha/2}\right),$$

then the exact $100(1-\alpha)\%$ confidence interval for $\mu$ is given by

$$\frac{\sum\limits_{i=1}^{k}\sqrt{n_i}\,u_i\,\bar{Y}_i\,/\,S_i}{\sum\limits_{i=1}^{k}\sqrt{n_i}\,u_i\,/\,S_i} \pm \frac{b_{1-\alpha/2}}{\sum\limits_{i=1}^{k}\sqrt{n_i}\,u_i\,/\,S_i}. \tag{11}$$

Let $t_\nu$ denote a $t$-distributed random variable with $\nu$ degrees of freedom, then it holds $\mathrm{Var}(t_\nu) = \nu/(\nu-2)$, $\nu > 2$, so that the distribution of $W_t$ essentially depends on the degrees of freedom of the $t$-test statistics. Fairweather [3] provided an approximation of the distribution of $W_t$ that can also be used to approximate the required quantile $b_{1-\alpha/2}$. Since $W_t$ is a linear combination of $t$-distributed random variables, the distribution of $W_t$ should resemble a scaled $t$-distribution, that is, we approximate the distribution of $W_t$ by a $c\, t_\nu$-distribution so that the second and fourth moment of both distributions coincide. The solution is given by $\nu = 4 + 1/\sum_{i=1}^{k}[u_i^2/(n_i-5)]$ and $c = \sqrt{(\nu-2)\,/\,(\nu A)}$ with $A = \sum_{i=1}^{k}(n_i-3)/(n_i-1)$, see [3]. Note that Fairweather's interval is always a genuine interval for $0 < \alpha < 0.5$.

Hartung and Knapp [6] used the $t$-test statistics $t_i$ from (9) and suggested two broad classes of exact $100(1-\alpha)\%$ confidence intervals for $\mu$. Let $F_{t_{n_i-1}}$ be the cumulative distribution function of the $t$-distribution with $(n_i-1)$ degrees of freedom. Then it holds

$$F_{t_{n_i-1}}(t_i) \sim U(0,1) \quad \text{and} \quad \Phi^{-1}[F_{t_{n_i-1}}(t_i)] \sim N(0,1),$$

where $U(0,1)$ stands for the uniform distribution on the unit interval and $\Phi^{-1}$ is the inverse of the cumulative distribution function $\Phi$ of the standard normal distribution.

Let us consider the weighted inverse normal combination statistic

$$Z(\mu) = \sum_{i=1}^{k} \sqrt{\frac{\gamma_i}{\sum_{j=1}^{k}\gamma_j}}\; \Phi^{-1}\left(F_{t_{n_i-1}}(t_i)\right) \tag{12}$$

with some positive weights $\gamma_i$, $i = 1,\ldots,k$. Clearly, $Z(\mu)$ is a standard normal random variable. One possible choice of positive weights is $\gamma_i = 1$, $i = 1,\ldots,k$. This means that the precision of each result is only represented through the cumulative distribution function $F_{t_{n_i-1}}$. Since the results of larger experiments are usually more precise, a natural choice of the weights $\gamma_i$ may be the sample size $n_i$ or the degrees of freedom $n_i - 1$.

The functions $F_{t_{n_i-1}}(\cdot)$ and $\Phi^{-1}(\cdot)$ are monotone increasing functions in their arguments $(\cdot)$, so that $Z(\mu)$ from (12) is a monotone decreasing function in $\mu$. Consequently, an exact $100(1-\alpha)\%$ confidence interval for $\mu$ is given by

$$[\mu_{L,Z}\ ;\ \mu_{U,Z}] \tag{13}$$

where the bounds $\mu_{L,Z}$ and $\mu_{U,Z}$ are the unique solutions of the equations

$$Z(\mu_{L,Z}) = \Phi^{-1}(1-\alpha/2) \quad \text{and} \quad Z(\mu_{U,Z}) = \Phi^{-1}(\alpha/2)\,.$$

A second class of exact confidence intervals for $\mu$ suggested by Hartung and Knapp [6] relies on the inverse $\chi^2$-method. Let $G_{\gamma_i}^{-1}$ denote the inverse of the cumulative distribution function $G_{\gamma_i}$ of a $\chi^2$-distribution with $\gamma_i$ degrees of freedom. The general inverse $\chi^2$-combination statistic is then given by

$$S(\mu) = \sum_{i=1}^{k} G_{\gamma_i}^{-1}\left(F_{t_{n_i-1}}(t_i)\right). \tag{14}$$

Clearly, $S(\mu)$ is a $\chi^2$-distributed random variable with $\gamma_\Sigma = \sum_{i=1}^k \gamma_i$ degrees of freedom. Since $F_{t_{n_i-1}}(\cdot)$ and $G_{\gamma_i}^{-1}(\cdot)$ are monotone increasing functions in their arguments $(\cdot)$, $S(\mu)$ is monotone decreasing in $\mu$. Consequently, an exact $100(1-\alpha)\%$ confidence interval for $\mu$ is given by

$$[\,\mu_{L,S}\,;\,\mu_{U,S}\,], \tag{15}$$

where the bounds $\mu_{L,S}$ and $\mu_{U,S}$ are the unique solutions of the equations

$$S(\mu_{L,S}) = \chi^2_{\gamma_\Sigma;1-\alpha/2} \quad \text{and} \quad S(\mu_{U,S}) = \chi^2_{\gamma_\Sigma;\alpha/2}$$

with $\chi^2_{\nu;\alpha}$ the $\alpha$-quantile of a $\chi^2$-distribution with $\nu$ degrees of freedom.

## 4 Generalized Confidence Intervals

The concept of *generalized P-values* was first introduced by Tsui and Weerahandi [15] to deal with the statistical testing problem in which nuisance parameters are present and it is difficult or impossible to obtain a non-trivial test with a fixed level of significance. Weerahandi [16] then introduced the concept of *generalized confidence interval* in this setting. Although a lot of exact confidence intervals for the common mean $\mu$ exist, the generalized confidence interval approach may be an alternative in the common mean problem since some of the exact confidence intervals do not always yield genuine intervals, see Sect. 1.

The general setup for constructing a generalized confidence interval is as follows: Let $\mathbf{X}$ be a random quantity having a density function $f(\mathbf{X}|\zeta)$, where $\zeta = (\theta, \eta)$ is a vector of unknown parameters, $\theta$ is the parameter of interest, and $\eta$ is a vector of nuisance parameters. Suppose we are interested in a confidence interval for $\theta$. Let $\mathbf{x}$ denote the observed value of $\mathbf{X}$ and consider the generalized variable $T(\mathbf{X};\mathbf{x},\zeta)$, which depends on the observed value $\mathbf{x}$ and the parameter vector $\zeta$, and satisfies the following requirements:

(a) The distribution of $T(\mathbf{X};\mathbf{x},\theta,\eta)$ does not depend on any unknown parameters.
(b) The observed value of $T(\mathbf{X};\mathbf{x},\theta,\eta)$ is free of the nuisance parameters.

Then, we say $T(\mathbf{X};\mathbf{x},\theta,\eta)$ is generalized pivotal quantity.

If $t_1$ and $t_2$ are such that

$$P(t_1 \le T(\mathbf{X};\mathbf{x},\theta,\eta) \le t_2) = 1-\alpha, \tag{16}$$

then, $\{\theta : t_1 \le T(\mathbf{X};\mathbf{x},\theta,\eta) \le t_2\}$ is a $100(1-\alpha)\%$ generalized confidence interval for $\theta$. For example, if the value of $T(\mathbf{X};\mathbf{x},\theta,\eta)$ at $\mathbf{X}=\mathbf{x}$ is $\theta$, then

$$[T(\mathbf{x};\alpha/2)\;,\;T(\mathbf{x};1-\alpha/2)]$$

is a $100(1-\alpha)\%$ confidence interval for $\theta$, where $T(\mathbf{x};\kappa)$ stands for the $\kappa$th quantile of $T(\mathbf{X};\mathbf{x},\theta,\eta)$.

Recall that we have independent samples from $k$ normal populations with common mean $\mu$ and possibly unequal variances $\sigma_i^2$, $i=1,\dots,k$. The sample sizes $n_i$, $i=1,\dots,k$, may differ from sample to sample. Let $\bar{Y}_i$ and $S_i^2$ be the sample mean and sample variance in the $i$th population. It is noted that $\bar{Y}_i$ and $S_i^2$ are stochastically independent with

$$\bar{Y}_i \sim N\left(\mu, \frac{\sigma_i^2}{n_i}\right), \quad U_i = \frac{(n_i-1)\,S_i^2}{\sigma_i^2} = \frac{V_i}{\sigma_i^2} \sim \chi^2_{n_i-1}, \quad i=1,\dots,k. \tag{17}$$

Let $\bar{y}_i$ and $s_i^2$ denote the observed values of $\bar{Y}_i$ and $S_i^2$, and $v_i$ stands for the observed value of $V_i$.

Krishnamoorthy and Lu [11] considered a weighted linear combination of sample generalized pivotal quantities. Within each sample, a generalized pivotal quantity for $\mu$ is given as

$$\begin{aligned} T_i &= \bar{y}_i - \left(\frac{\bar{Y}_i-\mu}{\sigma_i/\sqrt{n_i}}\right)\sqrt{\frac{\sigma_i^2\, v_i}{n_i\, V_i}} \\ &= \bar{y}_i - \frac{Z_i}{\sqrt{U_i}}\frac{\sqrt{v_i}}{\sqrt{n_i}}, \quad Z_i \sim N(0,1) \\ &= \bar{y}_i - t_i\frac{s_i}{\sqrt{n_i}}, \; i=1,\dots,k, \end{aligned} \tag{18}$$

with $t_i = \sqrt{n_i-1}\,Z_i/\sqrt{U_i} \sim t_{n_i-1}$. A general pivotal quantity for $\sigma_i^2$ is given as

$$R_i = \frac{\sigma_i^2}{V_i}\, v_i = \frac{v_i}{Q_i}, \quad Q_i = \frac{V_i}{\sigma_i^2} \sim \chi^2_{n_i-1}, \quad i=1,\dots,k. \tag{19}$$

Define $\bar{\mathbf{Y}} = (\bar{Y}_1,\dots,\bar{Y}_k)'$ and $\mathbf{V} = (V_1,\dots,V_k)'$ and let $\bar{\mathbf{y}}$ and $\mathbf{v}$ be the corresponding observed values. Then, the generalized pivotal quantity for the common mean $\mu$ is given as

$$T_{KL}\left(\bar{\mathbf{Y}},\mathbf{V};\bar{\mathbf{y}},\mathbf{v}\right) = \frac{\sum_{i=1}^{k} W_i\, T_i}{\sum_{j=1}^{k} W_j} \tag{20}$$

with

$$W_i = n_i\, Q_i/v_i = n_i\, R_i^{-1}.$$

The generalized pivotal quantity $T_{KL}$ fulfills the two conditions (A) and (B) above and the observed value of $T_{KL}$ is $\mu$. Consequently, $GCI_1(\mu) : \left(T_{KL;\alpha/2}, T_{KL;1-\alpha/2}\right)$ is a generalized confidence interval for $\mu$. Note that Krishnamoorthy and Lu [11] used two different $\chi^2$-random variables $U_i$ and $Q_i$ in the definitions of $T_i$ and $R_i$ even though they are related to the same sample sum of squares. As Krishnamoorthy and Lu [11] pointed out, the use of the same $\chi^2$-random variable in the generalized pivotal quantity produced confidence limits that are too liberal.

The algorithm for calculating $GCI_1(\mu)$ is as follows:
For given data $(\bar{y}_i, s_i^2, n_i)$, $i = 1, \dots k$:
For $j = 1, \dots, m$:

Generate $t_{n_1-1}, \dots, t_{n_k-1}$.

Generate $Q_i \sim \chi^2_{n_i-1}$, $i = 1, \dots, k$.

Compute $W_1, \dots, W_k$.

Compute $T_{KL,j} = \sum_{i=1}^k W_i \left(\bar{y}_i - t_{n_i-1}\, s_i/\sqrt{n_i}\right) / \sum_{j=1}^k W_j$ .

(end $j$ loop)
Compute the $\alpha/2$- and $(1-\alpha/2)$-quantile of $T_{KL,1}, \dots, T_{KL,m}$.

Then, $\left(T_{KL;\alpha/2}, T_{KL;1-\alpha/2}\right)$ is a $100(1-\alpha)\%$ generalized confidence interval on $\mu$.

Lin and Lee [12] first considered the best linear unbiased estimator for $\mu$ assuming that the variances $\sigma_i^2$, $i = 1, \dots, k$, are known. This estimator is given as, cf. (2),

$$\hat{\mu} = \frac{\sum\limits_{i=1}^k n_i\, \bar{Y}_i/\sigma_i^2}{\sum\limits_{j=1}^k n_j/\sigma_j^2} \tag{21}$$

with

$$\hat{\mu} \sim N\left(\mu,\, \left[\sum_{i=1}^k (n_i/\sigma_i^2)\right]^{-1}\right).$$

Consequently,

$$\sqrt{\sum_{i=1}^k (n_i/\sigma_i^2)}\ (\hat{\mu} - \mu) = Z \sim N(0,1).$$

The generalized pivotal quantity for $\mu$ is then given as

$$T_{LL}\left(\bar{\mathbf{Y}}, \mathbf{V}; \bar{\mathbf{y}}, \mathbf{v}\right) = \frac{\sum\limits_{i=1}^k \frac{n_i}{\sigma_i^2}\, \bar{y}_i\, \frac{V_i}{v_i}}{\sum\limits_{j=1}^k \frac{n_j}{\sigma_j^2}\, \frac{V_j}{v_j}} - \frac{\sqrt{\sum\limits_{i=1}^k n_i/\sigma_i^2}\ (\hat{\mu} - \mu)}{\sqrt{\sum\limits_{j=1}^k \frac{n_j}{\sigma_j^2}\, \frac{V_j}{v_j}}}$$

$$= \frac{\sum\limits_{i=1}^k \frac{n_i\, U_i}{v_i}\, \bar{y}_i}{\sum\limits_{j=1}^k \frac{n_j\, U_j}{v_j}} - \frac{Z}{\sqrt{\sum\limits_{j=1}^k \frac{n_j\, U_j}{v_j}}} = \frac{\sum\limits_{i=1}^k W_i\, \bar{y}_i}{\sum\limits_{j=1}^k W_j} - \frac{Z}{\sqrt{\sum\limits_{i=1}^k W_i}} \tag{22}$$

with

$$W_i = n_i\, U_i/v_i, i = 1, \dots, k.$$

The generalized pivotal quantity $T_{LL}$ fulfills the two conditions (A) and (B) and the observed value of $T_{LL}$ is $\mu$. Consequently, $GCI_2(\mu) : \left(T_{LL;\alpha/2}, T_{LL;1-\alpha/2}\right)$ is a generalized confidence interval for $\mu$.

The algorithm for calculating $GCI_2(\mu)$ is as follows:
For given data $(\bar{y}_i, s_i^2, n_i)$, $i = 1, \dots k$:
For $j = 1, \dots, m$:

Generate $Z \sim N(0,1)$.

Generate $U_i \sim \chi^2_{n_i-1}$, $i = 1, \dots, k$.

Compute $W_1, \dots, W_k$.

Compute $T_{LL,j} = \sum_{i=1}^k W_i \bar{y}_i / \sum_{j=1}^k W_j - Z/\sqrt{\sum_{i=1}^k W_i}$.

(end $j$ loop)
Compute the $\alpha/2$- and $(1-\alpha/2)$-quantile of $T_{LL,1}, \dots, T_{LL,m}$.

Then, $(T_{LL;\alpha/2}, T_{LL;1-\alpha/2})$ is a $100(1-\alpha)\%$ generalized confidence interval on $\mu$.

A new third approach also starts with the best linear unbiased estimator $\hat{\mu}$ from (21). Moreover, the statistic

$$\widehat{\mathrm{Var}}(\hat{\mu}) = \frac{1}{k-1} \left( \sum_{i=1}^k \frac{n_i}{\sigma_i^2} \right)^{-1} \sum_{i=1}^k \frac{n_i}{\sigma_i^2} \left( \bar{Y}_i - \frac{\sum_{j=1}^k n_j \bar{Y}_j / \sigma_j^2}{\sum_{\ell=1}^k n_\ell / \sigma_\ell^2} \right)^2 \tag{23}$$

is an unbiased estimator of the variance of $\hat{\mu}$ and stochastically independent of $\hat{\mu}$, see [5]. Hartung [5] also shows that

$$(k-1) \sum_{i=1}^k (n_i/\sigma_i^2)\ \widehat{\mathrm{Var}}(\hat{\mu}) \tag{24}$$

is a $\chi^2$-distributed random variable with $k-1$ degrees of freedom.

Consequently, $(\hat{\mu} - \mu)/\sqrt{\widehat{\mathrm{Var}}(\hat{\mu})}$ is a $t$-distributed random variable with $k-1$ degrees of freedom.

A new generalized pivotal quantity is then given by

$$\begin{aligned}
&T_{new}\left(\bar{\mathbf{Y}}, \mathbf{V}; \bar{\mathbf{y}}, \mathbf{v}\right) \\
&= \frac{\sum_{i=1}^k \frac{n_i U_i}{v_i} \bar{y}_i}{\sum_{j=1}^k \frac{n_j U_j}{v_j}} - t_{k-1} \sqrt{\frac{1}{k-1} \left( \sum_{i=1}^k \frac{n_i U_i}{v_i} \right)^{-1} \sum_{i=1}^k \frac{n_i U_i}{v_i} \left( \bar{y}_i - \frac{\sum_{j=1}^k (n_j U_j / v_j)\, \bar{y}_j}{\sum_{\ell=1}^k (n_\ell U_\ell / v_\ell)} \right)^2} \\
&= \frac{\sum_{i=1}^k W_i \bar{y}_i}{\sum_{j=1}^k W_j} - t_{k-1} \sqrt{\frac{1}{k-1} \left( \sum_{i=1}^k W_i \right)^{-1} \sum_{i=1}^k W_i \left( \bar{y}_i - \frac{\sum_{j=1}^k W_j \bar{y}_j}{\sum_{\ell=1}^k W_\ell} \right)^2}
\end{aligned} \tag{25}$$

with

$$W_i = n_i U_i / v_i, \ i = 1, \dots, k.$$

Again, the two conditions (A) and (B) above are fulfilled and the observed value of $T_{new}$ is $\mu$. Consequently, $GCI_3(\mu) : \left(T_{new;\alpha/2}, T_{new;1-\alpha/2}\right)$ is a generalized confidence interval for $\mu$.

The algorithm for calculating $GCI_3(\mu)$ is as follows:
For given data $(\bar{y}_i, s_i^2, n_i)$, $i = 1, \ldots k$:
For $j = 1, \ldots, m$:

Generate $t_{k-1}$.

Generate $U_i \sim \chi^2_{n_i-1}$, $i = 1 \ldots, k$.

Compute $W_1, \ldots, W_k$.

Compute
$$T_{new,j} = \sum_{i=1}^k W_i\, \bar{y}_i / \sum_{j=1}^k W_j - t_{k-1}\left[1/(k-1)\left(\sum_{i=1}^k W_i\right)^{-1} \sum_{i=1}^k W_i\left(\bar{y}_i - \sum_{j=1}^k W_j\, \bar{y}_j / \sum_{\ell=1}^k W_\ell\right)^2\right]^{1/2}.$$

(end $j$ loop)
Compute the $\alpha/2$- and $(1-\alpha/2)$-quantile of $T_{new,1}, \ldots, T_{new,m}$.

Then, $(T_{new;\alpha/2}, T_{new;1-\alpha/2})$ is a $100(1-\alpha)\%$ generalized confidence interval on $\mu$.

## 5 Real Data Examples

To illustrate the above presented exact and generalized confidence intervals, we use two small examples considered in Krishnamoorthy and Lu [11]. For calculating the intervals (13) and (15), we use two different sets of weights each. In the weighted inverse normal statistic (12), we consider $\gamma_i = 1\ \forall i$, that is, each sample gets the same weight, and $\gamma_i = n_i\ \forall i$, that is, the larger the experiment the larger the weight. In the general inverse $\chi^2$-combination statistic (14), we set $\gamma_i = 2\ \forall i$, that is, the combination method is then the well-known Fisher method, and again $\gamma_i = n_i\ \forall i$.

We use the following abbreviations for the confidence intervals:

$CI_1$ — Fairweather's confidence interval from (11)
$CI_2$ — Inverse normal confidence interval from (13) with $\gamma_i = 1\ \forall i$
$CI_3$ — Inverse normal confidence interval from (13) with $\gamma_i = n_i\ \forall i$
$CI_4$ — Inverse $\chi^2$-confidence interval from (15) with $\gamma_i = 2\ \forall i$
$CI_5$ — Inverse $\chi^2$-confidence interval from (15) with $\gamma_i = n_i\ \forall i$
$GCI_1$ — Generalized confidence interval using pivotal quantity (18)
$GCI_2$ — Generalized confidence interval using pivotal quantity (22)
$GCI_3$ — Generalized confidence interval using pivotal quantity (25)

The first example is originally from Meier [13]. In this example, four experiments are used to estimate the mean percentage of albumin in the plasma protein of normal human subjects. The observed means $\bar{y}_i$, the observed variances $s_i^2$, and the sample sizes $n_i$ of the four experiments are given in Table 1, so that the observed variance of $\bar{y}_i$ is $s_i^2/n_i$.

**Table 1** Percentage of albumin in plasma protein

| Experiment | $n_i$ | $\bar{y}_i$ | $s_i^2$ |
|---|---|---|---|
| A | 12 | 62.3 | 12.986 |
| B | 15 | 60.3 | 7.840 |
| C | 7 | 59.5 | 33.433 |
| D | 16 | 61.5 | 18.513 |

**Table 2** Results in the albumin example

| Method | Interval | Length |
|---|---|---|
| $CI_1$ | (59.90, 62.19) | 2.29 |
| $CI_2$ | (59.87, 62.18) | 2.31 |
| $CI_3$ | (59.94, 62.17) | 2.23 |
| $CI_4$ | (59.79, 62.31) | 2.52 |
| $CI_5$ | (59.91, 62.19) | 2.28 |
| $GCI_1$ | (59.82, 62.22) | 2.39 |
| $GCI_2$ | (59.92, 62.07) | 2.15 |
| $GCI_3$ | (59.42, 62.66) | 3.24 |

**Table 3** Selenium in nonfat milk powder

| Methods | $n_i$ | $\bar{y}_i$ | $s_i^2$ |
|---|---|---|---|
| Atomic absorption spectrometry | 8 | 105.0 | 85.711 |
| Neutron activation: | | | |
| 1. Instrumental | 12 | 109.75 | 20.748 |
| 2. Radiochemical | 14 | 109.5 | 2.729 |
| Isotope dilution mass spectrometry | 8 | 113.25 | 33.640 |

The resulting confidence intervals together with their lengths in the albumin example are displayed in Table 2. The exact confidence intervals are nearly identical. The shortest interval is the interval based on the inverse normal method with weights equal to the sample sizes ($CI_3$) followed by the interval based on the inverse $\chi^2$-method with sample sizes as weights ($CI_5$) and Fairweather's interval ($CI_1$). Only the interval calculated with the inverse $\chi^2$-method and equal weights ($CI_4$) is clearly wider than the other four exact confidence intervals.

The generalized confidence interval of Lin and Lee ($GCI_2$) is the shortest of all the intervals, the newly proposed generalized confidence interval ($GCI_3$) the widest one. The generalized confidence interval of Krishnamoorthy and Lu ($GCI_1$) is wider than all the exact confidence intervals except the widest exact interval.

The second example is quoted from [2] and deals with the problem of estimation of mean selenium in nonfat milk powder by combining the results of four methods. The observed means $\bar{y}_i$, the observed variances $s_i^2$, and the sample sizes $n_i$ of four different methods are given in Table 3, so that the observed variance of $\bar{y}_i$ is $s_i^2/n_i$.

**Table 4** Results in the milk powder example

| Method | Interval | Length |
|---|---|---|
| $CI_1$ | (108.53, 110.77) | 2.24 |
| $CI_2$ | (108.49, 110.86) | 2.37 |
| $CI_3$ | (108.57, 110.73) | 2.16 |
| $CI_4$ | (108.60, 112.47) | 3.87 |
| $CI_5$ | (108.63, 110.98) | 2.35 |
| $GCI_1$ | (108.65, 110.54) | 1.89 |
| $GCI_2$ | (108.73, 110.53) | 1.80 |
| $GCI_3$ | (107.84, 111.54) | 3.71 |

The results of the second example, confidence intervals plus lengths of the intervals, are displayed in Table 4. Again, the interval based on the inverse normal method with sample sizes as weights ($CI_3$) is the shortest of the exact confidence intervals, now followed by Fairweather's interval ($CI_1$) and then the interval based on the inverse $\chi^2$-method with sample sizes as weights ($GCI_5$).

The generalized confidence interval of Lin and Lee ($GCI_2$) is again the shortest of all the intervals, closely followed now by the generalized confidence interval of Krishnamoorthy and Lu ($GCI_1$). The newly proposed generalized confidence interval ($GCI_3$) is again rather wide compared to the other generalized confidence intervals but still shorter than the widest exact confidence interval ($CI_4$).

Based on the results of the two examples, the exact confidence intervals based on inverse normal and inverse $\chi^2$-method, both with samples sizes as weights ($CI_3$ and $CI_5$) , as well as Fairweather's interval ($CI_1$) may be recommended for practical use. The generalized confidence interval of Lin and Lee ($GCI_2$) may be recommended when this interval really keeps the nominal level. In the next section, some simulation results will be reported which provide further insight into the properties of the intervals.

## 6 Results of Simulation Study

In a simulation study, we investigated the eight confidence intervals used in the previous section with respect to their actual confidence levels and their expected lengths.

The setting of the simulation study is as follows: We considered $k = 3, 6$, and 9 populations with different sample sizes and error variances. For each number of population, we simulated ten different scenarios. The plans for $k = 3$ populations are displayed in Table 5. The first plan has small equal sample sizes and homogeneous error variances, whereas the second one has heterogeneous error variances instead. In the next two plans, we double the sample sizes of the first two plans. The other six plans are unbalanced. Plan 5 has small sample sizes and homogeneous error variances; plan 6 and 7, with the same sample sizes as plan 5, have heterogeneous error variances. In plan 6, the largest sample size is associated with the largest error

**Table 5** Simulation plans for $k = 3$ populations

| Plan | $(n_1, n_2, n_3)$ | $(\sigma_1^2, \sigma_2^2, \sigma_3^2)$ |
|---|---|---|
| 1 | (10,10,10) | (4,4,4) |
| 2 | (10,10,10) | (1,3,5) |
| 3 | (20,20,20) | (4,4,4) |
| 4 | (20,20,20) | (1,3,5) |
| 5 | (5,10,15) | (4,4,4) |
| 6 | (5,10,15) | (1,3,5) |
| 7 | (5,10,15) | (5,3,1) |
| 8 | (10,20,30) | (4,4,4) |
| 9 | (10,20,30) | (1,3,5) |
| 10 | (10,20,30) | (5,3,1) |

**Table 6** Estimated confidence coefficients and expected lengths of eight intervals on the common mean of $k = 3$ populations

| Estimated confidence coefficients | | | | | | | | |
|---|---|---|---|---|---|---|---|---|
| Plan | $CI_1$ | $CI_2$ | $CI_3$ | $CI_4$ | $CI_5$ | $GCI_1$ | $GCI_2$ | $GCI_3$ |
| 1 | 94.8 | 94.7 | 94.7 | 94.6 | 94.7 | 95.9 | 93.0 | 94.0 |
| 2 | 94.8 | 94.9 | 94.9 | 95.0 | 95.0 | 95.9 | 93.6 | 94.7 |
| 3 | 95.0 | 95.0 | 95.0 | 95.1 | 94.9 | 95.4 | 94.3 | 94.7 |
| 4 | 95.1 | 95.2 | 95.2 | 95.3 | 95.3 | 95.7 | 94.6 | 95.0 |
| 5 | 94.9 | 94.9 | 95.1 | 95.1 | 95.0 | 96.1 | 92.9 | 94.3 |
| 6 | 94.9 | 95.1 | 94.8 | 95.3 | 94.8 | 96.3 | 92.9 | 93.3 |
| 7 | 94.8 | 95.0 | 94.9 | 94.9 | 94.9 | 95.5 | 93.5 | 94.1 |
| 8 | 95.0 | 95.0 | 95.0 | 94.9 | 95.0 | 95.5 | 94.0 | 94.6 |
| 9 | 94.8 | 94.7 | 94.7 | 95.0 | 94.9 | 95.6 | 93.8 | 94.8 |
| 10 | 94.8 | 94.7 | 95.0 | 95.0 | 95.0 | 95.3 | 94.2 | 94.6 |
| Estimated expected lengths | | | | | | | | |
| Plan | $CI_1$ | $CI_2$ | $CI_3$ | $CI_4$ | $CI_5$ | $GCI_1$ | $GCI_2$ | $GCI_3$ |
| 1 | 1.532 | 1.505 | 1.505 | 1.600 | 1.524 | 1.678 | 1.468 | 2.580 |
| 2 | 1.138 | 1.136 | 1.136 | 1.250 | 1.160 | 1.181 | 1.051 | 1.890 |
| 3 | 1.041 | 1.038 | 1.038 | 1.110 | 1.046 | 1.097 | 1.027 | 1.915 |
| 4 | 0.772 | 0.776 | 0.776 | 0.862 | 0.785 | 0.765 | 0.725 | 1.364 |
| 5 | 1.548 | 1.551 | 1.509 | 1.684 | 1.530 | 1.733 | 1.480 | 2.637 |
| 6 | 1.309 | 1.264 | 1.320 | 1.347 | 1.346 | 1.464 | 1.217 | 2.083 |
| 7 | 1.030 | 1.089 | 0.992 | 1.300 | 1.013 | 1.050 | 0.943 | 1.808 |
| 8 | 1.063 | 1.065 | 1.039 | 1.160 | 1.046 | 1.106 | 1.029 | 1.928 |
| 9 | 0.867 | 0.859 | 0.900 | 0.921 | 0.911 | 0.917 | 0.842 | 1.538 |
| 10 | 0.726 | 0.743 | 0.680 | 0.900 | 0.687 | 0.674 | 0.646 | 1.259 |

variance and the smallest sample size with the smallest error variance. In plan 7, it is just the other way round. In plan 8, 9, and 10, we double the sample size of the three previous plans. For $k = 6$ populations, we replicated the plans for $k = 3$ populations once, and for $k = 9$ populations twice.

In Tables 6, 7, and 8, the estimated confidence coefficients (in %) and the estimated expected lengths of the eight intervals are displayed. We report the estimated

**Table 7** Estimated confidence coefficients and expected lengths of eight intervals on the common mean of $k = 6$ populations

| Estimated confidence coefficients | | | | | | | | |
|---|---|---|---|---|---|---|---|---|
| Plan | $CI_1$ | $CI_2$ | $CI_3$ | $CI_4$ | $CI_5$ | $GCI_1$ | $GCI_2$ | $GCI_3$ |
| 1 | 95.4 | 95.3 | 95.3 | 95.3 | 95.3 | 96.4 | 92.8 | 94.0 |
| 2 | 94.5 | 94.5 | 94.5 | 94.5 | 94.4 | 95.9 | 92.5 | 93.7 |
| 3 | 95.0 | 95.0 | 95.0 | 95.1 | 94.9 | 95.7 | 94.0 | 94.7 |
| 4 | 95.2 | 95.1 | 95.1 | 95.2 | 95.4 | 95.9 | 94.0 | 95.0 |
| 5 | 94.9 | 94.8 | 95.0 | 95.1 | 95.0 | 95.8 | 91.9 | 93.6 |
| 6 | 95.0 | 94.9 | 95.0 | 95.2 | 95.0 | 96.3 | 91.8 | 93.1 |
| 7 | 94.9 | 95.1 | 95.1 | 95.1 | 95.2 | 95.9 | 92.9 | 95.0 |
| 8 | 95.0 | 95.0 | 95.1 | 94.9 | 95.0 | 95.8 | 94.0 | 94.7 |
| 9 | 94.4 | 94.5 | 94.4 | 94.7 | 94.5 | 95.2 | 93.1 | 94.3 |
| 10 | 95.1 | 95.1 | 95.2 | 95.3 | 95.2 | 95.7 | 94.3 | 95.3 |
| Estimated expected lengths | | | | | | | | |
| Plan | $CI_1$ | $CI_2$ | $CI_3$ | $CI_4$ | $CI_5$ | $GCI_1$ | $GCI_2$ | $GCI_3$ |
| 1 | 1.069 | 1.054 | 1.054 | 1.152 | 1.074 | 1.211 | 1.033 | 1.273 |
| 2 | 0.792 | 0.786 | 0.786 | 0.880 | 0.805 | 0.844 | 0.727 | 0.902 |
| 3 | 0.731 | 0.729 | 0.729 | 0.796 | 0.736 | 0.782 | 0.721 | 0.897 |
| 4 | 0.542 | 0.543 | 0.543 | 0.609 | 0.549 | 0.546 | 0.507 | 0.631 |
| 5 | 1.080 | 1.080 | 1.052 | 1.192 | 1.071 | 1.261 | 1.045 | 1.301 |
| 6 | 0.913 | 0.876 | 0.915 | 0.956 | 0.937 | 1.064 | 0.845 | 1.018 |
| 7 | 0.721 | 0.753 | 0.689 | 0.896 | 0.703 | 0.757 | 0.661 | 0.866 |
| 8 | 0.746 | 0.748 | 0.729 | 0.825 | 0.736 | 0.790 | 0.724 | 0.906 |
| 9 | 0.607 | 0.601 | 0.629 | 0.658 | 0.637 | 0.658 | 0.589 | 0.725 |
| 10 | 0.510 | 0.519 | 0.476 | 0.615 | 0.481 | 0.480 | 0.452 | 0.578 |

confidence coefficients of the exact confidence intervals to demonstrate the accuracy of our simulation study. Each estimated confidence coefficient is based on 10,000 simulation runs, so that, using the central limit theorem, 95% confidence intervals around estimates between 94.6 and 95.4% cover the nominal confidence coefficient of $100(1-\alpha)\% = 95\%$.

The estimated confidence coefficients of the generalized confidence intervals are always in the same order given the number of populations and given a sampling plan. The generalized confidence interval of Krishnamoorthy and Lu ($GCI_1$) always produce actual confidence coefficients above the nominal confidence coefficient and is almost always significantly conservative, that is, the estimated confidence are larger than 95.4%.

The generalized confidence interval of Lin and Lee ($GCI_2$) always produce actual confidence coefficients below the nominal confidence coefficient and is (almost) always significantly liberal, that is, the estimated confidence coefficients are less than 94.6%. Consequently, this generalized confidence interval is not a suitable competitor to the exact confidence intervals in the scenarios covered by our simulation study.

**Table 8** Estimated confidence coefficients and expected lengths of eight intervals on the common mean of $k = 9$ populations

| Estimated confidence coefficients | | | | | | | | |
|---|---|---|---|---|---|---|---|---|
| Plan | $CI_1$ | $CI_2$ | $CI_3$ | $CI_4$ | $CI_5$ | $GCI_1$ | $GCI_2$ | $GCI_3$ |
| 1 | 95.5 | 95.5 | 95.5 | 95.6 | 95.4 | 96.5 | 92.9 | 94.1 |
| 2 | 94.7 | 94.7 | 94.7 | 94.7 | 94.7 | 95.9 | 92.0 | 93.7 |
| 3 | 95.0 | 95.0 | 95.0 | 95.2 | 95.0 | 95.7 | 93.8 | 94.6 |
| 4 | 94.8 | 94.8 | 94.8 | 95.2 | 95.0 | 95.5 | 93.6 | 94.5 |
| 5 | 95.0 | 95.2 | 95.3 | 95.0 | 95.0 | 96.2 | 92.2 | 94.1 |
| 6 | 95.0 | 94.8 | 95.0 | 94.7 | 94.8 | 96.3 | 91.1 | 92.9 |
| 7 | 94.4 | 94.5 | 94.5 | 94.8 | 94.7 | 95.3 | 92.3 | 95.1 |
| 8 | 95.3 | 95.0 | 95.1 | 95.4 | 95.2 | 95.8 | 93.9 | 94.4 |
| 9 | 94.9 | 94.8 | 95.1 | 94.8 | 94.9 | 95.8 | 93.2 | 93.9 |
| 10 | 95.1 | 95.1 | 95.1 | 95.0 | 94.9 | 95.8 | 94.3 | 95.1 |
| Estimated expected lengths | | | | | | | | |
| Plan | $CI_1$ | $CI_2$ | $CI_3$ | $CI_4$ | $CI_5$ | $GCI_1$ | $GCI_2$ | $GCI_3$ |
| 1 | 0.865 | 0.854 | 0.854 | 0.940 | 0.871 | 0.993 | 0.841 | 0.975 |
| 2 | 0.644 | 0.639 | 0.639 | 0.715 | 0.653 | 0.695 | 0.593 | 0.693 |
| 3 | 0.596 | 0.595 | 0.595 | 0.654 | 0.601 | 0.642 | 0.589 | 0.681 |
| 4 | 0.441 | 0.441 | 0.441 | 0.495 | 0.447 | 0.447 | 0.412 | 0.478 |
| 5 | 0.880 | 0.881 | 0.857 | 0.978 | 0.875 | 1.048 | 0.863 | 1.024 |
| 6 | 0.740 | 0.710 | 0.742 | 0.783 | 0.761 | 0.883 | 0.693 | 0.799 |
| 7 | 0.586 | 0.610 | 0.558 | 0.715 | 0.570 | 0.624 | 0.540 | 0.666 |
| 8 | 0.607 | 0.609 | 0.594 | 0.675 | 0.600 | 0.648 | 0.591 | 0.689 |
| 9 | 0.494 | 0.490 | 0.512 | 0.539 | 0.519 | 0.541 | 0.481 | 0.551 |
| 10 | 0.415 | 0.422 | 0.388 | 0.495 | 0.392 | 0.394 | 0.368 | 0.438 |

The estimated confidence coefficients of the newly proposed generalized confidence interval ($GCI_3$) always lie between the confidence coefficients of the other two generalized confidence intervals given a sampling plan. For small sample sizes, the interval ($GCI_3$) tends to be liberal, whereas for larger sample sizes, the interval seems to attain the nominal confidence coefficient, though nearly all the estimated confidence coefficients are less than 95%.

The generalized confidence interval of Lin and Lee ($GCI_2$) consistently produce the shortest average lengths. But since this interval does not attain the nominal level, we cannot recommend the use of this interval for the scenarios covered in our simulation study as already mentioned above.

Generally, the average lengths of all the intervals decrease when the number of populations or the sample sizes increase. Further note that the intervals $CI_2$ and $CI_3$ based on the inverse normal method are identical when the sample sizes are balanced. However, this is not true for the intervals $CI_4$ and $CI_5$ based on the inverse $\chi^2$-method when the sample sizes are balanced.

Of the exact confidence intervals, the interval $CI_4$ based on the inverse $\chi^2$-method with $\gamma_i = 2\ \forall i$, generally, produces the largest average lengths. Since we have four other exact confidence intervals with shorter average lengths, we cannot recommend interval $CI_4$ for practical use within the settings of our simulation study.

The picture of the other four exact confidence intervals with respect to expected length is not so clear. Though never producing the shortest average length, the interval based on the $\chi^2$-method with sample sizes as weights ($CI_5$) produces in some scenarios the second best result. The interval based on the inverse normal method with sample sizes as weights ($CI_3$) in most cases provides the shortest average length, but when larger error variances are associated with larger sample sizes, the interval based on the inverse normal methods with identical weights is better. The average lengths of Fairweather's interval $CI_1$ is in most cases close to the average lengths of the interval based on the inverse normal method with sample sizes as weights.

For $k = 6$ and $k = 9$ populations, the newly proposed generalized confidence interval $GCI_3$ has comparable average lengths like the generalized confidence interval $GCI_1$. The average lengths of $GCI_1$ are in most cases in the magnitude of the average lengths of the exact confidence intervals.

## 7 Final Remarks

In this paper, we have considered five exact and three generalized confidence intervals on the common mean of several independent normal populations. All the intervals have in common that they always produce genuine intervals. That is necessarily not the case for other exact intervals mentioned in the introduction. In a simulation study, we have investigated the performance of the intervals especially with respect to the expected length for a small or moderate number of populations with small sample sizes and homogeneous as well as heterogeneous error variances. Summarizing our findings quite generally, we can state that all the intervals considered in this paper can be recommended for practical use except the generalized confidence interval proposed by Lin and Lee ($GCI_2$) that turned out to be too liberal.

Note that the newly proposed generalized confidence interval $GCI_3$ is more robust than all the other intervals with respect to deviations from the model assumptions, for instance the occurrence of treatment-by-sample interactions, see [6], since the sample variances are only used for building the weights.

Typical common mean problems arise in interlaboratory trials. Identical parts of the same material of interest are sent to several laboratories for analyzing. The common mean of the several analyses is of interest. Since the laboratories do not work with the same precision, different population variances have to be taken into consideration.

More generally, common mean problems arise in meta-analysis when estimates from different experiments or studies have to be combined. If the meta-analysis is carried out retrospectively, all the intervals discussed in this paper can be applied. However, if the meta-analysis is carried out prospectively so that results of previous studies determine the sample sizes as well as the number of following studies, then only the exact confidence intervals based on the inverse normal or inverse $\chi^2$-combination methods are applicable. For example, a clear preference is given for a prospective meta-analysis when a new drug shall be approved for the market via meta-analysis, see [6].

A further field of applications lies in the adaptive extension of classical group sequential trials. Here a trial is performed consecutively on several independent stages and the results of previous stages determine the sample size of the next stage. Our exact confidence intervals based on the methods which combine $P$-values are applicable correctly, also in this adaptive sequential situation, see [7].

## References


[1] Cohen, A., Sackrowitz, H.B.: Testing hypotheses about the common mean of normal distributions. J. Stat. Plan. Infer. **9**, 207–227 (1984)

[2] Eberhardt, K.R., Reeve, C.P., Spiegelman, C.H.: A minimax approach to combining means, with practical examples. Chemometr. Intell. Lab. Syst. **5**, 129–148 (1989)

[3] Fairweather, W.R.: A method of obtaining an exact confidence interval for the common mean of several normal populations. Appl. Stat. **21**, 229–233 (1972)

[4] Graybill, F.A., Deal, R.B.: Combining unbiased estimators. Biometrics **15**, 543–550 (1959)

[5] Hartung, J.: An alternative method for meta-analysis. Biom. J. **41**, 901–916 (1999)

[6] Hartung, J., Knapp, G.: Models for combining results of different experiments: retrospective and prospective. Am. J. Math. Manag. Sci. **25**, 149–188 (2005)

[7] Hartung, J., Knapp, G.: Nested repeated confidence intervals in adaptive group sequential clinical trials. East-West J. Math. (in press)

[8] Hartung, J., Knapp, G., Sinha, B.K.: Statistical Meta-Analysis with Applications. Wiley, New York (2008)

[9] Hedges, L.V., Olkin, I.: Statistical Methods for Meta-Analysis. Academic, Boston (1985)

[10] Jordan, S.M., Krishnamoorthy, K.: Exact confidence intervals for the common mean of several normal populations. Biometrics **52**, 77–86 (1996)

[11] Krishnamoorthy, L., Lu, Y.: Inferences on the common mean of several normal populations based on the generalized variable method. Biometrics **59**, 237–247 (2003)

[12] Lin, S.-H., Lee, J.C.: Generalized inferences on the common mean of several normal populations. J. Stat. Plan. Infer. **134**, 568–582 (2005)

[13] Meier, P.: Variance of a weighted mean. Biometrics **9**, 59–73 (1953)

[14] Sinha, B.K.: Unbiased estimation of the variance of the Graybill-Deal estimator of the common mean of several normal populations. Can. J. Stat. **13**, 243–247 (1985)

[15] Tsui, K., Weerahandi, S.: Generalized *p*-values in significance testing of hypotheses in the presence of nuisance parameters. J. Am. Stat. Assoc. **84**, 602–607 (1989)
[16] Weerahandi, S.: Generalized confidence intervals. J. Am. Stat. Assoc. **88**, 899–905 (1993)
[17] Yu, Ph.L.H., Sun, Y., Sinha, B.K.: On exact confidence intervals for the common mean of several normal populations. J. Stat. Plan. Infer. **81**, 263–277 (1999)

# Locally Optimal Tests of Independence for Archimedean Copula Families

Jörg Rahnenführer

**Abstract** A multivariate distribution can be decoupled into its marginal distributions and a copula function, a distribution function with uniform marginals. Copulas are well suited for modelling the dependence between multivariate random variables independent of their marginal distributions. Applications range from survival analysis over extreme value theory to econometrics. In recent years, copulas have attracted increased attention in financial statistics, in particular regarding modelling issues for high-dimensional problems like value-at-risk or portfolio credit risk. The well studied subclass of Archimedean copulas can be expressed as a function of a one-dimensional generating function $\phi$. This class has become popular due to its richness in various distributional attributes providing flexibility in modelling. Here, we present locally optimal tests of independence for Archimedean copula families that are parameterized by a dependence parameter $\vartheta$, where $\vartheta = 0$ denotes independence of the marginal distributions. Under the general assumption of $L_2$-differentiability at $\vartheta = 0$ we calculate tangents of the underlying parametric families. For selected examples the optimal tests are calculated and connections to well-known correlation functionals are presented.

## 1 Introduction

Copula functions are multivariate distribution functions with uniformly distributed marginals. They are a powerful tool for modelling multivariate dependencies when marginal distributions are predetermined [5]. In financial statistics, copulas have recently become popular for modelling high-dimensional data sets related to pricing, risk management and credit risk analysis, see for example [3] for a recent book on these topics. In this context, Copula functions are used for implementing Monte Carlo analyses and for specifying non-linear dependencies between random variables that can not be captured by linear correlations.

Jörg Rahnenführer
Fakultät Statistik, Technische Universität Dortmund, D-44221 Dortmund, Germany
rahnenfuehrer@statistik.tu-dortmund.de

For example, individual risk models where dependence between contracts arises through mixtures are presented in [7]. Some of these models are generated by Archimedean copulas, a special subclass of copula families with convenient distributional properties. Archimedean copulas are also used to model portfolio allocations by exploiting separability properties of this subclass [11] and to construct models for data that may suffer from selectivity bias [20]. Especially for the class of Archimedean copulas, the resulting expressions for log-likelihood and score facilitate maximum likelihood estimation.

For all these models, besides estimating the multivariate distribution function, often a test for independence of the marginal distributions is of interest. Here, we present a formula for explicitly calculating locally optimal tests for arbitrary parameterized Archimedean copula families. We restrict the calculations to the bivariate case, an extension to higher dimensions is straightforward.

Similar results were first presented in two doctoral theses by Garralda-Guillem [4] and by Rahnenführer [19]. Later, Francois Verret and Christian Genest [9] presented a general theorem for constructing locally most powerful rank tests of independence for copula models. In this paper also additional interesting examples are discussed. However, the explicit formula for parameterized Archimedean copulas is not contained in any previous journal publication.

In the following, we first introduce bivariate copula functions and some basic properties. A detailed introduction to copulas can be found for example in [17]. Then we provide the construction of locally optimal tests for Archimedean copulas and the calculation of score functions. Finally, examples are presented for demonstrating the usefulness of this approach and for highlighting connections between dependence measures and corresponding optimal local tests.

## 2 Copula Families

We first introduce the notion of a copula function and state some well known convenient properties of copula families. In the following, let $F(x_1,x_2)$ be a two-dimensional right continuous distribution function on $\mathbb{R}^2$ with marginal distributions $F_1(x_1) := F(x_1,\infty)$ and $F_2(x_2) := F(\infty,x_2)$.

**Definition 1.** $C : [0,1]^2 \to [0,1]$ is called *copula* (copula function), if $C$ is a two-dimensional distribution function with marginal distributions that are uniformly distributed on the interval $[0,1]$:

$$
\begin{aligned}
C(x_1,0) = C(0,x_2) &= 0 \quad \forall\, 0 \le x_1,x_2 \le 1,\\
C(x_1,1) = x_1,\ C(1,x_2) &= x_2 \quad \forall\, 0 \le x_1,x_2 \le 1,\\
C(y_1,y_2) - C(y_1,x_2) - C(x_1,y_2) + C(x_1,x_2) &\ge 0 \quad \forall\, 0 \le x_1 \le y_1 \le 1\\
&\qquad\ \ \forall\, 0 \le x_2 \le y_2 \le 1.
\end{aligned}
$$

Any two-dimensional distribution function can be split into a copula function and its marginal distributions. For continuous distribution functions this separation is uniquely determined.

**Proposition 1.** *Let $F$ be a two-dimensional distribution function on $\mathbb{R}^2$ with marginal distributions $F_1$ and $F_2$. Then a copula $C_F(\cdot,\cdot)$ exists with*

$$F(x_1,x_2) = C_F(F_1(x_1),F_2(x_2)) \;\forall\, x_1,x_2 \in \mathbb{R}.$$

*If $F$ is continuous, then $C_F$ is uniquely defined by*

$$C_F(x_1,x_2) = F(F_1^{-1}(x_1),F_2^{-1}(x_2)) \;\forall\, x_1,x_2 \in [0,1].$$

**Proposition 2.** *Any copula $C$ fulfills the Lipschitz condition*

$$|C(y_1,y_2) - C(x_1,x_2)| \le |y_1 - y_2| + |x_1 - x_2| \;\forall\, (x_1,x_2),(y_1,y_2) \in [0,1]^2.$$

This property guarantees a certain amount of smoothness of copula functions. For proofs of Proposition 2 and 3 see for example [23], 7.51 and 7.52.

Copulas are especially useful for the interpretation of several well known correlation functionals that depend only on the copula function and not on the marginal distributions, see [23], Example 7.66.

**Definition 2.** For a fixed two-dimensional distribution function $F$ let the random variables $(X_1,X_2)$ and $(X_1',X_2')$ be distributed according to $F$.

$$\begin{aligned}\rho_S(F) &:= 12\int_0^1\int_0^1 (C_F(u_1,u_2) - u_1 u_2)\,du_2\,du_1\\ &= 12\int_0^1\int_0^1 \left(u_1 - \frac{1}{2}\right)\left(u_2 - \frac{1}{2}\right) dC_F(u_1,u_2)\\ &= \text{Corr}\,(F_1(X_1),F_2(X_2))\end{aligned}$$

is called *Spearman-functional. Corr* denotes the Spearman correlation coefficient.

$$\begin{aligned}\rho_F(F) &:= 4\int_0^1\int_0^1 C_F(u_1,u_2)\,dC_F(u_1,u_2) - 1\\ &= P\Big((X_1 - X_1')(X_2 - X_2') > 0\Big) - P\Big((X_1 - X_1')(X_2 - X_2') < 0\Big)\end{aligned}$$

is called *Fechner-Kendall-functional.*

$\rho_S$ is the correlation between $F_1(X_1)$ and $F_2(X_2)$, and $\rho_F$ is the difference between the probabilities of drawing concordant and discordant pairs, respectively, when drawing twice from the distribution function $F$.

The subclass of Archimedean copulas was first defined in [7] and includes several well known copula families.

**Definition 3.** A copula function $C(x_1,x_2)$ is called *Archimedean copula*, if a convex decreasing continuous function $\phi : [0,1] \to [0,\infty]$, $\phi(1) = 0$, exists with

$$C(x_1,x_2) = \phi^{-1}(\phi(x_1)+\phi(x_2)). \tag{1}$$

In this case $\phi$ is called generator.

For example, any function $\phi \in C^2[0,1]$ with $\phi(1) = 0$, $\phi'(x) < 0$ and $\phi''(x) > 0$ $\forall\, t \in (0,1)$ is a generator of an Archimedean copula.

Archimedean copula functions can be written as a mixture of powers of two distributions ([15], 4.2 or [7]):

$$C(x_1,x_2) = \int_0^\infty G_1(x_1)^\alpha G_2(x_2)^\alpha \, dM(\alpha),$$

where $G_i(x_i) = \exp(-\phi^{-1}(x_i))$, $i = 1,2$ and $M$ is a distribution function on $[0,\infty)$ with $M(0) = 0$. The construction of locally most powerful rank tests of independence for related models given by mixtures of one-parametric families of marginal distributions is presented in [12], §4 and §8. Further results on Archimedean copulas and related multivariate parametric families can be found in [8, 14] and [16]. A new Archimedean copula model for bivariate survival data is presented in [18].

## 3 Construction of Locally Optimal Tests of Independence

In this section we present an explicit representation of locally optimal tests for Archimedean copulas. We consider parametric families of copulas $C_\vartheta(x_1,x_2)$ with $\vartheta \in \mathbb{R}$. The parameter $\vartheta$ could be, for example, a monotonous transformation of the correlation coefficient $\rho$ of the one-dimensional marginal distributions. The value $\vartheta = 0$ represents independence, thus $C_0(x_1,x_2) = x_1 x_2$.

In this situation, testing for (one-sided) independence of two distributions that are coupled through the copula $C_\vartheta$ can be expressed as

$$H = \{\vartheta \le 0\} \quad \text{versus} \quad K = \{\vartheta > 0\}$$

or by respective expressions with reversed orientation.

The notion of locally optimal tests for this testing problem is based on the following smoothness condition for the underlying parametric family, see for example [2] or [21] for more details.

**Definition 4.** ($L_2$-differentiability): Let $E = (\Omega, \mathscr{A}, \{P_\vartheta : \vartheta \in \Theta\})$ be an experiment with a $\sigma$-finite dominating measure $\mu$, $\vartheta_0 \in \overset{\circ}{\Theta} \subset \mathbb{R}$ and let $|| \quad ||$ denote the $L_2(\mu)$-norm. $E$ is called *$L_2$-differentiable* in $\vartheta_0$, if there exists a function $g \in L_2(P_{\vartheta_0})$ with

$$\left\| 2\left(\frac{dP_{\vartheta_0+s}}{d\mu}\right)^{1/2} - 2\left(\frac{dP_{\vartheta_0}}{d\mu}\right)^{1/2} - sg\left(\frac{dP_{\vartheta_0}}{d\mu}\right)^{1/2} \right\| = o(s), \quad s \to 0.$$

The function $g$ is called tangent or score-function of the experiment. Under regularity conditions, if $\vartheta \mapsto f_\vartheta := \frac{dP_\vartheta}{dP_\mu}$ is differentiable with respect to $\vartheta$ in $\vartheta = \vartheta_0$, the tangent can be calculated as $g(x) = \frac{d}{d\vartheta} \ln f_\vartheta(x)|_{\vartheta=\vartheta_0}$.

Now assume the case where $\vartheta_0 = 0$ denotes independence. If a tangent in the sense of the preceding definition exists, the test statistic of the locally optimal test of independence based on $n$ independent pairs of observations is given by

$$T_n = \frac{1}{n} \sum_{i=1}^{n} g(x_{1i}, x_{2i}). \tag{2}$$

If densities $f_\vartheta$ (with respect to the uniform distribution on $[0,1]^2$) exist in a neighborhood of $\vartheta = 0$, the derivative can be calculated as

$$g(x) = \frac{d}{d\vartheta} \ln f_\vartheta(x)\bigg|_{\vartheta=0} = \frac{d}{d\vartheta} f_\vartheta(x)\bigg|_{\vartheta=0}.$$

Apparently $f_0 \equiv 1$ denotes the independence copula.

We now calculate the score function of a parametric Archimedean copula family $C_\vartheta(x_1,x_2)$ under appropriate differentiability conditions. In the following theorem, for better readability we replace $(x_1,x_2)$ with $(u,v)$. Let " $'$ " denote the partial derivative of $\phi_\vartheta(u)$ with respect to the argument $u$, and let " $\dot{}$ " denote the partial derivative with respect to the parameter $\vartheta$.

**Theorem 1.** *Let $C_\vartheta(u,v)$ be a family of Archimedean copulas with stochastic independence at $\vartheta = 0$. Assume that in $\vartheta_0$ the generator $\phi(\vartheta,u) := \phi_\vartheta(u)$ is twice differentiable with respect to the argument $u$ and once with respect to the parameter $\vartheta$, i.e. $\dot{\phi}_0''(u) := \dot{\phi}''(\vartheta,u)|_{\vartheta=0}$ exists. Then the score function is given by*

$$\begin{aligned} g(u,v) = {} & (uv)^2 \dot{\phi}_0''(uv) + 3uv\,\dot{\phi}_0'(uv) - u\,\dot{\phi}_0'(u) - v\,\dot{\phi}_0'(v) \\ & + \dot{\phi}_0(uv) - \dot{\phi}_0(u) - \dot{\phi}_0(v). \end{aligned} \tag{3}$$

*Proof.* According to (1) the family of copula functions can be written as

$$C_\vartheta(u,v) = \phi^{-1}(\vartheta, \phi(\vartheta,u) + \phi(\vartheta,v)).$$

The independence assumption at $\vartheta = 0$ translates into $\phi_0(u) + \phi_0(v) = \phi_0(uv)$. An immediate consequence of this formula and Definition 3 is $\phi_0(u) = a\ln(u)$ with

$a \in \mathbb{R}$, $a < 0$. We can set $a = -1$, since multiplying a generator $\phi$ of an Archimedean copula family with a positive constant yields exactly the same Copula family. It follows

$$\phi_0(u) = -\ln(u),\ \ \phi_0'(u) = -1/u,\ \ \phi_0^{-1}(u) = \exp(-u),\ \ (\phi^{-1})_0'(u) = -\exp(-u).$$

Applying a chain rule to the distribution function yields

$$\frac{d}{d\vartheta} C(u,v) = (\dot{\phi^{-1}})\Big(\vartheta, \phi(\vartheta,u) + \phi(\vartheta,v)\Big) \\ + (\phi^{-1})'\Big(\vartheta, \phi(\vartheta,u) + \phi(\vartheta,v)\Big) \cdot \frac{d}{d\vartheta}\Big(\phi(\vartheta,u) + \phi(\vartheta,v)\Big). \tag{4}$$

This requires the calculation of the term $(\dot{\phi^{-1}})(\vartheta,u)$:

$$\begin{aligned} &\phi(\vartheta, \phi^{-1}(\vartheta,u)) = u \\ \Longrightarrow\ &\dot{\phi}(\vartheta, \phi^{-1}(\vartheta,u)) + \phi'(\vartheta, \phi^{-1}(\vartheta,u)) \cdot (\dot{\phi^{-1}})(\vartheta,u) = 0 \\ \Longrightarrow\ &(\dot{\phi^{-1}})(\vartheta,u) = -\dot{\phi}(\vartheta, \phi^{-1}(\vartheta,u)) \Big/ \phi'(\vartheta, \phi^{-1}(\vartheta,u)) \\ \Longrightarrow\ &(\dot{\phi^{-1}})(\vartheta,u)\Big|_{\vartheta=0} = -\dot{\phi}_0(\phi_0^{-1}(u)) \Big/ \phi_0'(\phi_0^{-1}(u)). \end{aligned} \tag{5}$$

Combing equations (4) and (5) leads to

$$\begin{aligned} \frac{d}{d\vartheta} C(u,v)\Big|_{\vartheta=0} &\overset{(4)}{=} (\dot{\phi^{-1}})_0\Big(\phi_0(u) + \phi_0(v)\Big) \\ &\quad + (\phi^{-1})_0'\Big(\phi_0(u) + \phi_0(v)\Big) \cdot \Big(\dot{\phi}_0(u) + \dot{\phi}_0(v)\Big) \\ &\overset{(5)}{=} -\dot{\phi}_0(\phi_0^{-1}(\phi_0(u) + \phi_0(v))) \Big/ \phi_0'(\phi_0^{-1}(\phi_0(u) + \phi_0(v))) \\ &\quad - \phi_0^{-1}\Big(\phi_0(u) + \phi_0(v)\Big) \cdot \Big(\dot{\phi}_0(u) + \dot{\phi}_0(v)\Big) \\ &= -\dot{\phi}_0(uv) \Big/ (-1/(uv)) - uv\Big(\dot{\phi}_0(u) + \dot{\phi}_0(v)\Big) \\ &= uv\Big(\dot{\phi}_0(uv) - \dot{\phi}_0(u) - \dot{\phi}_0(v)\Big). \end{aligned} \tag{6}$$

From independence at $\vartheta = 0$ we get $c_0(u,v) \equiv 1$. Differentiation of the copula function with respect to $u$ and $v$ finally yields

$$\begin{aligned} g(u,v) &= \frac{d}{d\vartheta} c(u,v)\Big|_{\vartheta=0} \\ &= \frac{d}{du}\frac{d}{dv}\frac{d}{d\vartheta} C(u,v)\Big|_{\vartheta=0} \\ &\overset{(6)}{=} \frac{d}{du}\frac{d}{dv}\Big(uv\Big(\dot{\phi}_0(uv) - \dot{\phi}_0(u) - \dot{\phi}_0(v)\Big)\Big) \end{aligned}$$

$$
\begin{aligned}
&= \frac{d}{du}\left(u\left(\dot{\phi}_0(uv) - \dot{\phi}_0(u) - \dot{\phi}_0(v)\right) + u^2 v\,\dot{\phi}_0'(uv) - uv\,\dot{\phi}_0'(v)\right) \\
&= \dot{\phi}_0(uv) - \dot{\phi}_0(u) - \dot{\phi}_0(v) + u\left(\dot{\phi}_0'(uv)\,v - \dot{\phi}_0'(u)\right) + 2uv\,\dot{\phi}_0'(uv) \\
&\quad + u^2 v\,\dot{\phi}_0''(uv)\,v - v\,\dot{\phi}_0'(v) \\
&= (uv)^2\,\dot{\phi}_0''(uv) + 3uv\,\dot{\phi}_0'(uv) - u\,\dot{\phi}_0'(u) - v\,\dot{\phi}_0'(v) \\
&\quad + \dot{\phi}_0(uv) - \dot{\phi}_0(u) - \dot{\phi}_0(v).
\end{aligned}
$$

□

## 4 Examples of Score Functions and Locally Optimal Tests

We now explicitly calculate score functions for Archimedean copulas and describe the construction of locally optimal tests.

*Example 1 (Clayton-Copula (1978)).*

$$
\begin{aligned}
\phi_\vartheta(x) &= \frac{1}{\vartheta}(x^{-\vartheta} - 1), \;\; \vartheta > 0, \\
C_\vartheta(x_1, x_2) &= (x_1^{-\vartheta} + x_2^{-\vartheta} - 1)^{-1/\vartheta}.
\end{aligned}
$$

Straightforward calculations yield

$$
\dot{\phi}_0(x) = \frac{1}{2}\ln(x)^2, \quad \dot{\phi}_0'(x) = \frac{\ln(x)}{x}, \quad \dot{\phi}_0''(x) = \frac{1 - \ln(x)}{x^2}.
$$

From Theorem 1 we obtain

$$
\begin{aligned}
g(x_1, x_2) &= (x_1 x_2)^2 \frac{1 - \ln(x_1 x_2)}{(x_1 x_2)^2} + 3x_1 x_2 \frac{\ln(x_1 x_2)}{x_1 x_2} - x_1 \frac{\ln(x_1)}{x_1} - x_2 \frac{\ln(x_2)}{x_2} \\
&\quad + \frac{1}{2}\ln(x_1 x_2)^2 - \frac{1}{2}\ln(x_1)^2 - \frac{1}{2}\ln(x_2)^2 \\
&= 1 - \Big(\ln(x_1) + \ln(x_2)\Big) + 3\Big(\ln(x_1) + \ln(x_2)\Big) - \ln(x_1) - \ln(x_2) \\
&\quad + \frac{1}{2}(\ln(x_1) + \ln(x_2))^2 - \frac{1}{2}\ln(x_1)^2 - \frac{1}{2}\ln(x_2)^2 \\
&= 1 + \ln(x_1) + \ln(x_2) + \frac{1}{2} 2\ln(x_1)\ln(x_2) \\
&= \Big(1 + \ln(x_1)\Big)\Big(1 + \ln(x_2)\Big).
\end{aligned}
$$

*Example 2.* Let $\phi_\vartheta(x) = -\ln(x) - \vartheta(1 - x)$ with $\vartheta \in \mathbb{R}$, $\vartheta < 1$ be the generator of an Archimedean copula family. Then the score function is given by

$$
g(x_1, x_2) = (1 - 2x_1)(1 - 2x_2).
$$

For $\vartheta < 1$ the generator is a strictly decreasing and convex function. We can calculate $\dot\phi_0(x) = -(1-x)$, $\dot\phi_0'(x) = 1$ and $\dot\phi_0''(x) = 0$. Then Theorem 1 yields

$$\begin{aligned} g(x_1,x_2) &= 3x_1x_2 - x_1 - x_2 - (1 - x_1x_2) + (1 - x_1) + (1 - x_2) \\ &= (1-2x_1)(1-2x_2). \end{aligned}$$

The popular (non-Archimedean) Farlie-Gumbel-Morgenstern-Copula is given by

$$\begin{aligned} C_\vartheta(x_1,x_2) &= x_1x_2\Big(1 + \vartheta(1-x_1)(1-x_2)\Big), \quad |\vartheta| < 1, \\ f_\vartheta(x_1,x_2) &= 1 + \vartheta(1-2x_1)(1-2x_2). \end{aligned}$$

With $f_0 \equiv 1$, differentiation of the density function with respect to $\vartheta$ at $\vartheta = 0$ yields the score function $g(x_1,x_2) = (1-2x_1)(1-2x_2)$. Therefore this widely used Copula family has the same score function and thus is locally at $\vartheta = 0$ equivalent to the Archimedean copula from Example 2.

Finally, for completeness reasons, we describe the straightforward general construction of locally optimal tests, based on the preceding example.

In a first step, let the marginal distribution functions $F_1$ and $F_2$ of the random variables $X_1$ and $X_2$ be fixed. The test statistic of the locally optimal test then is given by

$$T_n = \frac{1}{n}\sum_{i=1}^{n}(1-2F_1(x_{1i}))(1-2F_2(x_{2i})),$$

compare with (2). Since $F_1(X_1)$ and $F_2(X_2)$ are both uniformly distributed on $[0,1]$, the distribution of $T_n$ is independent of $F_1$ and $F_2$. Hence, for $\vartheta = 0$ the critical value $c_\alpha$ of the test can be evaluated. For large $n$, one may substitute $c_\alpha$ by a normal critical value. Then the sequence

$$\varphi_n = \begin{cases} 1, & \sqrt{n}\,T_n/\sigma > u_{1-\alpha} \\ 0, & \sqrt{n}\,T_n/\sigma \le u_{1-\alpha} \end{cases}$$

with $\Phi(u_{1-\alpha}) = 1-\alpha$ and $\sigma = \int_o^1 (1-2u)^2\,du = 1/3$ is of asymptotic level $\alpha$. Recall from [21] that locally optimal tests are asymptotically efficient for local alternatives given by $L_2$-differentiable families. Obviously, for this example the statistic $T_n$ is just an empirical version of the Spearman-functional.

In a second step, let the marginal distribution functions $F_1$ and $F_2$ be unknown but fixed nuisance parameters. From $F_\vartheta(x_1,x_2) = C_\vartheta(F_1(x_1),F_2(x_2))$ we can conclude that the score function of the family $F_\vartheta$ is given by $g(F_1(x_1),F_2(x_2))$. Replacing $F_1$ and $F_2$ with their non-parametric estimators, the empirical distribution functions $\hat F_1$ and $\hat F_2$, naturally leads to rank tests. The resulting test statistic converges under suitable regularity conditions in distribution to the original test statistic. It follows that the tests are asymptotically equivalent. For a proof see [10], for further literature on these type of rank tests for independence see [1, 13, 22]. A more detailed discussion in the context of Archimedean copulas is presented in [9].

## References

[1] Behnen, K., Neuhaus, G.: Rank Tests with Estimated Scores and Their Application. Teubner, Stuttgart (1989)
[2] Bickel, P.J., Klaasen, C.A.J., Ritov, Y., Wellner, J.A.: Efficient and Adaptive Estimation for Semiparametric Models. Johns Hopkins University Press., Baltimore (1993)
[3] Cherubini, G., Vecchiato, W., Luciano, E.: Copula Methods in Finance. Wiley, New York (2004)
[4] Garralda-Guillem, A.I.: Dependencia y tests de rangos para leyes bidimensionales. Unpublished doctoral dissertation, Departamento de Matematica Aplicada, Facultad de Ciencias, Universidad de Granada, Granada, Spain (1997)
[5] Genest, C., MacKay, J.: The Joy of Copulas: Bivariate Distributions with Univariate Marginals. J. Am. Stat. Assoc. **40**, 280–283 (1986)
[6] Genest, C., MacKay, J.: Copules archimédiennes et familles de lois bidimensionnelles dont les marges sont données. Can. J. Stat. **14**, 145–159 (1986)
[7] Genest, C., Marceau, E., Mesfioui, M.: Compound Poisson Approximations for Individual Models with Dependent Risks. Insur. Math. Econ. **32**, 73–91 (2003)
[8] Genest, C., Rivest, L.: Statistical Inference Procedures for Bivariate Archimedean Copulas. J. Am. Stat. Assoc. **88**, 1034–1043 (1993)
[9] Genest, C., Verret, F.: Locally Most Powerful Rank Tests of Independence for Copula Models. J. Nonparametric Stat. **17**, 521–539 (2005)
[10] Hájek, J., Šidák, Z., Sen, P.K.: Theory of Rank Tests. Academic, San Diego (1999)
[11] Hennessy, D.A., Lapan, H.E.: The use of Archimedean Copulas to Model Portfolio Allocations. Math. Fin. **12**, 143–154 (2002)
[12] Janssen, A., Mason, D.M.: Non-Standard Rank Tests. Lecture Notes in Statistics **65**. Springer, Berlin (1990)
[13] Janssen, A., Rahnenführer, J.: A Hazard Based Approach to Dependence Tests for Bivariate Censored Models. Math. Method. Stat. **11**, 297–322 (2002)
[14] Joe, H.: Parametric Families of Multivariate Distributions with Given Margins. J. Multivariate Anal. **46**, 262–282 (1994)
[15] Joe, H.: Multivariate Models and Dependence Concepts. Chapman & Hall, London (1997)
[16] Nelsen, R.B.: Dependence and Order in Families of Archimedean Copulas. J. Multivariate Anal. **60**, 111–122 (1997)
[17] Nelsen, R.B.: An Introduction to Copulas. Springer, New York (1999)
[18] Oakes, D., Wang, A.T.: Copula Model Generated by Dabrowska's Association Measure. Biometrika **90**, 478–481 (2003)
[19] Rahnenführer, J.: Tests auf Unabhängigkeit in bivariaten Hazard-Modellen mit zensierten Daten. PhD thesis, dissertation.de - Verlag im Internet GmbH (1999)

[20] Smith, M.D.: Modelling Sample Selection Using Archimedean Copulas. Econom. J. **6**,1, 99–123 (2003)
[21] Strasser, H.: Mathematical Theory of Statistics. De Gruyter, Berlin (1985)
[22] Ruymgaart, F.H.: Asymptotic Theory of Rank Tests for Independence. Mathematical Centre Tracts **43**. Mathematical Centre, Amsterdam (1973)
[23] Witting, H., Müller–Funk, U.: Mathematische Statistik II. Teubner, Stuttgart (1995)

# Part III
# Design of Experiments and Analysis of Variance

*...no one believes an hypothesis except its originator but everyone believes an experiment except the experimenter.*

Anonymous

# Optimal Designs for Treatment-Control Comparisons in Microarray Experiments

Joachim Kunert, R.J. Martin, and Sabine Rothe

**Abstract** Two-colour microarray experiments form an important tool in modern molecular biology. But they are very expensive and so careful planning of experiments is necessary. In this paper we determine optimal approximate designs for microarray experiments when only treatment-control comparisons are of interest. Based on these results we construct near optimal finite designs and compare their efficiencies with those of the corresponding star designs, which are often used in microarray experiments.

## 1 Introduction

Two-colour microarray experiments enable simultaneous analysis of thousands of genes. Therefore they form an important tool in modern molecular biology. But microarray experiments are very expensive and so careful planning of experiments is necessary. A usual two-colour microarray experiment for comparison of $t$ *experimental conditions*, often called *tissues* or *treatments* uses $b$ arrays. An array is a small slide, spotted with thousands of genes. A mix of two tissues is hybridized to each array. The tissues are labelled with two fluorophores, usually red (Cy5) and green (Cy3) dye. After the hybridization process, the array is scanned to visualize the fluorescence intensities. Following the work of Bailey [1] we neglect the possibility of different dye effects and therefore the underlying model for this experiment can be seen as a block model for blocks of size 2, where every array can be senn as one block. Yang and Speed [8] argued that many important design issues have to be carefully considered during the whole experiment in order to achieve high precision. The probably most important point here is to determine which tissues should be allocated onto the same array.

Joachim Kunert
Fakultät Statistik, Technische Universität Dortmund, D-44221 Dortmund, Germany
kunert@statistik.tu-dortmund.de

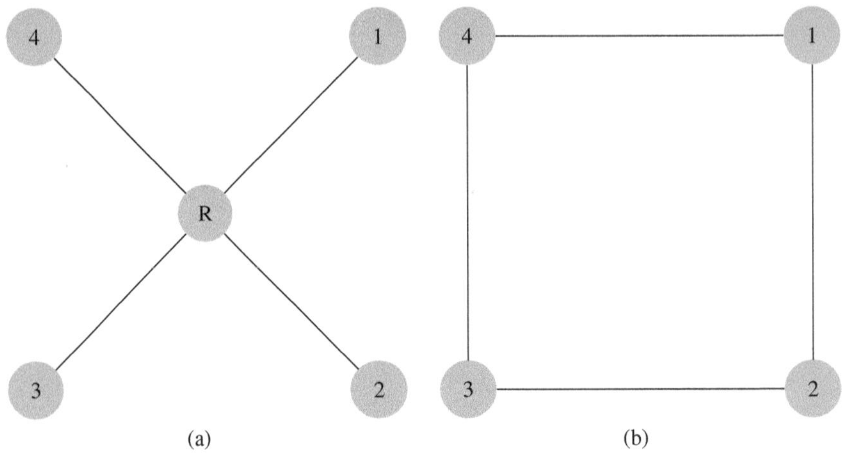


**Fig. 1** Reference design (**a**) and loop design (**b**) for $t = 4$ treatments and $b = 4$ arrays. Each line connecting two treatments indicates an array receiving these two treatments. The dye labelling is not illustrated. In the reference design, half of the available resources are used on the reference, which is not of interest. The loop design uses the same number of arrays, but no reference and therefore is more efficient

Some work has been done recently on the optimal design problem for microarray experiments. Kerr and Churchill [3] considered designs in common use and demonstrated the low efficiency of the ordinary reference design compared to a loop design. These two kinds of designs are illustrated in Fig. 1. If the underlying optimality criterion is the $A$– or $D$-criterion, it is well known that an optimal design for such experiments is the *balanced incomplete block design* (BIBD), see e.g. Shah and Sinha [7], Chap. 2. A BIBD for $t = 4$ treatments and $b = 6$ arrays is illustrated in Fig. 2 (a). But this design uses at least $t(t-1)/2$ arrays, which is a relatively large number and often not available. So Bailey [1] calculated optimal designs under the $A$- and $D$-criterion for small numbers of blocks.

Only relatively few papers deal with the problem of optimal designs for treatment-control comparisons. Here, merely the effects of some treatment contrasts, compared to a control are of interest. One could suppose that then a reference design which uses the control as the central treatment instead of a reference will perform better compared to the usual case where all treatment contrasts are of equal importance. Such a design is called a *star design.* An obvious optimality criterion for this situation is the $A$-criterion for treatment-control comparisons, see Bechhofer and Tamhane [2]. However, it is easily verified that under this criterion, a multiple star-design, as the one in Fig. 2 (b) is only as good as a BIBD for the same number of treatments and blocks. Since the BIBD remains optimal if all treatment contrasts are of equal interest, we would always recommend using the BIBD if the number of available arrays suffices. As we will see below, even for treatment-control comparisons, the star design is not the best design available.

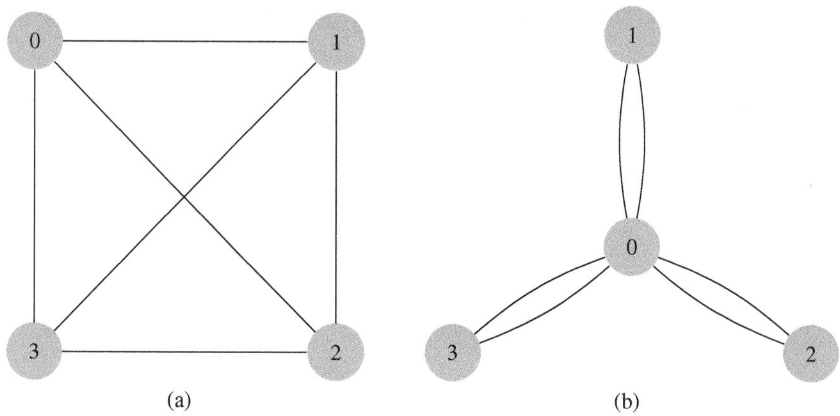


**Fig. 2** BIBD **(a)** and double star design **(b)** for $t = 3$ treatments, 1 control and $b = 6$ blocks. The labelling of treatments with fluorophores is not illustrated. The double star design and the BIBD have the same efficiencies under the $A$-criterium for treatment-control comparisons, but the BIBD is $A$-optimal if all treatment contrast are of equal interest

In this paper we determine optimal approximate designs for microarray experiments when only the treatment-control comparisons are of interest. Based on these results we construct appropriate finite designs and compare their efficiencies with those of the corresponding multiple star designs. In Sect. 2 we introduce the statistical model we use, following Bailey [1]. Optimal approximate designs for this model can be calculated with the approach proposed by Kunert et al [5] and will be described in Sect. 3. Resulting finite designs and their efficiencies, as well as a comparison with the efficiencies attained by the multiple star designs, follow in Sect. 4. We will end with a short summary and discussion in Sect. 5.

## 2 Statistical Model

Suppose there are $t \geq 2$ different treatments which we want to compare with a control 0. Like Bailey [1], we assume that after a normalisation process the $2b$ observations $y_{11}, \dots, y_{2b}$, where $b$ denotes the number of arrays available, can be described by a simple linear model

$$y = T\tau + B\beta + e,$$

where $\tau = (\tau_0, \dots \tau_t)^T$ and $\beta = (\beta_1, \dots \beta_b)^T$ are the vectors of treatment and array effects. Further, we make the usual assumption, that

$$E(e) = 0 \quad \text{and} \quad Var(e) = I_{2b}\sigma^2.$$

We are looking for $A$-optimal designs for such experiments, when only the treatment-control contrasts $\tau_i - \tau_0$, $1 \leq i \leq t$ are of interest. A design $d \in \Omega_{t+1,b,2}$ is a mapping

$$d\colon \{1,\dots,b\} \times \{1,2\} \to \{0,\dots,t\}$$

and $\Omega_{t+1,b,2}$ denotes the set of all designs for an experiment with $t+1$ treatments, $b$ blocks and 2 plots per block. The information matrix for estimating the treatment effects is then given by

$$C_d = T^T \omega^{\perp}(B) T,$$

where $\omega^{\perp}(B) = I_{2b} - B(B^T B)^{-} B^T$.

## 3 Determination of Optimal Approximate Designs and Application to Microarray Experiments

Kunert et al [5] constructed optimal and near optimal designs for treatment-control comparisons when the observations are dependent. In the special instance that $k=2$, their approach can be used for finding good designs for treatment-control comparisons in microarray experiments. We therefore follow the approach of Kunert et al [5] and partition the information matrix in the form

$$C_d = \begin{pmatrix} c_{d,ss} & c_{d,ns}^T \\ c_{d,ns} & D_d \end{pmatrix},$$

where $c_{d,ss} \in \mathbb{R}$ is a scalar, $c_{d,ns} \in \mathbb{R}^t$ is a $t$-vector and $D_d \in \mathbb{R}^{t \times t}$ is the $t$ by $t$ principle minor of $C_d$ formed by deleting the row and column corresponding to the control. A design d has supplemented balance, if $D_d$ is completely symmetric, $D_d = a_0 I_t - a_1 J_t$, with $a_0 = t a_1 + t^{-1} c_{d,ss}$. Then $c_{d,ns} = -D_d 1_t = -t^{-1} c_{d,ss} 1_t$ is a constant vector and $c_{d,ss} = 1_t^T D_d 1_t$.

Commonly used criteria for design optimality are mostly defined over functions of the information matrix $C_d$. We are here looking for a design minimizing the average pairwise variance of all treatment-control contrasts. This average variance is proportional to $\operatorname{tr}(D_d^{-1})$, the $A_{tc}$-value, say, of a design $d$, see Bechhofer and Tamhane [2]. For any design $d \in \Omega_{t+1,b,2}$ a simple lower bound for its $A_{tc}$-value is given by

$$l_d = \frac{t-1}{m_{d1}} + \frac{1}{m_{d2}},$$

where

$$m_{d2} = c_{d,ss}/t \quad \text{and} \quad m_{d_1} = (\operatorname{tr}(D_d) - m_{d2})/(t-1).$$

This bound is attained if $D_d$ is completely symmetric and hence the design has supplemented balance. The $A_{tc}$-efficiency of a design $d \in \Omega_{t+1,b,2}$ can be defined by comparing its $A_{tc}$-value to the minimum $A_{tc}$-value of all competing designs, or the minimum lower bound

$$l^* = \min_{d \in \Omega_{t+1,b,2}} l_d$$

for it. A design $d^*$ having supplemented balance and $m_{d^*1}$ and $m_{d^*2}$ such that

$$l_{d^*} = \frac{t-1}{m_{d^*1}} + \frac{1}{m_{d^*2}} = l^*$$

is thus $A_{tc}$-optimal. In many cases the overall lower bound $l^*$ cannot be attained, so the optimal design might have a higher $A_{tc}$-value.

Instead of minimizing the lower bound $l_d$ over all competing designs, Kunert et al [5] equivalently maximized

$$q_d = \frac{t-1}{l_d} = m_{d1} - \frac{m_{d1}^2}{(t-1)m_{d2} + m_{d1}}.$$

They then utilized the method of Kushner [6] for finding optimal designs (see also Kunert and Martin [4]). Hence, we consider the function

$$q_d(x) = (1+x)^2 m_{d1} + x^2(t-1)m_{d2},$$

which attains its minimum at

$$x^* = \frac{-m_{d1}}{(t-1)m_{d2} + m_{d1}}$$

with $\min q_d(x) = q_d$.

A design $d \in \Omega_{t+1,b,2}$ consists of $b$ sequences $s_{1d}, \ldots, s_{bd}$ of treatments and

$$m_{d1} = \sum_{i=1}^{b} m_1(s_{id}) \quad \text{and} \quad m_{d2} = \sum_{i=1}^{b} m_2(s_{id}),$$

where $m_j(s)$, $j = 1,2$, are the $m_{fj}$-values of a design $f \in \Omega_{t+1,1,k}$, that consists of only one block, occupied with sequence $s$. Two sequences $s$ und $v$ are equivalent, if $m_1(s) = m_1(v)$ and $m_2(s) = m_2(v)$. This, however, holds if and only if $s$ can be transformed into $v$ by relabelling the test-treatments or by reversing the order of treatments. Thus, the set of all sequences can be devided into $K$ equivalence classes. In microarray experiments we have merely $K = 3$ different equivalence classes of sequences with values $m_1(s_i)$ and $m_2(s_i)$, $i = 1,2,3$. Class 1 contains sequences with direct treatment-control comparisons. Class 2 contains comparisons between two different treatments and all sequences in which one treatment is compared with itself are gathered in class 3.

The value of $q_d$ only depends on the equivalence classes used and their proportions $\pi_{di}$, $i = 1,2,3$ in the design $d$. For each class $i$ of sequences we define the function

$$h_i(x) = (1+x)^2 m_1(s_i) + x^2(t-1)m_2(s_i),$$

where $s_i$ is an arbitrary sequence from class $i$. It then holds that

$$q_d(x) = b\sum_{i=1}^{3} \pi_{di} h_i(x).$$

We want to maximize $q_d = \min q_d(x)$. Define $y^* = \min_x \max_i h_i(x)$ and $x^*$ as the corresponding $x$-value. Then $y^* b = q^*$, say, is the maximal possible value of $q_d$ over all designs $d \in \Omega_{t+1,b,2}$. At the point $x^*$, either one of the $h_i$ wille have its minimum, or at least two of the $h_i$ will intersect. The corresponding classes are then to be used in a design with a maximum $q^*$. Define $h_i'(x) = \frac{\partial}{\partial x} h_i(x)$. Then the proportions of the $h_i$ in such a design are the values $\pi_{di}$, such that

$$b\sum_{i=1}^{3} \pi_{di} h_i'(x^*) = 0.$$

The values of $m_1(s)$ and $m_2(s)$ for the three equivalence classes, represented by $s_1 = (0,1)$ and $s_2 = (1,2)$ and $s_3 = (1,1)$ are

$$m_1(s_1) = \frac{1}{2t}, \qquad m_1(s_2) = \frac{1}{t-1}, \qquad m_1(s_3) = 0, \quad \text{and}$$
$$m_2(s_1) = \frac{1}{2t}, \qquad m_2(s_2) = 0, \qquad m_2(s_3) = 0.$$

This leads to the following polynomials for the sequence classes

$$h_1(x) = (1+x)^2 \frac{1}{2t} + x^2(t-1)\frac{1}{2t},$$
$$h_2(x) = (1+x)^2 \frac{1}{t-1} \quad \text{and} \quad h_3(x) = 0.$$

The functions $h_1$ and $h_2$ intersect at

$$x^* = -\frac{\sqrt{t+1}}{t-1+\sqrt{t+1}}.$$

Note that $y^* = h_1(x^*) = h_2(x^*) \geq 0 = h_3(x^*)$ and, therefore, $y^* = \max_i h_i(x^*)$. Since $(1+x^*) > 0$, $h_2'(x^*) = (1+x^*)\frac{2}{t-1} > 0$, and $h_1'(x^*) = (1+x^*)\frac{1-\sqrt{t+1}}{t} < 0$, so $h_1'(x^*)h_2'(x^*) < 0$ and $h_1'$ and $h_2'$ have opposite signs at $x^*$. In all,

$$y^* = \min_x \max_i h_i(x).$$

A design $d$ with a maximum $q_d$ therefore consists of sequences from classes 1 and 2 with appropriate proportions

$$\pi_{d1}^* = \frac{2t}{1+t+(t-1)\sqrt{t+1}}$$

for equivalence class 1 (treatment-control) and

$$\pi_{d2}^{*} = 1 - \pi_{d1}^{*}$$

for equivalence class 2 (treatment-treatment). The implied value for the global lower bound of a design $d \in \Omega_{t,b,2}$ is then

$$l^{*} = \frac{\left(t - 1 + \sqrt{t+1}\right)^{2}}{b}.$$

Note that this bound is attained if a design $d^{*}$ consists of exactly $b\pi_{d1}^{*}$ sequences from equivalence class 1 and of $b\pi_{d2}^{*}$ sequences from equivalence class 2 and additionally has supplemented balance.

# 4 Results

With the approach proposed in Sect. 3 we can now determine optimal approximate designs for given numbers of new treatments $t$ and blocks $b$. Thereby, an optimal design uses exactly $b_1^{*} = b\pi_{d1}^{*}$ sequences from equivalence class 1 and $b_2^{*} = b\pi_{d2}^{*}$ sequences from equivalence class 2. Additionally, those sequences must be such

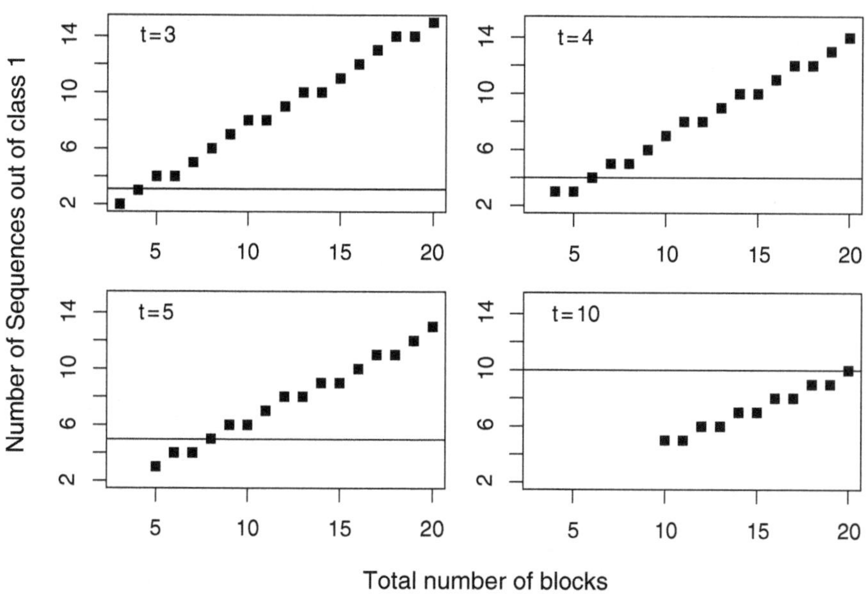


**Fig. 3** Rounded numbers of blocks that should be applied with sequences from equivalence class 1 for $t = 3, 4, 5, 10$ treatments and $t \leq b \leq 20$ blocks. The black line indicates the minimum number of sequences needed from class 1 to compare every treatment at least once directly with the control

that the resulting design has supplemented balance. Often, the calculated values of $\pi^*_{d1}$ and $\pi^*_{d2}$ for a design with a maximum overall lower bound $l^*$ are not rational. For every $b$, the optimal numbers of sequences used from the two classes are thus not integer numbers and the proposed designs do not exist. For the construction of efficient designs we therefore use proportions $\pi_{d1} \approx \pi^*_{d1}$ and $\pi_{d2} \approx \pi^*_{d2}$ such that $b_i = b\pi_{di} \in \mathbb{N}$, $i = 1,2$. Further, for given numbers $b_1$ and $b_2$, we can generally not construct a design that has supplemented balance. Instead, we choose a design that is as balanced as possible for given values of $b_1$ and $b_2$. With those two restrictions we are often not able to construct the proposed optimal designs. Nevertheless, we get designs with high $A_{tc}$-efficiencies.

For some combinations of $t$ and $b$, Fig. 3 shows the calculated numbers of sequences from equivalence class 1 that we use in our efficient designs. For relatively small numbers of blocks, those do not suffice for a complete star design. For

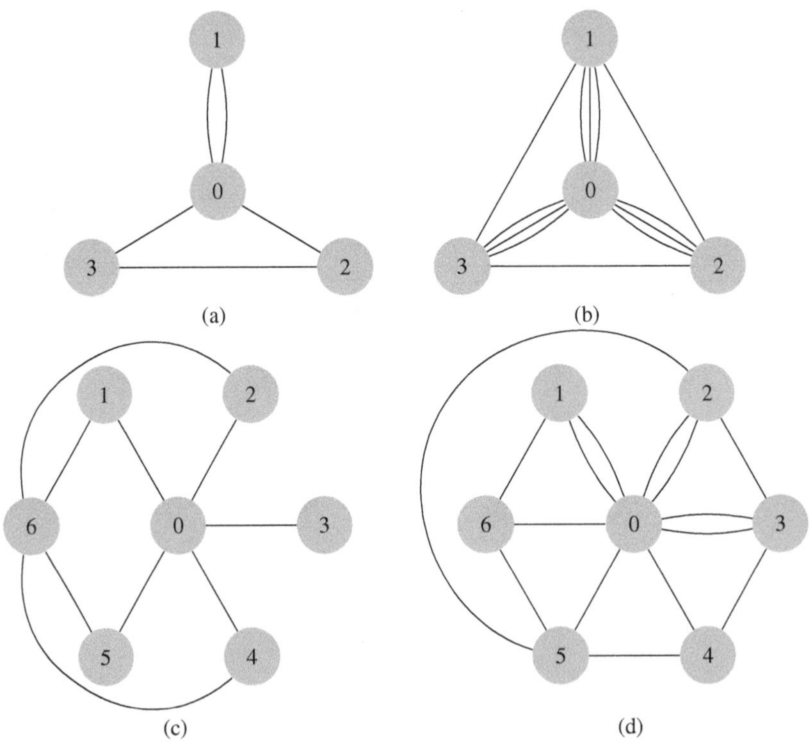


**Fig. 4** Designs for $t = 3$ treatments and $b = 5$ **(a)** and $b = 12$ **(b)** blocks and for $t = 6$ treatments and $b = 9$ **(c)** and $b = 15$ **(d)** blocks. The construction of the designs is based on the results of our approach. The first design has an $A_{tc}$-value of 3.67 with an overall lower bound $l^* = 3.2$ and therefore attains an $A_{tc}$-efficiency of 87%. The $A_{tc}$-value of the second design is equal to the overall bound $l^* = 1.33$ and the design is thus $A_{tc}$-optimal. The two designs for $t = 6$ new treatments attain $A_{tc}$-efficiencies of 81% and 90%, with $A_{tc}$-values equal to 8 and 4.32 and lower bounds $l^* = 6.50$ and $l^* = 3.90$

increasing numbers of treatments, the minimum number of blocks for which the calculated number $b_1$ allows the construction of a complete star design (with some further connections between two different treatments) also increases. Following our approach, efficient designs therefore do not compare every treatment directly with the control, if we only use small numbers of blocks. It is not until the total number of blocks reaches a specific bound that we construct a complete star design with additional comparisons between two different treatments.

Fig. 4 shows some efficient designs for different combinations of treatments and blocks, calculated with our approach. The $A_{tc}$-efficiencies of those four designs vary from 81% to 100%. Similar efficiencies are attained for many other designs based on our approach. But there are still rare cases in which the obtained results are not acceptable.

Our main goal in this paper is to compare the efficiencies of the designs attained with our approach with those of the corresponding multiple star designs. In Fig. 5 the relative efficiencies of the two types of designs are illustrated. Obviously, if we have the same number of treatments as blocks, the star designs perform better than the designs calculated with out approach. But for few treatments or increasing numbers of blocks, the efficiencies of the new designs increase, where the star designs improve only slightly. Hence, in most cases, our designs perform better than the multiple star designs. However, for larger numbers of treatments the number of blocks needed to get better than the star design increases. For $t = 10$ treatments

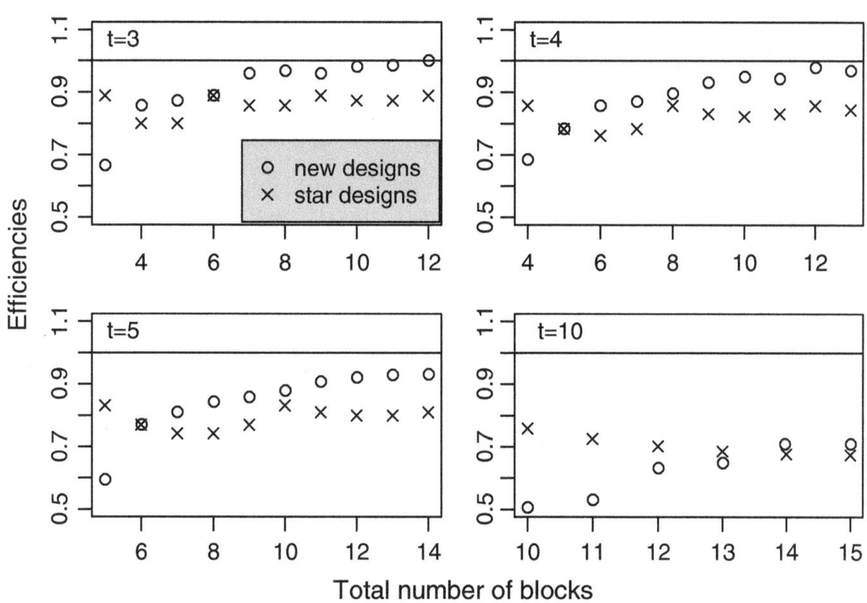


**Fig. 5** Efficiencies of the designs calculated with our approach and the corresponding multiple star designs for $t = 3, 4, 5, 10$ new treatments and different numbers of blocks. The new designs are a balance between having $b_1$ blocks 'close' to $b\pi_{d1}$ and $C_d$ 'close' to supplemented balance

it takes 14 blocks until our design is more efficient than the star design. Thus, the more treatments we have, the better it is to use a star design, as long as the number of blocks is small. With sufficiently many blocks, our designs are always preferable.

## 5 Summary and Discussion

In this paper, we have proposed an approach to determine optimal approximate designs for treatment-control comparisons in microarray experiments. Since, in general, these approximate designs cannot be constructed with a given number of arrays, we specified designs that are likely to attain high efficiencies. In contrast to the commonly used star design, most of the calculated designs use a certain proportion of direct treatment-treatment comparisons instead of simply comparing every treatment exclusively with the control. Those designs attain high efficiencies, mostly between 80% and 100%. In particular, they mostly perform better than the corresponding multiple star design.

## References


[1] Bailey, R.A.: Designs for two-colour microarray experiments. J. Roy. Stat. Soc. C–App. **56**, 365–394 (2007) doi: 10.1111/j.1467-9876.2007.00582.x

[2] Bechhofer, R.E., Tamhane, A.C.: Incomplete block designs for comparing treatments with a control: General theory. Technometrics **23**, 45–57 (1981)

[3] Kerr, M.K., Churchill, G.A.: Experimental design for gene expression microarrays. Biostatistics **2**, 183–201 (2001)

[4] Kunert, J., Martin, R.J.: On the determination of optimal designs for an interference model. Ann. Stat. **28**, 1728–1742 (2000)

[5] Kunert, J., Martin, R.J., Eccleston, J.A.: Optimal and near-optimal designs for test-control comparisons with dependent observations. Preprint (2008)

[6] Kushner, H.B.: Optimal repeated measurements designs: The linear optimality equations. Ann. Stat. **25**, 2328–2344 (1997)

[7] Shah, K.R., Sinha, B.K.: Theory of Optimal Designs. Springer, New York (1989)

[8] Yang, Y.H., Speed, T. Design issues for cdna microarray experiments. Nat. Rev. **3**, 579–588 (2002)

# Improving Henderson's Method 3 Approach when Estimating Variance Components in a Two-way Mixed Linear Model

Razaw al Sarraj and Dietrich von Rosen

**Abstract** A two-way linear mixed model, consisting of three variance components, $\sigma_1^2$, $\sigma_2^2$ and $\sigma_e^2$ is considered. The variance component estimators are estimated using a well known non-iterative estimation procedure, Henderson's method 3. For $\sigma_1^2$ we propose two modified estimators. The modification is carried out by perturbing the standard estimator, such that the obtained estimator is expected to perform better in terms of its mean square error.

## 1 Introduction

In an analysis of variance context, the most commonly used method for estimating the variance components has been through equating the observed and expected mean squares, and solving a set of linear equations. As long as the data are balanced the ANOVA estimators are known to have good statistical properties, i.e., the obtained estimators are unbiased and have minimum variance among all unbiased estimators which are quadratic functions of the observations, see Graybill and Hultquist [1]. However, since real world data often are always unbalanced, this method is no longer appealing. For instance, the uniformly minimum variance property is lost. Furthermore, whether data are balanced or unbalanced, there is nothing in the ANOVA methodology that would prevent negative estimates of the variance components to occur, LaMotte [4]. In a seminal paper Henderson [2] considered variance component estimation with unbalanced data. He presented three methods of estimation which later on, came to be known as Henderson's method 1, 2 and 3. The obtained estimators are unbiased and translation invariant.

However, since all three methods are variations of the general ANOVA method, they suffer from the weaknesses of it. In particular, the lack of uniqueness.

Dietrich von Rosen
Department of Energy and Technology, Box 7032, SE-750 07 Uppsala, Sweden
dietrich.von.rosen@et.slu.se

In this paper we were motivated by Kelly and Mathew's [3] work, where they improved the ANOVA estimators in a one-way variance component model. The model consists of two variance components, one is the random effect of interest, and the second is the error component. They modified the variance component estimator corresponding to the random effect such that the resulting estimator performed better than the unmodified ANOVA estimator in terms of the mean square error (MSE) criteria. If more components were to be included into the model, they were excluded by orthogonal projections. Hence, the model could always be dealt with as if it had two variance components.

Our aim is to modify the variance component estimators obtained by Henderson's method 3, in a two-way linear mixed model, i.e. a model with three variance components of which two components corresponding to the two random effects included in the model, and the third corresponds to the error component. Here, we want to emphasize that we are primarily interested in one of the variance components. We intend to modify this component and calculate its MSE. Thereafter, we compare it with the MSE of the unmodified one. This modified variance component estimator is expected to perform better in terms of the MSE criteria.

## 1.1 Quadratic Forms

Estimation of variance components for balanced and unbalanced data are based on quadratic forms $Y'AY$ where $A$ is a symmetric matrix, and

$$Y \sim N(\mu, V).$$

In particular the mean and the variance of $Y'AY$ are needed.

1. The mean of $Y'AY$, is equal to

$$\mathrm{E}(Y'AY) = \mathrm{tr}(AV) + \mu' A\mu, \tag{1}$$

   which is true even if $Y$ is not normally distributed.
2. The variance of $Y'AY$ is

$$\mathrm{D}[Y'AY] = 2\mathrm{tr}(AVAV) + 4(\mu' AVA\mu). \tag{2}$$

3. If $AV$ is idempotent, the distribution of $Y'AY$ is given by

$$Y'AY \sim \chi^2(r_A, \frac{1}{2}\mu' A\mu),$$

   where $\chi^2(r_A, \frac{1}{2}\mu' A\mu)$ is non-central chi-square distribution, with degrees of freedom equal to $r_A$, i.e., the rank of $A$, and the non-centrality parameter $\frac{1}{2}\mu' A\mu$.

## 1.2 Important Criteria for Deriving Estimators

Consider the following mixed linear model

$$Y = X\beta + Zu + e, \tag{3}$$

where $Y$ is the $N \times 1$ vector of observations, $X$ is a known $N \times m$ matrix, $\beta$ is an $m \times 1$ vector of unknown fixed effect parameters, and $e$ is an $N \times 1$ vector of random error with mean 0 and dispersion matrix $\sigma_e^2 I_N$. The term $Zu$ given in model (3) is a random term that can be partitioned conformably as

$$Zu = \begin{bmatrix} Z_1 \; Z_2 \; \dots \; Z_r \end{bmatrix} \begin{bmatrix} u_1 \\ u_2 \\ \vdots \\ u_r \end{bmatrix} = \sum_{i=1}^{r} Z_i u_i.$$

Thus, model (3) can be rewritten as

$$Y = X\beta + \sum_{i=1}^{r} Z_i u_i + e, \tag{4}$$

where $Z_i$ is $N \times n_i$ incidence matrix of known elements, $u_i$ is $n_i \times 1$ vector of random effects, with zero mean value and dispersion matrix $\sigma_i^2 I_{n_i}$, $i = 1, \cdots, r$. Further it is assumed that the $u_i$ and $e$ are uncorrelated random variables. Then from (4), $\mathrm{E}(Y) = X\beta$ and the dispersion matrix $\mathrm{V} = D[Y] = \Sigma_{i=1}^{r} Z_i Z_i' \sigma_i^2 + \sigma_e^2 I_N$. The parameters $\sigma_i^2$ and $\sigma_e^2$ are unknown. Since $Zu$ and $e$ are random effects, they can be combined into one random term. Thus (4) can be rewritten as $Y = X\beta + \Sigma_{i=0}^{r} Z_i u_i$ and the dispersion matrix $\mathrm{V} = \Sigma_{i=0}^{r} Z_i Z_i' \sigma_i^2$, where $u_0 = e$, $\sigma_0^2 = \sigma_e^2$ and $Z_0 = I_N$.

To generalize the idea of estimating a single variance component, we consider estimating a linear function of the variance components, $p_0\sigma_0^2 + p_1\sigma_2^2 + \cdots + p_r\sigma_r^2$, where $p_i$ are known, by a quadratic function $Y'AY$ of the random variable $Y$ in (4). The matrix $A$ should be chosen according to some suitable criteria.

1. Unbiasedness: If $Y'AY$ is unbiased for $\sum_{i=0}^{r} p_i \sigma_i^2$ for all $\sigma_i^2$, then under the restriction $X'AX = 0$,

$$\mathrm{E}(Y'AY) = \mathrm{tr}(AV) = \sum_{i=0}^{r} \mathrm{tr}(AZ_iZ_i')\sigma_i^2 = \sum_{i=0}^{r} p_i \sigma_i^2. \tag{5}$$

   i.e., an unbiased estimator is obtained if $p_i = \mathrm{tr}(AZ_iZ_i')$.
2. Translation Invariance: $Y'AY$ is translation invariant if it's value is not affected by any change in the fixed effect parameter for the model. If instead of $\beta$ we consider $\gamma = \beta - \beta_0$ as the unknown parameter, where $\beta_0$ is fixed. Then $Y'AY$ is translation invariant if $Y'AY = (Y - X\gamma)'A(Y - X\gamma)$ for all $\gamma$. Thus $AX = 0$. Since $AX = 0$ always implies $X'AX = 0$, we also have the unbiasedness condition satisfied. However, the reverse is not true i.e., unbiasedness does not imply invariance except when $A$ is n.n.d.

3. Minimum Variance: The variance of $Y'AY$ under a normality assumption equals

$$\mathrm{D}[Y'AY] = 2\mathrm{tr}[AVAV] + 4\beta' X' AVAX\beta. \tag{6}$$

Under unbiasedness i.e., $AX = 0$, the variance reduces to

$$\mathrm{D}[Y'AY] = 2\mathrm{tr}[AVAV].$$

The mean squared error, of $Y'AY$ equals

$$\mathrm{MSE}[Y'AY] = \mathrm{D}[Y'AY] + [\mathrm{Bias}(Y'AY)]^2. \tag{7}$$

Using the condition for translation invariance $AX = 0$ and unbiasedness $\mathrm{tr}[AZ_iZ_i'] = p_i$, (7) reduces to

$$\mathrm{MSE}[Y'AY] = D[Y'AY] = 2\mathrm{tr}[AVAV].$$

Both (6) and (7), under unbiasedness and invariance reduce to $2\mathrm{tr}(AVAV)$.

## *1.3 ANOVA-based Methods of Estimation*

This method is derived by equating the sums of squares in an analysis of variance table to their expected values. Let $\sigma^2$ be the vector of variance components to be estimated in some model, and let $s$ be a vector of sums of squares. Then taking the expected value

$$E(s) = C\sigma^2, \tag{8}$$

where $C$ is a non-singular matrix, the ANOVA estimator of $\widehat{\sigma}^2$ is based on (8) and is the solution to $s = C\widehat{\sigma}^2$, which equals

$$\widehat{\sigma}^2 = C^{-1}s. \tag{9}$$

The expression in (8) can be extended to include not only sums of squares but also any set of quadratic forms. Let $q = (q_1, q_2, \cdots, q_m)'$ be the $m \times 1$ vector of quadratic forms such that

$$E(q) = A\sigma^2, \tag{10}$$

where $\sigma^2 = (\sigma_1^2, \sigma_2^2, \ldots, \sigma_k^2)'$ is the vector of $k \times 1$ variance components and $A$ being an $m \times k$ matrix of known coefficients. Then, if $m = k$ and $A$ is non-singular, (10) will give $\widehat{\sigma}^2 = A^{-1}q$ as an unbiased estimator of $\sigma^2$, as in (9). In cases when there are more quadratic forms than there are variance components to estimate, the following formula gives an unbiased estimator: $\widehat{\sigma}^2 = (A'A)^{-1}A'q$, (see Searle et al. [9])

## 2 Henderson's Three Methods

Henderson [2] presented in his paper three methods of estimation of variance components, currently known as Henderson's method 1, 2 and 3. This paper is considered to be the landmark work of dealing with the problem of estimation of variance components for unbalanced data. For balanced data, variances are usually estimated using the minimum variance estimators based on the sums of squares, appearing in the analysis of variance table. For unbalanced data the situation is different; it is not always clear which mean squares should be used (see [7]) Henderson's methods are sometimes described as being three different ways of using the general ANOVA-method (see Searle [8]). They differ only in the different quadratics (not always sums of squares), used for a vector of any linearly independent quadratic forms of observations. All three methods involve calculations of mean squares, taking their expected values, equating them to the observed ones, and then solving the resulting equations in order to obtain the variance component estimators. Some of the merits of the methods is that they are easy to compute, they require no strong distributional assumptions, and by construction these methods yield unbiased estimators. However, the estimators can fall outside the parameter space, i.e., they can become negative. Moreover, the estimators are not unique, because when there are several random effects, the sums of squares for them can be computed in several ways, i.e, corrected for several combinations of other effects. When data are balanced, all three methods reduce to the usual ANOVA-method. (For a review of all three methods, see [6]). In our work, we will be concentrating on Henderson's method 3.

### 2.1 *Method 3*

This method can be used on mixed models with or without interactions. Instead of the sums of squares that method 1 and 2 use, method 3 uses reductions in sums of squares due to fitting sub-models of the full model, and then equating the reduced sums of squares to their respective expected values. The outcome will be a set of linear estimation equations, which have to be solved in order to obtain the variance component estimators. The drawback with this method is that sometimes more reduction sum of squares are available than necessary to estimate the variance component estimators (see [8]). In other words, occasionally more than one set of estimating equations for the variance components can be computed for one model. From each set we get different estimators of the variance components. Which set of estimators to prefer is not clear, i.e., the variance component estimators are not unique. We will consider the following two-way mixed model with no interaction,

$$Y = X\beta + Z_1u_1 + Z_2u_2 + e, \quad \text{(full model)} \tag{11}$$

where $\beta$ is the fixed parameter vector and $u_1$, $u_2$ are random effect parameters. For this model there are three variance components to estimate, i.e., the variance of the two random effects denoted by $\sigma_1^2$ and $\sigma_2^2$ respectively, and the third is the error variance component denoted by $\sigma_e^2$. We may obtain several sets of estimation equations. The sub-models which may give estimation equations are,

$$Y = X\beta + e, \tag{12}$$

$$Y = X\beta + Z_1 u_1 + e, \tag{13}$$

$$Y = X\beta + Z_2 u_2 + e. \tag{14}$$

Now we present some special notation for reduction sum of squares which was used by Searle [7, 8]. Let R(.) denote the reduction sum of squares. The sum of squares used for estimation corresponding to the sub-models (12), (13) and (14) can according to this notation be expressed as, $\mathrm{R}(\beta)$, $\mathrm{R}(\beta,u_1)$ and $\mathrm{R}(\beta,u_2)$, respectively. Another notation which will be needed before we write the possible set of equations is R(./.) which is the reduction sum of squares due to fitting the full model (11) minus that of the sub-model. For (11) two sets of estimation equations may be considered

$$\begin{cases} \mathrm{R}(u_1/\beta) \\ \mathrm{R}(u_2/\beta,u_1) \\ \mathrm{SSE} \end{cases} \qquad or \qquad \begin{cases} \mathrm{R}(u_2/\beta) \\ \mathrm{R}(u_1/u_2,\beta) \\ \mathrm{SSE} \end{cases}$$

where SSE denotes the residual sum of squares. For the first set of estimation equations we define the following partitioned matrices: $[X]$, $[X,Z_1]$ and $[X,Z_1,Z_2]$. Each reduction R(./.) can be expressed in the form $Y'AY$ for some symmetric matrix $A$. Define the projection matrix $P_w = w(w'w)^- w'$. Thus $P_w$ is an idempotent matrix, for more properties see Schott [5]. Assuming normality all the reduction sum of squares follow a non-central $\chi^2$ distribution and all these reduction sum of squares are independent of each other and of SSE, see [8]. We shall be using the first set of estimation equation in the first part of the work. In the second part, i.e., in Sect. 4, different reductions in sums of squares will be compared. For the first set of equations we need to define the following projection matrices,

$$P_x = X(X'X)^- X', \tag{15}$$

$$P_{x1} = (X,Z_1)((X,Z_1)'(X,Z_1))^-(X,Z_1)', \tag{16}$$

$$P_{x_{12}} = (X,Z_1,Z_2)((X,Z_1,Z_2)'(X,Z_1,Z_2))^-(X,Z_1,Z_2)'. \tag{17}$$

The reduction sums of squares R(./.) can now be obtained as.

$$\begin{aligned} \mathrm{R}(u_1/\beta) &= \mathrm{R}(u_1,\beta) - \mathrm{R}(\beta) \\ &= Y'(P_{x_1} - P_x)Y, \end{aligned}$$

$$\begin{aligned}\mathrm{R}(u_2/\beta,u_1) &= \mathrm{R}(\beta,u_1,u_2)-\mathrm{R}(\beta,u_1)\\ &= Y'(P_{x_{12}}-P_{x_1})Y,\end{aligned}$$

and

$$\mathrm{SSE} = Y'(I-P_{x_{12}})Y.$$

To apply the procedure, the expected values of the reduction sums of squares are computed. Thereafter the expected values are to be equated to their observed values and by solving the obtained equations the variance components are obtained. The expression for the expected value given in (1), can be used since the dispersion matrix, denoted by $V$ is $V=\sigma_1^2V_1+\sigma_2^2V_2+\sigma_e^2I$, where $V_1=Z_1Z_1'$ and $V_2=Z_2Z_2'$. The following is obtained

$$\mathrm{E}[\mathrm{R}(u_1/\beta)] = \mathrm{tr}(P_{x_1}-P_x)[\sigma_1^2V_1+\sigma_2^2V_2+\sigma_e^2I],$$

$$\mathrm{ER}(u_2/\beta,u_1) = \mathrm{tr}(P_{x_{12}}-P_{x_1})[\sigma_1^2V_1+\sigma_2^2V_2+\sigma_e^2I],$$

and

$$\mathrm{E[SSE]} = \mathrm{tr}(I-P_{x_{12}})[\sigma_1^2V_1+\sigma_2^2V_2+\sigma_e^2I].$$

The set of calculated reduction sum of squares may be arranged in a vector. Thereafter by equating these expected values to the observed ones we get

$$\begin{bmatrix} Y'(P_{x_1}-P_x)Y \\ Y'(P_{x_{12}}-P_{x_1})Y \\ Y'(I-P_{x_{12}})Y \end{bmatrix} = J\begin{bmatrix}\sigma_1^2\\ \sigma_2^2\\ \sigma_e^2\end{bmatrix},$$

where

$$J=\begin{bmatrix} \mathrm{tr}(P_{x_1}-P_x)V_1 & \mathrm{tr}(P_{x_1}-P_x)V_2 & \mathrm{tr}(P_{x_1}-P_x)I \\ \mathrm{tr}(P_{x_{12}}-P_{x_1})V_1 & \mathrm{tr}(P_{x_{12}}-P_{x_1})V_2 & \mathrm{tr}(P_{x_{12}}-P_{x_1})I \\ \mathrm{tr}(I-P_{x_{12}})V_1 & \mathrm{tr}(I-P_{x_{12}})V_2 & \mathrm{tr}(I-P_{x_{12}})I \end{bmatrix}.$$

Thus, the estimators of the variance components are

$$\begin{bmatrix}\widehat{\sigma}_1^2\\ \widehat{\sigma}_2^2\\ \widehat{\sigma}_e^2\end{bmatrix} = J^{-1}\begin{bmatrix} Y'(P_{x_1}-P_x)Y \\ Y'(P_{x_{12}}-P_{x_1})Y \\ Y'(I-P_{x_{12}})Y \end{bmatrix}.$$

However, since $P_{x_1}V_1=V_1$, $P_{x_{12}}V_2=V_2$ and $P_{x_{12}}V_1=V_1$, the $J$ matrix reduces to

$$J=\begin{bmatrix} \mathrm{tr}(P_{x_1}-P_x)V_1 & \mathrm{tr}(P_{x_1}-P_x)V_2 & \mathrm{tr}(P_{x_1}-P_x) \\ 0 & \mathrm{tr}(P_{x_{12}}-P_{x_1})V_2 & \mathrm{tr}(P_{x_{12}}-P_{x_1}) \\ 0 & 0 & \mathrm{tr}(I-P_{x_{12}}) \end{bmatrix}.$$

Let

$$
\begin{aligned}
&A=(P_{x_1}-P_x), \quad B=(P_{x_{12}}-P_{x_1}), \quad C=(I-P_{x_{12}}),\\
&a=tr(P_{x_1}-P_x)V_1, \quad b=tr(P_{x_{12}}-P_{x_1})V_2, \quad c=tr(I-P_{x12}),\\
&d=tr(P_{x_1}-P_x)V_2, \quad e=tr(P_{x_{12}}-P_{x_1}), \quad f=tr(P_{x_1}-P_x)),
\end{aligned}
\tag{18}
$$

we note that $A$, $B$ and $C$ are idempotent matrices. Using these notations the estimation equations can be written as

$$
\begin{bmatrix} \widehat{\sigma}_1^2 \\ \widehat{\sigma}_2^2 \\ \widehat{\sigma}_e^2 \end{bmatrix} = J^{-1} \begin{bmatrix} Y'AY \\ Y'BY \\ Y'CY \end{bmatrix}, \tag{19}
$$

The variance component estimator of $\sigma_1^2$, denoted by $\widehat{\sigma}_{u1}^2$ is:

$$
\begin{aligned}
\widehat{\sigma}_{u1}^2 = & \frac{\mathrm{tr}((P_{x_{12}}-P_{x_1})V_2)\mathrm{tr}(I-P_{x_{12}})Y'(P_{x_1}-P_x)Y}{\mathrm{tr}((P_{x_1}-P_x)V_1)\mathrm{tr}((P_{x_{12}}-P_{x_1})V_2)\mathrm{tr}(I-P_{x_{12}})} \\
& -\frac{\mathrm{tr}((P_{x_1}-P_x)V_2)\mathrm{tr}(I-P_{x_{12}})Y'(P_{x_{12}}-P_{x_1})Y}{\mathrm{tr}((P_{x_1}-P_x)V_1)\mathrm{tr}((P_{x_{12}}-P_{x_1})V_2)\mathrm{tr}(I-P_{x_{12}})} \\
& +\frac{kY'(I-P_{x_{12}})Y}{\mathrm{tr}((P_{x_1}-P_x)V_1)\mathrm{tr}((P_{x_{12}}-P_{x_1})V_2)\mathrm{tr}(I-P_{x_{12}})},
\end{aligned}
\tag{20}
$$

where $k=\mathrm{tr}((P_{x_1}-P_x)V_2)\mathrm{tr}(P_{x_{12}}-P_{x_1})-\mathrm{tr}(P_{x_1}-P_x)\mathrm{tr}((P_{x_{12}}-P_{x_1})V_2)$. Equation (20) simplifies to

$$
\begin{aligned}
\widehat{\sigma}_{u1}^2 = & \frac{Y'(P_{x_1}-P_x)Y}{\mathrm{tr}((P_{x_1}-P_x)V_1)} - \frac{\mathrm{tr}((P_{x_1}-P_x)V_2)Y'(P_{x_{12}}-P_{x_1})Y}{\mathrm{tr}((P_{x_1}-P_x)V_1)\mathrm{tr}((P_{x_{12}}-P_{x_1})V_2)} \\
& +\frac{kY'(I-P_{x_{12}})Y}{\mathrm{tr}((P_{x_1}-P_x)V_1)\mathrm{tr}((P_{x_{12}}-P_{x_1})V_2)\mathrm{tr}(I-P_{x_{12}})}.
\end{aligned}
\tag{21}
$$

Using the previous notations we can write $\widehat{\sigma}_{u1}^2$ as

$$
\widehat{\sigma}_{u1}^2 = \frac{Y'AY}{a} - \frac{d(Y'BY)}{ab} + \frac{k(Y'CY)}{abc}, \tag{22}
$$

where $A$, $B$, $C$, $b$, $c$ and $e$ are defined as in (18). Despite the fact that in our study we will focus on one of the variance components we also give the estimators of the two other components which may be calculated from (19);

$$
\widehat{\sigma}_{u_2}^2 = \frac{\mathrm{tr}(I-P_{x_{12}})Y'(P_{x_{12}}-P_{x_1})Y}{\mathrm{tr}((P_{x_{12}}-P_{x_1})V_2)\mathrm{tr}(I-P_{x_{12}})} - \frac{\mathrm{tr}(P_{x_{12}}-P_{x_1})Y'(I-P_{x_{12}})Y}{\mathrm{tr}((P_{x_{12}}-P_{x_1})V_2)\mathrm{tr}(I-P_{x_{12}})},
$$

$$
\widehat{\sigma}_e^2 = \frac{\mathrm{tr}((P_{x_1}-P_x)V_1)\mathrm{tr}((P_{x_{12}}-P_{x_1})V_2)Y'(I-P_{x_{12}})Y}{\mathrm{tr}((P_{x_1}-P_x)V_1)\mathrm{tr}((P_{x_{12}}-P_{x_1})V_2)\mathrm{tr}(I-P_{x_2})} = \frac{Y'(I-P_{x_{12}})Y}{\mathrm{tr}(I-P_{x_{12}})}.
$$

### 2.1.1 Mean Square Error of $\widehat{\sigma}^2_{u1}$

The mean square of $\widehat{\sigma}^2_{u1}$ equals its variance since $\widehat{\sigma}^2_{u1}$ is an unbiased estimator,

$$\begin{aligned} MSE(\widehat{\sigma}^2_{u1}) &= D[\widehat{\sigma}^2_{u1}] = D\left[\frac{Y'AY}{a} - \frac{d(Y'BY)}{ab} + \frac{k(Y'CY)}{abc}\right] \\ &= \frac{1}{a^2}D[Y'AY] + \frac{d^2}{a^2b^2}D[Y'BY] + \frac{k^2}{a^2b^2c^2}D[Y'CY] \\ &= \frac{2}{a}\mathrm{tr}[AV]^2 + \frac{2d^2}{a^2b^2}\mathrm{tr}[BV]^2 + \frac{2k^2}{a^2b^2c^2}\mathrm{tr}[CV]^2, \end{aligned} \tag{23}$$

Moreover since all the involved quadratic forms are uncorrelated, $V = \sigma_1^2V_1 + \sigma_2^2V_2 + \sigma_e^2 I$ and the MSE equals

$$\mathrm{D}[\widehat{\sigma}^2_{u_1}] = A_1 + A_2 + A_3, \tag{24}$$

where

$$\begin{aligned} A_1 = \frac{2}{a^2}[&\mathrm{tr}(AV_1AV_1)\sigma_1^4 + 2\mathrm{tr}(AV_1AV_2)\sigma_1^2\sigma_2^2 + \mathrm{tr}(AV_2AV_2)\sigma_2^4 + 2\mathrm{tr}(AV_1A)\sigma_1^2\sigma_e^2 \\ &+ 2\mathrm{tr}(AV_2A)\sigma_2^2\sigma_e^2 + \mathrm{tr}(A^2)\sigma_e^4], \end{aligned}$$

$$\begin{aligned} A_2 = \frac{2d^2}{a^2b^2}[&\mathrm{tr}(BV_1BV_1)\sigma_1^4 + 2\mathrm{tr}(BV_1BV_2)\sigma_1^2\sigma_2^2 + \mathrm{tr}(BV_2BV_2)\sigma_2^4 + 2\mathrm{tr}(BV_1B)\sigma_1^2\sigma_e^2 \\ &+ 2\mathrm{tr}(BV_2B)\sigma_2^2\sigma_e^2 + \mathrm{tr}(B^2)\sigma_e^4], \end{aligned}$$

$$\begin{aligned} A_3 = \frac{2k^2}{a^2b^2c^2}[&\mathrm{tr}(CV_1CV_1)\sigma_1^4 + 2\mathrm{tr}(CV_1CV_2)\sigma_1^2\sigma_2^2 + \mathrm{tr}(CV_2CV_2)\sigma_2^4 \\ &+ 2\mathrm{tr}(CV_1C)\sigma_1^2\sigma_e^2 + 2\mathrm{tr}(CV_2C)\sigma_2^2\sigma_e^2 + \mathrm{tr}(C^2)\sigma_e^4]. \end{aligned}$$

Thus, the following MSE is obtained:

$$\begin{aligned} \mathrm{MSE}(\widehat{\sigma}^2_{u1}) = &\left[\frac{2}{a^2}\mathrm{tr}(AV_1AV_1) + \frac{2d^2}{a^2b^2}\mathrm{tr}(BV_1BV_1) + \frac{2k^2}{a^2b^2c^2}\mathrm{tr}(CV_1CV_1)\right]\sigma_1^4 \\ &+ \left[\frac{2}{a^2}\mathrm{tr}(AV_2AV_2) + \frac{2d^2}{a^2b^2}\mathrm{tr}(BV_2BV_2) + \frac{2k^2}{a^2b^2c^2}\mathrm{tr}(CV_2CV_2)\right]\sigma_2^4 \\ &+ \left[\frac{4}{a^2}\mathrm{tr}(AV_1AV_2) + \frac{4d^2}{a^2b^2}\mathrm{tr}(BV_1BV_2) + \frac{4k^2}{a^2b^2c^2}\mathrm{tr}(CV_1CV_2)\right]\sigma_1^2\sigma_2^2 \\ &+ \left[\frac{4}{a^2}\mathrm{tr}(A^2V_1) + \frac{4d^2}{a^2b^2}\mathrm{tr}(B^2V_1) + \frac{4k^2}{a^2b^2c^2}\mathrm{tr}(C^2V_1)\right]\sigma_1^2\sigma_e^2 \\ &+ \left[\frac{4}{a^2}\mathrm{tr}(A^2V_2) + \frac{4d^2}{a^2b^2}\mathrm{tr}(B^2V_2) + \frac{4k^2}{a^2b^2c^2}\mathrm{tr}(C^2V_2)\right]\sigma_2^2\sigma_e^2 \\ &+ \left[\frac{2}{a^2}\mathrm{tr}(A^2) + \frac{2d^2}{a^2b^2}\mathrm{tr}(B^2) + \frac{2k^2}{a^2b^2c^2}\mathrm{tr}(C^2)\right]\sigma_e^4, \end{aligned}$$

since $\mathrm{tr}(CV_1)=0$, $\mathrm{tr}(CV_2)=0$ and $\mathrm{tr}(BV_1)=0$ . The above can be simplified to

$$\begin{aligned}\mathrm{MSE}(\widehat{\sigma}_{u1}^2) &= \left[\frac{2}{a^2}\mathrm{tr}(AV_1AV_1)\right]\sigma_1^4+\left[\frac{2}{a^2}\mathrm{tr}(AV_2AV_2)+\frac{2d^2}{a^2b^2}\mathrm{tr}(BV_2BV_2)\right]\sigma_2^4\\ &+\left[\frac{4}{a^2}\mathrm{tr}(AV_1AV_2)\right]\sigma_1^2\sigma_2^2+\left[\frac{4}{a^2}\mathrm{tr}(A^2V_1)\right]\sigma_1^2\sigma_e^2\\ &+\left[\frac{4}{a^2}\mathrm{tr}(AV_2A)+\frac{4d^2}{a^2b^2}\mathrm{tr}(BV_2B)\right]\sigma_2^2\sigma_e^2\\ &+\left[\frac{2}{a^2}\mathrm{tr}(A^2)+\frac{2d^2}{a^2b^2}\mathrm{tr}(B^2)+\frac{2k^2}{a^2b^2c^2}\mathrm{tr}(C^2)\right]\sigma_e^4. \end{aligned} \tag{25}$$

## 3 Perturbing Henderson's Equation

In this section, we modify the variance component estimators obtained by Henderson's method 3. This modification is carried out by perturbing the Henderson's estimation equation. Thus, the obtained variance component estimators are biased. Thereafter, by using some suitable criterion, for instance, the MSE, we evaluate the performance of the estimator by comparing it with the MSE of the unmodified estimator. For the estimation (19), we define a new class of estimators

$$\begin{bmatrix} c_1Y'AY \\ c_1d_1Y'BY \\ c_1d_2Y'CY \end{bmatrix} = J\begin{bmatrix}\sigma_1^2\\ \sigma_2^2\\ \sigma_e^2\end{bmatrix} \tag{26}$$

where $J$ is defined in Sect. (2.1), and $c_1\geq 0$, $d_1$ and $d_2$ are constants to be determined such that it would minimize the leading terms in the MSE of the estimator. The resulting estimator will perform better in terms of MSE since $c_1=d_1=d_2=1$ gives the same MSE. Thus, the modified variance component estimator of $\sigma_1^2$, denoted by $\widehat{\sigma}_{11}^2$ is

$$\widehat{\sigma}_{11}^2=\frac{c_1}{a}\left(Y'AY-\frac{d}{b}d_1Y'BY+\frac{k}{bc}d_2Y'CY\right), \tag{27}$$

where $A$, $B$, $C$, $a$, $b$, $c$ and $d$ are all defined in (18). The MSE of this modified variance component is

$$\mathrm{MSE}[\widehat{\sigma}_{11}^2]=\mathrm{D}[\widehat{\sigma}_{11}^2]+[\mathrm{E}(\widehat{\sigma}_{11}^2)-\sigma_1^2]^2. \tag{28}$$

Since now (19) is perturbed, the estimator is not unbiased, The variance in (27) equals

$$\mathrm{D}[\widehat{\sigma}_{11}^2]=\frac{c_1^2}{a^2}\mathrm{D}[Y'AY]+\frac{d^2c_1^2d_1^2}{a^2b^2}\mathrm{D}[Y'BY]+\frac{k^2c_1^2d_2^2}{a^2b^2c^2}D[Y'CY],$$

since $D[\widehat{\sigma}_{11}^2]$ has the same structure as (25). Hence the variance of the modified estimator $\widehat{\sigma}_{11}^2$ can be written

$$
\begin{aligned}
D[\widehat{\sigma}_{11}^2] = & \left[\frac{2c_1^2}{a^2}\mathrm{tr}(AV_1AV_1)\right]\sigma_1^4 + \left[\frac{2c_1^2}{a^2}tr(AV_2AV_2) + \frac{2d^2c_1^2d_1^2}{a^2b^2}tr(BV_2BV_2)\right]\sigma_2^4 \\
& + \left[\frac{4c_1^2}{a^2}tr(AV_1AV_2)\right]\sigma_1^2\sigma_2^2 + \left[\frac{4c_1^2}{a^2}tr(A^2V_1)\right]\sigma_1^2\sigma_e^2 \\
& + \left[\frac{4c_1^2}{a^2}tr(AV_2A) + \frac{4d^2c_1^2d_1^2}{a^2b^2}tr(BV_2B)\right]\sigma_2^2\sigma_e^2 \\
& + \left[\frac{2c_1^2}{a^2}tr(A^2) + \frac{2d^2c_1^2d_1^2}{a^2b^2}tr(B^2) + \frac{2k^2c_1^2d_2^2}{a^2b^2c^2}tr(C^2)\right]\sigma_e^4. \qquad (29)
\end{aligned}
$$

Now we will calculate the bias part of (27), and thus the expectation of $\widehat{\sigma}_{11}^2$ is needed:

$$
\begin{aligned}
E[\widehat{\sigma}_{11}^2] &= \frac{c_1}{a}E(Y'AY) - \frac{dc_1}{ab}d_1E(Y'BY) + \frac{c_1kd_2}{abc}E(Y'CY) \\
&= \frac{c_1}{a}tr[A(\sigma_1^2V_1 + \sigma_2^2V_2 + \sigma_e^2I)] \\
&\quad - \frac{dc_1d_1}{ab}tr[B(\sigma_1^2V_1 + \sigma_2^2V_2 + \sigma_e^2I)] \\
&\quad + \frac{c_1kd_2}{abc}tr[C(\sigma_1^2V_1 + \sigma_2^2V_2 + \sigma_e^2I)],
\end{aligned}
$$

which can be simplified to

$$
\begin{aligned}
E[\widehat{\sigma}_{11}^2] = & \left[\frac{c_1}{a}tr(AV_1) - \frac{dc_1d_1}{ab}tr(BV_1) + \frac{c_1kd_2}{abc}tr(CV_1)\right]\sigma_1^2 \\
& + \left[\frac{c_1}{a}tr(AV_2) - \frac{dc_1d_1}{ab}tr(BV_2) + \frac{c_1kd_2}{abc}tr(CV_2)\right]\sigma_2^2 \\
& + \left[\frac{c_1}{a}tr(A) - \frac{dc_1d_1}{ab}tr(B) + \frac{c_1kd_2}{abc}tr(C)\right]\sigma_e^2. \qquad (30)
\end{aligned}
$$

Thus, the squared bias can be written

$$
\begin{aligned}
(E[\widehat{\sigma}_{11}^2] - \sigma_1^2)^2 = & \Bigg[\left(\frac{c_1}{a}tr(AV_1) - 1\right)\sigma_1^2 + \left(\frac{c_1}{a}tr(AV_2) - \frac{dc_1d_1}{ab}tr(BV_2)\right)\sigma_2^2 \\
& + \left(\frac{c_1}{a}tr(A) - \frac{dc_1d_1}{ab}tr(B) + \frac{c_1kd_2}{abc}tr(C)\right)\sigma_e^2\Bigg]^2. \qquad (31)
\end{aligned}
$$

If we substitute the variance and biased part back into (28), we get the following:

$$\begin{aligned}
\mathrm{MSE}(\widehat{\sigma}_{11}^2) = & \left[\frac{2c_1^2}{a^2}tr(AV_1AV_1)\right]\sigma_1^4 + \left[\frac{4c_1^2}{a^2}tr(AV_1AV_2)\right]\sigma_1^2\sigma_2^2 \\
& + \left[\frac{2c_1^2}{a^2}tr(AV_2AV_2) + \frac{2d^2c_1^2d_1^2}{a^2b^2}tr(BV_2BV_2)\right]\sigma_2^4 \\
& + \left[\frac{4c_1^2}{a^2}tr(A^2V_1)\right]\sigma_1^2\sigma_e^2 \\
& + \left[\frac{4c_1^2}{a^2}tr(A^2V_2) + \frac{4d^2c_1^2d_1^2}{a^2b^2}tr(B^2V_2)\right]\sigma_2^2\sigma_e^2 \\
& + \left[\frac{2c_1^2}{a^2}tr(A^2) + \frac{2d^2c_1^2d_1^2}{a^2b^2}tr(B^2) + \frac{2k^2c_1^2d_2^2}{a^2b^2c^2}tr(C^2)\right]\sigma_e^4 \\
& + \left[\left(\frac{c_1}{a}tr(AV_1) - 1\right)\sigma_{u1}^2 + \left(\frac{c_1}{a}tr(AV_2) - \frac{dc_1d_1}{ab}tr(BV_2)\right)\right]\sigma_2^2 \\
& + \left[\frac{c_1}{a}tr(A) - \frac{dc_1d_1}{ab}tr(B) + \frac{c_1kd_2}{abc}tr(C)\sigma_e^2\right]^2. \qquad (32)
\end{aligned}$$

We write the latter expression as below. First let

$$r = \frac{c_1}{a}tr(AV_2) - \frac{dc_1d_1}{ab}tr(BV_2).$$

Rewriting it gives the following:

$$r = \frac{c_1d}{a} - \frac{dc_1d_1}{a},$$

where from (18) we have $tr(AV_2) = d$ and $tr(BV_2) = b$. Moreover, let

$$t = \frac{c_1}{a}tr(A) - \frac{dc_1d_1}{ab}tr(B) + \frac{c_1kd_2}{ab}. \qquad (33)$$

Hence, the following mean square error is obtained for the modified estimator $\widehat{\sigma}_{u11}^2$:

$$\begin{aligned}
\mathrm{MSE}(\widehat{\sigma}_{11}^2) = & \left[\frac{2c_1^2}{a^2}tr(AV_1AV_1) + (c_1-1)^2\right]\sigma_1^4 \\
& + \left[\frac{4c_1^2}{a^2}tr(AV_1AV_2) + 2(c_1-1)r\right]\sigma_1^2\sigma_2^2 \\
& + \left[\frac{2c_1^2}{a^2}tr(AV_2AV_2) + \frac{2d^2c_1^2d_1^2}{a^2b^2}tr(BV_2BV_2) + r^2\right]\sigma_2^4 \\
& + \left[\frac{4c_1^2}{a^2}tr(A^2V_1) + 2(c_1-1)t\right]\sigma_1^2\sigma_e^2
\end{aligned}$$

$$+\left[\frac{4c_1^2}{a^2}tr(A^2V_2)+\frac{4d^2c_1^2d_1^2}{a^2b^2}tr(B^2V_2)+2rt\right]\sigma_2^2\sigma_e^2$$
$$+\left[\frac{2c_1^2}{a^2}tr(A^2)+\frac{2d^2c_1^2d_1^2}{a^2b^2}tr(B^2)+\frac{2k^2c_1^2d_2^2}{a^2b^2c^2}tr(C^2)+t^2\right]\sigma_e^4. \quad (34)$$

## 3.1 Mean Square Error Comparison

In this section we compare the mean square errors of the modified $\widehat{\sigma}_{11}^2$ and the unmodified estimator $\widehat{\sigma}_{u1}^2$, given by (34) and (25), respectively. We will investigate if $\mathrm{MSE}(\widehat{\sigma}_{11}^2) \le \mathrm{MSE}(\widehat{\sigma}_{u1}^2)$. To do so we compare all coefficients of $\sigma_1^4$, $\sigma_2^4$ and $\sigma_e^4$ and all their cross combinations which appeared in (34) and (25). We will investigate a number of inequalities. If they hold, then the coefficients of the modified estimator $\widehat{\sigma}_{11}^2$ are less than the coefficients of the unmodified one $\widehat{\sigma}_{u1}^2$.

From the terms corresponding to $\sigma_1^4$ in (34) and (25) it follows that we have to investigate if

$$\frac{2c_1^2}{a^2}\mathrm{tr}(AV_1AV_1)+(c_1-1)^2 \le \frac{2}{a^2}\mathrm{tr}(AV_1AV_1). \quad (35)$$

From the terms corresponding to $\sigma_2^4$ we obtain that

$$\frac{2c_1^2}{a^2}\mathrm{tr}(AV_2AV_2)+\frac{2d^2c_1^2d_1^2}{a^2b^2}\mathrm{tr}(BV_2BV_2)+r^2$$
$$\le \frac{2}{a^2}\mathrm{tr}(AV_2AV_2)+\frac{2d^2}{a^2b^2}\mathrm{tr}(BV_2BV_2), \quad (36)$$

should be studied, where $r=(\frac{c_1d}{a}-\frac{dc_1d_1}{a})$ and by assumption $c_1>0$. Corresponding to $\sigma_e^4$ we will study the inequality

$$\frac{2c_1^2}{a^2}\mathrm{tr}(A^2)+\frac{2d^2c_1^2d_1^2}{a^2b^2}\mathrm{tr}(B^2)+\frac{2k^2c_1^2d_2^2}{a^2b^2c}+t^2$$
$$\le \frac{2}{a^2}\mathrm{tr}(A^2)+\frac{2d^2}{a^2b^2}\mathrm{tr}(B^2)+\frac{2k^2}{a^2b^2c} \quad (37)$$

where $k=d\mathrm{tr}(B)-b\mathrm{tr}(A)$ and $t$ is defined in (33).

Now the cross combination coefficients of (25) and (34) will be compared. We have first the coefficients of $\sigma_1^2\sigma_2^2$.

$$\frac{4c_1^2}{a^2}\mathrm{tr}(AV_1AV_2)+2(c_1-1)r \le \frac{4}{a^2}\mathrm{tr}(AV_1AV_2), \quad (38)$$

where

$$(c_1-1)r=(c_1-1)(\frac{c_1d}{a}-\frac{dc_1d_1}{a})=\frac{d}{a}(1-d_1)(c_1^2-c_1).$$

Corresponding to $\sigma_1^2\sigma_e^2$ we investigate

$$\frac{4c_1^2}{a^2}\mathrm{tr}(A^2V_1)+2(c_1-1)t \leq \frac{4}{a^2}tr(A^2V_1), \tag{39}$$

where

$$(c_1-1)t=(c_1-1)\left(\frac{c_1}{a}\mathrm{tr}(A)-\frac{dc_1d_1}{ab}tr(B)+\frac{c_1kd_2}{ab}\right),$$

and $A$, defined in (18), is an idempotent matrix. Finally we also study the coefficients corresponding to $\sigma_2^2\sigma_e^2$,

$$\frac{4c_1^2}{a^2}\mathrm{tr}(AV_2)+\frac{4d^2c_1^2d_1^2}{a^2b^2}tr(B^2V_2)+2rt \leq \frac{4}{a^2}\mathrm{tr}(A^2V_2)+\frac{4d^2}{a^2b^2}tr(B^2V_2), \tag{40}$$

where

$$\begin{aligned}2rt &= 2\left(\frac{c_1d}{a}-\frac{dc_1d_1}{a}\right)\left(\frac{c_1}{a}\mathrm{tr}(A)-\frac{dc_1d_1}{ab}\mathrm{tr}(B)+\frac{c_1kd_2}{ab}\right)\\ &= \frac{2c_1^2d}{a^2}(1-d_1)\left(\mathrm{tr}(A)-\frac{dd_1}{b}\mathrm{tr}(B)+\frac{k}{b}d_2\right).\end{aligned}$$

In order to find appropriate values of $c_1$, $d_1$ and $d_2$ we have chosen to minimize the leading terms in (34), i.e., the terms that involve the coefficients of $\sigma_1^4$, $\sigma_2^4$ and $\sigma_e^4$, respectively. When minimizing the coefficient of $\sigma_1^4$ in (34) the following equation is obtained,

$$\frac{\partial}{\partial c_1}\left[\frac{2c_1^2}{a^2}tr(AV_1AV_1)+(c_1-1)^2\right]=0,$$

with a solution given by

$$c_1=\frac{1}{\frac{2}{a^2}tr(AV_1AV_1)+1}. \tag{41}$$

Moreover, minimizing the coefficient of $\sigma_2^4$ gives

$$\frac{\partial}{\partial d_1}\left[\frac{2c_1^2}{a^2}tr(AV_2AV_2)+\frac{2d^2c_1^2d_1^2}{a^2b^2}tr(BV_2BV_2)+\left(\frac{c_1d}{a}-\frac{dc_1d_1}{a}\right)^2\right]=0,$$

which implies

$$d_1=\frac{1}{\frac{2}{b^2}tr(BV_2BV_2)+1}. \tag{42}$$

Finally, when minimizing the coefficient of the error variance component $\sigma_e^4$ we have to solve

$$\frac{\partial}{\partial d_2}\left[\frac{c_1^2}{a^2}tr(A^2)+\frac{2d^2c_1^2d_1^2}{a^2b^2}tr(B^2)+\frac{2k^2c_1^2d_2^2}{a^2b^2c}\right.$$
$$+\frac{c_1^2}{a^2}(tr(A))^2-\frac{2dc_1^2d_1}{a^2b}tr(A)tr(B)+\frac{2c_1^2kd_2}{a^2b}tr(A)$$
$$\left.-\frac{2dc_1^2kd_1d_2}{a^2b^2}tr(B)+\frac{c_1^2k^2d_2^2}{a^2b^2}+\frac{d^2c_1^2d_1^2}{a^2b^2}(tr(B))^2\right]=0.$$

The minimum is obtained when

$$d_2=\frac{\frac{d}{b}d_1tr(B)-tr(A)}{(\frac{k}{b})(\frac{2}{c}+1)}. \tag{43}$$

It has been verified that if $c_1$, $d_1$ and $d_2$ satisfy the minimum of the coefficients $\sigma_1^4$, $\sigma_2^4$ and $\sigma_e^4$, respectively, in (34). It follows that (35) and (36) hold for the given values in (41) and (42), respectively. Concerning (37), omitting $a^2$ and simplifying, the left hand side can be written as

$$c_1^2tr(A)+\frac{d^2c_1^2d_1^2}{b^2}\mathrm{tr}(B)+\frac{k^2c_1^2d_2^2}{b^2c}+\frac{1}{2}c_1^2\left(\mathrm{tr}(A)-\frac{d}{b}d_1\mathrm{tr}(B)+\frac{kd_2}{b}\right)^2. \tag{44}$$

However, since $c_1$ and $d_1$ given by (41) and (42) respectively, are less than 1 it is enough to study when

$$\frac{k^2c_1^2d_2^2}{b^2c}+\frac{1}{2}c_1^2\left(\mathrm{tr}(A)-\frac{d}{b}d_1\mathrm{tr}(B)+\frac{kd_2}{b}\right)^2\leq\frac{k^2}{b^2c} \tag{45}$$

The following is obtained after substituting $d_2$ defined in (43) into the left hand side of (45)

$$\frac{k^2c_1^2}{b^2c}\frac{(\frac{d}{b}d_1\mathrm{tr}(B)-\mathrm{tr}(A))^2}{(\frac{k}{b})^2(\frac{2}{c}+1)^2}+\frac{c_1^2}{2}\left(\mathrm{tr}(A)-\frac{d}{b}d_1tr(B)+\frac{k}{b}\frac{\frac{d}{b}d_1tr(B)-\mathrm{tr}(A)}{(\frac{k}{b})(\frac{2}{c}+1)}\right)^2 \tag{46}$$

which can be simplified to,

$$c_1^2\left(\frac{d}{b}d_1\mathrm{tr}(B)-\mathrm{tr}(A)\right)^2\left[\frac{c}{(2+c)^2}-\frac{2}{(2+c)^2}\right]. \tag{47}$$

Hence, for (37) to hold the following must be satisfied

$$\left(\frac{d}{b}d_1\mathrm{tr}(B)-\mathrm{tr}(A)\right)^2\leq\left(\frac{d}{b}\mathrm{tr}(B)-\mathrm{tr}(A)\right)^2. \tag{48}$$

Therefore we have two cases to consider, either

$$\mathrm{tr}(A) \leq \frac{d}{b} d_1 \mathrm{tr}(B), \tag{49}$$

or

$$\mathrm{tr}(A) > \frac{d}{b} d_1 \mathrm{tr}(B), \tag{50}$$

which have to be treated separately. If (49) holds, then (48) is always satisfied. If instead (50) is true we will return one step and suppose $d_1 = 1$. Then, obviously (36) and (48) will hold. Observe that $d_1 = 1$ means that we should not perturb (26) with respect to $d_1$.

Moreover, (38) is always satisfied since,

$$(c_1 - 1)r = \frac{d}{a}(1 - d_1)(c_1^2 - c_1) \leq 0. \tag{51}$$

Concerning (39), we study the second term in the left hand side,

$$(c_1 - 1)t = (c_1 - 1)\left(\frac{c_1}{a} tr(A) - \frac{dc_1 d_1}{ab} tr(B) + \frac{c_1 k d_2}{ab}\right).$$

Substituting $d_2$, defined in (43), yields

$$(c_1 - 1)\left(\frac{c_1}{a} tr(A) - \frac{dc_1 d_1}{ab} tr(B) + \frac{c_1 k}{ab} \frac{\frac{d}{b} d_1 tr(B) - tr(A)}{(\frac{k}{b})(\frac{2}{c} + 1)}\right),$$

giving

$$\frac{1}{a}(c_1^2 - c_1)\left(tr(A) - \frac{dc_1}{b} d_1 tr(B) + \frac{\frac{d}{b} d_1 tr(B) - tr(A)}{\frac{2}{c} + 1}\right).$$

Thus, for (39), we have from (18) that $\mathrm{tr}(AV_1) = a$ which implies that (39) can be written as

$$\frac{2c_1^2}{a} + \frac{1}{a}(c_1^2 - c_1)\left(tr(A) - \frac{dc_1}{b} d_1 tr(B) + \frac{\frac{d}{b} d_1 tr(B) - tr(A)}{\frac{2}{c} + 1}\right) \leq \frac{2}{a}.$$

Hence, if (49) is true (39) will hold if

$$2c_1^2 + (c_1^2 - c_1)\left(\mathrm{tr}(A) - \frac{d}{b} d_1 \mathrm{tr}(B)\right)\left(\frac{2}{2 + c}\right) \leq 2, \tag{52}$$

and we obtain the additional condition

$$\mathrm{tr}(A) \geq \frac{d}{b} d_1 \mathrm{tr}(B) - \frac{(2 + c)(1 + c_1)}{c_1}. \tag{53}$$

If (50) holds, then it's obvious that (53) will be true. Finally, we check the inequality (40). Since from (18) we have $\mathrm{tr}(AV_2) = d$ and $\mathrm{tr}(BV_2) = b$ we rewrite (40) as

$$\frac{4c_1^2 d}{a^2} + \frac{4d^2 c_1^2 d_1^2}{a^2 b^2} + \frac{2c_1^2 d}{a^2}(1-d_1)\left(\mathrm{tr}(A) - \frac{dd_1}{b}\mathrm{tr}(B) + \frac{k}{b}d_2\right) \leq \frac{4d}{a^2} + \frac{4d^2}{a^2 b}.$$

It is enough to investigate the third term in the left hand side:

$$\frac{c_1^2 d}{a^2}(1-d_1)\left(\mathrm{tr}(A) - \frac{dd_1}{b}\mathrm{tr}(B) + \frac{k}{b}d_2\right).$$

As previously, after substituting $d_2$ and omitting identical terms from both sides, (40) can be written as,

$$2c_1^2 + \frac{2dc_1^2 d_1^2}{b} + c_1^2(1-d_1)\left(\mathrm{tr}(A) - \frac{d}{b}d_1 \mathrm{tr}(B)\right)\left(\frac{2}{2+c}\right) \leq 2 + \frac{2d}{b}. \quad (54)$$

Thus, (40) is satisfied under (49). Moreover, if $d_1 = 1$ as assumed if (50) holds, then (40) is also valid. The above results can be summarized in the following proposition

**Proposition 1.** *Let the variance component estimator corresponding to the first random effect $\widehat{\sigma}_{u1}^2$ in the model defined in (11) be modified as in (27), where $c_1$, $d_1$ and $d_2$ are chosen as in (41), (42) and (43), respectively. Then (35)– (40) are sufficient conditions for $MSE(\widehat{\sigma}_{11}^2) \leq MSE(\widehat{\sigma}_{u1}^2)$.*

Moreover, for the two cases that emerged from (48) we have the following theorem

**Theorem 1.** *Given the model defined in (11), let $MSE(\widehat{\sigma}_{u1}^2)$ be the mean square error of the unmodified estimator given in (25) and let $MSE(\widehat{\sigma}_{11}^2)$ be the mean square error of the modified estimator given in (34).*

*(i) If (49) and (53) hold, $MSE(\widehat{\sigma}_{11}^2) \leq MSE(\widehat{\sigma}_{u1}^2)$.*
*(ii)If (50) and $d_1 = 1$, $MSE(\widehat{\sigma}_{11}^2) \leq MSE(\widehat{\sigma}_{u1}^2)$.*

## 4 Conclusion

The problem of modifying the variance component estimator obtained by using Henderson's method 3, has been the focus of our work.

För a two-way linear mixed model, consisting of three variance components, $\sigma_1^2$, $\sigma_2^2$, and $\sigma_e^2$, we have perturbed the Henderson's estimation equations. The main aim, was to modify the standard unbiased estimator, corresponding to one of the random effects, by multiplying the estimator with some coefficients that are chosen to minimize the leading terms $\sigma_1^2$, $\sigma_2^2$, and $\sigma_e^2$ in the mean square error equation.

Two modified variance component estimators are proposed; each appropriate under certain given conditions. Our proposed estimators are easy to compute and have smaller MSE than the unmodified one. Moreover, the conditions under which each of the proposed estimators are valid, are easy to investigate. For instance, in practical application if the unbiasedness condition is not of major concern, our proposed estimators should be considered.

**Acknowledgement** We wish to thank professor Thomas Mathew for many helpful discussions.

## References

[1] Graybill, F.A., Hultquist, R.A.: Theorems concerning Eisenhart's Model II. Ann. Math. Stat. **32**, 261–269 (1961)
[2] Henderson, C.R.: Estimation of variance and covariance components. Biometrics **9**, 226–252 (1953)
[3] Kelly, J.R., Mathew, T.: Improved nonnegative estimation of variance components in some mixed models with unbalanced data. Technometrics **36**, 171–181 (1994)
[4] LaMotte, L.R.: On non-negative quadratic unbiased estimation of variance components. J. Am. Stat. Assoc. **68**, 728–730 (1973)
[5] Schott, J.R.: Matrix analysis for statistics. Wiley, New York (1997)
[6] Searle, S.R.: Another look at Henderson's methods of estimating variance components. Biometrics **24**, 749–778 (1968)
[7] Searle, S.R.: Linear models. Wiley, New York (1971)
[8] Searle, S.R.: Linear models for unbalanced data. Wiley, New York (1987)
[9] Searle, S.R., Casella, G., McCulloch, C.E.: Variance components. Wiley, New York (1992)

# Implications of Dimensionality on Measurement Reliability

Kimmo Vehkalahti, Simo Puntanen, and Lauri Tarkkonen

**Abstract** We study some topics of the reliability of measurement, especially certain implications of multidimensionality and unidimensionality. We consider these aspects within a measurement framework focusing on one hand on the dimensionality of the measurement model and on the other hand on the dimensionality of the measurement scale. Working through theorems and examples we compare two reliability estimators, namely Cronbach's alpha and Tarkkonen's rho. It seems that there is not much use for Cronbach's alpha. It is based on unidimensional models and scales, while the models and scales used in practice are multidimensional. Tarkkonen's rho seems to work well in multidimensional studies, giving support to the real purpose of reliability estimation which seems to have been lost for a quite long time.

## 1 Introduction

Measurement brings uncertainty in all statistical research. Assessing its quality requires two concepts: *validity* and *reliability*. The problems of validity can seldom be solved statistically, whereas the reliability is clearly a statistical question, being closely related to the variance of the measurement. Therefore a measurement model is needed to assess the reliability, since the variance of the measurement error must be estimated. The models to be used should be multidimensional with flexible assumptions in order to be applicable in practical applications. In social sciences and behavioral sciences, where the measurements are usually far from stable, the applied research has traditionally concentrated on unidimensional models. Recently, that tradition has been questioned, and the need of multidimensionality recognized (Lucke [11], ten Berge and Sočan [18]).

In this paper we study some topics of reliability, especially certain implications of the multidimensionality and unidimensionality. Section 2 presents a framework for

Simo Puntanen
Department of Mathematics and Statistics, University of Tampere, FI-33014 Tampere Finland
Simo.Puntanen@uta.fi

the study, Sect. 3 introduces the concept of reliability, Sect. 4 reveals implications through theorems and examples, and Sect. 5 closes with a short discussion.

## 2 Framework for multidimensional measurement

In order to assess the validity and reliability of multivariate measurements one must have means for setting up the relations between the latent constructs and the measured variables. Several approaches have been suggested in the literature, such as the factor analysis and the structural equations model (Bollen [3]).

Tarkkonen and Vehkalahti [17] have introduced a measurement framework which consists of four multidimensional parts: 1) the measurement model, 2) the measurement scale, 3) the second-order scale, and 4) the external validity criteria (Tarkkonen and Vehkalahti [17], Vehkalahti et al [20], Vehkalahti [19], Tarkkonen [16]). Here, we focus on the parts 1) and 2) of that framework. The questions of validity are often connected to the substantial theory. However, within the framework we will briefly touch the issues of structural validity and predictive validity.

### 2.1 Measurement model

In Fig. 1, the frame on the left illustrates the measurement model. Three types of concepts exist in that frame: true scores, measurement errors, and observed variables

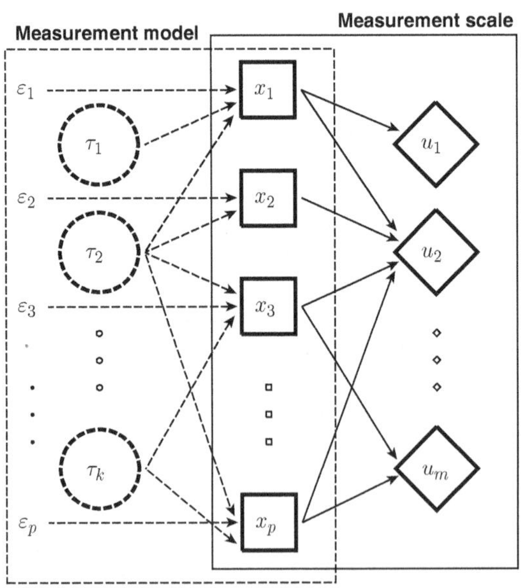


**Fig. 1** Main frames of measurement framework: measurement model and measurement scale

(or items). Except for the items $x_1,\ldots,x_p$, we use dash lines, because the concepts are hypothetical. Here, we are asking questions, such as: How many dimensions are there? How to measure them as well as possible?

We define the model as follows. Let $\boldsymbol{x} = (x_1,\ldots,x_p)'$ measure $k$ ($< p$) unobservable true scores $\boldsymbol{\tau} = (\tau_1,\ldots,\tau_k)'$ with unobservable measurement errors $\boldsymbol{\varepsilon} = (\varepsilon_1,\ldots,\varepsilon_p)'$. It is essential that $k$ is less than $p$, since we are willing to solve the problem of measurement in a *reduced true score space*. The measurement model is

$$\boldsymbol{x} = \boldsymbol{\mu} + \boldsymbol{B\tau} + \boldsymbol{\varepsilon},$$

where $\boldsymbol{\mu}$ is the expectation of $\boldsymbol{x}$ and $\boldsymbol{B} \in \mathbb{R}^{p\times k}$ specifies the relationship between $\boldsymbol{x}$ and $\boldsymbol{\tau}$, illustrated in Fig. 1 as arrows from true scores to items. We assume that $\mathrm{E}(\boldsymbol{\varepsilon}) = \boldsymbol{0}$ and $\mathrm{cov}(\boldsymbol{\tau},\boldsymbol{\varepsilon}) = \boldsymbol{0}$. Denoting $\mathrm{cov}(\boldsymbol{\tau}) = \boldsymbol{\Phi}$ and $\mathrm{cov}(\boldsymbol{\varepsilon}) = \boldsymbol{\Psi}$ we have

$$\mathrm{cov}(\boldsymbol{x}) = \boldsymbol{\Sigma} = \mathrm{cov}(\boldsymbol{B\tau}) + \mathrm{cov}(\boldsymbol{\varepsilon}) = \boldsymbol{B\Phi B}' + \boldsymbol{\Psi}, \tag{1}$$

where we assume that $\boldsymbol{\Sigma}$ and $\boldsymbol{\Phi}$ are positive definite and $\boldsymbol{B}$ has full column rank. We may use the triplet $\mathscr{M} = \{\boldsymbol{x}, \boldsymbol{B\tau}, \boldsymbol{B\Phi B}' + \boldsymbol{\Psi}\}$ to denote the measurement model shortly. The model $\mathscr{M}$ is rather general, but it is not identifiable. The parameters of the model are the $pk + k(k+1)/2 + p(p+1)/2$ (unique) elements of the matrices $\boldsymbol{B}$, $\boldsymbol{\Phi}$, and $\boldsymbol{\Psi}$, respectively. In general, there are too many parameters, since $\boldsymbol{\Sigma}$ has only $p(p+1)/2$ (unique) elements. The identifiability of the model is obtained by imposing suitable assumptions on the true scores and the measurement errors.

At least in a basic exploratory setting it is typical to assume that $\mathrm{cov}(\boldsymbol{\varepsilon}) = \boldsymbol{\Psi}_d = \mathrm{diag}(\psi_1^2,\ldots,\psi_p^2)$. (We note that throughout this paper notation $\boldsymbol{K}_d$ indicates either that $\boldsymbol{K}$ is a diagonal matrix or that $\boldsymbol{K}_d$ is a diagonal matrix comprising the diagonal elements of $\boldsymbol{K}$.) Then, the more restricted model $\mathscr{M}_d = \{\boldsymbol{x}, \boldsymbol{B\tau}, \boldsymbol{B\Phi B}' + \boldsymbol{\Psi}_d\}$ conforms with the factor analysis model, where the common factors are *directly associated with the true scores* and the rest is interpreted as measurement errors. This is a more straight-forward approach compared to the usual considerations, see, e.g., Alwin [1, ch. 4]. Assuming multinormality the parameters can be estimated using the maximum likelihood method.

### 2.1.1 Structural validity

Structural validity is a property of the measurement model. It is important, because the model forms the core of the framework and hence affects the quality of all scales created. Similarly with other questions of validity, knowledge of the theory and practice of the application is necessary. However, some statistical considerations may be useful as well.

The lack of the structural validity can be revealed by testing hypotheses on the dimension of $\boldsymbol{\tau}$ and on the (estimated) effects of $\boldsymbol{\tau}$ on $\boldsymbol{x}$ (the matrix $\boldsymbol{B}$). The latter may be called *true score images* (or *factor images*), since they reflect the unobservable true scores (or factors). Of course, the factors can not have reliability, but we can think of the factor images as a special measurement scale (see Sect. 2.2). Then, the reliabilities of the factor images may be used to assess the structural validity of the model. An appropriate factor rotation is essential to fine-tune the factor images. For a skilled researcher, *graphical rotation*, implemented in SURVO MM software (Mustonen [13]), is an ideal choice.

In an exploratory setting, we may use the residuals of the model to tune the dimensionality, i.e., the number of the factors. As soon as we decide it, the reliabilities of the observed items will be identified. This may sound like a confirmatory factor analysis. To some extent, most studies are both exploratory and confirmatory (Jöreskog [8]). Indeed, our approach could be called *semi-confirmatory*: perhaps exploratory but based on a measurement model, creating a sound base for working with the scales.

## 2.2 Measurement scale

In further analyses, the items are best used by creating multivariate measurement scales $\boldsymbol{u} = (u_1, \ldots, u_m)'$ as linear combinations $\boldsymbol{u} = \boldsymbol{A}'\boldsymbol{x}$, where $\boldsymbol{A} \in \mathbb{R}^{p \times m}$ is a weight matrix. Tarkkonen and Vehkalahti [17] give various criteria for choosing the weights, but generally it is assumed that $\boldsymbol{A}$ has full column rank and $\boldsymbol{B}'\boldsymbol{a}_i \neq \boldsymbol{0}, i = 1, \ldots, m$, where $\boldsymbol{a}_i$ is the $i$th column vector of $\boldsymbol{A}$. It is possible to create any number of scales regardless of the dimension of the model, so $m$ does not have to be equal to $k$. It can be as well equal to one, for example.

In Fig. 1, the measurement scale is illustrated by the frame on the right. As we see, the items belong to both frames: they are the only observable part of the model, and they are used to construct the scales, such as factor scores, psychological test scales, plain sums, indexes or any other linear combinations of the items. The weights may be predetermined values according to a theory. A special case mentioned in Sect. 2.1.1 is the true score (or factor) images, defined as $\boldsymbol{A} = \boldsymbol{B}$, that is, the weights are the coefficients of the measurement model. This scale may be useful in assessing the structural validity of the model.

Using (1) we obtain the covariances of the scale in the form

$$\operatorname{cov}(\boldsymbol{u}) = \boldsymbol{A}'\boldsymbol{\Sigma}\boldsymbol{A} = \boldsymbol{A}'\boldsymbol{B}\boldsymbol{\Phi}\boldsymbol{B}'\boldsymbol{A} + \boldsymbol{A}'\boldsymbol{\Psi}\boldsymbol{A}, \tag{2}$$

separately for the effects of the true scores and the measurement errors. We are most interested in the variances, which are needed to estimate the reliability. We will get back to this in Sect. 3.1.

### 2.2.1 Predictive validity

Predictive validity is assessed by the correlations of the scale and an external criterion $\boldsymbol{y}$. It is typical to further condense the data by creating second-order scales $\boldsymbol{z} = \boldsymbol{W}'\boldsymbol{u} = \boldsymbol{W}'\boldsymbol{A}'\boldsymbol{x}$, where $\boldsymbol{W}$ is a weight matrix. These scales are often results of regression analysis, discriminant analysis, or other multivariate statistical methods. In a quite general case, the predictive validity would then be assessed by the canonical correlations between $\boldsymbol{z}$ and $\boldsymbol{y}$.

Consider the linear regression model $y = \beta_0 + \boldsymbol{\beta}'\boldsymbol{u} + \theta$, where $y$ is the response variable, $\beta_0$ is the intercept, $\boldsymbol{\beta} = (\beta_1, \ldots, \beta_m)'$ is the vector of the regression coefficients, $\boldsymbol{u}$ is the vector of the predictors, a scale such as the factor scores, and $\theta$ is a model error. Now, the criterion $y$ is a scalar, the second-order scale is given by the prediction scale $z = \hat{\boldsymbol{\beta}}'\boldsymbol{u}$, and the predictive validity is the multiple correlation.

## 3 Reliability

Reliability is a property of the measurement scale. It is a well-established concept defined in the test theory of psychometrics (Lord and Novick [10]), where the measurements are assumed to consist of unobserved true scores and measurement errors as $x = \tau + \varepsilon$, where $\mathrm{E}(\varepsilon) = 0$ and $\mathrm{cov}(\tau, \varepsilon) = 0$. It follows that $\mathrm{var}(x) = \sigma_x^2 = \sigma_\tau^2 + \sigma_\varepsilon^2$. The reliability of $x$ is defined as the ratio $\sigma_\tau^2/\sigma_x^2$, but since

$$\mathrm{cov}(x, \tau) = \mathrm{cov}(\tau + \varepsilon, \tau) = \mathrm{cov}(\tau, \tau) = \mathrm{var}(\tau) = \sigma_\tau^2,$$

it can also be seen as the squared correlation between $x$ and $\tau$:

$$\rho_{x\tau}^2 = \frac{[\mathrm{cov}(x, \tau)]^2}{\mathrm{var}(x)\,\mathrm{var}(\tau)} = \frac{(\sigma_\tau^2)^2}{\sigma_x^2 \sigma_\tau^2} = \frac{\sigma_\tau^2}{\sigma_x^2}.$$

The notation $\rho_{x\tau}^2$ is used in the literature, because the true score is often taken as a scalar in psychometrics. The point in the definition is the ratio of the variances, but either $\sigma_\tau^2$ or $\sigma_\varepsilon^2$ must be estimated to obtain a reliability estimate. In general, this depends on the assumptions both on the model and the scale.

### *3.1 Multidimensional case*

The measurement framework of Sect. 2 will now be completed with an estimator of reliability, which we have suggested to be called *Tarkkonen's rho*, since the idea was originally proposed by Tarkkonen [16]. According to the definition of reliability, the estimator is obtained as a ratio of the variances, that is, the diagonal elements of the matrices in (2). In the general form Tarkkonen's rho is a reliability matrix (Vehkalahti et al [20], Tarkkonen and Vehkalahti [17], Vehkalahti [19])

$$\boldsymbol{\rho_u} = \mathrm{diag}\left(\frac{\boldsymbol{a}_1'\boldsymbol{B\Phi B}'\boldsymbol{a}_1}{\boldsymbol{a}_1'\boldsymbol{\Sigma a}_1}, \ldots, \frac{\boldsymbol{a}_m'\boldsymbol{B\Phi B}'\boldsymbol{a}_m}{\boldsymbol{a}_m'\boldsymbol{\Sigma a}_m}\right) = (\boldsymbol{A}'\boldsymbol{B\Phi B}'\boldsymbol{A})_d \times [(\boldsymbol{A}'\boldsymbol{\Sigma A})_d]^{-1}. \quad (3)$$

For making more detailed assumptions on the measurement errors, we can write (3) with $\boldsymbol{\Psi}$ explicitly present in the form

$$\begin{aligned}\boldsymbol{\rho_u} &= \text{diag}\left(\left[1+\frac{\boldsymbol{a}_1'\boldsymbol{\Psi}\boldsymbol{a}_1}{\boldsymbol{a}_1'\boldsymbol{B\Phi B}'\boldsymbol{a}_1}\right]^{-1},\ldots,\left[1+\frac{\boldsymbol{a}_m'\boldsymbol{\Psi}\boldsymbol{a}_m}{\boldsymbol{a}_m'\boldsymbol{B\Phi B}'\boldsymbol{a}_m}\right]^{-1}\right)\\ &= \{\boldsymbol{I}_m + (\boldsymbol{A}'\boldsymbol{\Psi}\boldsymbol{A})_d \times [(\boldsymbol{A}'\boldsymbol{B\Phi B}'\boldsymbol{A})_d]^{-1}\}^{-1},\end{aligned} \tag{4}$$

where $\boldsymbol{I}_m$ is an identity matrix of order $m$. The reliabilities of various measurement scales are obtained by (3) or (4) by substituting the matrix $\boldsymbol{A}$ with the actual weight matrix of the scale. For example, the factor scores have $\boldsymbol{A} = \boldsymbol{\Sigma}^{-1}\boldsymbol{B}$.

## 3.2 *Unidimensional case*

Although alternative reliability estimators based on factor analysis have been suggested (Werts et al [22], Heise and Bohrnstedt [6], McDonald [12]), the reliability studies have been predominantly unidimensional. This is mostly due to historical reasons. For over fifty years, the most common (if not only) estimator that has been applied in practice is called *Cronbach's alpha* (Cronbach [5]). In the following we briefly summarize its historical roots since it helps to understand the implications of the original assumptions. For more comprehensive reviews, see, e.g., Blinkhorn [2] or Weiss and Davison [21].

### 3.2.1 Kuder–Richardson formula 20

Cronbach's alpha is essentially based on its predecessor, *Kuder–Richardson formula 20*, or KR-20, where "20" refers to the number of the formula in the original paper by Kuder and Richardson [9]. The formula can be derived as follows.

Let $\boldsymbol{x}$ and $\boldsymbol{y}$ be $p$-dimensional random vectors with a $(2p \times 2p)$ covariance matrix

$$\text{cov}\begin{pmatrix}\boldsymbol{x}\\ \boldsymbol{y}\end{pmatrix} := \begin{pmatrix}\boldsymbol{\Sigma}_{xx} & \boldsymbol{\Sigma}_{xy}\\ \boldsymbol{\Sigma}_{yx} & \boldsymbol{\Sigma}_{yy}\end{pmatrix},$$

where $\boldsymbol{\Sigma}_{xx} = \boldsymbol{\Sigma}_{yy}$ and $\boldsymbol{\Sigma}_{xy} = \boldsymbol{\Sigma}_{yx}$. Then, denoting symbolically,

$$\boldsymbol{\Sigma}_{xx} = \begin{pmatrix}\sigma_1^2 & \cdots & \rho_{ij}\sigma_i\sigma_j\\ \vdots & \ddots & \vdots\\ \rho_{ij}\sigma_i\sigma_j & \cdots & \sigma_p^2\end{pmatrix} \quad \text{and} \quad \boldsymbol{\Sigma}_{xy} = \begin{pmatrix}\rho_{11}\sigma_1^2 & \cdots & \rho_{ij}\sigma_i\sigma_j\\ \vdots & \ddots & \vdots\\ \rho_{ij}\sigma_i\sigma_j & \cdots & \rho_{pp}\sigma_p^2\end{pmatrix},$$

where $\sigma_i^2 = \text{var}(x_i)$, $\rho_{ij} = \text{cor}(x_i, y_j)$, and $\rho_{ii}$ is the reliability of $x_i$. Note that $\boldsymbol{\Sigma}_{xy} = \boldsymbol{\Sigma}_{xx} - (\boldsymbol{I}_p - \boldsymbol{\rho}_d)\boldsymbol{\Sigma}_d$, where $\boldsymbol{\rho}_d = \text{diag}(\rho_{11}, \ldots, \rho_{pp})$, and $\boldsymbol{\Sigma}_d = \text{diag}(\sigma_1^2, \ldots, \sigma_p^2)$. It is assumed that $\rho_{ij} = \rho_{xx}$ and $\sigma_i = \sigma_j = \sigma_x$ for all $i, j$, which implies that also

$\rho_{ii} = \rho_{xx}$ for all $i$. This means that $x_i$ and $y_j$ are *parallel measurements*, that is, $\boldsymbol{x}$ and $\boldsymbol{y}$ have an intraclass correlation structure with a covariance matrix

$$\operatorname{cov}\begin{pmatrix}\boldsymbol{x}\\ \boldsymbol{y}\end{pmatrix} = \sigma_x^2 \begin{pmatrix}\boldsymbol{\Sigma} & \rho_{xx}\mathbf{1}\mathbf{1}'\\ \rho_{xx}\mathbf{1}\mathbf{1}' & \boldsymbol{\Sigma}\end{pmatrix} := \begin{pmatrix}\boldsymbol{\Sigma}_{xx} & \boldsymbol{\Sigma}_{xy}\\ \boldsymbol{\Sigma}_{yx} & \boldsymbol{\Sigma}_{yy}\end{pmatrix},$$

where $\boldsymbol{\Sigma} = (1-\rho_{xx})\boldsymbol{I}_p + \rho_{xx}\mathbf{1}\mathbf{1}'$ and $\mathbf{1}$ is the vector of ones.

The *parallel model* has $x_i = \tau_i + \varepsilon_i$ and $y_j = \tau_j + \varepsilon_j$ with $\operatorname{cov}(\varepsilon_i, \varepsilon_j) = 0$ and $\operatorname{cov}(\tau_i, \varepsilon_i) = \operatorname{cov}(\tau_j, \varepsilon_j) = 0$. It is assumed that $\operatorname{var}(\varepsilon_i) = \operatorname{var}(\varepsilon_j)$ and $\operatorname{cov}(\tau_i, \tau_j) = \operatorname{cov}(\tau_i, \tau_i) = \operatorname{var}(\tau_i)$. Denoting $\operatorname{var}(x) = \operatorname{var}(x_i)$ and $\operatorname{var}(\tau) = \operatorname{var}(\tau_i)$ we obtain

$$\rho_{xx} = \frac{\operatorname{cov}(\tau_i+\varepsilon_i, \tau_j+\varepsilon_j)}{\operatorname{var}(x)} = \frac{\operatorname{var}(\tau)}{\operatorname{var}(x)} = \frac{\sigma_\tau^2}{\sigma_x^2} = \rho_{x\tau}^2,$$

that is, the item reliabilities, the common correlations of the intraclass correlation structure, are in accordance with the definition of the reliability. We note that the strict assumptions of equal variances of the measurement errors (and thus equal variances of the items) are not required in the definition of the reliability, they are merely properties of the parallel model.

Consider two measurement scales, $u = \mathbf{1}'\boldsymbol{x}$ and $v = \mathbf{1}'\boldsymbol{y}$. The variance of $u$ is

$$\sigma_u^2 = \operatorname{var}(\mathbf{1}'\boldsymbol{x}) = \mathbf{1}'\boldsymbol{\Sigma}\mathbf{1} = p(p-1)\sigma_x^2\rho_{xx} + p\sigma_x^2, \tag{5}$$

and the correlation between $u$ and $v$ can be written, using (5), as

$$\rho_{uv} = \operatorname{cor}(\mathbf{1}'\boldsymbol{x}, \mathbf{1}'\boldsymbol{y}) = \frac{\mathbf{1}'\boldsymbol{\Sigma}_{xy}\mathbf{1}}{\operatorname{var}(\mathbf{1}'\boldsymbol{x})} = \frac{\sigma_x^2 p^2 \rho_{xx}}{\sigma_x^2 p[1+(p-1)\rho_{xx}]} = \frac{p\rho_{xx}}{1+(p-1)\rho_{xx}}, \tag{6}$$

which is known as the *Spearman–Brown formula*. Solving $\rho_{xx}$ from (5) and substituting it in (6) leads to KR-20 in the form

$$\rho_{uv} = \frac{p}{p-1}\left(1 - \frac{p\sigma_x^2}{\sigma_u^2}\right). \tag{7}$$

With the earlier methods of reliability estimation the item reliabilities $\rho_{xx}$ had been a nuisance. In the derivation of KR-20 they were hidden by algebraic manipulations, as the aim was to provide quick methods for practical needs. However, it was clear that the figures obtained would be underestimates if the strict assumptions were not met (Kuder and Richardson [9, p. 159]).

### 3.2.2 Cronbach's alpha

The principal advantages claimed for KR-20 were ease of calculation and conservatism. The method was also criticized, because the magnitude of the underestimate was unknown, and even negative values could be obtained. One of the critics was *Lee J. Cronbach*, who stated that "while conservatism has advantages in research, in this case it leads to difficulties" (Cronbach [4, p. 487]) and that "the Kuder–Richardson formula is not desirable as an all-purpose substitute for the usual techniques" (Cronbach [4, p. 488]). In 1951, he presented "the more general formula" (Cronbach [5, p. 299])

$$\alpha = \frac{p}{p-1}\left(1 - \frac{\sum_{i=1}^{p}\sigma_{x_i}^2}{\sigma_u^2}\right), \tag{8}$$

and adviced that we should take it "as given, and make no assumptions regarding it" (Cronbach [5, p. 299]). It is easy to see that Cronbach's alpha (8) resembles KR-20 (7), as only the term $p\sigma_x^2$ is replaced by $\sum_{i=1}^{p}\sigma_{x_i}^2$. This loosens the strict assumption of the equal variances made in the derivation of KR-20, but the problem inherited from the KR-20 remains: as there is no trace of the true scores or measurement errors in the formula, any assumptions of them are easy to forget.

## 4 Implications of dimensionality

In the following we focus on two particular estimators of reliability, namely Tarkkonen's rho and Cronbach's alpha. The former is interesting because of its generality, while the latter is interesting because of its popularity. We show that under certain models and conditions Cronbach's alpha is a special case of Tarkkonen's rho, and that the multidimensionality, which is the starting point for Tarkkonen's rho, is beyond the scope of Cronbach's alpha.

We will investigate the properties of the estimators through theoretical comparisons and numerical examples. The examples are based on using the matrix interpreter of SURVO MM (Mustonen [13]).

### *4.1 Dimensionality of the model*

We will support the theoretical comparisons with numerical examples based on four simple measurement models (see Table 1). Without losing generality, we assume that the items $x_1, x_2, \ldots, x_7$ are standardized, that is, $\mathrm{E}(\boldsymbol{x}) = \mathbf{0}$ and $\mathrm{cov}(\boldsymbol{x}) = \mathrm{cor}(\boldsymbol{x}) = \boldsymbol{\Sigma}$ in each case. Then it is obvious that $\boldsymbol{\Sigma}$ is known as soon as $\boldsymbol{B}$ is known. Hence we can specify a measurement model by choosing the elements of the matrix $\boldsymbol{B}$ so that $\boldsymbol{B}$ has full column rank and $\boldsymbol{\Sigma}$ is positive definite.

We will refer to the models by indexing $\boldsymbol{B}$, $\boldsymbol{\Sigma}$, and other related matrices by numbers from 1 to 4. All these models can be considered as population models, as we do not consider any sampling variation. However, it would be possible to study the sampling properties of the reliability estimators by conducting Monte Carlo simulations based on these specifications, see Vehkalahti et al [20].

**Table 1** The specifications of the measurement models used in the examples

| | **B1** | **B2** | **B3** | | **B4** | |
|---|---|---|---|---|---|---|
| | $\tau$ | $\tau$ | $\tau_1$ | $\tau_2$ | $\tau_1$ | $\tau_2$ |
| $x_1$ | .9 | .9 | .9 | 0 | .9 | 0 |
| $x_2$ | .9 | .9 | .9 | 0 | .9 | 0 |
| $x_3$ | .9 | .8 | .9 | 0 | .8 | -.5 |
| $x_4$ | .9 | .8 | .9 | 0 | .8 | -.5 |
| $x_5$ | .9 | .5 | 0 | .9 | -.5 | .8 |
| $x_6$ | .9 | .5 | 0 | .9 | 0 | .9 |
| $x_7$ | .9 | .5 | 0 | .9 | 0 | .9 |

Models ***B*1** and ***B*2** are unidimensional. Using the notation established in Sect. 3 we can summarize three variants of the unidimensional model that has dominated the test theory of psychometrics:

$$\mathscr{M}_1 = \{\boldsymbol{x},\, \mathbf{1}\tau,\, \sigma_\tau^2\mathbf{1}\mathbf{1}' + \sigma_\varepsilon^2\boldsymbol{I}_p\},$$
$$\mathscr{M}_2 = \{\boldsymbol{x},\, \mathbf{1}\tau,\, \sigma_\tau^2\mathbf{1}\mathbf{1}' + \boldsymbol{\Psi}_d\},\ \text{and}$$
$$\mathscr{M}_3 = \{\boldsymbol{x}, \boldsymbol{b}\tau,\, \sigma_\tau^2\boldsymbol{b}\boldsymbol{b}' + \boldsymbol{\Psi}_d\},\ \text{where } \boldsymbol{b} \in \mathbb{R}^p.$$

$\mathscr{M}_1$ is the parallel model, $\mathscr{M}_2$ is the *$\tau$-equivalent model* (Novick and Lewis [14]), and $\mathscr{M}_3$ is the *congeneric model* (Jöreskog [7]). Obviously, all these are special cases of the general model $\mathscr{M} = \{\boldsymbol{x}, \boldsymbol{B}\boldsymbol{\tau}, \boldsymbol{B}\boldsymbol{\Phi}\boldsymbol{B}' + \boldsymbol{\Psi}\}$.

Clearly, ***B*1** represents $\mathscr{M}_1$ or $\mathscr{M}_2$, and ***B*2** represents $\mathscr{M}_3$. The two-dimensional models ***B*3** and ***B*4** represent the simplest examples of multidimensional models. For further simplicity in these models we assume that the true scores are uncorrelated.

## *4.2 Dimensionality of the scale*

The general, multidimensional measurement scale given in Sect. 2.2 is important as such, because it connects the measurement framework with multivariate statistical models, such as regression or discriminant analysis. The special cases of the scale are the weighted sum $u = \boldsymbol{a}'\boldsymbol{x}$, where $\boldsymbol{a} \in \mathbb{R}^p$ and the unweighted sum $u = \mathbf{1}'\boldsymbol{x}$. The latter is a traditional scale in psychometrics. We will consider these together with different measurement models. As the multidimensional scale is definitely beyond the scope of Cronbach's alpha, the theoretical comparisons between Tarkkonen's rho and Cronbach's alpha are conducted only with the unidimensional scales. However, the examples will also show how the dimensionality of the scale works with Tarkkonen's rho.

We begin from the unweighted sum, and then proceed to the weighted sum. We find it instructive to go through the unweighted sum first, although the results will follow immediately as special cases of the weighted sum. In each case, we present theorems and examples, revealing implications of the dimensionality behind Tarkkonen's rho and Cronbach's alpha, which we here denote respectively by

$$\rho_{uu}(\mathbf{1}) = \frac{\mathbf{1}'\boldsymbol{B\Phi B}'\mathbf{1}}{\mathbf{1}'\boldsymbol{\Sigma}\mathbf{1}} \quad \text{and} \quad \alpha(\mathbf{1}) = \frac{p}{p-1}\left(1 - \frac{\operatorname{tr}(\boldsymbol{\Sigma})}{\mathbf{1}'\boldsymbol{\Sigma}\mathbf{1}}\right),$$

where $\operatorname{tr}(\cdot)$ is the trace. In the case of the weighted sum we will denote them by

$$\rho_{uu}(\boldsymbol{a}) = \frac{\boldsymbol{a}'\boldsymbol{B\Phi B}'\boldsymbol{a}}{\boldsymbol{a}'\boldsymbol{\Sigma}\boldsymbol{a}} \quad \text{and} \quad \alpha(\boldsymbol{a}) = \frac{p}{p-1}\left(1 - \frac{\boldsymbol{a}'\boldsymbol{\Sigma}_d\boldsymbol{a}}{\boldsymbol{a}'\boldsymbol{\Sigma}\boldsymbol{a}}\right).$$

We note that $\rho_{uu}(\mathbf{1})$ and $\rho_{uu}(\boldsymbol{a})$ follow as special cases from (3) for unidimensional scales. We also note that although Cronbach's alpha has been generalized for the weighted scale, it is obvious that replacing the unit weights by arbitrary weights violates the original assumptions and may lead to doubtful results.

### 4.3 Case of the unweighted sum

When the scale is the unweighted sum $u = \mathbf{1}'\boldsymbol{x}$ and the measurement model is $\mathscr{M}_2$, Tarkkonen's rho and Cronbach's alpha are algebraically identical. Under these circumstances we have $\boldsymbol{\Sigma} = \sigma_\tau^2\mathbf{1}\mathbf{1}' + \boldsymbol{\Psi}_d$, and therefore

$$\sigma_u^2 = \mathbf{1}'\boldsymbol{\Sigma}\mathbf{1} = \sigma_\tau^2\mathbf{1}'\mathbf{1}\mathbf{1}'\mathbf{1} + \mathbf{1}'\boldsymbol{\Psi}_d\mathbf{1} = p^2\sigma_\tau^2 + \operatorname{tr}(\boldsymbol{\Psi}_d).$$

In this simple case it is easy to show that

$$\rho_{uu}(\mathbf{1}) = \frac{\mathbf{1}'\boldsymbol{B\Phi B}'\mathbf{1}}{\mathbf{1}'\boldsymbol{\Sigma}\mathbf{1}} = \frac{p^2\sigma_\tau^2}{\mathbf{1}'\boldsymbol{\Sigma}\mathbf{1}} = \frac{p}{p-1}\left(\frac{p^2\sigma_\tau^2 - p\sigma_\tau^2}{\mathbf{1}'\boldsymbol{\Sigma}\mathbf{1}}\right) = \frac{p}{p-1}\left(1 - \frac{\operatorname{tr}(\boldsymbol{\Sigma})}{\mathbf{1}'\boldsymbol{\Sigma}\mathbf{1}}\right) = \alpha(\mathbf{1}).$$

To investigate the conditions for the equality, we begin with a theorem.

**Theorem 1.** *Let $\boldsymbol{V}$ be a symmetric nonnegative definite $p \times p$ matrix. Then*

$$\mathbf{1}'\boldsymbol{V}\mathbf{1} \geq \frac{p}{p-1}[\mathbf{1}'\boldsymbol{V}\mathbf{1} - \operatorname{tr}(\boldsymbol{V})], \tag{9}$$

*i.e. (assuming $\mathbf{1}'\boldsymbol{V}\mathbf{1} \neq 0$),*

$$1 \geq \frac{p}{p-1}\left(1 - \frac{\operatorname{tr}(\boldsymbol{V})}{\mathbf{1}'\boldsymbol{V}\mathbf{1}}\right).$$

*The equality in* (9) *is obtained if and only if $\boldsymbol{V} = \delta^2\mathbf{1}\mathbf{1}'$ for some $\delta \in \mathbb{R}$.*

*Proof.* Let us rewrite (9) as $(p-1)\mathbf{1}'\boldsymbol{V}\mathbf{1} \geq p\mathbf{1}'\boldsymbol{V}\mathbf{1} - p\operatorname{tr}(\boldsymbol{V})$, that is,

$$\frac{\mathbf{1}'\boldsymbol{V}\mathbf{1}}{\mathbf{1}'\mathbf{1}} \leq \operatorname{tr}(\boldsymbol{V}) = \lambda_1 + \lambda_2 + \cdots + \lambda_p, \tag{10}$$

where $\lambda_1 \geq \lambda_2 \geq \cdots \geq \lambda_p \geq 0$ are the eigenvalues of $\boldsymbol{V}$; we will denote $\lambda_i = \mathrm{ch}_i(\boldsymbol{V})$. Since

$$\max_{z \neq 0} \frac{z'\boldsymbol{V}z}{z'z} = \lambda_1 = \mathrm{ch}_1(\boldsymbol{V}),$$

the inequality (10) indeed holds. The equality in (10) means that

$$\lambda_1 \geq \frac{\mathbf{1}'\boldsymbol{V}\mathbf{1}}{\mathbf{1}'\mathbf{1}} = \lambda_1 + \lambda_2 + \cdots + \lambda_p,$$

which holds if and only if $\lambda_2 = \cdots = \lambda_p = 0$ and vector $\mathbf{1}$ is the eigenvector of $\boldsymbol{V}$ with respect to $\lambda_1$, i.e., $\boldsymbol{V}$ is of the form $\boldsymbol{V} = \lambda_1 \mathbf{1}\mathbf{1}'$. □

Puntanen and Styan [15, pp. 137–138] have proved Theorem 1 using orthogonal projectors. Another proof appears in Vehkalahti [19, Lemma 4.1].

**Corollary 1.** *Theorem 1 implies immediately that*

$$1 \geq \alpha(\mathbf{1}) = \frac{p}{p-1}\left(1 - \frac{\mathrm{tr}(\boldsymbol{\Sigma})}{\mathbf{1}'\boldsymbol{\Sigma}\mathbf{1}}\right) \tag{11}$$

*for any $p \times p$ nonnegative definite matrix $\boldsymbol{\Sigma}$ (for which $\mathbf{1}'\boldsymbol{\Sigma}\mathbf{1} \neq 0$), and that the equality is obtained in* (11) *if and only if $\boldsymbol{\Sigma} = \delta^2\mathbf{1}\mathbf{1}'$ for some $\delta \in \mathbb{R}$.*

In the following we use Theorem 1 to prove that the equality of $\rho_{uu}(\mathbf{1})$ and $\alpha(\mathbf{1})$ requires the model to be unidimensional, either $\mathscr{M}_2$ or $\mathscr{M}_1$. In other cases $\rho_{uu}(\mathbf{1})$ is shown to be greater than $\alpha(\mathbf{1})$.

**Theorem 2.** *Consider the measurement model $\mathscr{M}_d = \{\boldsymbol{x}, \boldsymbol{B\tau}, \boldsymbol{B\Phi B}' + \boldsymbol{\Psi}_d\}$. Then*

$$\rho_{uu}(\mathbf{1}) \geq \alpha(\mathbf{1}),$$

*where the equality is obtained if and only if $\boldsymbol{B\Phi B}' = \delta^2\mathbf{1}\mathbf{1}'$ for some $\delta \in \mathbb{R}$, i.e., $\boldsymbol{\Sigma} = \delta^2\mathbf{1}\mathbf{1}' + \boldsymbol{\Psi}_d$, where $\boldsymbol{\Psi}_d = \mathrm{diag}(\psi_1^2, \ldots, \psi_p^2)$.*

*Proof.* To prove that $\rho_{uu}(\mathbf{1}) \geq \alpha(\mathbf{1})$, we have to show that

$$\frac{\mathbf{1}'\boldsymbol{B\Phi B}'\mathbf{1}}{\mathbf{1}'\boldsymbol{\Sigma}\mathbf{1}} \geq \frac{p}{p-1}\left(\frac{\mathbf{1}'\boldsymbol{\Sigma}\mathbf{1} - \mathrm{tr}(\boldsymbol{\Sigma})}{\mathbf{1}'\boldsymbol{\Sigma}\mathbf{1}}\right). \tag{12}$$

Since $\boldsymbol{\Psi}_d$ is a diagonal matrix, we have

$$\mathbf{1}'\boldsymbol{\Sigma}\mathbf{1} - \mathrm{tr}(\boldsymbol{\Sigma}) = \mathbf{1}'(\boldsymbol{B\Phi B}' + \boldsymbol{\Psi}_d)\mathbf{1} - \mathrm{tr}(\boldsymbol{B\Phi B}' + \boldsymbol{\Psi}_d) = \mathbf{1}'\boldsymbol{B\Phi B}'\mathbf{1} - \mathrm{tr}(\boldsymbol{B\Phi B}'),$$

and hence (12) is equivalent to

$$\mathbf{1}'\boldsymbol{B\Phi B}'\mathbf{1} \geq \frac{p}{p-1}[\mathbf{1}'\boldsymbol{B\Phi B}'\mathbf{1} - \mathrm{tr}(\boldsymbol{B\Phi B}')]. \tag{13}$$

In view of Theorem 1, (13) holds for every $\boldsymbol{B}$ (and every nonnegative definite $\boldsymbol{\Phi}$) and the equality is obtained if and only if $\boldsymbol{B\Phi B}' = \delta^2\mathbf{1}\mathbf{1}'$ for some $\delta \in \mathbb{R}$. □

The equality of $\rho_{uu}(\mathbf{1})$ and $\alpha(\mathbf{1})$ indeed requires the model to be unidimensional. Since the condition for the equality depends only on $\boldsymbol{B\Phi B}'$, the true score part of the covariance matrix, it applies similarly to the models $\mathscr{M}_2$ and $\mathscr{M}_1$. This well-known condition is called $\tau$-equivalence (Novick and Lewis [14]).

The following example shows how the reliability estimates for the unweighted sum are obtained under the model $\boldsymbol{B1}$ by using the matrix interpreter of SURVO MM (Mustonen [13]). The matrix commands begin with `MAT` and the rest is mostly comments written freely around them. The results are saved in matrices, which are given names after the word `MAT`. Their numerical values extracted from the matrices (by means of the *editorial computing*) display the equality of $\rho_{uu}(\mathbf{1})$ and $\alpha(\mathbf{1})$:

```
Computing rhouu_1=MAT_Rhouu(1,1) and alpha_1=MAT_Alpha(1,1)
MAT One=CON(p,1) / vector of ones, dimension p=7
MAT Ip=IDN(p,p)  / identity matrix of order p
MAT S1=B1*B1'+Ip-DIAG(B1*B1') / SIGMA constructed
MAT Rhouu=(One'*B1*B1'*One)*INV(One'*S1*One)
MAT Alpha=(p/(p-1))*(1-(TRACE(S1)*INV(One'*S1*One)))
The numerical results are:
    rhouu_1=0.96757679180887
    alpha_1=0.96757679180887
```

Similarly we obtain the results for the model $\boldsymbol{B2}$. According to Theorem 2 the estimates are different, because the elements of $\boldsymbol{B2}$ are not equal. They are also lower than before, because the elements are not as high as in $\boldsymbol{B1}$.

```
MAT S2=B2*B2'+Ip-DIAG(B2*B2') / SIGMA constructed
MAT Rhouu=(One'*B2*B2'*One)*INV(One'*S2*One)
MAT Alpha=(p/(p-1))*(1-(TRACE(S2)*INV(One'*S2*One)))
The numerical results are:
    rhouu_1=0.87755847953216
    alpha_1=0.86817738791423
```

## 4.4 Case of the weighted sum

It is instructive to study a slightly more general case, that is, the case of the weighted sum $u = \boldsymbol{a}'\boldsymbol{x}$. We will show that Cronbach's alpha is equal to Tarkkonen's rho only in a rare special case. Before proceeding onwards, we prove the following theorem:

**Theorem 3.** *Let* $\boldsymbol{V}$ *be a symmetric nonnegative definite* $p \times p$ *matrix with* $\boldsymbol{V}_d = \mathrm{diag}(\boldsymbol{V})$ *being positive definite. Then*

$$\max_{\boldsymbol{a}\neq\boldsymbol{0}} \frac{\boldsymbol{a}'\boldsymbol{V}\boldsymbol{a}}{\boldsymbol{a}'\boldsymbol{V}_d\boldsymbol{a}} = \mathrm{ch}_1(\boldsymbol{V}_d^{-1/2}\boldsymbol{V}\boldsymbol{V}_d^{-1/2}) = \mathrm{ch}_1(\boldsymbol{R}_V),$$

*where* $\boldsymbol{R}_V = \boldsymbol{V}_d^{-1/2}\boldsymbol{V}\boldsymbol{V}_d^{-1/2}$*, i.e.,* $\boldsymbol{R}_V$ *can be considered as a correlation matrix. Moreover,*

$$\frac{\boldsymbol{a}'\boldsymbol{V}\boldsymbol{a}}{\boldsymbol{a}'\boldsymbol{V}_d\boldsymbol{a}} \leq p \quad \textit{for all } \boldsymbol{a} \in \mathbb{R}^p, \tag{14}$$

*where the equality is obtained if and only if* $\boldsymbol{V} = \delta^2 \boldsymbol{q}\boldsymbol{q}'$ *for some* $\delta \in \mathbb{R}$ *and some* $\boldsymbol{q} = (q_1, \ldots, q_p)'$, *and* $\boldsymbol{a}$ *is a multiple of* $\tilde{\boldsymbol{a}} = (1/q_1, \ldots, 1/q_p)'$.

*Proof.* We first note that

$$\max_{a \neq 0} \frac{\boldsymbol{a}'\boldsymbol{V}\boldsymbol{a}}{\boldsymbol{a}'\boldsymbol{V}_d\boldsymbol{a}} = \max_{a \neq 0} \frac{\boldsymbol{a}'\boldsymbol{V}_d^{1/2}\boldsymbol{V}_d^{-1/2}\boldsymbol{V}\boldsymbol{V}_d^{-1/2}\boldsymbol{V}_d^{1/2}\boldsymbol{a}}{\boldsymbol{a}'\boldsymbol{V}_d^{1/2}\boldsymbol{V}_d^{1/2}\boldsymbol{a}}$$

$$= \max_{z \neq 0} \frac{\boldsymbol{z}'\boldsymbol{V}_d^{-1/2}\boldsymbol{V}\boldsymbol{V}_d^{-1/2}\boldsymbol{z}}{\boldsymbol{z}'\boldsymbol{z}}$$

$$= \mathrm{ch}_1(\boldsymbol{V}_d^{-1/2}\boldsymbol{V}\boldsymbol{V}_d^{-1/2})$$

$$= \mathrm{ch}_1(\boldsymbol{R}_V) := \mu_1.$$

It is obvious that the largest eigenvalue $\mu_1$ of a $p \times p$ correlation matrix $\boldsymbol{R}_V$ is $\leq p$ and clearly $\mu_1 = p$ if and only if $\boldsymbol{R}_V = \boldsymbol{1}\boldsymbol{1}'$, i.e., $\boldsymbol{V}$ must be of the form $\boldsymbol{V} = \delta^2 \boldsymbol{q}\boldsymbol{q}'$ for some $\delta \in \mathbb{R}$ and $\boldsymbol{q} = (q_1, \ldots, q_p)' \in \mathbb{R}^p$. It is easy to conclude that if $\boldsymbol{V} = \delta^2 \boldsymbol{q}\boldsymbol{q}'$, then the equality in (14) is obtained if and only if $\boldsymbol{a}$ is a multiple of $\tilde{\boldsymbol{a}} = \boldsymbol{V}_d^{-1/2}\boldsymbol{1} = \frac{1}{\delta}(1/q_1, \ldots, 1/q_p)'$. □

**Corollary 2.** *Using the notation above,*

$$\alpha(\boldsymbol{a}) = \frac{p}{p-1}\left(1 - \frac{\boldsymbol{a}'\boldsymbol{\Sigma}_d\boldsymbol{a}}{\boldsymbol{a}'\boldsymbol{\Sigma}\boldsymbol{a}}\right) \leq \frac{p}{p-1}\left(1 - \frac{1}{\mathrm{ch}_1(\boldsymbol{R}_\Sigma)}\right) \text{ for all } \boldsymbol{a} \in \mathbb{R}^p.$$

*Moreover,*

$$\alpha(\boldsymbol{a}) \leq 1 \quad \text{for all } \boldsymbol{a} \in \mathbb{R}^p. \tag{15}$$

*The equality in* (15) *is obtained if and only if* $\boldsymbol{\Sigma} = \delta^2 \boldsymbol{q}\boldsymbol{q}'$ *for some* $\delta \in \mathbb{R}$ *and some* $\boldsymbol{q} = (q_1, \ldots, q_p)'$, *and* $\boldsymbol{a}$ *is a multiple of* $\tilde{\boldsymbol{a}} = (1/q_1, \ldots, 1/q_p)'$.

*Proof.* The proof comes at once from Theorem 3. □

Finally we use Theorem 3 to prove that the equality of $\rho_{uu}(\boldsymbol{a})$ and $\alpha(\boldsymbol{a})$ requires the model to be unidimensional, either $\mathcal{M}_2$ or $\mathcal{M}_1$. This is a straight-forward generalization of Theorem 1. In a rather peculiar special case, however, the model may be $\mathcal{M}_3$, but then the scale weights will depend completely on the model weights. This result can be seen only with the weighted scale. Similarly as before, $\rho_{uu}(\boldsymbol{a})$ exceeds $\alpha(\boldsymbol{a})$ in all other circumstances.

**Theorem 4.** *Consider the measurement model* $\mathcal{M}_d = \{\boldsymbol{x}, \boldsymbol{B}\boldsymbol{\tau}, \boldsymbol{B}\boldsymbol{\Phi}\boldsymbol{B}' + \boldsymbol{\Psi}_d\}$. *Then,*

$$\alpha(\boldsymbol{a}) \leq \rho_{uu}(\boldsymbol{a}) \quad \text{for all } \boldsymbol{a} \in \mathbb{R}^p,$$

*and the equality is obtained if and only if* $\boldsymbol{B}\boldsymbol{\Phi}\boldsymbol{B}' = \delta^2 \boldsymbol{q}\boldsymbol{q}'$ *for some* $\delta \in \mathbb{R}$ *and some* $\boldsymbol{q} = (q_1, \ldots, q_p)'$, *i.e.,* $\boldsymbol{\Sigma} = \delta^2 \boldsymbol{q}\boldsymbol{q}' + \boldsymbol{\Psi}_d$, *and* $\boldsymbol{a}$ *is a multiple of* $\tilde{\boldsymbol{a}} = (1/q_1, \ldots, 1/q_p)'$.

*Proof.* Our claim is

$$\frac{p}{p-1}\left(1-\frac{\boldsymbol{a}'\boldsymbol{\Sigma}_d\boldsymbol{a}}{\boldsymbol{a}'\boldsymbol{\Sigma}\boldsymbol{a}}\right)\leq\frac{\boldsymbol{a}'\boldsymbol{B}\boldsymbol{\Phi}\boldsymbol{B}'\boldsymbol{a}}{\boldsymbol{a}'\boldsymbol{\Sigma}\boldsymbol{a}},$$

i.e.,

$$\frac{p}{p-1}(\boldsymbol{a}'\boldsymbol{\Sigma}\boldsymbol{a}-\boldsymbol{a}'\boldsymbol{\Sigma}_d\boldsymbol{a})\leq\boldsymbol{a}'\boldsymbol{B}\boldsymbol{\Phi}\boldsymbol{B}'\boldsymbol{a}. \tag{16}$$

Substituting $\boldsymbol{\Sigma}=\boldsymbol{B}\boldsymbol{\Phi}\boldsymbol{B}'+\boldsymbol{\Psi}_d$, (16) becomes

$$p[\boldsymbol{a}'\boldsymbol{B}\boldsymbol{\Phi}\boldsymbol{B}'\boldsymbol{a}+\boldsymbol{a}'\boldsymbol{\Psi}_d\boldsymbol{a}-\boldsymbol{a}'(\boldsymbol{B}\boldsymbol{\Phi}\boldsymbol{B}')_d\boldsymbol{a}-\boldsymbol{a}'\boldsymbol{\Psi}_d\boldsymbol{a}]\leq(p-1)\boldsymbol{a}'\boldsymbol{B}\boldsymbol{\Phi}\boldsymbol{B}'\boldsymbol{a},$$

which simplifies into the form

$$\frac{\boldsymbol{a}'\boldsymbol{B}\boldsymbol{\Phi}\boldsymbol{B}'\boldsymbol{a}}{\boldsymbol{a}'(\boldsymbol{B}\boldsymbol{\Phi}\boldsymbol{B}')_d\boldsymbol{a}}\leq p.$$

The proof is now completed using Theorem 3. □

The equality of $\rho_{uu}(\boldsymbol{a})$ and $\alpha(\boldsymbol{a})$ indeed requires the model to be unidimensional. Since the condition for the equality again depends only on the true score part of the covariance matrix, it applies similarly to the models $\mathcal{M}_2$ and $\mathcal{M}_1$. The weighted scale reveals that the equality holds also in the case of the model $\mathcal{M}_3$, the congeneric model (Jöreskog [7]), but only in a rare special case where the scale weights are inverses of the model weights. Again $\rho_{uu}(\boldsymbol{a})$ exceeds $\alpha(\boldsymbol{a})$ in all other circumstances.

In the following examples the weighted sum is represented by the factor scores (by regression method).

```
Computing rhouu_a=MAT_Rhouu(1,1) and alpha_a=MAT_Alpha(1,1)
MAT A2=INV(S2)*B2 / factor score coefficients
MAT Ip=IDN(p,p)   / identity matrix of order p=7
MAT S2=B2*B2'+Ip-DIAG(B2*B2') / SIGMA constructed
MAT Rhouu=(A2'*B2*B2'*A2)*INV(A2'*S2*A2)
MAT Alpha=(p/(p-1))*(1-(A2'*DIAG(S2)*A2)*INV(A2'*S2*A2))
The numerical results are:        to be compared with these:
    rhouu_a=0.92898671096345      rhouu_1=0.87755847953216
    alpha_a=0.81147318009654      alpha_1=0.86817738791423
```

With the model ***B*1** the weighted scale does not have any effect on the estimates, since the factor score coefficients are constant for each $x_i$. With ***B*2** the discrepancy between them becomes quite clear when compared to the results of the unweighted sum: Tarkkonen's rho increases but Cronbach's alpha decreases. This is a clear sign of the well-known underestimation problem of Cronbach's alpha.

## *4.5 Multidimensional models*

Let us take a look at the reliability estimation under the multidimensional models ***B*3** and ***B*4**. First, we compute the reliability matrix $\boldsymbol{\rho}_u$ in the case of the model ***B*3** using Eq. (4) given in Sect. 3.1.

```
MAT RLABELS x TO B3 / sets names x1 etc. for rows
MAT CLABELS t TO B3 / sets names t1,t2 for columns
MAT Ip=IDN(p,p)      / identity matrix of order p=7
MAT Ik=IDN(k,k)      / identity matrix of order k=2
MAT S3=B3*B3'+Ip-DIAG(B3*B3') / SIGMA constructed
MAT Psi_d=DIAG(S3-B3*B3') / cov.matrix of meas.errors
MAT A3=INV(S3)*B3          / factor score coefficients
MAT Rho_u=INV(Ik+DIAG(A3'*Psi_d*A3)*INV(DIAG(A3'*B3*B3'*A3)))
MAT LOAD Rho_u 1.23456789012345 CUR+2

MATRIX Rho_u
///                    t1                t2
t1         0.94460641399417 0.00000000000000
t2         0.00000000000000 0.92748091603053
```

Since ***B*3** is truely a multidimensional model, although very simple, it is beyond the scope of Cronbach's alpha. Of course it is possible to compute the reliability of the unweighted sum, but the dimensionality will make Cronbach's alpha quite low:

```
MAT One=CON(p,1) / p=7 for computing alpha_1=MAT_Alpha(1,1)
MAT Alpha=(p/(p-1))*(1-(One'*DIAG(S3)*One)*INV(One'*S3*One))
and we have alpha_1=0.78822984244671
```

In practice, it is typical to compute Cronbach's alpha by selecting only the "best" items for the sum. Here, the items $x_1$ to $x_4$ would obviously represent the first dimension while $x_5$ to $x_7$ would be the choice for the second dimension. The following example shows how this would be done for the first dimension. The other one would follow similarly:

```
MAT One=CON(p,1) / p=4 for computing alpha_1=MAT_Alpha(1,1)
MAT S=S3(x1:x4,x1:x4) / pick a sub-matrix of S3
MAT Alpha=(p/(p-1))*(1-(One'*DIAG(S)*One)*INV(One'*S*One))
and we have alpha_1=0.94460641399417
```

The result is equal to the first element of the reliability matrix $\boldsymbol{\rho_u}$. This is, however, due to the very simple structure of the model ***B*3**. The selection of the items becomes arbitrary as soon as the structure gets more complicated.

A bit more realistic variant of ***B*3** is provided by model ***B*4**, where the items $x_3, x_4$, and $x_5$ are related with both dimensions (see Table 1). It does not cause any difficulties for Tarkkonen's rho, since repeating the previous computations for ***B*4** we obtain the reliability matrix $\boldsymbol{\rho_u}$ again. The figures are quite similar, because ***B*3** and ***B*4** do not differ remarkably:

```
MATRIX Rho_u
///                    t1                t2
t1         0.94374792204044 0.00000000000000
t2         0.00000000000000 0.93354550709381
```

However, these tiny changes in the model cause additional problems with Cronbach's alpha, as the selection of the items becomes more difficult. In addition, some of the elements are negative, and will distort the estimates dramatically:

```
.......... Select positive elements only:
MAT One=CON(p,1) / p=4 for computing alpha_1=MAT_Alpha(1,1)
MAT S=S4(x1:x4,x1:x4) / pick a sub-matrix of S4
MAT Alpha=(p/(p-1))*(1-(One'*DIAG(S)*One)*INV(One'*S*One))
and we have alpha_1=0.92806484295846
.......... Select also x5:
MAT One=CON(p,1) / p=5 for computing alpha_1=MAT_Alpha(1,1)
MAT S=S4(x1:x5,x1:x5) / pick a sub-matrix of S4
MAT Alpha=(p/(p-1))*(1-(One'*DIAG(S)*One)*INV(One'*S*One))
and we have alpha_1=0.56768558951965
```

Negative elements that arise naturally in multidimensional models cause severe problems with Cronbach's alpha. On the contrary, Tarkkonen's rho works well with positive and negative elements, and there is no need to select items in order to increase the reliability. All items that have been measured can also be used.

### 4.6 Conclusions

It seems clear that Tarkkonen's rho works well with multidimensional models and scales. However, any kind of multidimensionality seems to be a problem for Cronbach's alpha. This discrepancy is easy to show with simple examples. To make fair comparisons, one has to stick with unidimensional models and scales. Some conclusions can be drawn, and none of them seems to be in favor of Cronbach's alpha.

First, with the unweighted sum we can conclude that

$$\alpha(\mathbf{1}) \leq \rho_{uu}(\mathbf{1}) \leq 1,$$

where the assumptions of Tarkkonen's rho ensure that $\rho_{uu}(\mathbf{1}) > 0$. However, $\alpha(\mathbf{1})$ may tend negative because the strict assumptions made in the derivation of KR-20 (7), and mostly inherited in $\alpha(\mathbf{1})$, are easily violated. Negative estimates do not make sense, of course, as reliability is by definition a ratio of variances.

In a bit more general setting we can also conclude that

$$\alpha(\boldsymbol{a}) \leq \rho_{uu}(\boldsymbol{a}) \leq 1 \quad \text{for all } \boldsymbol{a} \in \mathbb{R}^p.$$

Again the assumptions ensure that $\rho_{uu}(\boldsymbol{a}) > 0$, but $\alpha(\boldsymbol{a})$ may tend negative. An additional reason for this is that the original formula (KR-20) was derived only for an unweighted sum, not for arbitrary weighted sums.

## 5 Discussion

For a long time, research has focused on examining the properties of Cronbach's alpha without questioning its original assumptions. This is supported by an excellent review by Weiss and Davison [21, pp. 630–635], or more recent studies basically in about any journals of psychometrics. The motivation of estimating the reliability and using the estimates for assessing the quality of measurements has been obscured, because the estimates have been so poor. As Weiss and Davison [21, p. 633] put it: "Somewhere during the three-quarter century history of classical test theory the real purpose of reliability estimation seems to have been lost".

What is then the "real purpose of reliability estimation"? First of all, it is important for a measurement scale to have a high reliability, because the further analyses could then be based mainly on the true variation instead of random measurement errors. One topic that has been nearly forgotten because of the poor reliability estimates, is the correction for attenuation, which is badly needed, for example, in survey research (Alwin [1], ch. 4).

Of course, the questions of validity should usually have the highest priority. Unfortunately, the strict assumptions behind Cronbach's alpha have lead in the opposite direction: maximizing the reliability of the scale by discarding any "unsuitable" items. The applied criterion, the *internal consistency*, requires all items to be equally good indicators of the trait under study. Some statistical program packages even support this procedure by reporting *Cronbach's alpha if item deleted* statistics. Combined with the underestimation problems of Cronbach's alpha, it is clear that this approach has unfavourable consequences.

Most empirical problems are multidimensional, and it is difficult to develop items that measure only one dimension. Indeed, the most striking problem of Cronbach's alpha is its built-in assumption of unidimensionality. On the contrary, Tarkkonen's rho seems to be well-suited for multidimensional reliability studies.

**Acknowledgement** The authors are grateful to Jarkko M. Isotalo and Maria Valaste for helpful discussions and to Jarmo Niemelä for LaTeX assistance.

## References

[1] Alwin, D.F.: Margins of Error: A Study of Reliability in Survey Measurement. Wiley, Hoboken, New Jersey (2007)
[2] Blinkhorn, S.F.: Past imperfect, future conditional: Fifty years of test theory. Brit. J. Math. Stat. Psy. **50**, 175–185 (1997)
[3] Bollen, K.A.: Structural Equations with Latent Variables. Wiley, New York (1989)
[4] Cronbach, L.J.: On estimates of test reliability. J. Educ. Psychol. **34**, 485–494 (1943)
[5] Cronbach, L.J.: Coefficient alpha and the internal structure of tests. Psychometrika **16**, 297–334 (1951)
[6] Heise, D.R., Bohrnstedt, G.W.: Validity, invalidity and reliability. In: Borgatta, E.F., Bohrnstedt, G.W. (eds.) Sociological Methodology, pp. 104–129. Jossey-Bass, San Francisco (1970)

[7] Jöreskog, K.G.: Statistical analysis of sets of congeneric tests. Psychometrika **36**, 109–133 (1971)
[8] Jöreskog, K.G.: Factor analysis and its extensions. In: Cudeck, R., MacCallum, R.C. (eds.) Factor Analysis at 100: Historical Developments and Future Directions, pp. 47–77. Lawrence Erlbaum, Mahwah, New Jersey (2007)
[9] Kuder, G.F., Richardson, M.W.: The theory of the estimation of test reliability. Psychometrika **2**, 151–160 (1937)
[10] Lord, F.M., Novick, M.R.: Statistical Theories of Mental Test Scores. Addison-Wesley, London (1968)
[11] Lucke, J.F.: The $\alpha$ and the $\omega$ of congeneric test theory: An extension of reliability and internal consistency to heterogeneous tests. Appl. Psych. Meas. **29**, 65–81 (2005)
[12] McDonald, R.P.: The theoretical foundations of principal factor analysis, canonical factor analysis, and alpha factor analysis. Brit. J. Math. Stat. Psy. **23**, 1–21 (1970)
[13] Mustonen, S.: SURVO MM: Computing environment for creative processing of text and numerical data. http://www.survo.fi/mm/english.html (2001)
[14] Novick, M.R., Lewis, C.: Coefficient alpha and the reliability of composite measurements. Psychometrika **32**, 1–13 (1967)
[15] Puntanen, S., Styan, G.P.H.: Matrix tricks for linear statistical models: our personal Top Thirteen. Research report A 345, Dept. of Mathematics, Statistics & Philosophy, University of Tampere, Tampere, Finland (2003)
[16] Tarkkonen, L.: On Reliability of Composite Scales. No. 7 in Statistical Studies. Finnish Statistical Society, Helsinki, Finland (1987)
[17] Tarkkonen, L., Vehkalahti, K.: Measurement errors in multivariate measurement scales. J. Multivariate Anal. **96**, 172–189 (2005)
[18] ten Berge, J.M.F., Soĉan, G.: The greatest lower bound to the reliability of a test and the hypothesis of unidimensionality. Psychometrika **69**, 611–623 (2004)
[19] Vehkalahti, K.: Reliability of Measurement Scales. No. 17 in Statistical Research Reports. Finnish Statistical Society, Helsinki, Finland (2000)
[20] Vehkalahti, K., Puntanen, S., Tarkkonen, L.: Effects of measurement errors in predictor selection of linear regression model. Comput. Stat. Data An. **52**, 1183–1195 (2007)
[21] Weiss, D.J., Davison, M.L.: Test theory and methods. Annu. Rev. Psychol. **32**, 629–658 (1981)
[22] Werts, C.E., Rock, R.D., Linn, R.L., Jöreskog, K.G.: A general method of estimating the reliability of a composite. Educ. Psychol. Meas. **38**, 933–938 (1978)

# Part IV
# Linear Models and Applied Econometrics

*All models are wrong, but some are useful.*

George E.P. Box

# Robust Moment Based Estimation and Inference: The Generalized Cressie–Read Estimator

Ron C. Mittelhammer and George G. Judge

**Abstract** In this paper a range of information theoretic distance measures, based on Cressie-Read divergence, are combined with mean-zero estimating equations to provide an efficient basis for semi parametric estimation and testing. Asymptotic properties of the resulting semi parametric estimators are demonstrated and issues of implementation are considered.

## 1 Introduction

For a range of statistical models when the functional form of the likelihood is known, the likelihood concept is appealing for estimation and inference purposes from both a sampling theory and Bayesian perspective. If insufficient information about the underlying data sampling process is available for specifying the functional form of the likelihood function, parametric maximum likelihood (ML) methods are inapplicable. In econometrics, because information about the underlying data sampling process is usually partial and incomplete, much estimation and inference over the last two decades has proceeded under formulations that are semi parametric in the sense that the joint probability distribution of the data is unspecified apart from a finite set of moment conditions or conditional moments restrictions.

A way of avoiding an explicit functional specification of the likelihood is to use an estimation method that is still likelihood based, but that does not assume a specific parametric family of probability distributions for the underlying data sampling process. One such possibility is the general "empirical likelihood" concept where, given observations $\{y_1, y_2, \ldots, y_n\}$, a possible sample distribution function is chosen from the multinomial family that assigns probability weight $w_i$ to observation $y_i$. Under the empirical likelihood concept, empirical likelihood weights, $\mathbf{w}$, supported

George G. Judge
University of California-Berkeley, 207 Giannini Hall, Berkeley, CA 94720
judge@are.berkeley.edu

on the sample of observed data outcomes are used to reduce the infinite dimensional problem of nonparametric likelihood estimation to a finite dimensional one. This represents a generalization of the Empirical Distribution Function in which $w_i = n^{-1}, \forall_i$.

Regarding extremum metrics from which to derive empirical likelihood weights, the general [6] family of power divergence statistics represents a flexible family of pseudo-distance measures leading to associated estimators. The Cressie–Read (CR) statistic contains a parameter $\lambda$ that indexes a set of empirical goodness-of-fit (empirical divergence) measures and estimation criteria. As $\lambda$ varies the resulting empirical likelihood estimators exhibit qualitatively different sampling behavior ([1]; [2]; [17] [16]; [27]). Within this context the purpose of this paper is to investigate the statistical implications of generalizing the empirical likelihood principle to encompass the entire range of distance measures contained in the Cressie–Read family. Using empirical moments as constraints, an empirical likelihood solution basis is demonstrated based on the CR-optimized $(\lambda, \mathbf{w})$ combination-estimator for any particular data sampling process, and corresponding statistical implications of basing parameter estimation on the generalized CR $(\lambda, \mathbf{w})$ are assessed. The resulting CR$(\lambda, \mathbf{w})$ formulations can be used to help avoid the use of tenuous model assumptions, and to provide optimal estimation and testing methods for semiparametric models based on mean zero estimating equations.

In line with the objectives noted above, in 2 a basic statistical model is specified and specific estimators within the Cressie–Read family of estimators are identified and developed. In 3 the generally user-specific parameter in the Cressie–Read statistic is considered free to vary outside of the limiting cases usually considered and the estimation and inference theory underlying this generalization is developed. In the last section econometric implications of this generalization are considered and conclusions are drawn regarding the use of the generalized Cressie–Read distance measure (CRDM) in empirical practice.

## 2 Preliminaries-statistical Models, Estimators, and Inference Procedures

Consider a structural equation that is contained within a system of structural equations and that exhibits the semiparametric linear statistical model form $\mathbf{Y} = \mathbf{X}\beta + \varepsilon$. Assume a vector of sample outcomes $\mathbf{y} = (y_1, y_2, \cdots, y_n)'$ associated with this linear model is observed, where $\mathbf{X}$ is a $(n \times k)$ matrix of stochastic explanatory variables, $\varepsilon$ is an unobservable random noise vector with mean vector $\mathbf{0}$ and covariance matrix $\sigma^2 \mathbf{I}_n$, and $\beta \in \mathbf{B}$ is a $(k \times 1)$ vector of unknown parameters. If one or more of the right hand side $\mathbf{X}_{.j}$'s is correlated with the equation noise, then $E\left[n^{-1}\mathbf{X}'\varepsilon\right] \neq 0$ or plim $\left[n^{-1}\mathbf{X}'\varepsilon\right] \neq \mathbf{0}$ and traditional Gauss–Markov based procedures such as the least squares (LS) estimator, or equivalently the method of moments (MOM) estimator defined by $\hat{\beta}_{mom} = \arg_{\beta \in B}\left[n^{-1}\mathbf{X}'(\mathbf{Y} - \mathbf{X}\beta) = \mathbf{0}\right]$, are biased and inconsistent, with unconditional expectation and probability limit given by $E\left[\hat{\beta}\right] \neq \beta$ and plim $\left[\hat{\beta}\right] \neq \beta$. Given a sampling process characterized by nonorthogonality of $\mathbf{X}$ and $\varepsilon$, in order to avoid the use of strong distributional assumptions it is conventional to introduce additional information in the form of a $(n \times m)$, $m \geq k$, random matrix $\mathbf{Z}$ of instrumental variables whose elements are correlated with $\mathbf{X}$ but uncorrelated with $\varepsilon$. This information is introduced into the statistical model by specifying the sample analog moment condition

$$\mathbf{h}(\mathbf{Y},\mathbf{X},\mathbf{Z};\beta) = n^{-1}\left[\mathbf{Z}'(\mathbf{Y}-\mathbf{X}\beta)\right] \xrightarrow{p} \mathbf{0}, \tag{1}$$

relating to the underlying population moment condition derived from the orthogonality of instruments, $\mathbf{Z}$, and model noise defined by

$$E\left[\mathbf{Z}'(\mathbf{Y}-\mathbf{X}\beta)\right] = \mathbf{0}. \tag{2}$$

When the usual moment regularity conditions are fulfilled, this IV estimator is consistent, asymptotically normal distributed, and is an optimal estimating function (OptEF) estimator ([8]; [13]; [26]).

If the vector of moment conditions overdetermines the model parameters, other estimation procedures are available. For example, one possibility is the estimator formed by following [8]; [10]; [14], and applying the optimal estimating function (OptEF) transformation $\left(\mathbf{X}'\mathbf{Z}(\mathbf{Z}'\mathbf{Z})^{-1}\mathbf{Z}'\mathbf{X}\right)^{-1}\mathbf{X}'\mathbf{Z}(\mathbf{Z}'\mathbf{Z})^{-1}$ to the moment conditions in (2). The GMM estimator ([12]; and [15]) that minimizes a quadratic form in the sample moment information is another popular estimator that makes use of the information in (2).

## 2.1 *Empirical Likelihood (EL) Type Estimators*

In contrast to traditional instrument-moment based estimators, the empirical likelihood (EL) approach ([21],[22],[23]; [25]; [11]; [4]; [24]; [26]; and [18]) allows the investigator to employ likelihood methods for model estimation and inference without having to choose a specific parametric family of probability densities on which to base the likelihood function. As noted in 1, one possibility for an extremum metric-estimation criterion is the general CR [6] and [29], power divergence family of statistics

$$I(\mathbf{w},\mathbf{q},\lambda) = \frac{1}{\lambda(\lambda+1)}\sum_{i=1}^{n} w_i\left[\left(\frac{w_i}{q_i}\right)^{\lambda} - 1\right], \tag{3}$$

where $\lambda$ is a parameter that indexes members of the CR family, and the $q_i$'s are interpreted as reference probabilities that satisfy $q_i \in (0,1), \forall i$ and $\sum_{i=1}^{n} q_i = 1$. In a linear structural model estimation context based on instrumental variables, if we use (3) as the goodness-of-fit criterion and (1) as the moment-estimating function information, the CRDM estimation problem can be formulated as the following extremum-type estimator for $\beta$ for any given choice of the $\lambda$ parameter:

$$\hat{\beta}(\lambda) = \arg\max_{\beta \in B} \left[ \ell_E(\beta;\lambda) = \max_{w} \left\{ -I(\mathbf{w},\mathbf{q},\lambda) \mid \sum_{i=1}^{n} w_i \mathbf{z}_{i.}'(y_i - \mathbf{x}_{i.}\beta) = \mathbf{0}, \right.\right.$$
$$\left.\left. \sum_{i=1}^{n} w_i = 1, w_i \geq 0 \ \forall i \right\} \right] \quad (4)$$

where $\ell_E(\beta;\lambda)$ can be interpreted as an empirical likelihood function parameterized by the parameter $\lambda$. It is important to note that the family of power divergence statistics is symmetric in the choice of which set of probabilities are considered as the first and second arguments of the function (3). In particular, regardless of whether the statistic is designated as $\mathbf{I}(\mathbf{w},\mathbf{q},\lambda)$ or $\mathbf{I}(\mathbf{q},\mathbf{w},\lambda)$ precisely the same family of divergence measures is defined across the range of possible values $\lambda \in (-\infty,\infty)$. This point, which is discussed by Österreicher (2002) and Österreicher and Vajda (2003), is demonstrated in the Appendix as Proposition 1.

## *2.2 Three Main Variants of* $\mathbf{I}(\mathbf{w},\mathbf{q},\lambda)$

Three main variants of $\mathbf{I}(\mathbf{w},\mathbf{q},\lambda)$, and the associated empirical likelihood functions $\ell_E(\beta;\lambda)$, have emerged and received explicit attention in the econometrics literature. All are based on the reference distribution specification $\mathbf{q} = n^{-1}\mathbf{1}_n$, where $\mathbf{1}_n$ denotes an $n \times 1$ vector of unit values, and we adopt this reference distribution specification for the remainder of the paper. Note this choice of the reference distribution is tantamount to choosing the classical empirical distribution function (EDF) as the target empirical likelihood function. We utilize the abbreviated notation $CR(\lambda) \equiv I(\mathbf{w},\mathbf{q},\lambda)$, where the arguments $\mathbf{w}$ and $\mathbf{q}$ are tacitly understood to be evaluated at relevant vector values. In the two special cases where $\lambda = 0$ *or* $-1$, the notations CR(0) and CR($-1$) are to be interpreted as the continuous limits $\lim_{\lambda\to 0} CR(\lambda)$ and $\lim_{\lambda\to -1} CR(\lambda)$, respectively.

The specification CR(-1) leads to the traditional empirical log-likelihood (EL) objective function, $n^{-1}\sum_{i=1}^{n}\ln(w_i)$, and the maximum empirical likelihood estimate of $\beta$, while the specification CR(0) leads to the empirical exponential likelihood objective function, $-\sum_{i=1}^{n} w_i \ln(w_i)$, and the Maximum Empirical Exponential Likelihood estimate of $\beta$ (see Proposition 2 of the appendix for derivations of these two results). Finally, CR(1) defines the log Euclidean or least squares likelihood function $\frac{n}{2}\left(-\sum_{i=1}^{n}\left(w_i{}^2 - \frac{1}{n}\right)\right) \propto \left(\mathbf{w} - n^{-1}\mathbf{1}_n\right)'\left(\mathbf{w} - n^{-1}\mathbf{1}_n\right)$ leading to the Maximum Log Euclidean Likelihood or least squares empirical likelihood estimate of $\beta$. This estimator is related to an updating variant of the GMM estimator, where the unknown covariance matrix is handled internally ([3]).

### 2.2.1 Estimation

If the traditional EL criterion CR(-1) is used, the estimation objective involves finding the feasible weights $\hat{\mathbf{w}}$ that maximizes the joint empirical log likelihood assigned to the observed set of sample observations, conditional on the moment constraints. In the sense of objective function analogies, the EL approach is the closest to the classical maximum likelihood approach. The CR(0) criterion of maximizing $-\sum_{i=1}^{n} w_i \ln(w_i)$ is equivalent to defining an estimator by *minimizing* the Kullback-Leibler (KL) information criterion $\sum_{i=1}^{n} w_i \ln\left(w_i/n^{-1}\right)$ and using the maximum entropy principle of Jaynes (1957). Interpreted in the KL context, this estimation objective finds the feasible weights $\hat{\mathbf{w}}$ that define the minimum value of all possible *expected* log-likelihood ratios consistent with the structural moment constraints ([7]; and [26]). The expectations are based on the $\hat{\mathbf{w}}$ distribution and the log-likelihood ratio has the restricted (by moment constraints) likelihood in the numerator and the unrestricted (i.e., uniform distribution) likelihood in the denominator. The CR(1) solution seeks feasible weights $\hat{\mathbf{w}}$ that minimize the Euclidean distance of $\mathbf{w}$ from the uniform probability distribution, the square of this Euclidean distance being $\left(\mathbf{w}-n^{-1}\mathbf{1}_n\right)'\left(\mathbf{w}-n^{-1}\mathbf{1}_n\right)$. The optimum weights, subject to moment and adding up constraints, are necessarily nonnegative valued by the functional characteristics of the estimation objective represented by the CR(0) and CR(-1) cases, but negative weights are not ruled out by the CR(1) specification in the absence of explicitly imposed nonnegativity constraints.

Under the usual regularity conditions assumed when establishing the asymptotics of traditional structural equation estimators, all of the preceding EL-like estimators of $\beta$ obtained by optimizing the $w_i$'s , for fixed choices of $\lambda$, are consistent, asymptotically normally distributed, and asymptotically efficient relative to the optimal estimating function (OptEF) estimator ([1]). The solution to the constrained optimization problem (4) yields an optimal estimate, $\hat{\mathbf{w}}(\lambda)$ and $\hat{\beta}(\lambda)$, that cannot, in general, be expressed in closed form and thus must be obtained using numerical methods. Note further that for the typical application in which the reference distribution $q_i = n^{-1}\,\forall i$, *any* of the estimation objective functions contained in the Cressie–Read family achieve *unconstrained* (by moment equations) optima when the empirical probability distribution is given by $\mathbf{w} = n^{-1}\mathbf{1}_n$.

### 2.2.2 Inference

EL-type inference methods, including hypothesis testing and confidence region estimation, bear a strong analogy to inference methods used in traditional ML and GMM approaches. [21],[22] showed that an analog of Wilks' Theorem for likelihood ratios, specifically -2ln(LR) $\overset{a}{\sim} \chi^2_j$ under $H_o$, hold for the empirical likelihood CR(-1) approach, where $j$ denotes the number of functionally independent restrictions on the parameter space. Baggerly [1] demonstrated that this calibration remains applicable when the likelihood is replaced with any properly scaled member of the Cressie–Read family of power divergence statistics (3). In this context, the empirical likelihood ratio (LR) for testing the linear combinations hypothesis $\mathbf{c}\beta = \mathbf{r}$ when $\text{rank}(\mathbf{c}) = j$, is given for the CR(-1) case by

$$LR_{CR(-1)}(\mathbf{y}) = \frac{\max_{\beta}\left[\ell_E(\beta, \lambda \to -1)\,\text{s.t. } \mathbf{c}\beta = \mathbf{r}\right]}{\max_{\beta} \ell_E(\beta, \lambda \to -1)} \tag{5}$$

and

$$-2\ln\left(LR_{CR(-1)}(\mathbf{y})\right) \overset{a}{\sim} \chi_j^2 \tag{6}$$

under $H_o$ when $m \geq k$. An analogous pseudo-LR approach can be applied, mutatis mutandis, to other members of the Cressie–Read family.

To place this approach to estimation and inference in a more general context, it is important to relate the CR($\lambda$) approach to the score test proposed by [28]. Although his quadratic inference function form of test statistic was in a parametric context, its recent impact has been within semiparametric inference based on estimating equations. In econometric type problems where there are more estimating equations than unknowns, the quadratic inference function expressed as a function of the parameters provides an optimal basis for estimation and testing. As noted by [15], the Rao quadratic inference formulation is also closely related to the empirical likelihood-CR($\lambda$) methods where an empirical likelihood objective function is used to create a multinomial-type likelihood. In the following sections we demonstrate some of the statistical implications of extending the quadratic inference methodology to the [6] family of distance-divergence measures.

## 3 Global Minimum Discrepancy (GMD) Approach to Estimation and Inference

The definition of the estimator implied by the optimization problem in (4) is conditional on a choice of the arbitrary parameter $\lambda$ in the CR family of divergence statistics. This parameter indexes functional forms for divergence measures in a family of divergence measures, and indexing empirical likelihood functions defined in (4) is analogous to the way parameters index functional forms for likelihood measures in a parametric family of likelihood functions. This suggests that one might consider, in the context of (4), optimizing with respect to the choice of the $\lambda$ parameter when defining the estimator for the data-probability weights and the parameter vector $\beta$ of a data sampling model (see for example [5]). We consider such an idea in this section.

### *3.1 The GMD Measure*

The use of the CR($\lambda$) statistic in defining an empirical likelihood-type function, conditional on $\lambda$, was discussed in 2.2. We now define an empirical likelihood function based on the CR statistic that is not conditional on the choice of the arbitrary $\lambda$ parameter value:

$$\ell_E(\beta) = \max_{\lambda}\left[\ell_E(\beta;\lambda)\right] = \max_{\omega,\lambda}\left\{-I\left(\mathbf{w}, n^{-1}\mathbf{1}_n, \lambda\right) \mid \sum_{i=1}^{n} w_i z_{i.}'(y_i - \mathbf{x}_{i.}\beta) = \mathbf{0}, \sum_{i=1}^{n} w_i = 1, w_i \geq 0 \ \forall i\right\} \quad (7)$$

Moving in the direction of defining a global minimum discrepancy (GMD) measure, it is apparent that the optimization problem is segmentable in the $\lambda$ parameter, so that (7) can be optimized with respect to $\lambda$ at the outset, for a given $\mathbf{w}$, as

$$I_*\left(\mathbf{w}, n^{-1}\mathbf{1}_n\right) = \min_{\lambda} I\left(\mathbf{w}, n^{-1}\mathbf{1}_n, \lambda\right) = \min_{\lambda}\left[\frac{1}{\lambda(\lambda+1)}\sum_{i=1}^{n} w_i\left[(nw_i)^{\lambda} - 1\right]\right]. \quad (8)$$

The first order conditions for the optimization are given by

$$\frac{\partial CR(\lambda)}{\partial \lambda} = \frac{\left[\sum_{i=1}^{n}(nw_i)^{\lambda} w_i\left[(\lambda(\lambda+1)\ln(nw_i) - (2\lambda+1))\right]\right] + 2\lambda + 1}{\left[\lambda(\lambda+1)\right]^2} = 0. \quad (9)$$

Consequently, the first order conditions will be met iff the numerator of the derivative expression is zero.

The solution for $\lambda$ in (9) is not available in closed form, but can be straightforwardly solved numerically on a computer, and characterizes a global optimal solution, because:

1. CR($\lambda$) is strictly convex in $\lambda$, for all feasible $\mathbf{w}$, except on a set of measure zero that amounts to only a singleton exception, and
2. There will always exist a finite choice of $\lambda$ that will satisfy the first order condition (9).

The strict convexity of (9) in $\lambda$ result follows because it can be shown[1] that the second order derivative defined by

$$\frac{\partial^2 CR(\lambda)}{\partial \lambda^2} = \left[\sum_{i=1}^{n}(nw_i)^{\lambda} w_i\left[\frac{(\ln(nw_i))^2}{\lambda(\lambda+1)} - \frac{2(2\lambda+1)\ln(nw_i)}{(\lambda(\lambda+1))^2} + \frac{6\lambda(\lambda+1)+2}{(\lambda(\lambda+1))^3}\right]\right] - \frac{6\lambda(\lambda+1)+2}{(\lambda(\lambda+1))^3} \quad (10)$$

is such that $\frac{\partial^2 CR(\lambda)}{\partial\lambda^2} > 0 \ \forall\lambda$ when $w_i \neq n^{-1}$ for some i, while $\frac{\partial^2 CR(\lambda)}{\partial\lambda^2} = 0 \ \forall\lambda$ when $\mathbf{w} = n^{-1}\mathbf{1}_n$. Thus the CR measure is convex in the $\lambda$ parameter, and strictly

[1] For any choice of n, these results can be directly numerically verified by solving for the unique global minimum of $\left(\partial^2 CR(\lambda)\right)/\left(\partial\lambda^2\right)$ with respect to the choice of $\mathbf{w}, \lambda$, leading to $\mathbf{w} = n^{-1}\mathbf{1}_n$ and $\lambda$ arbitrary.

convex when $\mathbf{w} \neq n^{-1}\mathbf{1}_n$, ensuring that any solution to the first order condition in (9) defines a well-defined minimum with respect to the choice of $\lambda$.[2] The existence result follows from the convexity result, along with the fact that CR is continuous in $\lambda$ and that for any $\mathbf{w} \neq n^{-1}\mathbf{1}_n$, $\lim_{\lambda \to -\infty} \mathrm{CR}(\lambda) = \lim_{\lambda \to \infty} \mathrm{CR}(\lambda) = +\infty$. Therefore, for any finite value $\eta > \min_\lambda \mathrm{CR}(\lambda)$, there exists finite $\lambda_1$ and $\lambda_2$ such that $\eta = \mathrm{CR}(\lambda_1) = \mathrm{CR}(\lambda_2)$, and by Rolle's Theorem, $\frac{\partial \mathrm{CR}(\lambda)}{\partial \lambda} = 0$ for some finite $\lambda \in [\lambda_1, \lambda_2]$.

Given the conditional-on-**w** optimal solution for $\lambda$ defined above, the global optimal solution to (7) can be obtained by choosing **w** optimally. The existence of such a global optimal **w** is assured by Weierstrass's theorem because the feasible space of choices for **w** is closed and bounded (and also convex), and the objective function is continuous (as well as differentiable).

## *3.2 Rationale*

The objective underlying the definition of the GMD measure can be interpreted as choosing the discrepancy measure, among all of the discrepancy measures available in the CR family, that results in the assignment of the least discrepancy value to any given *subject* probability distribution, **w**, and *reference* probability distribution, **q**. The discrepancy measure chosen via the GMD principle thus attempts to rationalize any specification of **w**, conditional on **q**, by judging the subject distribution **w** to be closer to the reference distribution **q** (in terms of discrepancy value) than would be the case for any other choice of discrepancy measure in the CR family. As such, $I_*(\mathbf{w}, \mathbf{q})$ minimizes the influence of the choice of discrepancy measure on the actual choice of subject distribution to be paired with the reference distribution in the estimation of the probability weights and parameter values in the data sampling model. It follows that $I_*(\mathbf{w}, \mathbf{q})$ can be thought of as being the least influential, or most neutral discrepancy measure to be applied to any pair of subject and reference distributions in the CR family. In the absence of any informative prior information suggesting that some discrepancy measures are relatively more appropriate than others, this would appear to be a defensible way to proceed in attempting to reconcile data weights with a reference distribution, together with any estimating equation or moment information that the data observations were required to satisfy.

Explicit extremum metrics used in estimation and inference are often chosen in an ad-hoc manner. We proceed by specifying a general family of extremum metrics that all measure divergence between subject and reference probability distributions and then solve, based on the data, for the subject distribution **w** and particular metric that results in **w** being overall closest to the reference distribution **q**. When **q** is based on the EDF of the data, as has most often been the case in the literature, this principle leads to adopting an empirical likelihood function that is as close to EDF weights

[2] Note that the exception $\mathbf{w} = n^{-1}\mathbf{1}_n$ is generally relegated to infeasibility whenever the moment conditions or estimating equations overidentify the parameters of a model being estimated, making even this singleton exception moot in applications.

as the moment constraints will allow. Choosing an integer value for $\lambda$ in (4), as is traditional, would appear to use information one usually does not posses, and our approach avoids the need for making an arbitrary choice.

### 3.3 GMD Estimation and Inference

The GMD estimator of $\beta$ will be defined by solving the maximum empirical likelihood-type problem of the form

$$\hat{\beta} = \underset{\beta \in B}{\arg\max} \left[ \ell_E(\beta) = \max_{w} \left\{ -I_*(\mathbf{w}, \mathbf{q}) \mid \sum_{i=1}^{n} w_i \mathbf{z}_{i.}' (y_i - \mathbf{x}_{i.}\beta) = \mathbf{0}, \sum_{i=1}^{n} w_i = 1, w_i \geq 0; \forall i \right\} \right] \quad (11)$$

It can be anticipated that the typical first order asymptotic results continue to hold. In particular, given that the estimating equations are valid, then $w_i \to n^{-1}\ \forall i$ and the estimator of $\beta$ will have the same first order asymptotics as the GMM estimator based on the moments $n^{-1} \sum_{i=1}^{n} \mathbf{z}_{i.}' (y_i - \mathbf{x}_{i.}\beta) = 0$, which applies as well to all of the specific members of the CRMD family of estimators.

Baggerly [1] has shown that the same asymptotic chi-square calibration holds for every choice of the parameter $\lambda$. Thus inference can proceed by an appropriate application of the empirical likelihood ratio statistic defined in (5). Given that the probability distribution of the unconditional empirical likelihood statistic is a mixture, over the distribution of $\hat{\lambda}$, of the associated appropriately scaled likelihood ratio statistics, the asymptotic chi-square calibration is maintained for the statistic based on the GMD approach.

## 4 Econometric Implications

Given a structural equation statistical model and a corresponding sample of data, we cast the estimation problem as one of how to best estimate the response coefficients when one's prior knowledge consists only of the expected values of moment functions of the sample information. Rather than choosing the estimation criterion in an ad hoc manner, we ask the basic question of how to make best use of the sample information in an information theoretic context and use the Cressie–Read family of divergence statistics as a basis for identifying a family of potential distance measures. Thus the measure of goodness-distance measure chosen, between a reference probability distribution q and the probability distribution **p** being approximated, is thus a data based decision. The CR measure of divergence includes within its family of estimators the empirical likelihood (EL), the Kullback–Leibler (KL)-maximum entropy, and the log Euclidean likelihood estimation alternatives.

Some of the attractive characteristics of the CR divergence measures are that first and second derivatives of the CR function with respect to $\lambda$ exist and the derivatives are smoothly differentiable and permit a definition of the Hessian of the CR

Statistic (see the Appendix, Definition 1). The CR as a function of $\lambda$ is strictly convex and has a well-defined minimum for a given $\mathbf{w}$ vector. A global minimum value also exists relative to closed and bounded choices of both $\mathbf{w}$ and $\lambda$ , and if $\lambda$ is confined to a fixed interval, the entire set of $\mathbf{w}$ and $\lambda$ values is in fact closed and bounded. If there are no data-moment constraints on the choices of the $\mathbf{w}$ and $\lambda$ arguments, the global optimum is not unique and a resulting ridge of $\lambda$ values are all optimal. Implicit differentiation can be used to define the appropriate derivatives of the probabilities with respect to the explanatory variables. The derivatives are flexible in the sense that they are functions of the sample data and not dependent on parameters other than data-determined Lagrange multipliers. The optimal member of the class of CR-based estimators avoids tuning parameter choices or estimation of nuisance parameters such as unknown covariance components in the case of the traditional GMM estimator. Finally, from an asymptotic standpoint the range of estimators from the CR($\lambda$) family under the usual regularity conditions are consistent and asymptotically normal and efficient.

Econometric and statistical ventures are by necessity conditional in nature. In this research we have, in the spirit of Occam's Razor, attempted to reduce the conditions required for obtaining a superior performing solution for this estimation and inference problem.

Finally, we note that [9], building on the work of [30], [31], and [32], have established a close relationship between the maximum entropy principle CR(0) and the problem of minimizing worst case expected loss. Using a game theoretic interpretation, they demonstrate that the $P^*$ distribution that maximizes entropy over a class of $\Gamma$ distributions also minimizes the worst case expected logarithmic score (log loss). In decision theory terminology this means $P^*$ is a robust Bayes or $\Gamma$ minimax rule when the loss is measured by a log score. In the important case where $\Gamma$ is described by mean value constraints, the challenge is to extend their results to our CR($\lambda$) context of $\mathbf{p}$ and $\mathbf{q}$ distributions and the distance measures of relative entropy, discrepancy and divergence.

**Acknowledgements** We gratefully acknowledge the helpful comments and insights provided by Keith Baggerly, Marian Grendar, and Guido Imbens.

## 5 Appendix: Propositions, Proofs, and Definitions

**Proposition 1.** *Subject-Reference Distribution Symmetry in the CR Family of Power Divergence Statistics*

Let $\mathbf{w}$ and $\mathbf{q}$ denote two n-element probability distributions, such that $\mathbf{w} \gg \mathbf{0}$, $\mathbf{q} \gg \mathbf{0}$, and $\mathbf{1}_n'\mathbf{w} = \mathbf{1}_n'\mathbf{q} = 1$. Define the family of CR divergence statistics, indexed by the parameter $\lambda$, alternatively as

$$I(\mathbf{w},\mathbf{q},\lambda) = \frac{1}{\lambda(\lambda+1)} \sum_{i=1}^{n} w_i \left[ \left( \frac{w_i}{q_i} \right)^{\lambda} - 1 \right] \tag{12}$$

or

$$I(\mathbf{q},\mathbf{w},\lambda)=\frac{1}{\lambda(\lambda+1)}\sum_{i=1}^{n} q_i\left[\left(\frac{q_i}{w_i}\right)^{\lambda}-1\right]. \tag{13}$$

Then

a) $I(\mathbf{w},\mathbf{q},\alpha)=I(\mathbf{q},\mathbf{w},-(1+\alpha))\,\forall\alpha\neq 0 \text{ or } -1,$

and

b) $\lim\limits_{\alpha\to 0 \text{ or } -1} I(\mathbf{w},\mathbf{q},\alpha)=\lim\limits_{\alpha\to 0 \text{ or} -1} I(\mathbf{q},\mathbf{w},-(1+\alpha)).$

*Proof. Part (a)*: Evaluating the two discrepancy measures at $\alpha$ and $-(1+\alpha)$, respectively, when $\alpha\neq 0$ *or* $-1$ yields

$$I(\mathbf{w},\mathbf{q},\alpha)=\frac{1}{\alpha(\alpha+1)}\sum_{i=1}^{n} w_i\left[\left(\frac{w_i}{q_i}\right)^{\alpha}-1\right]=\frac{1}{\alpha(\alpha+1)}\left(\left[\sum_{i=1}^{n} w_i^{\alpha+1}q_i^{-\alpha}\right]-1\right) \tag{14}$$

$$\begin{aligned}I(\mathbf{q},\mathbf{w},-(1+\alpha))&=\frac{1}{-(1+\alpha)(1-(1+\alpha))}\sum_{i=1}^{n} q_i\left[\left(\frac{q_i}{w_i}\right)^{-(1+\alpha)}-1\right]\\&=\frac{1}{\alpha(\alpha+1)}\left(\left[\sum_{i=1}^{n} w_i^{\alpha+1}q_i^{-\alpha}\right]-1\right)\end{aligned} \tag{15}$$

which demonstrates the validity of part a).

*Part (b)*: First examine the case where $\alpha\to 0$. Applying L'Hopital's rule to evaluate the limits yields

$$\begin{aligned}\lim_{\alpha\to 0} I(\mathbf{w},\mathbf{q},\alpha)&=\lim_{\alpha\to 0}\frac{\sum_{i=1}^{n} w_i\left[\left(\frac{w_i}{q_i}\right)^{\alpha}-1\right]}{\alpha(\alpha+1)}\\&=\lim_{\alpha\to 0}\left[\frac{\sum_{i=1}^{n} w_i\left(\frac{w_i}{q_i}\right)^{\alpha}\ln\left(\frac{w_i}{q_i}\right)}{2\alpha+1}\right]=\sum_{i=1}^{n} w_i\ln\left(\frac{w_i}{q_i}\right)\end{aligned} \tag{16}$$

and

$$\begin{aligned}\lim_{\alpha\to 0} I(\mathbf{q},\mathbf{w},-(1+\alpha))&=\lim_{\gamma\to -1} I(\mathbf{q},\mathbf{w},\gamma)\\&=\lim_{\gamma\to 0}\frac{\sum_{i=1}^{n} q_i\left[\left(\frac{q_i}{w_i}\right)^{\gamma}-1\right]}{\gamma(\gamma+1)}\\&=\lim_{\gamma\to -1}\left[\frac{\sum_{i=1}^{n} q_i\left(\frac{q_i}{w_i}\right)^{\gamma}\ln\left(\frac{q_i}{w_i}\right)}{2\gamma+1}\right]=\sum_{i=1}^{nw_i}\ln\left(\frac{w_i}{q_i}\right)\end{aligned} \tag{17}$$

which demonstrates the result when $\alpha\to 0$.

Now examine the case where $\alpha \to -1$. Again applying L'Hopital's rule to evaluate the limits yields

$$\lim_{\alpha\to-1} I(\mathbf{w},\mathbf{q},\alpha) = \lim_{\alpha\to-1} \frac{\sum_{i=1}^{n} w_i \left[\left(\frac{w_i}{q_i}\right)^{\alpha} - 1\right]}{\alpha(\alpha+1)}$$
$$= \lim_{\alpha\to-1} \left[\frac{\sum_{i=1}^{n} w_i \left(\frac{w_i}{q_i}\right)^{\alpha} \ln\left(\frac{w_i}{q_i}\right)}{2\alpha+1}\right] = \sum_{i=1}^{n} q_i \ln\left(\frac{q_i}{w_i}\right) \tag{18}$$

and

$$\lim_{\alpha\to-1} I(\mathbf{q},\mathbf{w},-(1+\alpha)) = \lim_{\gamma\to0} I(\mathbf{q},\mathbf{w},\gamma)$$
$$= \lim_{\gamma\to0} \frac{\sum_{i=1}^{n} q_i \left[\left(\frac{q_i}{w_i}\right)^{\gamma} - 1\right]}{\gamma(\gamma+1)}$$
$$= \lim_{\alpha\to0} \left[\frac{\sum_{i=1}^{n} q_i \left(\frac{q_i}{w_i}\right)^{\gamma} \ln\left(\frac{q_i}{w_i}\right)}{2\gamma+1}\right] = \sum_{i=1}^{n} q_i \ln\left(\frac{q_i}{w_i}\right) \tag{19}$$

which demonstrates the result when $\alpha \to -1$.

For a discussion of the symmetry issue in a somewhat different context see [19].

**Proposition 2.**

$$\lim_{\gamma\to0} I\left(\mathbf{w}, n^{-1}\mathbf{1}_n, \gamma\right) = \lim_{\gamma\to0} \left[\frac{\sum_{i=1}^{n} w_i \left[\left(\frac{w_i}{n^{-1}}\right)^{\gamma} - 1\right]}{\gamma(\gamma+1)}\right] = \sum_{i=1}^{n} w_i \ln(w_i) + \ln(n) \tag{20}$$

*Proof.* Applying L'Hopital's rule to the ratio of terms yields

$$\lim_{\gamma\to0} \left[\frac{\sum_{i=1}^{n} w_i (n w_i)^{\gamma} \ln(n w_i)}{2\gamma+1}\right] = \sum_{i=1}^{n} w_i \ln(w_i) + \ln(n) \tag{21}$$

because $\sum_{i=1}^{n} w_i = 1$.

**Proposition 3.**

$$\lim_{\gamma\to-1} I\left(\mathbf{w}, n^{-1}\mathbf{1}_n, \gamma\right) = \lim_{\gamma\to-1}\left[\frac{\sum\limits_{i=1}^{n} w_i\left[\left(\frac{w_i}{n^{-1}}\right)^{\gamma}-1\right]}{\gamma(\gamma+1)}\right] = -\sum_{i=1}^{n} n^{-1}\ln(w_i) - \ln(n) \tag{22}$$

*Proof.* Applying L'Hopital's rule to the ratio of terms yields

$$\lim_{\gamma\to-1}\left[\frac{\sum\limits_{i=1}^{n} w_i\,(n\,w_i)^{\gamma}\ln(n\,w_i)}{2\gamma+1}\right] = -\sum_{i=1}^{n} n^{-1}\ln(w_i) - \ln(n) \tag{23}$$

because $\sum\limits_{i=1}^{n} w_i = 1$.

**Definition 1.** Hessian of $I\left(\mathbf{w}, n^{-1}\mathbf{1}_n, \lambda\right)$

Let $\xi \equiv \begin{bmatrix}\mathbf{w}\\ \lambda\end{bmatrix}$. Then

$$\frac{\partial^2 I\left(\mathbf{w}, n^{-1}\mathbf{1}_n, \lambda\right)}{\partial\xi\partial\xi\prime} = \begin{bmatrix} \frac{\partial^2 I\left(\mathbf{w}, n^{-1}1_n, \lambda\right)}{\partial\mathbf{w}\partial\mathbf{w}'} & \frac{\partial^2 I\left(\mathbf{w}, n^{-1}1_n, \lambda\right)}{\partial\mathbf{w}\partial\lambda} \\ \frac{\partial^2 I\left(\mathbf{w}, n^{-1}\mathbf{1}_n, \lambda\right)}{\partial\lambda\partial\mathbf{w}\prime} & \frac{\partial^2 I\left(\mathbf{w}, n^{-1}\mathbf{1}_n, \lambda\right)}{\partial\lambda^2} \end{bmatrix} \tag{24}$$

where

$$\frac{\partial^2 I\left(\mathbf{w}, n^{-1}\mathbf{1}_n, \lambda\right)}{\partial\mathbf{w}\,\partial\mathbf{w}'} = n^{\lambda}\begin{bmatrix} w_1^{\lambda-1} & 0 & \cdots & 0 \\ 0 & w_2^{\lambda-1} & \cdots & 0 \\ 0 & 0 & \ddots & \vdots \\ 0 & 0 & \cdots & w_2^{\lambda-1} \end{bmatrix} = n^{\lambda}\left(\mathbf{I}_n \odot \mathbf{w}^{\lambda-1}\right) \tag{25}$$

$$\frac{\partial^2 I\left(\mathbf{w}, n^{-1}\mathbf{1}_n, \lambda\right)}{\partial\lambda^2} = \sum_{i=1}^{n} n^{\lambda} {w_i}^{\lambda+1}\left[\frac{(\ln(nw_i))^2}{\lambda(\lambda+1)} - \frac{2(2\lambda+1)\ln(nw_i)}{(\lambda(\lambda+1))^2} + \frac{6\lambda(\lambda+1)+2}{(\lambda(\lambda+1))^3}\right] - \frac{6\lambda(\lambda+1)+2}{(\lambda(\lambda+1))^3}, \tag{26}$$

and

$$\frac{\partial^2 I\left(\mathbf{w}, n^{-1}\mathbf{1}_n, \lambda\right)}{\partial\mathbf{w}\partial\lambda} = \left(\frac{n^{\lambda}}{\lambda^2}\right)\left[\mathbf{w}^{\lambda} \odot \left(\lambda\ln(n\mathbf{w}) - \mathbf{1}_n\right)\right] \tag{27}$$

## References

[1] Baggerly, K.: Empirical likelihood as a goodness of fit measure. Biometrika **85**, 535–547 (1998)
[2] Baggerly, K.: Studentized empirical likelihood and maximum entropy empirical t. Working paper, Department of Statistics, Rice University, Houston, Texas (2001)
[3] Brown, B., Chen, S.: Combined least squares empirical likelihood. Ann. Inst. Stat. Math. **60**, 697–714 (1998)
[4] Corcoran, S.: Empirical exponential family likelihood using several moment conditions. Stat. Sinica **10**, 545–557 (2000)
[5] Cotofrei, P.: A possible generalization of the empirical likelihood computer sciences. Technical report, University of "A.I. Cuza," Iasi, Romania (2003)
[6] Cressie, N., Read, T.: Multinomial goodness of fit tests. J. Roy. Statist. Soc. B **46**, 440–464 (1984)
[7] Golan, A., Judge, G.G., Miller, D.: Maximum Entropy Econometrics. Wiley, New York (1996)
[8] Godambe, V.: An optimum property of regular maximum likelihood estimation. Ann. Math. Stat. **31**, 1208–1212 (1960)
[9] Grunwald, P., David, A.: Game theory, maximum entropy, minimum discrepancy and robust bayesian decision theory. Ann. Stat. **32**, 1367–1433 (2004)
[10] Judge, G.G., Griffith, W.E., Hill, R.C., Lütkepohl, H., Lee, T.-C.: The Theory and Practice of Econometrics. Wiley, New York (1985)
[11] Imbens, G.W., Spady, R.H., Johnson, P.: Information theoretic approaches to inference in moment condition models. Econometrica **66**, 333–357 (1998)
[12] Hansen, L.: Large sample properties of generalized method of moments estimators. Econometrica **50**, 1029–1054 (1982)
[13] Heyde, C.: Quasilikelihood and optimality of estimating functions: some current and unifying themes. Bull. Int. Stat. Inst. **1**, 19–29 (1989)
[14] Heyde, C., Morton, R.: Multiple roots in general estimating equations. Biometrika **85**, 954–959 (1998)
[15] Lindsay, B., Qu, A.: Inference functions and quadratic score tests. Stat. Sci. **18**, 394–410 (2003)
[16] Mittelhammer, R., Judge, G.G.: Robust empirical likelihood estimation of models with nonorthogonal noise components. Volume in Honor of Henri Theil. To appear in: J. Agr. Appl. Econ. (2008)
[17] Mittelhammer, R., Judge, G.G.: Endogeneity and Moment Based Estimation under Squared Error Loss. In: Wan, A., Ullah, A. (eds.) Handbook of Applied Econometrics and Statistical Inference, Dekker, New York (2001)
[18] Newey, W., Smith, R.: Asymptotic bias and equivalence of GMM and GEL estimators. MIT Working paper, MIT, USA (2000)
[19] Österreicher, F.: Csiszar's f-divergencies-basic properties. Working paper, Institute of Mathematics, University of Salzburg, Austria (2002)
[20] Österreicher, F., Vajda, I: A new class of metric divergencies on probability spaces and its applicability in statistics. Ann. Inst. Stat. Math. **55**, 639–653 (2003)

[21] Owen, A.: Empirical likelihood ratio confidence intervals for a single functional. Biometrika **75**, 237-249 (1988)
[22] Owen, A.: Empirical likelihood for linear models. Ann. Stat. **19**, 1725–1747 (1991)
[23] Owen, A.: Empirical Likelihood, Chapman & Hall, New York (2001)
[24] Qin, J.: Combining parametric and empirical likelihood data. Biometrika **87**, 484–490 (2000)
[25] Qin, J., Lawless, J.: Empirical likelihood and general estimating equations. Ann. Stat. **22**, 300–325 (1994)
[26] Mittelhammer, R., Judge, G.G., Miller, D.: Econometric Foundations. Cambridge University Press, Cambridge (2000)
[27] Mittelhammer, R., Judge, G.G., Schoenberg, R.: Empirical Evidence Concerning the Finite Sample Performance of EL-Type Structural Equation Estimation and Inference Methods. In: Festschrift in Honor of Thomas Rothenberg, Cambridge University Press, Cambridge, UK (2003)
[28] Rao, C.R: Tests of significance in multivariate analysis. Biometrika **35**, 58–79 (1948)
[29] Read, T., Cressie, N.: Goodness of fit statistics for discrete multivariate data. Springer, New York (1988)
[30] Topsoe, F.: Informational theoretical optimization techniques. Kybernetika **15**, 8–27 (1979)
[31] Walley, P.: Statistical Reasoning with Imprecise Probabilities. Chapman & Hall, New York (1991)
[32] Wolpert, D., Wolf, D.: Estimating functions of probability distributions from a finite set of data. Phys. Rev. E **6**, 6841–6852 (1995)

# More on the F-test under Nonspherical Disturbances

Walter Krämer and Christoph Hanck

**Abstract** We show that the F-test can be both liberal and conservative in the context of a particular type of nonspherical behaviour induced by spatial autocorrelation, and that the conservative variant is more likely to occur for extreme values of the spatial autocorrelation parameter. In particular, it will wipe out the progressive one as the sample size increases.

## 1 Introduction and Summary

The robustness of the F-test to nonspherical disturbances has concerned applied statisticians for many decades. The present paper considers the F-test in the context of the linear regression model

$$y = X\beta + u = X^{(1)}\beta^{(1)} + X^{(2)}\beta^{(2)} + u, \tag{1}$$

where $y$ and $u$ are $T \times 1$, $X$ is $T \times K$ and nonstochastic of rank $K < T$, $\beta$ is $K \times 1$, and the disturbance vector $u$ is multivariate normal with mean zero and (possibly) nonscalar covariance matrix $V$. The design matrix is partioned into $X^{(1)}(T \times q)$ and $X^{(2)}(T \times (K-q))$ and the null hypothesis to be tested is $H_0 : \beta^{(1)} = b^{(1)}$.

The standard F-test assumes that $V = \sigma^2 I$ and rejects for large values of

$$F = \frac{(\tilde{u}'\tilde{u} - \hat{u}'\hat{u})/q}{\hat{u}'\hat{u}/(T-K)}, \tag{2}$$

where $\hat{u} = y - X\hat{\beta}$, $\hat{\beta} = (X'X)^{-1}X'y$, $\tilde{u} = y - X^{(1)}b^{(1)} - X^{(2)}\tilde{\beta}^{(2)}$, $\tilde{\beta}^{(2)} = (X^{(2)\prime} X^{(2)})^{-1}X^{(2)\prime}(y - X^{(1)}b^{(1)})$. Its null distribution is central F with $q$ and $T-K$ degrees

Christoph Hanck
Department Quantitative Economics, Universiteit Maastricht, NL-6211 LM Maastricht, Netherlands
c.hanck@ke.unimaas.nl

of freedom and the problem to be studied here is the robustness of this null distribution to deviations from $V = \sigma^2 I$.

Vinod [8] and Kiviet [4] address this problem for a given disturbance covariance matrix $V$, and derive bounds for the size of the test when the design matrix $X$ varies across all $T \times K$ matrices of rank $K$, while Banerjee and Magnus [2] and Hillier and King [3] consider the test statistics themselves. Below we follow Krämer [5, 6] and Krämer et al. [7] by fixing $X$ and letting $V$ vary across certain subsets of possible disturbance covariance matrices which are likely to occur in practice. This seems the more natural approach, as $X$ is always known in applications, whereas $V$ is an unknown $T \times T$ parameter matrix.

The subset of disturbance covariance matrices under study here is implicitly defined by the spatial autoregressive scheme

$$u = \rho W u + \varepsilon, \tag{3}$$

where $\varepsilon$ is a $T \times 1$ normal random vector with mean zero and scalar covariance matrix $\sigma_\varepsilon^2 I$, and $W$ is some known $T \times T$-matrix of nonnegative spatial weights with $w_{ii} = 0$ $(i = 1, \ldots, T)$. Although there are many other patterns of spatial dependence which have been suggested in the literature (see Anselin and Florax [1] for an overview), the one defined by (3) is by far the most popular, so it seems worthwhile to investigate the behaviour of parameter estimates and tests when the regression disturbances "misbehaves" according to this particular scheme.

Below we build on Krämer [6], who shows that the size of the test can tend to both one and zero as the parameter $\rho$ varies across its allowable range. While Krämer [6] is silent on the respective empirical relevance of the two extreme cases, we show here that the conservative variant is far more likely to occur in practice, and will wipe out the liberal one as sample size increases.

## 2 The Null Distribution under Spatial Autocorrelation

The coefficient $\rho$ in (3) measures the degree of correlation, which can be both positive and negative. There is no disturbance autocorrelation where $\rho = 0$. Below we focus on the empirically more relevant case of positive disturbance correlation, where

$$0 \leq \rho < \frac{1}{\lambda_{max}}$$

and where $\lambda_{max}$ is the Frobenius-root of $W$ (i.e. the unique positive real eigenvalue such that $\lambda_{max} \geq |\lambda_i|$ for arbitrary eigenvalues $\lambda_i$). The disturbances are then given by

$$u = (I - \rho W)^{-1} \varepsilon, \tag{4}$$

so $V := Cov(u) = \sigma_\varepsilon^2 \left[(I - \rho W)(I - \rho W)'\right]^{-1}$ and $V = \sigma_\varepsilon^2 I$ whenever $\rho = 0$.

The behaviour of the test statistic (2) when disturbances are given by (3) is best seen by first rewriting it as

$$F = \frac{u'(M^{(2)} - M)u/q}{u'Mu/(T-K)}, \tag{5}$$

where $M = I - X(X'X)^{-1}X'$ and $M^{(2)} = I - X^{(2)}(X^{(2)'}X^{(2)})^{-1}X^{(2)'}$. Let $F^{\alpha}_{q,T-K}$ be the $(1-\alpha)$ quantile of the central F-distribution with $q$ and $T-K$ degrees of freedom, respectively, where $\alpha$ is the nominal size of the test. Then

$$\begin{aligned}
P(F \geq F^{\alpha}_{q,T-K}) &= P(u'(M^{(2)} - M)u - \frac{q}{T-K}F^{\alpha}_{q,T-K}u'Mu \geq 0) \\
&= P(u'(M^{(2)} - dM)u \geq 0) \\
&\quad (\text{where } d = 1 + \frac{q}{T-K}F^{\alpha}_{q,T-K}) \\
&= P(\eta'(I-\rho W)'(M^{(2)} - dM)(I-\rho W)\eta \geq 0) \\
&\quad (\text{where } \eta = \frac{1}{\sigma_{\varepsilon}}\varepsilon \sim N(0,I)) \\
&= P(\sum_{i=1}^{T} \lambda_i \xi_i^2 \geq 0) \\
&= P((1-\rho\lambda_{max})^2 \sum_{i=1}^{T} \lambda_i \xi_i^2 \geq 0),
\end{aligned} \tag{6}$$

where the $\xi_i^2$ are iid $\chi^2_{(1)}$ and the $\lambda_i$ are the eigenvalues of $(I-\rho W)'(M^{(2)} - dM)(I-\rho W)$, and therefore also of $V(M^{(2)} - dM)$.

The limiting rejection probability as $\rho \to 1/\lambda_{max}$ depends upon the limiting behaviour of $(1-\rho\lambda_{max})^2 V$. We confine ourselves to the case where $W$ is symmetric, which appears to be the more important one in practice. This will for instance occur if spatial dependence follows the $j$-ahead-and-$j$-behind or the equal-weight criteria (see section 3 below). Then $W$ admits a spectral decomposition

$$W = \sum_{i=1}^{T} \lambda_i \omega_i \omega_i', \tag{7}$$

where we have without loss of generality arranged the eigenvalues $\lambda_i$ in increasing order, and

$$V = \sum_{i=1}^{T} \frac{\sigma_{\varepsilon}^2}{(1-\rho\lambda_i)^2} \omega_i \omega_i' \tag{8}$$

is the resulting spectral decomposition of $V$, which always exists as $V$ is symmetric. The point of our argument now is that

$$\lim_{\rho \to 1/\lambda_{max}} (1-\rho\lambda_{max})^2 V = \sigma_{\varepsilon}^2 \omega_T \omega_T', \tag{9}$$

a matrix of rank 1. Therefore, all limiting eigenvalues of

$$(1-\rho\lambda_{max})^2 V(M^{(2)}-dM) \tag{10}$$

are zero except one, which is given by

$$c_T = tr(\omega_T \omega_T'(M^{(2)}-dM)) = \omega_T'(M^{(2)}-dM)\omega_T. \tag{11}$$

This constant $c_T$ is crucial for our analysis. It determines whether the F-test will eventually be conservative or liberal. If $c_T$ is positive, the rejection probability of the F-test will tend to 1 as $\rho$ approaches $1/\lambda_{max}$. The test is then liberal in the extreme, at least for values of $\rho$ close to the edge of the parameter space.

If $c_T$ is negative, the rejection probability will tend to zero, and the test will eventually be extremely conservative. And if $c_T = 0$, the limiting behaviour of the test cannot be determined from the limiting behaviour of the eigenvalues of $(1-\rho\lambda_{max})^2 V(M^{(2)}-dM)$ (which are all zero). Section 3 now sheds some light on which of these cases is more likely to occur in empirical applications.

## 3 Exact Rejection Probabilities in Finite Samples

The first important point to make is that the crucial constant $c_T$ depends only on $X$ and $W$ and the nominal size of the test, all of which are known. Therefore, $c_T$ is known as well and can guide the user in interpreting a test: If $c_T < 0$, one has to beware of a loss in power, and if $c_T > 0$, one has to beware of spurious rejections.

The following argument shows that the first problem is far more likely to occur in practice: Rewrite the critical constant as

$$c_T = \omega_T' M^{(2)} \omega_T - \omega_T' \frac{T-K+q}{T-K} F^{\alpha}_{q,T-K} M \omega_T. \tag{12}$$

Then it is easily seen that in general $c_T < 0$ (i.e. except in very contrived cases). This follows from the fact that

$$\frac{T-K+q}{T-K} F^{\alpha}_{q,T-K} \rightarrow \chi_q^{2,\alpha}/q \tag{13}$$

as $T \rightarrow \infty$, which is larger than 2 for moderate values of $\alpha$ and $q$. (It takes the values 3.84, 2.99 and 2.60 for $\alpha = 0.05$ and $q$ =1, 2 und 3, respectively). This will in general be more than enough to counterbalance the fact that $\omega_T' M^{(2)} \omega_T > \omega_T' M \omega_T$.

Of course one can always construct a weighting matrix and regressor matrices $W, X$ and $X^{(2)}$ such that $\omega_T' M^{(2)} \omega_T = 1$ and $\omega_T' M \omega_T = 0$ and therefore $c_T > 0$. For instance, let $\iota = (1,\ldots,1)'$ be a $(T \times 1)$-vector and choose $X^{(2)}$ orthogonal to $\iota$. E.g., for $T$ even, pick $X^{(2)} = (1,-1,1,-1,\ldots,1,-1)'$. Let

$$X = \left[\iota \vdots X^{(2)}\right] \tag{14}$$

(that is, test $H_0 : \beta^{(1)} = 0$) and let

$$W = W^{EW} = (w_{ij}^{EW}) = \begin{cases} 1 & \text{for } i \neq j \\ 0 & \text{for } i = j \end{cases}, \tag{15}$$

the equal weight matrix. This and similar cases will however rarely happen in the natural course of events, and will become ever more unlikely as sample size increases.

Figure 1 gives an example where $W$ is derived from the queen-criterion (see Figure 2 for an illustration of the criterion with $N = 9$):

There is a square of cells, and all cells around a given cell obtain a weight of 1. The sample size is then a square number. We choose $X$ to have $K = 2$ and $T = 16$ or 25 such that, for $T = 16$, the second column is the (normalized) eigenvector

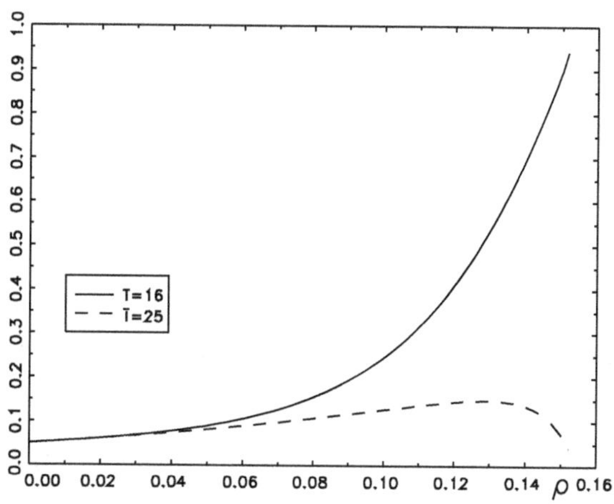


**Fig. 1** Rejection probabilities for the queen matrix

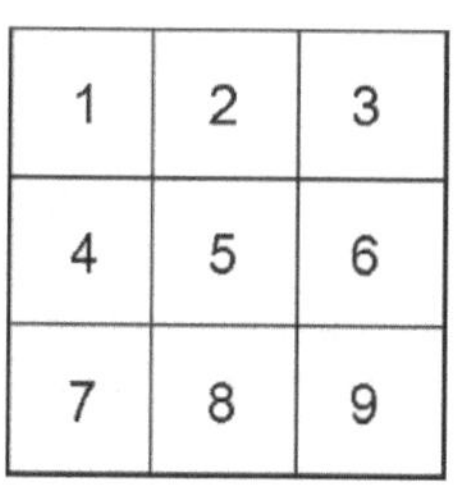


$$W = \begin{pmatrix} 0\,1\,0\,1\,1\,0\,0\,0\,0 \\ 1\,0\,1\,1\,1\,1\,0\,0\,0 \\ 0\,1\,0\,0\,1\,1\,0\,0\,0 \\ 1\,1\,0\,0\,1\,0\,1\,1\,0 \\ 1\,1\,1\,1\,0\,1\,1\,1\,1 \\ 0\,1\,1\,0\,1\,0\,0\,1\,1 \\ 0\,0\,0\,1\,1\,0\,0\,1\,0 \\ 0\,0\,0\,1\,1\,1\,1\,0\,1 \\ 0\,0\,0\,0\,1\,1\,0\,1\,0 \end{pmatrix}$$

**Fig. 2** An example of the queen matrix

corresponding to the largest eigenvalue of $W$ (which happens to be $\lambda_{max} = 5.85$), and the first column is any $(16 \times 1)$-vector orthogonal to $\omega_{16}$. For $t > 16$, $x_{t_1} = 1$ and $x_{t_2} = t - 16$. Then we have $c_{16} = 0.316$ and $c_{25} = -0.850$, and so our theoretical result predicts that the rejection probabilities will tend to one as $\rho \to 1/\lambda_{max}$ for $T = 16$ and will tend to zero as $\rho \to 1/\lambda_{max}$ for $T = 25$. Figure 1 shows that this is indeed the case.

The case $c_T = 0$, where our analysis does not apply, will occur for instance whenever $\omega_T$ is in the column space of $X^{(2)}$. The most important special case is when $W$ is row-normalized and therefore $\omega_T = \frac{1}{\sqrt{T}}(1,\ldots,1)'$ and where in addition $X^{(2)}$ contains an intercept. However, row-normalization will often destroy the symmetry of $W$, so this case is not covered by our discussion above.

**Acknowledgement** Research supported by Deutsche Forschungsgemeinschaft (DFG) under SFB 475. We are grateful to Anurag Banerjee for providing us the Gauss routines for our calculations.

## References

[1] Anselin, L., Florax, R. (eds.): New directions in spatial econometrics. Springer, Berlin (1995)
[2] Banerjee, A.N., Magnus, J.: On the sensitivity of the usual t- and F-tests to covariance misspecifications. J. Econometrics **95**, 157–176 (2000)
[3] Hillier, O.H., King, M.L: Linear regression with correlated errors: bounds on coefficient estimates and t-values. In: King, M.L., Giles, D.E.A.(eds.) Specification Analysis in the Linear Model, pp. 74–80. Routledge and Kegan-Paul, London (1982)
[4] Kiviet, J.F.: Effects of ARMA errors on tests for regression coefficients: comments on Vinod's article, improved and additional results. J. Am. Stat. Assoc. **75**, 333–358 (1980)
[5] Krämer, W.: On the robustness of the F-test to autocorrelation among disturbances. Econ. Lett. **30**, 37–40 (1989)
[6] Krämer, W.: The robustness of the F-test to spatial autocorrelation among regression disturbances. Statistica **63**, 435–440 (2003)
[7] Krämer, W., Kiviet, J., Breitung, J.: The null distribution of the $F$-test in the linear regression model with autocorrelated disturbances. Statistica **50**, 503–509 (1990)
[8] Vinod, H.D.: Effects of ARMA errors on the significance tests for regression coefficients. J. Am. Stat. Assoc. **71**, 929–933 (1976)

# Optimal Estimation in a Linear Regression Model using Incomplete Prior Information

Helge Toutenburg, Shalabh, and Christian Heumann

**Abstract** For the estimation of regression coefficients in a linear model when incomplete prior information is available, the optimal estimators in the classes of linear heterogeneous and linear homogeneous estimators are considered. As they involve some unknowns, they are operationalized by substituting unbiased estimators for the unknown quantities. The properties of resulting feasible estimators are analyzed and the effect of operationalization is studied. A comparison of the heterogeneous and homogeneous estimation techniques is also presented.

## 1 Introduction

Postulating the prior information in the form of a set of stochastic linear restrictions binding the coefficients in a linear regression model, Theil and Goldberger [3] have developed an interesting framework of the mixed regression estimation for the model parameters; see e.g., Srivastava [2] for an annotated bibliography of earlier developments and Rao et al. [1] for some recent advances. Such a framework assumes that the variance covariance matrix in the given prior information is known. This specification may not be accomplished in many practical situations where the variance covariance may not be available for one reason or the other. Even if available, its accuracy may be doubtful and consequently its credibility may be sufficiently low. One may then prefer to discard it and treat it as unknown. Appreciating such circumstances, Toutenburg et al. [4] have introduced the method of weakly unbiased estimation for the regression coefficients and have derived the optimal estimators in the classes of linear homogeneous as well as linear heterogeneous estimators through the minimization of risk function under a general quadratic loss structure. Unfortunately, the thus obtained optimal estimators are not functions of

Helge Toutenburg
Institut für Statistik, Universität München, D-80799 München, Germany
toutenb@stat.uni-muenchen.de

observations alone. They involve the coefficient vector itself, which is being estimated, besides the scaling factor of the disturbance variance covariance matrix. Consequently, as acknowledged by Toutenburg et al. [4], such estimators have no practical utility.

In this paper, we apply a simple operationalization technique for obtaining the feasible versions of the optimal estimators. The technique essentially involves replacement of unknown quantities by their unbiased and/or consistent estimators. Such a substitution generally destroys the optimality and superiority properties. A study of the damage done to the optimal properties is the subject matter of our investigations. It is found that the process of operationalization may often alter the conclusions that are drawn from the performance of optimal estimators that are not friendly with users due to involvement of unknown parameters.

The plan of presentation is as follows. In Sect. 2, we describe the model and present the estimators for the vector of regression coefficients. Their properties are discussed in Sect. 3. Some numerical results about the behaviour of estimators in finite samples are reported in Sect. 4. Some summarizing remarks are then presented in Sect. 5. In the last, the Appendix gives the derivation of main results.

## 2 Estimators for Regression Coefficients

Consider the following linear regression model:

$$y = X\beta + \varepsilon \,, \tag{1}$$

where $y$ is a $n \times 1$ vector of $n$ observations on the study variable, $X$ is a $n \times p$ matrix of $n$ observations on the $p$ explanatory variables, $\beta$ is a $p \times 1$ vector of regression coefficients and $\varepsilon$ is a $n \times 1$ vector of disturbances.

In addition to the observations, let us be given some incomplete prior information in the form of a set of stochastic linear restrictions binding the regression coefficients:

$$r = R\beta + \phi \,, \tag{2}$$

where $r$ is a $m \times 1$ vector, $R$ is a full row rank matrix of order $m \times p$ and $\phi$ is a $m \times 1$ vector of disturbances.

It is assumed that $\varepsilon$ and $\phi$ are stochastically independent. Further, $\varepsilon$ has mean vector 0 and variance covariance matrix $\sigma^2 W$ in which the scalar $\sigma$ is unknown but the matrix $W$ is known. Similarly, $\phi$ has mean vector 0 and variance covariance matrix $\sigma^2 V$.

When $V$ is available, the mixed regression estimator of $\beta$ proposed by Theil and Goldberger [3] is given by

$$\begin{aligned} b_{MR} &= (S + R'V^{-1}R)^{-1}(X'W^{-1}y + R'V^{-1}r) \\ &= b + S^{-1}R'(RS^{-1}R' + V)^{-1}(r - Rb) \,, \end{aligned} \tag{3}$$

where $S$ denotes the matrix $X'W^{-1}X$ and $b = S^{-1}X'W^{-1}y$ is the generalized least squares estimator of $\beta$.

In practice, $V$ may not be known all the time and then the mixed regression estimator cannot be used. Often, $V$ may be given but its accuracy and credibility may be questionable. Consequently, one may be willing to assume $V$ as unknown rather than known. In such circumstances, the mixed regression estimator (3) cannot be used.

For handling the case of unknown $V$, Toutenburg et al. [4] have pioneered the concept of weakly unbiasedness and utilized it for the estimation of $\beta$. Accordingly, an estimator $\hat{\beta}$ is said to be weakly–$(R,r)$–unbiased with respect to the stochastic linear restrictions (2) when the conditional expectation of $R\hat{\beta}$ given $r$ is equal to $r$ itself, i.e.,

$$E(R\hat{\beta} \mid r) = r \tag{4}$$

whence it follows that the unconditional expectation of $R\hat{\beta}$ is $R\beta$.

It may be observed that the unbiasedness of $\hat{\beta}$ for $\beta$ implies weakly–$(R,r)$–unbiasedness of $\hat{\beta}$ but its converse may not be necessarily always true.

Taking the performance criterion as

$$R_A(\hat{\beta},\beta) = E(\hat{\beta}-\beta)'A(\hat{\beta}-\beta)\,, \tag{5}$$

that is, the risk associated with an estimator $\hat{\beta}$ of $\beta$ under a general quadratic loss function with a positive definite loss matrix $A$, Toutenburg et al. [4] have discussed the minimum risk estimator of $\beta$; see also Rao et al. [1] for an expository account.

The optimal estimator in the class of linear and weakly unbiased heterogeneous estimators for $\beta$ is given by

$$\hat{\beta}_1 = \beta + A^{-1}R'(RA^{-1}R')^{-1}(r - R\beta) \tag{6}$$

while the optimal estimator in the class of linear and weakly unbiased homogeneous estimators is

$$\hat{\beta}_2 = \frac{\beta'X'W^{-1}y}{\sigma^2 + \beta'S\beta}\left[\beta + A^{-1}R'(RA^{-1}R')^{-1}\left(\frac{\sigma^2\beta'S\beta}{\beta'S\beta}r - R\beta\right)\right]. \tag{7}$$

Clearly, $\hat{\beta}_1$ and $\hat{\beta}_2$ are not estimators in true sense owing to involvement of $\beta$ itself besides $\sigma^2$ which is also unknown. As a consequence, they have no practical utility.

A simple solution to operationalize $\hat{\beta}_1$ and $\hat{\beta}_2$ is to replace the unknown quantities by their estimators. Such a process of operationalization generally destroys the optimality of estimators.

If we replace $\beta$ by its generalized least squares estimator $b$ and $\sigma^2$ by its unbiased estimator

$$s^2 = \left(\frac{1}{n-p}\right)(y - Xb)'W^{-1}(y - Xb), \tag{8}$$

we obtain the following feasible versions of $\hat{\beta}_1$ and $\hat{\beta}_2$:

$$\tilde{\beta}_1 = b + A^{-1}R'(RA^{-1}R')^{-1}(r - Rb) \tag{9}$$

$$\tilde{\beta}_2 = \frac{b'Sb}{s^2 + b'Sb}\left[b + A^{-1}R'(RA^{-1}R')^{-1}\left(\frac{s^2 + b'Sb}{b'Sb}r - Rb\right)\right]. \tag{10}$$

It may be remarked that Toutenburg, Toutenburg et al. ([4], Sect. 4) have derived a feasible and unbiased version of the estimator $\hat{\beta}_1$ such that it is optimal in the class of linear homogeneous estimators. This estimator is same as $\tilde{\beta}_1$. It is thus interesting to note that when the optimal estimator in the class of linear heterogeneous estimators is operationalized, it turns out to have optimal performance in the class of linear homogeneous estimators.

## 3 Comparison of Estimators

It may be observed that a comparison of the estimator $\tilde{\beta}_1$ with $\hat{\beta}_1$ and $\tilde{\beta}_2$ with $\hat{\beta}_2$ will furnish us an idea about the changes in the properties due to the process of operationalization. Similarly, if we compare $\hat{\beta}_1$ and $\hat{\beta}_2$ with $\tilde{\beta}_1$ and $\tilde{\beta}_2$, it will reveal the changes in the properties of the optimal estimator and its feasible version in the classes of linear heterogeneous and linear homogeneous estimators.

### *3.1 Linearity*

First of all, we may observe that both the estimators $\hat{\beta}_1$ and $\tilde{\beta}_1$ are linear and thus the process of operationalization does not alter the linearity of estimator. This is not true when we consider the optimal estimator $\hat{\beta}_2$ in the class of linear homogeneous estimators and its feasible version $\tilde{\beta}_2$. Further, from (9) and (10), we notice that

$$\tilde{\beta}_2 = \frac{1}{s^2 + b'Sb}[b'Sb\tilde{\beta}_1 + s^2A^{-1}R'(RA^{-1}R')^{-1}r] \tag{11}$$

so that $\tilde{\beta}_2$ is a weighted average of $\tilde{\beta}_1$ and $A^{-1}R'(RA^{-1}R')^{-1}r$ while such a result does not hold in case of $\hat{\beta}_2$.

### *3.2 Unbiasedness*

From (9) and (10), we observe that

$$R\tilde{\beta}_1 = R\tilde{\beta}_2 = r \tag{12}$$

whence it is obvious that both the estimators $\tilde{\beta}_1$ and $\tilde{\beta}_2$ are weakly–$(R,r)$–unbiased like $\hat{\beta}_1$ and $\hat{\beta}_2$. Thus the operationalization does not disturb the property of weakly unbiasedness.

Next, let us consider the traditional unbiasedness property. It is easy to see that the optimal estimator $\hat{\beta}_1$ and its feasible version $\tilde{\beta}_1$ in the class of linear heterogeneous estimators are unbiased while the optimal estimators $\hat{\beta}_2$ and its feasible version $\tilde{\beta}_2$ in the class of homogeneous estimators are generally not unbiased. This may serve as an interesting example to demonstrate that weakly unbiasedness does not necessarily imply unbiasedness. Thus, with respect to the criterion of unbiasedness, no change arises due to operationalization.

## *3.3 Bias Vector*

Let us examine the bias vectors of the estimators $\hat{\beta}_2$ and $\tilde{\beta}_2$.

It is easy to see that the bias vector of $\hat{\beta}_2$ is given by

$$\begin{aligned} B(\hat{\beta}_2) &= E(\hat{\beta}_2 - \beta) \\ &= -\frac{\sigma^2}{\sigma^2 + \beta' S \beta} A^{-1} M \beta \,, \end{aligned} \tag{13}$$

where

$$M = A - R'(RA^{-1}R')^{-1}R. \tag{14}$$

The exact expression for the bias vector of $\tilde{\beta}_2$ is impossible to derive without assuming any specific distribution for the elements of disturbance vector $\varepsilon$. It may be further observed that even under the specification of distribution like normality, the exact expression will be sufficiently intricate and any clear inference will be hard to deduce. We therefore consider its approximate expression using the large sample asymptotic theory. For this purpose, it is assumed that explanatory variables in the model are at least asymptotically cooperative, i.e., the limiting form of the matrix $n^{-1}X'W^{-1}X$ as $n$ tends to infinity is a finite and nonsingular matrix. We also assume that $\varepsilon$ follows a multivariate normal distribution.

**Theorem I:** If we write $Q = n^{-1}S$, the bias vector of $\tilde{\beta}_2$ to order $O(n^{-1})$ is given by

$$\begin{aligned} B(\tilde{\beta}_2) &= E(\tilde{\beta}_2 - \beta) \\ &= -\frac{\sigma^2}{n\beta' Q \beta} A^{-1} M \beta + \frac{\sigma^2}{n^2 \beta' Q \beta}\left[p + (p-1)\frac{\sigma^2}{\beta' Q \beta}\right] A^{-1} M \beta \end{aligned} \tag{15}$$

which is derived in the Appendix.

A similar expression for the optimal estimator to order $O(n^{-2})$ can be straightforwardly obtained from (13) as follows:

$$
\begin{aligned}
B(\hat{\beta}_2) &= -\frac{\sigma^2}{n\beta'Q\beta}\left(1+\frac{\sigma^2}{n\beta'Q\beta}\right)^{-1} A^{-1}M\beta \\
&= -\frac{\sigma^2}{n\beta'Q\beta}A^{-1}M\beta + \frac{\sigma^4}{n^2(\beta'Q\beta)^2}A^{-1}M\beta.
\end{aligned} \tag{16}
$$

If we compare the optimal estimators $\hat{\beta}_2$ and its feasible version $\tilde{\beta}_2$ with respect to the criterion of bias to order $O(n^{-1})$ only, it follows from (15) and (16) that both the estimators are equally good. This implies that operationalization does not alter the asymptotic bias to order $O(n^{-1})$.

When we retain the term of order $O(n^{-2})$ also in the bias vector, the two estimators are found to have different bias vectors and the effect of operationalization precipitates.

Let us now compare the estimators $\hat{\beta}_2$ and $\tilde{\beta}_2$ according to the length of their bias vectors. If we consider terms upto order $O(n^{-3})$ only, we observe from (15) and (16) that

$$
[B(\hat{\beta}_2)]'[B(\hat{\beta}_2)] - [B(\tilde{\beta}_2)]'[B(\tilde{\beta}_2)] = \frac{2\sigma^2}{n^3\beta'Q\beta}\left[p+(p-2)\frac{\sigma^2}{\beta'Q\beta}\right]\beta'MA^{-2}M\beta.
$$

It is thus surprising that the feasible estimator $\tilde{\beta}_2$ is preferable to the optimal estimator with respect to the criterion of the bias vector length to the given order of approximation in the case of two or more explanatory variables in the model. If $p = 1$, this result continues to hold true provided that $\beta'Q\beta$ is greater than $\sigma^2$. Thus it is interesting to note that operationalization of optimal estimator improves the performance with respect to the bias vector length criterion.

## *3.4 Conditional Risk Function*

From Toutenburg et al. ([4], p. 530), the conditional risk function of $\hat{\beta}_1$, given $r$ is

$$
\begin{aligned}
R_A(\hat{\beta}_1,\beta \mid r) &= E[(\hat{\beta}_1-\beta)'A(\hat{\beta}_1-\beta) \mid r] \\
&= (r-R\beta)'(RA^{-1}R')^{-1}(r-R\beta).
\end{aligned} \tag{17}
$$

Similarly, the conditional risk function of $\hat{\beta}_2$ given $r$ can be easily obtained:

$$
\begin{aligned}
R_A(\hat{\beta}_2,\beta \mid r) &= E[(\hat{\beta}_2-\beta)'A(\hat{\beta}_2-\beta) \mid r] \\
&= (r-R\beta)'(RA^{-1}R')^{-1}(r-R\beta) \\
&\quad + \frac{\sigma^2}{\sigma^2+n\beta'Q\beta}\left[\beta'M\beta + \left(1+\frac{\sigma^2}{n\beta'Q\beta}\right)r'(RA^{-1}R')^{-1}r\right].
\end{aligned} \tag{18}
$$

Using the result

$$\frac{\sigma^2}{\sigma^2+n\beta'Q\beta} = \frac{\sigma^2}{n\beta'Q\beta}\left(1+\frac{\sigma^2}{n\beta'Q\beta}\right)^{-1}$$
$$= \frac{\sigma^2}{n\beta'Q\beta} - \frac{\sigma^4}{n^2(\beta'Q\beta)^2} + O(n^{-3}), \tag{19}$$

we can express

$$R_A(\hat{\beta}_2,\beta \mid r) = (r-R\beta)'(RA^{-1}R')^{-1}(r-R\beta)$$
$$+\frac{\sigma^2}{n\beta'Q\beta}[\beta'M\beta + r'(RA^{-1}R')^{-1}r] - \frac{\sigma^4\beta'M\beta}{n^2(\beta'Q\beta)^2} + O(n^{-3}). \tag{20}$$

For the feasible estimator $\tilde{\beta}_1$ , it can be easily seen that the conditional risk function of $\tilde{\beta}_1$ given $r$ is given by

$$R_A(\tilde{\beta}_1,\beta \mid r) = E[(\tilde{\beta}_1-\beta)'A(\tilde{\beta}_1-\beta) \mid r]$$
$$= (r-R\beta)'(RA^{-1}R')^{-1}(r-R\beta) + \frac{\sigma^2}{n} trMQ^{-1}. \tag{21}$$

As the exact expression for the conditional risk of the estimator $\tilde{\beta}_2$ is too complex to permit the deduction of any clear inference regarding the performance relative to other estimators, we consider its asymptotic approximation under the normality of disturbances. This is derived in Appendix.

**Theorem II:** The conditional risk function of the estimator $\tilde{\beta}_2$ given $r$ to order $O(n^{-2})$ is given by

$$R_A(\tilde{\beta}_2,\beta \mid r) = E[(\tilde{\beta}_2-\beta)'A(\tilde{\beta}_2-\beta) \mid r]$$
$$= (r-R\beta)'(RA^{-1}R')^{-1}(r-R\beta) + \frac{\sigma^2}{n} trMQ^{-1}$$
$$-\frac{\sigma^4}{n^2\beta'Q\beta}\left[2trMQ^{-1} - 5\left(\frac{\beta'M\beta}{\beta'Q\beta}\right)\right]. \tag{22}$$

It is obvious from (17) and (21) that the operationalization process leads to an increase in the conditional risk. Similarly, comparing $\hat{\beta}_2$ and $\tilde{\beta}_2$ with respect to the criterion of the conditional risk given $r$ to order $O(n^{-1})$, we observe from (20) and (22) that the operationalization process results in an increase in the conditional risk when

$$trMQ^{-1} > \frac{\beta'M\beta}{\beta'Q\beta} + \frac{r'(RA^{-1}R')r}{\beta'Q\beta}. \tag{23}$$

The opposite is true, i.e., operationalization reduces the conditional risk when the inequality (23) holds true with a reversed sign.

If we compare the exact expressions (17) and (18) for the conditional risk function given $r$, it is seen that the estimator $\hat{\beta}_1$ is uniformly superior to $\hat{\beta}_2$. This result remains true, as is evident from (21) and (22), for their feasible versions also when the criterion is the conditional risk given $r$ to order $O(n^{-2})$ and

$$trMQ^{-1} < 2.5\left(\frac{\beta' M\beta}{\beta' Q\beta}\right) \tag{24}$$

while the opposite is true, i.e., $\tilde{\beta}_2$ has smaller risk than $\tilde{\beta}_1$ when

$$trMQ^{-1} > 2.5\left(\frac{\beta' M\beta}{\beta' Q\beta}\right). \tag{25}$$

The conditions (24) and (25) have little usefulness in actual practice because they cannot be verified due to involvement of $\beta$. However, we can deduce sufficient conditions that are simple and easy to check.

Let $\lambda_{\min}$ and $\lambda_{\max}$ be the minimum and maximum eigen values of the matrix $M$ in the metric of $Q$, and $T$ be the total of all the eigenvalues. Now, it is seen that the condition (24) is satisfied so long as

$$T < 2.5\lambda_{\min} \tag{26}$$

which is a sufficient condition for the superiority of $\tilde{\beta}_1$ over $\tilde{\beta}_2$.

Similarly, for the superiority of $\tilde{\beta}_2$ over $\tilde{\beta}_1$, the following sufficient condition can be deduced from (25):

$$T > 2.5\lambda_{\max}. \tag{27}$$

We thus observe that the optimal estimator $\hat{\beta}_1$ is uniformly superior to $\hat{\beta}_2$ with respect to both the criteria of conditional and unconditional risks. The property of uniform superiority is lost when they are operationalized for obtaining feasible estimators. So much so that the superiority result may take an opposite turn at times.

Further, we notice that the reduction in the conditional risk of $\hat{\beta}_1$ over $\hat{\beta}_2$ is generally different in comparison to the corresponding reduction in the conditional risk when their feasible versions are considered. The change in the conditional risk performance of the optimal estimators starts appearing in the term of order $O(n^{-1})$. When their feasible versions are compared, the leading term of the change in risk is of order $O(n^{-2})$. This can be attributed to the process of operationalization.

### *3.5 Unconditional Risk Function*

Now let us compare the estimators under the criterion of the unconditional risk function.

It can be easily seen from (17), (18), (20), (21) and (22) that the unconditional risk functions of the four estimators are given by

$$R_A(\hat{\beta}_1,\beta) = E(\hat{\beta}_1-\beta)'A(\hat{\beta}_1-\beta) = \sigma^2 trV(RA^{-1}R')^{-1} \tag{28}$$

$$\begin{aligned}R_A(\hat{\beta}_2,\beta) &= E(\hat{\beta}_2-\beta)'A(\hat{\beta}_2-\beta)\\ &= \sigma^2 trV(RA^{-1}R')^{-1}\\ &\quad+\frac{\sigma^2}{n\beta'Q\beta}\left[\beta'A\beta+\sigma^2 trV(RA^{-1}R')^{-1}-\frac{\sigma^2}{\sigma^2+n\beta'Q\beta}\beta'M\beta\right]\\ &= \sigma^2 trV(RA^{-1}R')^{-1}+\frac{\sigma^2}{n\beta'Q\beta}\left[\beta'A\beta+\sigma^2 trV(RA^{-1}R')^{-1}\right]\\ &\quad-\frac{\sigma^4\beta'M\beta}{n^2(\beta'Q\beta)^2}+O(n^{-3})\end{aligned} \tag{29}$$

$$R_A(\tilde{\beta}_1,\beta) = E(\tilde{\beta}_1-\beta)'A(\tilde{\beta}_1-\beta) = \sigma^2 trV(RA^{-1}R')^{-1}+\frac{\sigma^2}{n}trMQ^{-1} \tag{30}$$

$$\begin{aligned}R_A(\tilde{\beta}_2,\beta) &= E(\tilde{\beta}_2-\beta)'A(\tilde{\beta}_2-\beta)\\ &= \sigma^2 trV(RA^{-1}R')^{-1}+\frac{\sigma^2}{n}trMQ^{-1}\\ &\quad-\frac{\sigma^4}{n^2\beta'Q\beta}\left[2trMQ^{-1}-5\left(\frac{\beta'M\beta}{\beta'Q\beta}\right)\right]+O(n^{-3}).\end{aligned} \tag{31}$$

Looking at the above expressions, it is interesting to note that the relative performance of one estimator over the other is same as observed under the criterion of the conditional risk given $r$.

## 4 Simulation Study

We conducted a simulation experiment to study the performance of the estimators $\tilde{\beta}_1$ and $\tilde{\beta}_2$ with respect to the ordinary least squares estimator $b$. The sample size was fixed at $n=30$. The design matrix $X$ contained an intercept term and six covariates which were generated from multivariate normal distribution with variance 1 and equal correlation of 0.4. The mean vector of the covariates was $(-2,-2,-2,2,2,2)$. The true response vector (without the error term $\varepsilon$) was then calculated as $\tilde{y}=X\beta$ with the $7\times 1$ true parameter vector $\beta=(10,10,10,10,-1,-1,-1)$. The restriction matrix $R$ was generated as a $3\times 7$ matrix containing uniform random numbers. The true restriction vector (without the error term $\phi$) was calculated as $\tilde{r}=R\beta$. Then in a loop with 5,000 replications, in every replication, new error terms $\varepsilon$ and $\phi$ were added in $\tilde{y}$ and $\tilde{r}$ to get $y$ and $r$ respectively. The errors were generated independently from normal random variables with variances $\sigma^2 = 40$ for $\varepsilon_i$, $i = 1,\ldots,n$ and $\sigma^2/c$ for $\phi_j$, $j = 1,2,3$. The factor $c$ controls the accuracy of the prior information compared to the noise in the data. If $c$ is high, the prior information is more accurate than the case when $c$ is low. Note that $c < 1$ means that the prior information is more noisy than the data which indicates that it is probably useless in practice. In fact we only expect the proposed estimators to be better than $b$ if $c$ is considerably larger than 1. For comparison of the estimators, we calculated the measure

$$\text{MRMSE} = \frac{1}{5000}\sum_{k=1}^{5000}\sqrt{\frac{1}{7}(\hat{\beta}-\beta)'(\hat{\beta}-\beta)}\,,$$

(mean of root mean squared errors) where $\hat{\beta}$ stands for one of the estimators $b$, $\tilde{\beta}_1$ or $\tilde{\beta}_2$. Figure 1 shows the distribution of the root mean squared errors $\sqrt{\frac{1}{7}(\hat{\beta}-\beta)'(\hat{\beta}-\beta)}$ for each estimator based on 5,000 replications with $c = 100$. This means that the prior information was not perfect but very reliable ($\sigma^2/c = 40/100 = 0.4$). A considerable gain can be observed by using one of the new proposed estimators while there is no noticeable difference between $\tilde{\beta}_1$ and $\tilde{\beta}_2$. The *MRMSE*s in that run were 2.64 for $b$ and 1.85 for $\tilde{\beta}_1$ and $\tilde{\beta}_2$. The picture changes when we decrease $c$. For example when $c = 4$ (which means that the standard error of $\phi_j$ is half of the standard error of the noise in the data), then the *MRMSE* were 3.09 for $b$ and 2.65 for $\tilde{\beta}_1$ and $\tilde{\beta}_2$. Figure 2 shows the corresponding boxplots. But a general conclusion is not possible since the results clearly also depend on the matrices $X$, $R$ and vector $\beta$ itself.

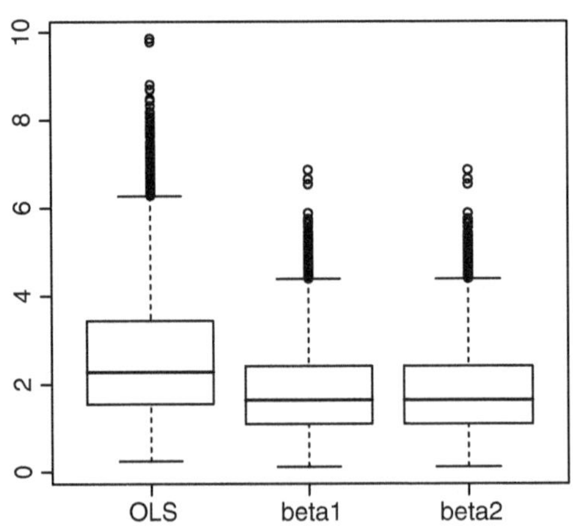


**Fig. 1** Boxplot of root mean squared errors of the three estimators with $c = 100$

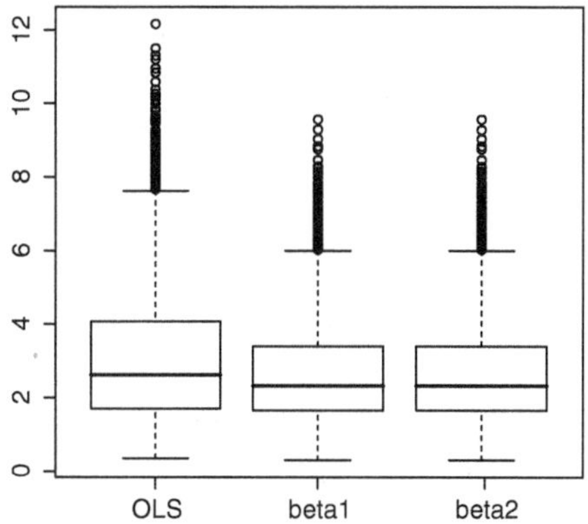


**Fig. 2** Boxplot of root mean squared errors of the three estimators with $c = 4$

## 5 Some Summarizing Remarks

We have considered the minimum risk approach for the estimation of coefficients in a linear regression model when incomplete prior information specifying a set of linear stochastic restrictions with unknown variance covariance matrix is available. In the linear and weakly unbiased heterogeneous and homogeneous classes of estimators, the optimal estimators obtained by Toutenburg et al. [4] as well as their feasible versions are presented. Properties of these four estimators are then discussed.

Analyzing the effect of operationalizing the optimal estimators, we have observed that the property of linearity is retained only in case of heterogeneous estimation. So far as the property of weakly unbiasedness is concerned, the process of operationalization has no influence. But when the traditional unbiasedness is considered, it is seen that the optimal heterogeneous estimator remains unbiased while the optimal homogeneous estimator is generally biased. This remains true when their feasible versions are considered. In other words, the process of operationalizations does not bring any change in the performance of estimators.

Looking at the direction and magnitude of bias, we have found that the optimal estimator and its feasible version in the case of homogeneous estimation have identical bias vectors to order $O(n^{-1})$ implying that the operationalization process has no effect on the bias vector in large samples. But when the sample size is not large enough and the term of order $O(n^{-2})$ is no more negligible, the effect of operationalization appears. If we compare the optimal estimator and its feasible version with respect to the criterion of the length of the bias vector to order $O(n^{-3})$, it is seen that the operationalization improves the performance provided that there are two or more explanatory variables in the model. This result remains true in the case of one explanatory variable also under a certain condition.

Examining the risk functions, it is observed that the relative performance of one estimator over the other remains unaltered whether the criterion is conditional risk given $r$ or the unconditional risk.

When we compare the risk functions of the optimal heterogeneous estimator and its feasible version, it is found that the process of operationalization invariably increases the risk. Such is not the case when we compare the optimal homogeneous estimator and its feasible version. Here the operationalization may lead to a reduction in risk at times; see the condition (23).

Next, it is observed that the optimal heterogeneous estimator has always smaller risk in comparison to the optimal homogeneous estimator. When they are operationalized in a bid to obtain feasible estimators, the property of uniform superiority is lost. We have therefore obtained sufficient conditions for the superiority of one feasible estimator over the other. An important aspect of these conditions is that they are simple and easy to check in practice.

Further, we have observed the magnitude of change in the risk of one optimal estimator over the other optimal estimator is generally different when their feasible versions are considered. In case of optimal estimators, the change occurs at the level of order $O(n^{-1})$ but when the feasible estimators are compared, this level is of order $O(n^{-2})$. This brings out the impact of operationalization process.

Finally, it may be remarked that if we consider the asymptotic distribution of the estimation error, i.e., the difference between the estimator and the coefficient vector, both the optimal estimators as well as their feasible versions have same asymptotic distribution. Thus the process of operationalization does not show any impact on the asymptotic properties of estimators. It may alter the performance of estimators when the number of observations is not sufficiently large. The difference in the performance of estimators is clear in finite samples through simulation experiment.

## Appendix

If we define

$$\begin{aligned} z &= \frac{1}{n^{1/2}} X'W^{-1}\varepsilon\,, \\ u &= \frac{1}{\sigma^2 n^{1/2}} \varepsilon' W^{-1}\varepsilon - n^{1/2}\,, \\ v &= \frac{1}{\sigma^2} \varepsilon' W^{-1} X S^{-1} X' W^{-1}\varepsilon\,, \end{aligned}$$

we can write

$$\begin{aligned} b'Sb &= \beta' S\beta + 2\beta' X' W^{-1}\varepsilon + \varepsilon' W^{-1} X S^{-1} X' W^{-1}\varepsilon \\ &= n\beta' Q\beta + 2n^{1/2}\beta' z + \sigma^2 v \qquad (32) \\ s^2 &= \frac{1}{(n-p)}(y - Xb)' W^{-1}(y - Xb) \\ &= \sigma^2\left[1 + \frac{u}{n^{1/2}} - \frac{v}{n}\right] + O_p(n^{-3/2}). \qquad (33) \end{aligned}$$

Using these, we can express

$$\begin{aligned}\frac{s^2}{s^2+b'Sb} &= \frac{\sigma^2}{n\beta'Q\beta}\left[1+\frac{u}{n^{1/2}}-\frac{v}{n}+O_p(n^{-3/2})\right]\\ &\quad *\left[1+\frac{2\beta'z}{n^{1/2}\beta'Q\beta}+\frac{\sigma^2(1+v)}{n\beta'Q\beta}+O_p(n^{-3/2})\right]^{-1}\\ &= \frac{\sigma^2}{n\beta'Q\beta}\left[1+\frac{u}{n^{1/2}}-\frac{v}{n}+O_p(n^{-3/2})\right]\\ &\quad *\left[1-\frac{2\beta'z}{n^{1/2}\beta'Q\beta}-\sigma^2\left(1+v-\frac{4\beta'zz'\beta}{\sigma^2\beta'Q\beta}\right)+O_p(n^{-3/2})\right]\\ &= \frac{\sigma^2}{n\beta'Q\beta}+\frac{\sigma^2}{n^{3/2}\beta'Q\beta}\left(u-\frac{2\beta'z}{\beta'Q\beta}\right)\\ &\quad -\frac{\sigma^2}{n^2\beta'Q\beta}\left(v+\frac{2u\beta'z+\sigma^2+\sigma^2 v}{\beta'Q\beta}-\frac{4\beta'zz'\beta}{(\beta'Q\beta)^2}\right)+O_p(n^{-5/2}).\end{aligned} \tag{34}$$

Utilizing these results, we can express

$$\begin{aligned}(\tilde{\beta}_2-\beta) &= (\tilde{\beta}_1-\beta)-\frac{s^2}{s^2+b'Sb}A^{-1}Mb\\ &= \xi_0+\frac{1}{n^{1/2}}\xi_{1/2}+\frac{1}{n}\xi_1+\frac{1}{n^{3/2}}\xi_{3/2}+\frac{1}{n^2}\xi_2+O_p(n^{-5/2}),\end{aligned} \tag{35}$$

where

$$\begin{aligned}\xi_0 &= A^{-1}R'(R'A^{-1}R')^{-1}(r-R\beta)\\ \xi_{1/2} &= A^{-1}MQ^{-1}z\\ \xi_1 &= -\frac{\sigma^2}{\beta'Q\beta}A^{-1}M\beta\\ \xi_{3/2} &= -\frac{\sigma^2}{\beta'Q\beta}\left[\left(u-\frac{2\beta'z}{\beta'Q\beta}\right)A^{-1}M\beta+A^{-1}MQ^{-1}z\right]\\ \xi_2 &= \frac{\sigma^2}{\beta'Q\beta}\left[\left(v+\frac{2u\beta'z+\sigma^2+\sigma^2 v}{\beta'Q\beta}-\frac{4\beta'zz'\beta}{(\beta'Q\beta)^2}\right)A^{-1}M\beta\right.\\ &\quad \left.-\left(u-\frac{2\beta'z}{\beta'Q\beta}\right)A^{-1}MQ^{-1}z\right].\end{aligned}$$

By virtue of normality of $\varepsilon$, it is easy to see that

$$\begin{aligned}E(\xi_0\mid r) &= \xi_0\,,\ \ E(\xi_0)=0\,,\\ E(\xi_{1/2}\mid r) &= E(\xi_{1/2})=0\,,\\ E(\xi_1\mid r) &= E(\xi_1)=\xi_1\,,\end{aligned}$$

$$E(\xi_{3/2} \mid r) = E(\xi_{3/2}) = 0\,,$$

$$E(\xi_2 \mid r) = E(\xi_2) = \frac{\sigma^2}{\beta' Q \beta}\left[p + (p-1)\frac{\sigma^2}{\beta' Q \beta}\right] A^{-1} M \beta.$$

Using these results, we obtain from (35) the expression (15) of Theorem I.

Next, we observe from (35) that the conditional risk function of $\tilde{\beta}_2$ to order $O(n^{-2})$ is given by

$$\begin{aligned}
R_A(\tilde{\beta}_2, \beta \mid r) &= E[(\tilde{\beta}_2 - \beta)' A (\tilde{\beta}_2 - \beta) \mid r] \\
&= \xi_0' A \xi_0 + \frac{2}{n^{1/2}} E(\xi_0' A \xi_{1/2}) \\
&\quad + \frac{1}{n} E(\xi_{1/2}' A \xi_{1/2} + 2\xi_0' A \xi_1) + \frac{2}{n^{3/2}} E[\xi_0' A \xi_{3/2} + \xi_{1/2}' A \xi_1] \\
&\quad + \frac{1}{n^2} E(\xi_1' A \xi_1 + 2\xi_0' A \xi_2 + 2\xi_{1/2}' A \xi_{3/2}) + O_p(n^{-5/2}).
\end{aligned} \tag{36}$$

Now it can be easily seen that

$$\begin{aligned}
E(\xi_0' A \xi_{1/2} \mid r) &= 0\,, \\
E(\xi_{1/2}' A \xi_{1/2} \mid r) &= \sigma^2 tr M Q^{-1}\,, \\
E(\xi_0' A \xi_1 \mid r) &= 0\,, \\
E(\xi_0' A \xi_{3/2} \mid r) &= 0\,, \\
E(\xi_{1/2}' A \xi_1 \mid r) &= 0\,, \\
E(\xi_1' A \xi_1 \mid r) &= \frac{\sigma^4}{(\beta' Q \beta)^2} \beta' M \beta\,, \\
E(\xi_0' A \xi_2 \mid r) &= 0\,, \\
E(\xi_{1/2}' A \xi_{3/2} \mid r) &= \frac{\sigma^4}{\beta' Q \beta}\left[-tr M Q^{-1} + 2\left(\frac{\beta' M \beta}{\beta' Q \beta}\right)\right],
\end{aligned}$$

where repeated use has been made of the results $RA^{-1}M = 0$ and $MA^{-1}M = M$.

Substituting these results in (36), we obtain the result stated in Theorem II.

## References

[1] Rao, C.R., Toutenburg, H., Shalabh, Heumann, C.: Linear Models and Generalizations: Least Squares and Alternatives (3rd ed.). Springer, New York (2008)

[2] Srivastava, V.K.: Estimation of linear single-equation and simultaneous equation models under stochastic linear constraints: An annotated bibliography. Int. Stat. Rev. **48**, 79–82 (1980)

[3] Theil, H., Goldberger, A.S.: On pure and mixed estimation in econometrics. Int. Econ. Rev. **2**, 65–78 (1961)
[4] Toutenburg, H., Trenkler, G., Liski, E.P.: Optimal estimation methods under weakened linear restrictions. Comput. Stat. Data An. **14**, 527–536 (1992)

# Minimum Description Length Model Selection in Gaussian Regression under Data Constraints

Erkki P. Liski and Antti Liski

**Abstract** The normalized maximum likelihood (*NML*) formulation of the stochastic complexity Rissanen ([10]) contains two components: the maximized log likelihood and a component that may be interpreted as the parametric complexity of the model. The stochastic complexity for the data, relative to a suggested model, serves as a criterion for model selection. The calculation of the stochastic complexity can be considered as an implementation of the minimum description length principle (*MDL*) (cf. Rissanen [12]). To obtain an *NML* based model selection criterion for the Gaussian linear regression, Rissanen [11] constrains the data space appropriately. In this paper we demonstrate the effect of the data constraints on the selection criterion. In fact, we obtain various forms of the criterion by reformulating the shape of the data constraints. A special emphasis is placed on the performance of the criterion when collinearity is present in data.

## 1 Introduction

The variable selection problem is most familiar in the Gaussian regression context. Suppose that the response variable $\boldsymbol{y}$ and the potential explanatory variables $\boldsymbol{x}_1,\ldots,\boldsymbol{x}_K$ are vectors of $n$ observations. The problem of variable selection arises when one wants to decide which variables to include into the model. If we let $\gamma$ index the subsets of $\boldsymbol{x}_1,\ldots,\boldsymbol{x}_K$ and let $k_\gamma$ be the size of the $\gamma$th subset, then the problem is to select and fit a model of the form

$$\boldsymbol{y} = \mathbf{X}_\gamma \boldsymbol{\beta}_\gamma + \boldsymbol{\varepsilon}, \tag{1}$$

where $\mathbf{X}_\gamma$ is an $n \times k_\gamma$ regression matrix corresponding to the $\gamma$th subset, $\boldsymbol{\beta}_\gamma$ is the $k_\gamma \times 1$ vector of unknown regression coefficients and $\boldsymbol{\varepsilon} \sim \mathrm{N}_n(\mathbf{0}, \sigma^2 \boldsymbol{I})$.

Erkki P. Liski
University of Tampere, Tampere, Finland
Erkki.Liski@uta.fi

Let $\hat{\boldsymbol{\theta}}_\gamma = (\hat{\boldsymbol{\beta}}_\gamma, \hat{\sigma}^2_\gamma)$ denote the *ML* estimates

$$\hat{\boldsymbol{\beta}}_\gamma = (\mathbf{X}'_\gamma \mathbf{X}_\gamma)^{-1}\mathbf{X}'_\gamma \mathbf{y} \quad \text{and} \quad \hat{\sigma}^2_\gamma = RSS_\gamma / n \tag{2}$$

of $\boldsymbol{\beta}_\gamma$ and $\sigma^2$ from the model (1), where $RSS_\gamma = \|\mathbf{y} - \hat{\mathbf{y}}_\gamma\|^2$ is the residual sum of squares and $\hat{\mathbf{y}}_\gamma = \boldsymbol{X}_\gamma \hat{\boldsymbol{\beta}}_\gamma$ is the vector of fitted values. Here we assume that $\boldsymbol{X}_\gamma$ is of full column rank.

The two most well-known methods for model selection are the *Akaike information criterion* or *AIC* (Akaike [1], Burnham [3]) and the *Bayesian information criterion* or *BIC* (Schwarz [13]). The *Akaike information criterion* is defined by

$$AIC(\gamma) = -2\log f(\mathbf{y}; \hat{\boldsymbol{\theta}}_\gamma) + 2k_\gamma,$$

where $f(\mathbf{y}; \boldsymbol{\theta}_\gamma)$ is the density function of $\mathbf{y}$. The corresponding *BIC* criterion is

$$BIC(\gamma) = -2\log f(\mathbf{y}; \hat{\boldsymbol{\theta}}_\gamma) + k_\gamma \log n.$$

The *MDL* principle for statistical model selection is based on the idea to capture regular features in data by constructing a model in a certain class which permits the shortest description of the data and the model itself. Rissanen's [10, 11] *MDL* approach to modeling utilizes ideas of coding theory. The expression

$$-\log \hat{f}(\mathbf{y}; \gamma) = -\log f(\mathbf{y}; \hat{\boldsymbol{\theta}}_\gamma) + \log C(\gamma) \tag{3}$$

defines the "shortest code length" for the data $\mathbf{y}$ that can be obtained with the model $\gamma$ and it is called the *stochastic complexity* of $\mathbf{y}$, given $\gamma$.

Under certain conditions $\log C(\gamma)$ has the estimate ([10])

$$\log C(\gamma) = \log \frac{n}{2\pi} + \log \int |\mathbf{J}(\boldsymbol{\theta}_\gamma)|^{1/2} \, \mathrm{d}\boldsymbol{\theta}_\gamma + o(1), \tag{4}$$

where $|\mathbf{J}(\boldsymbol{\theta}_\gamma)|$ is the determinant of the Fisher's information matrix. Since the last term $o(1)$ in (4) goes to zero as $n \to \infty$ and the second term is constant, asymptotically $\log C(\gamma)$ behaves like the first term. Thus we see the asymptotic connection with the *BIC*. For some important models $\log C(\gamma)$ can be calculated exactly, for example by using the *NML* technique. In statistical literature the *MDL* principle is often confused with a particular implementation of it as the selection criterion *BIC* (For discussion see [6], p. 552). In fact, the stochastic complexity (3) has the adaptation property that it behaves more like *AIC* when the number of parameters is getting large compared with the number of observations.

## 2 Selection by Stochastic Complexity

Assume that $\mathbf{y}$ follows the Gaussian linear model (1) with $\boldsymbol{\theta}_\gamma = (\boldsymbol{\beta}_\gamma, \sigma^2)$. Here we consider the family of models

$$\mathcal{M}_\gamma = \{f(\mathbf{y}; \boldsymbol{\theta}_\gamma) : \gamma \in \Gamma\} \tag{5}$$

defined by the normal densities $f(\boldsymbol{y};\boldsymbol{\theta}_\gamma)$, where $\Gamma$ denotes a set of subsets of $\boldsymbol{x}_1,\ldots,\boldsymbol{x}_K$, i.e. the set of models we wish to consider.

After observing $\boldsymbol{y}$ we may determine the maximum likelihood (*ML*) estimate $\hat{\boldsymbol{\theta}}_\gamma = \hat{\boldsymbol{\theta}}_\gamma(\boldsymbol{y})$ of $\boldsymbol{\theta}_\gamma$ such that $f(\boldsymbol{y};\hat{\boldsymbol{\theta}}_\gamma) = \max_{\boldsymbol{\theta}_\gamma} f(\boldsymbol{y};\boldsymbol{\theta}_\gamma)$. Rissanen [10] introduced the *NML* function

$$\hat{f}(\boldsymbol{y};\gamma) = \frac{f(\boldsymbol{y};\hat{\boldsymbol{\theta}}_\gamma(\boldsymbol{y}))}{C(\gamma)} \quad \text{with} \quad C(\gamma) = \int f(\boldsymbol{y};\hat{\boldsymbol{\theta}}_\gamma(\boldsymbol{y}))\,\mathrm{d}\boldsymbol{y}, \tag{6}$$

where $\hat{f}(\boldsymbol{y};\gamma)$ is a density function, provided that $C(\gamma)$ is bounded. The *NML* density function provides a general technique to apply the *MDL* (minimum description length) principle. Therefore the derivation of the *NML* density is a crucial step in the practical implementation of the *MDL* principle.

For each model $\gamma \in \Gamma$ we have an *NML* density (6) which depends on $\gamma$. In the sequel, $\hat{f}(\boldsymbol{y};\gamma)$ refers to the *NML* density of the model $\gamma$, and $C(\gamma)$ denotes the corresponding normalizing constant. Now the stochastic complexity (3) can be calculated by using the *NML* density:

$$-\log \hat{f}(\boldsymbol{y};\gamma) = -\log f(\boldsymbol{y};\hat{\boldsymbol{\theta}}_\gamma(\boldsymbol{y})) + \log C(\gamma).$$

The last term in the (3) is called *the parametric complexity* of the model. According to the *MDL* principle we seek to find the index value $\gamma = \hat{\gamma}$ that minimizes the stochastic complexity (3). The basics of the *MDL* theory are presented in the recent books by Grünwald [6] and Rissanen [12].

Since the following development will be for a fixed $\gamma$, we may drop the subindex $\gamma$ for a while without loss of clarity. It turns out that the *NML* function (6) for the normal distribution is undefined, since the normalizing constant $C$ is not bounded. Hence Rissanen [11] suggested the constrained data space

$$\mathscr{Y}(s,R) = \{\boldsymbol{y} : \hat{\boldsymbol{\beta}}'\mathbf{X}'\mathbf{X}\hat{\boldsymbol{\beta}} \le nR,\ \hat{\sigma}^2 \ge s\}, \tag{7}$$

where $s > 0$ and $R > 0$ are given positive constants. Then the *NML* density under the constraints (7) will be

$$\hat{f}(\boldsymbol{y};s,R) = f(\boldsymbol{y};\hat{\boldsymbol{\theta}}(\boldsymbol{y}))/C(s,R), \tag{8}$$

where now the normalizing constant $C(s,R)$ depends on two hyperparameters $s$ and $R$.

To get rid of these hyperparameters Rissanen [11] applied another level of normalization. Maximizing the function (8) with respect of $R$ and $s$ yields the *ML* estimates $\hat{R} = \|\hat{\boldsymbol{y}}\|^2/n$ and $\hat{s} = \hat{\sigma}^2$. The maximized *NML* function *mNML* is obtained by substituting these estimates into (8) in place of $s$ and $R$. Then the function *mNML* is normalized. In this second stage normalization the data space is constrained such that

$$\mathscr{Y} = \{\boldsymbol{y} : nR_1 \le \|\hat{\boldsymbol{y}}\|^2 \le nR_2,\ s_1 \le \hat{\sigma}^2 \le s_2\}, \tag{9}$$

where $0 < R_1 < R_2$ and $0 < s_1 < s_2$ are given positive constants. By normalizing the function $\hat{f}(\mathbf{y};\hat{s},\hat{R})$ we obtain the normalized *mNML* function $\hat{f}(\mathbf{y})$, say. Finally the stochastic complexity (3) takes the form

$$-\log \hat{f}(\mathbf{y}) = \frac{n-k}{2}\log\hat{\sigma}^2 + \frac{k}{2}\log\hat{R} - \log\Gamma(\frac{n-k}{2}) - \log\Gamma(\frac{k}{2}) + c, \tag{10}$$

where $\Gamma(\cdot)$ denotes the gamma function and $c = \frac{n}{2}\log(n\pi) + \log[\log\frac{s_2}{s_1}\log\frac{R_2}{R_1}]$ is the same for all models, and hence it can be ignored. More details can be found in Rissanen [11, 12].

## 3 The Effect of Data Constraints

For the Gaussian density $f(\mathbf{y};\boldsymbol{\theta})$ the numerator in (8) takes a simple form

$$f(\mathbf{y};\hat{\boldsymbol{\theta}}) = (2\pi\hat{\sigma}^2 e)^{-\frac{n}{2}},$$

but the normalizing constant $C(s,R)$ will essentially depend on two hyperparameters $s$ and $R$. The estimator $\hat{\boldsymbol{\theta}} = (\hat{\boldsymbol{\beta}},\hat{\sigma}^2)$ is a sufficient statistic for $\boldsymbol{\theta} = (\boldsymbol{\beta},\sigma^2)$ under the model (1). By sufficiency the density $f(\mathbf{y};\boldsymbol{\theta})$ belonging to the family (5) can be written as

$$f(\mathbf{y};\boldsymbol{\theta}) = f(\mathbf{y}|\hat{\boldsymbol{\theta}})g(\hat{\boldsymbol{\theta}};\boldsymbol{\theta}), \tag{11}$$

where the conditional density $f(\mathbf{y}|\hat{\boldsymbol{\theta}})$ does not depend on the unknown parameter vector $\boldsymbol{\theta}$. The *ML* estimators $\hat{\boldsymbol{\beta}}$ and $\hat{\sigma}^2$, given in (2), are independent. Therefore

$$g(\hat{\boldsymbol{\beta}},\hat{\sigma}^2;\boldsymbol{\beta},\sigma^2) = g_1(\hat{\boldsymbol{\beta}};\boldsymbol{\beta},\sigma^2)g_2(\hat{\sigma}^2;\sigma^2), \tag{12}$$

where $g_1(\hat{\boldsymbol{\beta}};\boldsymbol{\beta},\sigma^2)$ and $g_2(\hat{\sigma}^2;\sigma^2)$ are the densities of the *ML* estimators $\hat{\boldsymbol{\beta}}$ and $\hat{\sigma}^2$, respectively. Substituting $\hat{\boldsymbol{\beta}}$ and $\hat{\sigma}^2$ into (12) in place of $\boldsymbol{\beta}$ and $\sigma^2$, respectively, yields (cf. [11], [12], p. 115)

$$g_1(\hat{\boldsymbol{\beta}};\hat{\boldsymbol{\beta}},\hat{\sigma}^2)g_2(\hat{\sigma}^2;\hat{\sigma}^2) = A_{n,k}(\hat{\sigma}^2)^{-\frac{k}{2}-1}, \tag{13}$$

where

$$A_{n,k} = \frac{|\mathbf{X}'\mathbf{X}|^{1/2}(\frac{n}{2e})^{n/2}}{(2\pi)^{k/2}\Gamma(\frac{n-k}{2})}.$$

Utilizing the factorization (11) and the result (13) we get the normalizing constant $C(s,R)$ under the constraint (7) corresponding to (8) as follows:

$$
\begin{aligned}
C(s,R) &= \int_{\mathscr{T}(s,R)} \Big[ \int_{\mathscr{Y}(\hat{\boldsymbol{\theta}})} f(\mathbf{y}|\hat{\boldsymbol{\theta}})\,\mathrm{d}\mathbf{y} \Big] \tilde{g}(\hat{\sigma}^2)\,\mathrm{d}\hat{\boldsymbol{\theta}} \\
&= A_{v,k} \int_s^{\infty} (\hat{\sigma}^2)^{-\frac{k}{2}-1}\,\mathrm{d}\hat{\sigma}^2 \int_{\mathscr{B}(R)} \mathrm{d}\hat{\boldsymbol{\beta}} \\
&= A_{v,k} V_k \frac{2}{k} \left(\frac{R}{s}\right)^{k/2},
\end{aligned} \tag{14}
$$

where $\mathscr{T}(s,R) = \{\hat{\boldsymbol{\theta}} : \hat{\sigma}^2 \geq s,\ \hat{\boldsymbol{\beta}}'\mathbf{Q}\hat{\boldsymbol{\beta}} \leq nR\}$ and $\mathbf{Q}$ is a $k \times k$ positive definite matrix. Integrating the inner integral in the first line of (14) over $\mathscr{Y}(\hat{\boldsymbol{\theta}}) = \{\mathbf{y} :\ \hat{\boldsymbol{\theta}} = \hat{\boldsymbol{\theta}}(\mathbf{y})\}$ for a fixed value of $\hat{\boldsymbol{\theta}}$ gives unity. In the last line of (14)

$$
V_k R^{k/2} = \frac{\pi^{k/2} n R^{k/2}}{\frac{k}{2}\Gamma(\frac{k}{2})|\mathbf{Q}|^{1/2}}
$$

is the volume of an ellipsoid

$$
\mathscr{B}(\mathbf{Q},R) = \{\hat{\boldsymbol{\beta}} : \hat{\boldsymbol{\beta}}'\mathbf{Q}\hat{\boldsymbol{\beta}} \leq nR\} \tag{15}
$$

(cf. [4], p. 120).

The form of the stochastic complexity under the ellipsoidal constraint (15) takes the form

$$
-\log \hat{f}(\mathbf{y}) = \frac{n-k}{2}\log\hat{\sigma}^2 + \frac{k}{2}\log\hat{R} - \log\Gamma\!\left(\frac{n-k}{2}\right) - \log\Gamma\!\left(\frac{k}{2}\right) + \frac{1}{2}\log\frac{|\mathbf{X}'\mathbf{X}|}{|\mathbf{Q}|}, \tag{16}
$$

where $\hat{R} = \hat{\boldsymbol{\beta}}'\mathbf{Q}\hat{\boldsymbol{\beta}}/n$. The constant $c$, given in (10), is not essential in model comparison, and hence it is omitted. If we choose the constraint $\mathscr{B}(\mathbf{X}'\mathbf{X},R)$ in (15), then $\log\frac{|\mathbf{X}'\mathbf{X}|}{|\mathbf{Q}|} = 0$ and the stochastic complexity (16) takes the form (10). This is the constraint Rissanen [11, 12] uses. It is now clear that the matrix $\mathbf{Q}$ in the ellipsoidal constraint (15) has an essential effect on the stochastic complexity.

## 4 Effects of Collinearity

If we apply Stirling's approximation

$$
\Gamma(x+1) \approx (2\pi)^{1/2}(x+1)^{x+1/2}e^{-x-1}
$$

to $\Gamma$-functions in (16), omit the terms that do not depend on $\gamma$ or $k_\gamma$ and multiply (16) by 2, just for convenience, we have the *NML* criterion function of the form

$$
MDL(\gamma,\mathbf{Q}) = n\log S_\gamma^2 + k_\gamma \log F(\mathbf{Q})_\gamma + \log[k_\gamma(n-k_\gamma)] + \log\frac{|\mathbf{X}_\gamma'\mathbf{X}_\gamma|}{|\mathbf{Q}|}, \tag{17}
$$

where

$$S_\gamma^2 = \frac{RSS_\gamma}{n-k_\gamma} \quad \text{and} \quad F(\mathbf{Q})_\gamma = \frac{\hat{\boldsymbol{\beta}}'_\gamma \mathbf{Q} \hat{\boldsymbol{\beta}}_\gamma}{k_\gamma S_\gamma^2}.$$

In the special case $\mathbf{Q} = \mathbf{X}'\mathbf{X}$ the criterion (17) takes the form

$$MDL(\gamma, \mathbf{X}'\mathbf{X}) = n \log S_\gamma^2 + k_\gamma \log F_\gamma + \log[k_\gamma(n-k_\gamma)], \tag{18}$$

where

$$F_\gamma = \frac{\mathbf{y}'\mathbf{y} - RSS_\gamma}{kS_\gamma^2}$$

is the usual $F$-statistic. The formulation (18) was presented in Liski [8], and also Hansen and Yu [7] considered it in the context of a slightly different criterion.

Consider the set of models $\mathscr{M}_k$, where $k = k_\gamma$ and $RSS = RSS_\gamma$ for all $\gamma \in \mathscr{M}_k$. Then clearly the criterion (18) does not discriminate the models in $\mathscr{M}_k$. Assume that we have a satisfactory set of explanatory variables $\{x_1, \ldots, x_{k-1}\}$ and we try to add new variables $x_k$ and $x_{k+1}$. Consider a situation when both the model $\{x_1, \ldots, x_{k-1}, x_k\}$, say $\gamma_1$, and $\{x_1, \ldots, x_{k-1}, x_{k+1}\}$, say $\gamma_2$, yield the same, or a very close, residual sum of squares $RSS$, i.e. the models lie in $\mathscr{M}_k$. Hence, in terms of the $MDL$ criterion (18), the two models are indistinguishable.

Assume that due to the collinearity between $x_1, \ldots, x_{k-1}, x_k$, for example, the model yields large standard errors and low $t$-statistics for the estimates of the regression coefficients. On the other hand, the model with explanatory variables $x_1, \ldots, x_{k-1}, x_{k+1}$ may still have satisfactory $t$-statistics. Clearly, this second model would be better, if our interest is also in regression coefficients, not only in prediction. However, the $MDL$ criterion (18) fails to identify it. Note that $AIC$ and $BIC$ criteria have this same property.

For a collinear model $\gamma$ the determinant $|\mathbf{X}'_\gamma \mathbf{X}_\gamma| \approx 0$ and the $ML$ estimates of the regression coefficients become unstable, which may lead to a large value of $\|\hat{\boldsymbol{\beta}}_\gamma\|^2$ (cf. Belsley [2], for example). Let us further consider the set of models $\mathscr{M}_k$ and take $\mathbf{Q} = \mathbf{I}$ in (17). Then in the criterion $MDL(\gamma, \mathbf{I})$

$$F(\mathbf{I})_\gamma = \frac{\|\hat{\boldsymbol{\beta}}_\gamma\|^2}{kS_\gamma^2}$$

and the last term in (17) is $\log|\mathbf{X}'_\gamma \mathbf{X}_\gamma|$. Due to collinearity, $\log|\mathbf{X}'_{\gamma_1} \mathbf{X}_{\gamma_1}| < \log|\mathbf{X}'_{\gamma_2} \mathbf{X}_{\gamma_2}|$, but on the other hand $\|\hat{\boldsymbol{\beta}}_{\gamma_1}\|^2$ tends to be larger than $\|\hat{\boldsymbol{\beta}}_{\gamma_2}\|^2$. Thus the criterion (17) with $\mathbf{Q} = \mathbf{I}$ responds to the collinearity, but the message is not quite clear, since the two terms have opposite effects. If we use the criterion $MDL(\gamma, (\mathbf{X}'\mathbf{X})^2)$, then

$$F((\mathbf{X}'_\gamma \mathbf{X}_\gamma)^2) = \frac{\|\mathbf{X}'_\gamma \mathbf{y}\|^2}{kS_\gamma^2}$$

and the last term in (17) is $-\log|\mathbf{X}'_\gamma \mathbf{X}_\gamma|$. Now clearly the last term penalises the collinearity.

**Table 1** Five best-fitting subsets of two and three variables, and two models of four variables for the STEAM data

| Variables | $RSS_\gamma$ | $MDL(\gamma,\mathbf{X}'\mathbf{X})$ | $MDL(\gamma,\mathbf{I})$ | $MDL(\gamma,(\mathbf{X}'\mathbf{X})^2)$ |
|---|---|---|---|---|
| 1, 7 | 8.93 | 4.251 | −0.818 | 10.618 |
| 5, 7 | 9.63 | 5.904 | 0.611 | 12.318 |
| 2, 7 | 9.78 | 6.258 | 1.226 | 12.631 |
| 4, 7 | 15.60 | 16.511 | 11.342 | 22.893 |
| 7, 9 | 15.99 | 17.051 | 11.680 | 23.486 |
| 4, 5, 7 | 7.34 | 7.744 | −0.278 | 17.357 |
| 1, 5, 7 | 7.68 | 8.696 | −0.066 | 18.977 |
| 1, 7, 9 | 8.61 | 11.087 | 2.847 | 21.221 |
| 1, 4, 7 | 8.69 | 11.283 | 3.276 | 21.011 |
| 5, 7, 8 | 8.71 | 11.321 | 3.121 | 21.291 |
| 2, 4, 5, 7 | 7.162 | 14.671 | −18.112 | −15.699 |
| 1, 2, 5, 7 | 7.156 | 14.656 | −3.367 | −0.954 |

## *An Example: STEAM Data*

As an example we consider the STEAM data set ([5], p. 616, [9], p. 69) which contains 25 observations on 10 variables. The response $y$ is *pounds of steam used monthly* (the variable 1 in Draper and Smith), and the other nine variables constitute the set of potential explanatory variables. We center and scale the explanatory variables which does not affect the fitted model but $\mathbf{X}_\gamma'\mathbf{X}_\gamma$ is the correlation matrix. Here the $MDL(\gamma,\mathbf{X}'\mathbf{X})$ increases monotonously as the function of $RSS_\gamma$ when $k_\gamma = k$ is fixed. However, $MDL(\gamma,\mathbf{I})$ and $MDL(\gamma,(\mathbf{X}'\mathbf{X})^2)$ respond to collinearity. The two and three variable sets of explanatory variables given in Table 1 are not collinear. Therefore also $MDL(\gamma,\mathbf{I})$ and $MDL(\gamma,(\mathbf{X}'\mathbf{X})^2)$ put the models almost in same order as $MDL(\gamma,\mathbf{X}'\mathbf{X})$. However, the four variable models $\{x_2,x_4,x_5,x_7\}$ and $\{x_1,x_2,x_5,x_7\}$ have practically the same value of $MDL(\gamma,\mathbf{X}'\mathbf{X})$, but both $MDL(\gamma,\mathbf{I})$ and $MDL(\gamma,(\mathbf{X}'\mathbf{X})^2)$ strongly prefer $\{x_2,x_4,x_5,x_7\}$ to $\{x_1,x_2,x_5,x_7\}$. This is because the variables $x_1,x_2,x_5,x_7$ are much more collinear (the determinant of the correlation matrix $|\mathbf{R}| = 0.033$) than the variables $x_2,x_4,x_5,x_7$ ($|\mathbf{R}| = 0.299$). The length $\|\hat{\boldsymbol{\beta}}\|$ has larger value for the model $\{x_1,x_2,x_5,x_7\}$ than for $\{x_2,x_4,x_5,x_7\}$ which has an effect on the criterion $MDL(\gamma,\mathbf{I})$. Especially the size of the coefficient $\hat{\beta}_2$ and the intercept increase dramatically whereas the coefficients $\hat{\beta}_5$ and $\hat{\beta}_7$ remain practically same.

## References

[1] Akaike, H.: Information theory as an extension of the maximum likelihood principle. In: Petrov, B.N., Csaki, F. (eds.) Second International Symposium on Information Theory, pp. 267–281. Akademiai Kiado, Budapest (1973)

[2] Belsley, H.: Conditioning Diagnostics. Wiley, New York (1991)

[3] Burnham, K.P., Anderson, D.R.: Model Selection and Multi-model Inference. Springer, New York (2002)

[4] Cramer, H.: Mathematical Methods of Statistics. Princeton University Press, Princeton (1946)

[5] Draper, N.R., Smith, H.: Applied Regression Analysis, 2nd edn. Wiley, New York (1981)

[6] Grünwald, P.D.: The Minimum Description Length Principle. MIT, London (2007)

[7] Hansen, A.J., Yu, B.: Model Selection and the Principle of Minimum Description Length. J. Am. Stat. Assoc. **96**, 746–774 (2001)

[8] Liski, E.P.: Normalized ML and the MDL Principle for Variable Selection in Linear Regression. In: Festschrift for Tarmo Pukkila on his 60th Birthday, pp. 159–172, Tampere, Finland (2006)

[9] Miller, A.: Subset Selection in Regression, 2nd edn. Chapman & Hall/CRC, New York (2002)

[10] Rissanen, J.: Fisher Information and Stochastic Complexity. IEEE Trans. Inf. Theory, **IT-42, 1**, 40–47 (1996)

[11] Rissanen, J.: MDL Denoising. IEEE Trans. Inf. Theory, **IT-46, 1**, 2537–2543 (2000)

[12] Rissanen, J.: Information and Complexity and in Statistical Modeling. Springer, New York (2007)

[13] Schwarz, G.: Estimating the Dimension of a Model. Ann. Stat. **6**, 461–464 (1978)

# Self-exciting Extreme Value Models for Stock Market Crashes

Rodrigo Herrera and Bernhard Schipp

**Abstract** We demonstrate the usefulness of Extreme value Theory (EVT) to evaluate magnitudes of stock market crashes and provide some extensions. A common practice in EVT is to compute either unconditional quantiles of the loss distribution or conditional methods linking GARCH models to EVT. Our approach combines self-exciting models for exceedances over a given threshold with a marked dependent process for estimating the tail of loss distributions. The corresponding models allow to adopt ex-ante estimation of two risk measures in different quantiles to assess the expected frequency of different crashes of important stock market indices. The paper concludes with a backtesting estimation of the magnitude of major stock market crashes in financial history from one day before an important crash until one year later. The results show that this approach provides better estimates of risk measures than the classical methods and is moreover able to use available data in a more efficient way.

## 1 Introduction

The characterization of stock market changes and especially extreme negative events are of a profound importance to risk management. In financial markets, these extreme price movements correspond to market corrections and also to stock market crashes. On October 19, 1987, known as Black Monday, the S&P500 index fell by 20,5% of its value in a single session. The market simply gapped down at the opening and it did not stop falling until more than US$ 25,1 trillion in equity value had been erased.

Every market crash induces significant losses to the economy as a whole and its impact should be minimized as much as possible. This paper is motivated by

Bernhard Schipp
Fakultät Wirtschaftswissenschaften, Technische Universiät Dresden, D-01062 Dresden, Germany
Bernhard.Schipp@tu-dresden.de

the following question: can we forecast the magnitude of a crash? This particular question was raised in [14]. He asked what is the worst scenario that the daily sample period from 1 January 1960 to 16 October 1987 would allow us to predict?.

Assuming that yearly maximum values are i.i.d random variables [14] has estimated the Value at Risk $VaR_{0.01\%}$ of the S&P500 index at a level of 7.4% level with a 95% confidence interval ranging from 4.9 to 24%. Matthys and Beirlant [12] have estimated the $VaR_{0.01\%}$ of the S&P500 index with 5.28 and an 95% confidence interval ranging from 4.24 to 8%). Novak [17] have evaluated the $VaR_{0.01\%}$ and the Expected Shortfall $ES_{0.01\%}$ for the same data at 18.08 and 24.18% respectively.

Obviously the results differ by large. A possible explanation is that [14] used yearly data and the blocks method approach, which is less accurate than other extreme value approaches as the Peaks Over Threshold (POT) method. An important difference between the work of [17] and [14] is that the first author has used a non-parametric approach.

The main question now is whether it is possible to develop (in an objective and non arbitrary manner) good approaches to risk measurement such as Value at Risk ($VaR$) or Expect Shortfall ($ES$), especially in times of extreme market movements.

Several authors as [6, 11, 12, 14, 16, 17] have argued that EVT allows us to take explicitly into account rare events contained in the tails. This approach offers three main advantages over classical methods such as conditional models of the volatility (GARCH, EWMA or SV processes), historical simulations or normal distribution approach.

First, as it is a parametric method, high probability values can be estimated from, for example, out-of-sample $VaR$ computations. Second, extreme value methods focus on modelling the tail behaviour of a loss distribution using only extreme value rather than the whole data set. Third, as we only assume a particular model for returns exceeding a high threshold, we allow the data to speak for themselves to fit the distribution tails, which are not influenced by the center of the returns distribution.

The standard Peaks over threshold (POT) model in EVT, which describes the appearance of extremes in i.i.d data, subsumes elegantly the models for maximum values and the Generalized Pareto distribution (GPD) models for excess losses. However, the characteristics of financial return series such as clustered extremes and serial dependence typically violate the assumptions of POT model. These problems are often addressed by the application of a declustering method, and then the standard model is fitted to cluster maximum values only.

In this paper we use a self-exciting version of POT model introduced preliminarily in [16, 15, 2] where the data will be fitted in a single step and will not involve a prefiltering of data. The main aims of the models are based on marked point processes combined with self-exciting processes where the intensity of arrival of extreme events depend on the marks, which allows more realistic models.

Point processes of this kind have proven to be an efficient tool to model earthquake occurrences. Corresponding models (Hawkes or ETAS, Epidemic Type Aftershock Sequence) are considered to be standard branching models. For our study we examine the behaviour of different daily stock indices from many years before a crash up to one day before a crash, as has previously been made in [14, 17]. We estimate how the value of the returns of an index would change under the worst scenario. This definition of worst scenario corresponds to a worst possible movement in approximately 40 years as in the above studies.

In particular we consider the Crash of October 1987, the Hong Kong Crashes of 1997, and the NASDAQ Crash of April 2000.

This paper is organized as follows. In Sect. 2 we outline relevant aspects of the classical POT model, and then in Sect. 3 we describe the Self-exciting POT model. Section 4 presents a preliminary analysis of the data used for the applied illustrations in the paper. Results of the modelling of some important historical crashes and a Backtesting simulation for each return are given in the Sects. 5 and 6 respectively. Conclusions and discussions are resumed in Sect. 7.

## 2 The Classical Peaks over Threshold Method

In this section we summarize the results from the POT method which serve as a basis for our approach to modelling. Relevant references on the subject of extreme value theory include [6, 16, 7].

The POT method is based on the limit law for excess distributions (see for example [6]). The predominant aim is to fit a generalized Pareto distribution (GPD) to excesses over a high threshold of a random variable, under the condition that sufficient data are available above the chosen threshold. As a convention in this paper, a negative value is treated as a positive number and extreme events take place when losses are part of the right tail of the distribution.

Suppose that $X_1,\dots,X_n$ are i.i.d with distribution function $F \in MDA(H_\xi)$ for some $\xi \in \mathbb{R}$, where $H_\xi$ is a non-degenerate limiting distribution and $\lim_{n\to\infty} \overline{F}(c_n x + d_n) = -\ln H_\xi(x)$ holds for normalizing sequences $c_n$ and $d_n$. The excess distribution function of $X$ is given by

$$F_u(x) = P(X - u \leq x \mid X > u),\ x \geq 0.$$

This relation can be rewritten as

$$\bar{F}(u+x) = \bar{F}(u)\bar{F_u}(x).$$

Now by definition a GPD $G_{\xi,\beta}$ with parameters $\xi \in \mathbb{R}$ and $\beta > 0$ has distribution tail

$$\bar{G}_{\xi,\beta}(x) = \begin{cases} \left(1+\xi\frac{x}{\beta}\right) & if\ \xi \neq 0, \\ \exp(-x/\beta) & if\ \xi = 0, \end{cases} \tag{1}$$

and $x \in D(\xi,\beta)$

$$D(\xi,\beta) = \begin{cases} [0,\infty) & if\ \xi \geq 0, \\ [0,-\beta/\xi] & if\ \xi < 0. \end{cases}$$

Equation (1) provides a log likelihood for estimation of $\beta$ and $\xi$ equals

$$l((\xi,\beta),\mathbf{X}) = -n\ln\beta - (\xi^{-1}+1)\sum_{i=1}^{n}\ln(1+\xi x_i/\beta), \tag{2}$$

where the arguments of the above function have to satisfy the domain restriction $x_i \in D(\xi,\beta)$.

As $F \in MDA(H_\xi)$ then for an appropriate positive function $\beta$

$$\lim_{u\uparrow x_F}\sup_{0<x<x_F-u}\left|\bar{F}_u(x) - \bar{G}_{\xi,\beta}(x)\right| = 0,$$

where $x_F$ is the right endpoint.

Thus, for high threshold $u$ one expects that the excess distribution $F_u$ can be well approximated by a GPD

$$\bar{F}_u(x) \approx \bar{G}_{\xi,\beta}(x)$$

Based on this result, tail of $F$ can be estimated with

$$\bar{F}(x) = \bar{F}_n(u)\bar{G}_{\hat{\xi},\hat{\beta}}(x) \approx \frac{N_u}{n}\left(1+\hat{\xi}\frac{x}{\hat{\beta}}\right), \tag{3}$$

where $\bar{F}_n(u)$ is the empirical distribution function in $u$, $N_u$ is the number of exceedances about $u$ and $\hat{\xi} = \hat{\xi}_{N_u}, \hat{\beta} = \hat{\beta}_{N_u}$.

By inverting (3) an estimator of the quantile $x_p$ results immediately:

$$\hat{x}_p = u + \frac{\hat{\beta}}{\hat{\xi}}\left(\left(\frac{n}{N_u}(1-p)\right)^{-\hat{\xi}} - 1\right).$$

This method of analyzing the extreme values has the advantage of being straightforward to implement, but there is a number of disadvantages when considering broader features of this distribution, such as non stationary effects, trends and seasonality in the model.

A stronger approach to model specification is via a point process representation of the exceedances introduced by [19]. The idea is to view all events exceeding a given level $x$ as a bidimensional point process. The intensity of this process is defined for all Borel sets of the form $A = (t_1,t_2)\times(y,\infty)$ where $t_1$ and $t_2$ are time coordinates and $x \geq u$ is a given high threshold of the process. If the process is stationary and satisfies a condition that there are asymptotically no clusters among extremes, then its limiting form is a non-homogeneous Poisson process and the intensity at a point $(t,x)$ is given by

$$\lambda(t,x) = \frac{1}{\sigma}\left(1+\xi\frac{x-\mu}{\sigma}\right)_+^{-1/\xi-1}, \tag{4}$$

where $y_+ = \max(x,0)$ and $\mu,\sigma,\xi$ represent respectively a location parameter, scale parameter and shape parameter. Therefore, the intensity measure of the set $A$ for any $x \geq u$ may be expressed in the form of an one-dimensional homogeneous Poisson process with rate $\tau(x) = -\ln H_{\xi,\mu,\beta}$.

$$\Lambda(A) = \int_{t_1}^{t_2} \int_{x}^{\infty} \lambda(t,y) d_y dt = -(t_2 - t_1) \ln H_{\xi,\mu,\sigma}.$$

Now the tail of the excess over the threshold $u$, denoted $\bar{F}_u(x)$, can be calculated as the ratio of the rates of exceeding the levels $u+x$ and $u$ as follows

$$\bar{F}_u(x) = \frac{\tau(u+x)}{\tau(u)} = \left(1 + \frac{\xi x}{\sigma + \xi(u-\mu)}\right)^{-1/\xi} = \bar{G}_{\xi,\beta}(x),$$

where $\beta = \sigma + \xi(u-\mu)$ is simply a scaling parameter. Note that this is the same model described informally at the beginning of this section.

A useful reparametrization of (4) for the next section can be rewritten in terms of $\tau(u) = -\ln H_{\xi,\mu,\beta}(u)$ and $\beta = \sigma + \xi(u-\mu)$.

$$\lambda(t,x) = \frac{\tau}{\beta}\left(1 + \xi\frac{x-u}{\beta}\right)^{-1/\xi-1}, \tag{5}$$

where $\xi \in \mathbb{R}$ and $\tau, \beta > 0$.

The aim of presenting the two dimensional point process derivation of the POT method is for introducing easily the new methods, which have as basis the point process representation.

In the following section we describe some of the best-known and more successful attempts.

# 3 Self-exciting Peaks over Threshold (SEPOT) Method

The problem with the theory outlined in the previous section is that it assumes that the underlying series is independent, which is unrealistic in most applications.

Serial dependence and volatility cluster play an important role in most applications on returns of financial series, and so exceedances of a high threshold for daily financial return series do not necessarily occur according to a homogeneous Poisson process. Therefore, the application direct of the POT method is nonviable.

However, under weak conditions the POT representation may be applied to the maximum value of each cluster. The problem here is the identification of independent clusters of exceedances over a high threshold. This is because of the fact that the choice of declustering scheme often has a significant impact on estimates of cluster characteristics.

Possible algorithms are given by the run method, the block method or the interval method (see [1]). In particular, the interval estimator introduced by [8] proposes an automatic declustering scheme that is justified by an asymptotic result for the arrival times between threshold exceedances. The scheme relies on the estimation of *extremal index* prior to declustering, which can be interpreted as the reciprocal of the mean cluster size. However, this method consists of a two step procedure and the cluster behaviour of the extremes is lost.

The methodology introduced in this section takes advantage of the structure of the model, thus allowing the existing (dependent) data to be used more efficiently.

In the early 1970s, [9] introduced a family of what he called "self-exciting" or "mutually exciting" models, which became both pioneering examples of the conditional intensity methodology. The models have been greatly improved and extended by [18], whose ETAS model has been successfully used to elucidate the detailed structure of aftershock sequences.

In these models a recent spate of threshold exceedances causes the instantaneous risk of a threshold exceedances at a particular time to be higher. As [16] suggest the structure of these processes, which has traditionally been used in the modelling of earthquakes, would also seem appropriate for modelling market shocks and the tremors that follow these.

Basically, the Hawkes process is a model for clustering. Consider a simple, temporal point process $N$ on $[0,\infty)$ adapted to a filtration $\mathscr{H}_t$, which denote the entire history of the process up to time $t$ denoted as $\mathscr{H}_t = \{t_i : t_i < t\}$. Assuming it exists, the conditional intensity $\lambda_{\mathscr{H}_t}(t)$ is defined as the unique non decreasing, $\mathscr{H}$-predictable process such that $N([0,t)) - \int \lambda_{\mathscr{H}_t}(t)dt$ is a martingale.

Since the finite-dimensional distributions of such a point process are uniquely determined by its conditional intensity (see [5]), one way to model a point process is via its conditional intensity.

In self-exciting point processes, the conditional intensity is given by

$$\lambda_{\mathscr{H}_t}(t) = \mu + \sum_{i:t_i<t} g(t - t_i.) \tag{6}$$

where $\mu > 0$ is a short term clustering component, $g(v) \geq 0$ represents the contribution to the conditional intensity satisfying $\int_0^y g(v)dv < 1$, and the sum is taken over all events $\{t_i\}$ occurring before the current time $t$. The process has both epidemic and branching process interpretations. In practice, $g(v)$ is usually given a simple parametric form, such as a finite sum of Laguerre polynomials, or the Pareto-type form used in Ogata's ETAS model, by analogy with Omori's law.

Given a sequence $X_1,\ldots,X_n$ consider an stationary process$(T_k, Z_k)$ with state space $\chi = (0,n] \times (u,\infty)$ for $k = 1,\ldots,N_u$, where $T_k$ are the times of occurrence of the process, $Z_k$ are the marks of the excess,i.e, $Z_k = (X_n - u) > 0$ and $N_u$ are the number of exceedances. Let $\mathscr{H}_t$ denote the entire history of the process up to time $t$ as has be above defined.

We define a point process of exceedances $N(\cdot) = \sum_{i=1}^n I_{\{(T_i,X_i)\in\cdot\}}$ with a self-exiting structure given by a conditional intensity as (6) but with marks of the exceedances among the threshold $u$.

$$\lambda_{\mathscr{H}_t}(t) = \mu + \phi \sum_{k:T_k<t} g(t - T_k; Z_k),$$

where $\tau > 0$, $\phi \geq 0$ and $g(t,z)$ is a self-excitement function and monotonic decreasing that contributes an amount to the intensity originating from a previous exceedance. For our study we use two definitions, in the first model we consider a simple Hawkes model of the form

$$g(t,z) = (1 + \delta z)\exp(-\gamma t),\ \delta,\gamma > 0, \tag{7}$$

(see [9]), which is a generalized Poisson cluster process associating to cluster centres of a branching process of descendants.

The second is an ETAS model introduced by [10],

$$g(t,z) = (1 + t\gamma^{-1})^{-(\rho+1)}(1 + \delta z),\ \delta,\gamma,\rho > 0, \tag{8}$$

which can be viewed as a generalization of the modified Omori law, which takes into account the secondary aftershock sequences triggered by all events. In this model, all extreme events are simultaneously mainshocks, aftershocks and possible foreshocks. An observed aftershock sequence in the ETAS model is the result of the activity of all events triggering events triggering themselves other events, and so on, taken together. In particular, it could be considered as an extension of the Hawkes model.

The above specifications can be extended to other more complicated models such as generalized Hawkes processes or a marked self-exciting Stress Release Process (see [5], Chap. 6.). However, this would increase a lot the complexity of the model. For this reason, we opted for keeping the models as simple as possible.

Following the approach outlined in Sect. 2 we modify (5) by incorporating a model for dependence of the frequencies and sizes of the events over a high threshold $u$.

We begin replacing $\tau$ and $\beta$ in (5) by $\tau(t) = \tau + \phi w(t)$ and $\beta(t) = \beta + \eta w(t)$, where $w(t) = \sum_{k:T_k<t} g(t - T_k; Z_k)$ and $\eta$ models the hypothesis that the marks are predictable when $\eta > 0$. Note that the only thing that we are doing is to make $\tau$ and $\beta$ depend on historical exceedances according to a common self-exciting function.

It implicates that the exceedances of the threshold $u$ occur according to a one dimensional self-exciting process with conditional intensity described by the equation

$$\lambda_{\mathscr{H}_t}(t,x) = \frac{(\tau + \phi w(t))}{(\beta + \eta w(t))}\left(1 + \xi \frac{x - u}{\beta + \eta w(t)}\right)^{-1/\xi}.$$

Thus the conditional rate of crossing the threshold $u$ at time $t$ is defined as follows

$$\tau_{\mathscr{H}_t}(t,x) = \int_x^{\infty} \lambda(t,y)dy = (\tau + \phi w(t))\left(1 + \xi \frac{x - u}{\beta + \eta w(t)}\right)^{-1/\xi},$$

Therefore, the implied distribution of the excess losses when exceedances take place is generalized Pareto

$$\begin{aligned} P(X_k > u+x \mid T_k = t, \mathscr{H}_t) &= \frac{\tau_{\mathscr{H}_t}(t, x+u)}{\tau_{\mathscr{H}_t}(t, x)} \\ &= \bar{G}_{\xi, \beta+\eta w(t)}(x). \end{aligned}$$

For such a model, it is straightforward to write down a likelihood function, and hence, to find maximum likelihood estimators. The maximum likelihood inference involves maximization of

$$L(\theta; \{(T_1, Z_1), \ldots, (T_{N_u}, Z_{N_u})\}) = \left[\prod_{i=1}^{N_u} \lambda_{\mathscr{H}_t}(T_i)\right] \left[\prod_{i=1}^{N_u} f(Z_i \mid T_i)\right] \exp\left(-\int_0^n \lambda_g(u)\, du\right),$$

where $\theta$ is the vector of parameters to be estimated, $\lambda_{\mathscr{H}_t}(T_i)$ is the conditional intensity, $f(Z_i \mid T_i)$ is the density of the marks and $\lambda_g(u)$ is the intensity of the ground process associated to the time of the exceedances. For a rigorous and detailed exposition on the maximum likelihood estimation of marked self-exciting process trough conditional intensities, we refer to [5] page 246.

Note that the model with unpredictable marks can be obtained if $\eta = 0$. By comparing a model with $\eta = 0$ and a model with $\eta > 0$ we can formally test the hypothesis that the marks are unpredictable using a likelihood ratio test.

In particular, for our quantitative analysis we fitted eight different sub models derived from the (8) and (7). *Model a* is a Hawkes model without predictable marks ($\eta = 0$) and without influence of the marks on the estimation of the self-exciting function ($\delta = 0$). *Model b* is a Hawkes model without predictable marks and $\delta > 0$. *Model c* is an ETAS model without predictable marks and without influence of the marks. *Model d* is an ETAS without predictable marks. *Model e* is a Hawkes model without influence of the marks $\delta = 0$ and predictable marks $\eta > 0$. *Model f* is a Hawkes model with $\delta, \eta > 0$. *Model g* is an ETAS model with predictable marks and $\delta = 0$. *Model h* is the general formulation for an ETAS model with $\delta, \eta > 0$.

## 3.1 Measures of Extreme Risk

The measurement of market risk to which financial institutions are exposed has become an important instrument for market regulators, portfolio managers and for internal risk control. In this paper we will concentrate on two measures which attempt to describe the tail of a loss distribution, VaR and expected shortfall.

Value-at-risk (VaR) has become a standard measure used in financial risk management due to its conceptual simplicity, computational facility, and ready applicability.

In this paper, Value at Risk (VaR) is defined as the $q$th quantile of a distribution $F, VaR_q = F^{-1}(q)$. Thus, as in the classical POT model we can obtain a VaR estimator at level $\alpha$ by solving the equation $\tau(t_f,x) = 1-\alpha$, where $t_f$ denotes the conditional exceedances intensity at a time point just after $t$ for which the level is $1-\alpha$.

For our model this is only possible if $\tau + \phi w(t_f) > 1-\alpha$ and the resulting VaR estimator is

$$\mathrm{VaR}_\alpha^t = u + \frac{\beta + \eta w(t_f)}{\xi}\left(\left(\frac{1-\alpha}{\tau + \phi w(t_f)}\right)^{-\xi} - 1\right). \tag{9}$$

Many authors claim that VaR has several conceptual problems as its coherence as a risk measure. To alleviate the problems inherent in VaR, some practitioners have been turning their attention towards expected shortfall and away from VaR.

In the case of expected shortfall (ES), it is defined as the average of all losses which are greater or equal than VaR, i.e. the average loss in the worst $(1-q)\%$ cases $ES_q = E\left[X \mid X > VaR_q\right]$.

Exploiting the fact that the excess distribution above the higher threshold is also GPD with the same shape but different scale parameter, the $\mathrm{VaR}_\alpha^t$ can also be rewritten as

$$F_{VaR_q}(y) = G_{\xi,\beta+\eta w(t_f)+\xi(\mathrm{VaR}_\alpha^t - u)}(y) \tag{10}$$

As a consequence of (10) and provided that $\xi < 1$, the mean of $F_{VaR_q}(y)$ is $(\beta + \eta w(t_f) + \xi(\mathrm{VaR}_\alpha^t - u))\,/\,(1-\xi)$, the expected shortfall is then

$$ES_q = \frac{\mathrm{VaR}_\alpha^t}{1-\xi} + \frac{\beta + \eta w(t_f) - \xi u}{1-\xi}. \tag{11}$$

Thus, the $ES_q$ and $\mathrm{VaR}_\alpha^t$ are estimated by substituting the data based estimate for everything which is unknown in the last two expressions to obtain the approaches.

We shall remember the convention in this paper, where a loss is a positive number and a profit is a negative number

## 4 Preliminary Data Analysis

Our data set consists of daily returns defined by $r_t = -\ln(p_t/p_{t-1})$, where $p_t$ denotes the value of the index at day $t$, from the S&P 500 index over a sample period from 4 January to 16 October 1987, one day before the crash of 1987. The Hang Seng Index over a sample period from 2 January 1986 to 17 October 1997 and the NASDAQ Composite Index over a sample period between 1 January 1980 and 13 March 2000.

We use these sample periods for backtesting the estimation of the different risk measures in each index until one year after. We update new information on a daily basis that becomes available for the parameter estimates previously obtained. Thus we dynamically adjust quantiles, that allow us to improve as accurately as possible the estimation of the risk measures.

The crashes that we are going to study are the crash in October, 1987 (in particular on Monday 19), The Hong Kong crash of 1997 and The Dot com Crash 2000.

In Table 1 of the Appendix, we find some descriptive statistics of the daily returns on the above series. The mean return is close to zero for all of the four series. However, they differ considerably in terms of standard deviation, skewness and kurtosis. In particular, the returns of the Hang Seng index exhibit a high kurtosis compared with the other two series.

The assumption of normally distributed returns is strongly rejected by all series through the Jarque-Bera test. Other assumptions such as the null hypothesis that the returns series are iid random variables as well that the returns have a unit root are strongly rejected.

Figure 1 shows on the top the evolution of the indices for the sample period plus a year more to observe the crashes that we want to quantify, on the middle QQ

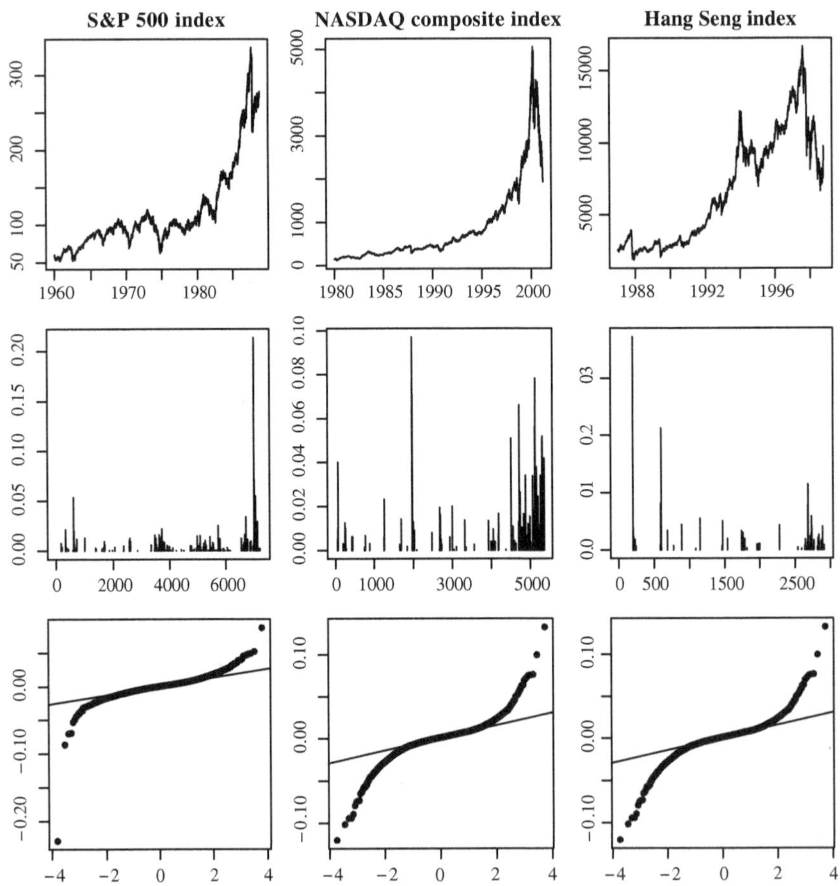


**Fig. 1** On the top the evolution of the indices S&P 500, NASDAQ and Hang Seng for the sample period. In the low panel, quantiles of the respective return distribution against those of normal distribution. In the middle panel the exceedances or losses over a defined threshold $u$

plots for the empirical return distributions relative to the normal distribution, which provides evidence for the fat-tailed property of the returns. An important point is the choice of the threshold, which implies a balance between bias and variance. On the one hand, if a too high threshold is selected the bias decreases. On the other hand, by taking a lower threshold, the variance decreases but the approximation becomes poorer. Cotter and Dowd [4] suggest that the points where the QQ plots change shape provide us with natural estimates of the tail threshold $u$. These leads us to select thresholds for S&P 500 equal to 3.3% of the sample and for the NASDAQ and Hang Seng indices a 3% of the sample is a fair approximation for the threshold $u$.

## 5 Quantitative Analysis of Stock Market Crashes

Subsequently, we calculate the VaR and ES for these three famous stock market crashes one day before a crash will take place. For this we estimate the eight models proposed in Sect. 3. Next, through likelihood ratio tests we formally test which model is more appropriate for each time series. The confidence intervals for the risk measures were obtained using a semi-parametric bootstrap proposed by [4] based on the maximum likelihood estimates of the SEPOT models. The obvious alternative is to bootstrap from the distribution of sample returns and re-estimate the SEPOT parameters for each resample. However, some of these resample estimates could be degenerate. This semi-parametric method used here tries to avoid this problem in a simple way.

We first take 10,000 bootstrap resamples, each of which consists of 20,000 uniform random variables. For each sample of the uniform random variable, we calculate the empirical quantile in question. Then we use this empirical quantile to calculate the cumulative probability $\hat{\alpha}$ of the uniform random variable of the sample. With this estimate and the parameter obtained in each model fitted we calculate 10,000 resamples estimates of the risk measures. For a 95% confidence interval we take the $20000 \times 0.025 = 500th$ and $20000 \times 0.975 = 19500th$ largest resample estimates of the risk measures.

Moreover, we test the null hypothesis of estimating correctly the Risk measures at time $t_i$ against the alternative that the method systematically underestimates the returns $r_{t_{i+1}}$. Thus, the indicator for a violation at time $t_i$ is Bernoulli $I_t := 1_{\{r_{t_{i+1}} > \{VaR_{\alpha,t_i}, ES_{\alpha,t_i}\}\}} \sim Be(1-\alpha)$. How it is described by [13], $I_t$ and $I_s$ are independent for $t, s \in T$, then

$$\sum_{t_i \in T} \sim B(n, 1-\alpha). \tag{12}$$

Expression 12 is a two-tailed test that is asymptotically distributed as binomial. We perform the null hypothesis that it is a method that correctly estimates the risk measures against the alternative that the method has a systematic estimation error and gives too few or too many violations.

## *5.1 Stock Market Crash 1987*

The first stock market crash that we want to analyze is the crash on October 19, 1987, a date that is also known as Black Monday. The S&P 500 dropped by 20.4%, falling from 282.7 to 225.06. That is almost double as the 12% loss experienced by investors in 1929 on Black Tuesday. Thus, this crash was the greatest single-day loss that Wall Street had ever suffered in continuous trading up to that point.

On October 16, 1987, the S&P 500 index failed by 5.25%, the second largest fall since 1960, only in this week the index downed by 9.21%. However, this was nothing compared to the subsequent Monday.

In this first *ex-ante* estimation of the risk measures defined in (9) and (11) we investigate a "worst case scenario" as in [14]. This worst case scenario is defined as a 40-year return level, i.e., a level which, on average, should only be exceeded in one year every forty years.

These results for the S&P 500 index are summarized in the Table 4. With the help of a likelihood ratio test, we compare the different models and choose the best model for this index. The best fitted model is the *model h*, which provides the higher VaR and ES level in comparison to the studies of [14] and [12]. In particular, the model gives evidence for the predictable marks ($\eta$) and influence of the marks in the self-exciting process.

The *VaR* estimates for the Monday 19, 1987 is 12.31% with confidence interval (8.42, 16.48%) and an *ES* from 15.46% with confidence interval (10.74, 20.51%). Notwithstanding we have made a more realistic estimation for the *VaR* and *ES* in comparison with the studies of [14] and [12] we did not manage to reach the magnitude of crash on Monday 19 for a return level of 40 years. The negative log-return on October 19 clearly stands out in the upper right of the upper left plot (1). The weakness can occur because we used a semi-parametric bootstrap for the construction of the confidence interval.

According to our model the magnitude of the crash on Monday 19 corresponds to an event in 130 years $VaR_{0.003\%} = 15.9\%\,(12.3\%, 20.4\%)$. Anyway the event on October 19, 1897, is according to our theory of extreme values and therefore, far from impossible at all.

An *ex-ante* forecast of the Risk measures *VaR* and *ES* at the 0.9999th-quantile for the choose model is presented in Fig. 2. Although the daily quantile forecast is quite volatile, the *model h* provides rather stable quantile forecasts across volatile return periods. Note that following the method given in [13], it is possible to develop a binomial test of the success of self-exciting marked point process estimation based on the number of violation.

Table 2 presents backtesting results based on population quantiles 0.99th, 0.999th, and 0.9999th, i.e., an event in 100 days, an event in 4 years and an event in 40 years respectively. In all cases the *model h* estimates correctly estimates the conditional VaR for all the quantiles, the null hypothesis is rejected whenever the p-value of the binomial test is less than 5%.

For the backtest in the next section we will use one year sample period from 16 October 1987 to 16 October 1988. In this period we find three important movements

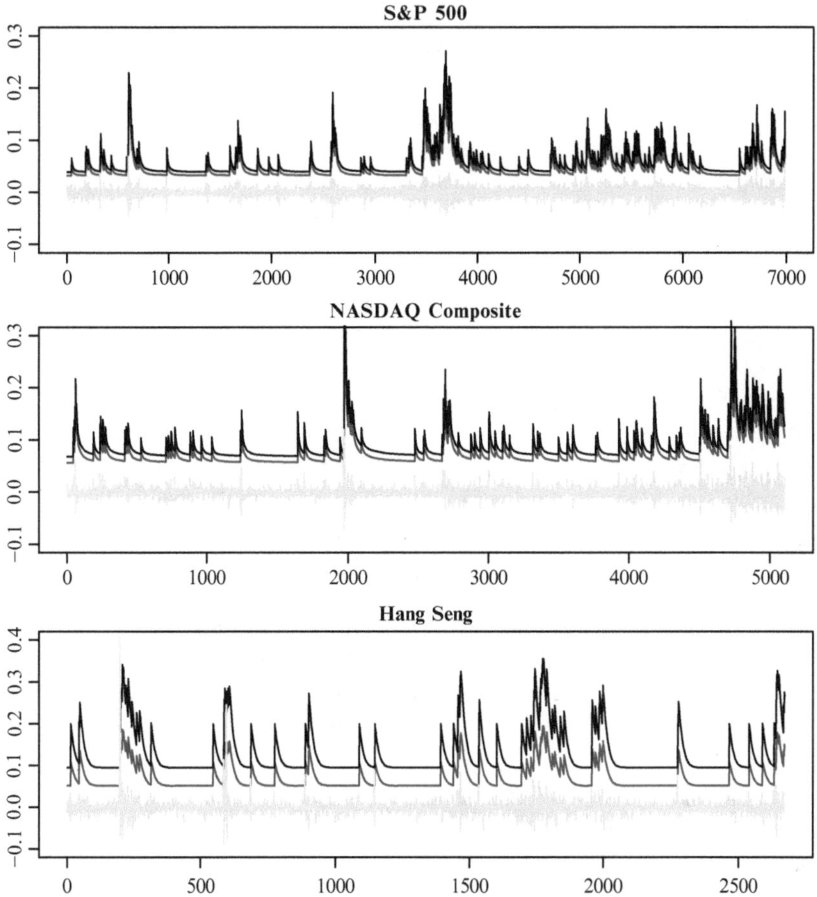


**Fig. 2** *Ex-ante* estimation of the risk measures *VaR* and *ES* for the 0.9999th quantile for the S&P 500 returns with the best model fitted (*model h*), NASDAQ returns with a *VaR* and *ES* for the 0.9999th quantile (*model h*), and the Hang Seng returns with a *VaR* and *ES* for the 0.999th quantile (*model a*). The light-grey line is the returns of the sample period, the dark grey line is the *VaR* estimations, and the black line is the *ES*

on October 19, 1987 the black Monday with 22.89 %, 26 October 1987 with a loss of 8.64% and 1 August 1988 with a 7%. Moreover, we found circa 24 exceedances in this period over the threshold $u = 0.0153$.

## 5.2 *The NASDAQ Crash 2000*

Our second investigation concerns the NASDAQ crash in March and April, 2000. This index dropped precipitously between March 14 and April 14 with a cumulative loss of approx. Fifty percent counted from its all-time high of 5,133 reached on

March 10, 2000. The drop was mostly driven by the so-called "New Economy" stocks which have risen nearly four-fold over 1998 and 1999 compared to a gain of only 50% for the S&P 500 index.

We try to estimate the worst case scenario for this index at the beginning on March 14, 2000. The heaviest drops were on April 3 and April 12, 2000, with 7.94 and 7.32% respectively. The results for the sample period are displayed in the Table 5.

The best model fitted is a model with predictable marks, i.e., the *model h* with a log-likelihood −66.53. In Fig. 2 we show the *ex-ante* estimation at the 0.9999 quantile for the VaR as well as the ES for the *model h* for the sample of study. On March 14, 2000 the NASDAQ dropped by a 4,2% , our estimation for the worst case covers without problems this movement.

Detailed results of the estimation for the period of crisis between March 14, 2000, and March 14, 2001, is presented in Sect. 6. In a preliminary analysis for this year and particularly between March 14 and April 14, 2000 we find three important movements. On 14 April, the log-return was 10.1%, one day before the drop was 7.94% and a week later on April 25 the log-return index lost 7.32%. Moreover, 67 extremes of the total data are found in the backtest sample period, which are defined over the threshold $u = 2.33$.

The violations corresponding to the model fitted are resumed in Table 2. Our model correctly estimates the conditional VaR for all the quantiles, the null hypothesis is rejected whenever the p-value of the binomial test is less than 5 percent. In all quantiles our model is closer to the expected number of violations and in all cases it performs very well.

### *5.3 The Hong Kong Crash 1997*

The third analysis is the Hong Kong crash of 1997. The Hong Kong market has the second largest stock market in Asia and the 7th largest in the world. However, in the last twenty years we can identify three major crashes. The first crash was synchronous to the worldwide October 1987 crash. The second crash began on February 4, 1994 and ended on March 3, 1994, with a cumulative loss of 19.4%. The third crash was on October 23, 1997, when the Hang Seng index lost more than 10% of its value in the biggest one-day fall in its history.

For our analysis we use the sample period from 2 January 1986 to 17 October 1997 and for the backtest we use one year sample period, from 18 October 1997 to 18 October 1998.

A great difference between this series and the others is that it has resisted about three crashes during one decade. The number of extreme movements is much more ample in relation to the other series, which would affect considerably the estimation of high quantiles. To gain some feeling for the problem, let us focus on the sample period from 2 January 1986 to 17 October 1997, where we found an extreme log-return over 40% on November 5,1987, an extreme log-return over 24% on June

5, 1981, and four extreme log-returns over 10%, 14.7% on October 28, 1997, 11.7% on November 4,1987, 11.4% on Mai 22, 1989 and 10.9% on October 23,1997.

Note that we will calculate the conditional risk measures for two of these last values in the backtest sample period. The results for a high threshold $u$ for the SE-POT models are summarized in Table 6. Our results show that the best fitted model is a Hawkes model without predictable marks and without influence of the marks, for the exceedances of the sample. The fitting exercise for example for different thresholds $u$ leads to estimations of parameters and risk measures that do not vary too much for the sample period.

The binomial test for the period before the backtest is displayed in Table 2. The results confirm that the model predicts the quantiles of interest well at a p-value level of 0.05.

The results of the best models for each one of the series at the worst case level are displayed in Fig. 2.

### *Goodness of Fit for the Models*

We provide also some goodness of fit for the best models of each index. In particular, we use the W-statistics to assess our success in modelling the temporal behaviour of the exceedances of the threshold $u$. The W-statistics are defined to be

$$W = \xi^{-1} \ln \left( 1 + \xi \frac{x-u}{\beta + \eta w(t)} \right).$$

This statistic states that if the GPD parameter model is correct, then the residuals are approximately independent unit exponential variables. The corresponding QQ-plots, displayed in Fig. 4 in the Appendix, do not show a substantial deviation from an exponential distribution.

Furthermore, to check that there is no further time series structure the autocorrelation function (ACF) for the residuals (left panel) and for their squares (right panel) are also included. Both autocorrelations are negligible at nearly all lags.

## 6 Backtesting

In order to assess the accuracy of our models to compute VaR and ES, we backtested the methods on the three return series described earlier by the following procedure. Let $T = \{t_1, \dots t_n\}$ be the time interval for the back test sample. On each day in the back test, we fitted the best self-exciting model that we selected before. Then we re-estimated the risk measures for each returns series according to expressions (9) and (11).

The risk measures estimate in $t_i$ $VaR_{\alpha,t_i}$ are compared in each case with $r_{t_i}$, the log-negative return for $q = \{0.99, 0.999, 0.9999\}$. A violation is said to take place whenever $r_{t_{i+1}} > VaR_{\alpha,t_i}$. The results are shown in the Fig. 3. It shows clearly the very satisfying overall performance of the methodology proposed in this paper for all the backtesting quantiles experiments. A more detailed analysis is realized to test the null hypothesis, that the models estimated well the conditionals risk measures.

Following, we construct a formal test in similar lines to the binomial test of quantile violation as in the Sect. 5 to verify the potential of the models fitted to estimate the conditionals *VaR* and *ES* for different levels of quantiles is conducted in this section and they are displayed in Table 3. The results of our backtesting procedure, which dynamically adjusts quantiles incorporating the new information daily allows us to conclude statistically that the models estimated compute well the different risk measures.

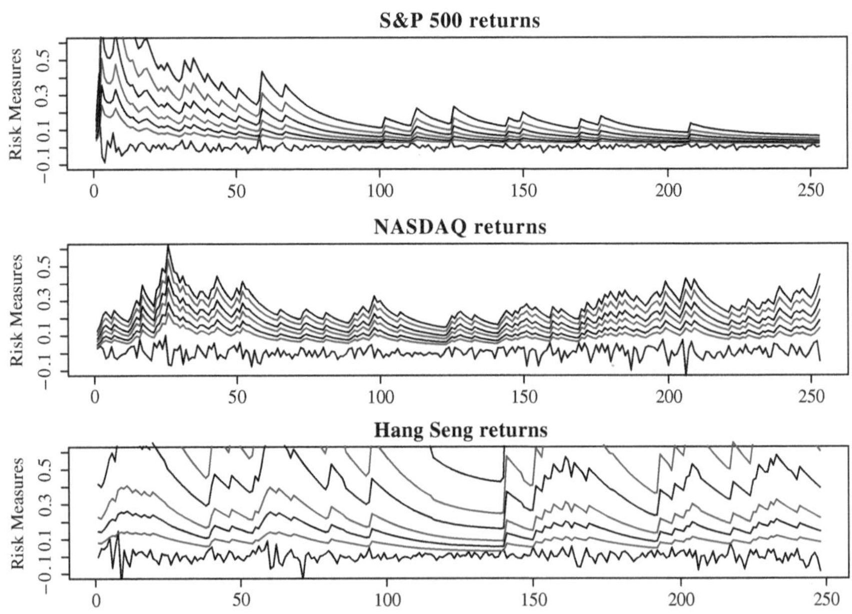


**Fig. 3** Backtest estimation of the risk measures *VaR* and *ES* for the $\{0.99, 0.999, 0.9999\}$-th quantile for the S&P 500 returns with the best model fitted (*model h*), NASDAQ returns with a *VaR* and *ES* for the $\{0.99, 0.999, 0.9999\}$-th quantile (*model h*), and the Hang Seng returns with a *VaR* and *ES* for the$\{0.99, 0.999, 0.9999\}$-th quantile (*model a*). The grey line is the returns of the sample period, the light-grey line is the *VaR* estimations, and the black line is the *ES*

# 7 Conclusions

It is by now well known that returns from financial assets exhibit heavy tails and clusters at the extremes. As a consequence the normal distribution is regarded as an inappropriate model to characterize the return distribution. Especially, extreme market conditions tend to happen more frequently than expected on the basis of a normal distribution.

In this paper we concentrate on the estimation of the tail of the distribution of financial return time series, and some popular risk measures such as *VaR* and *ES*. We propose an extension of the classical POT to model cluster behaviour through self-exciting processes for the exceedance times and a marked point process for the exceedances themselves.

The idea of a Peaks over threshold model with self-exciting structures is relatively new and was recently explored in [16, 2].

We fitted eight different models to return data from three important crashes in the history, the crash of 1987, the NASDAQ crash 2000, and the Hong Kong crash in 1997. Maximum likelihood methods are used to calculate the parameters, where the self-exciting approach can follow a Hawkes model or an ETAS model for the point process. At the same time the exceedances over a defined threshold were modelled with a generalized Pareto approximation suggested by extreme value theory.

In contrast to the classical model approaches such as proposed in [14, 17, 12], we apply a direct treatment to the effects of temporal dependence especially in periods of high volatility, which may cause large losses to cluster over time. The models proposed in this paper are a refinement designed to account for such stylized factors and provide a strong alternative to the models applied until now.

Two directions for future research emerge from the results. An extension to non stationary series could be made as in [3], and other flexible forms for the self-exciting function could be used incorporating other characteristics of the series such as trend of increasingly exceedances or different regimes as after shocks.

# Appendix

**Table 1** Summary statistics for the stock index returns. Asymptotic p-values are shown in the brackets. *,**,*** denote statistical significance at the 1, 5 and 10 level respectively

| | S&P 500 | NASDAQ | Hang Seng |
|---|---|---|---|
| N° obs. | 6985 | 5105 | 2681 |
| Std. dev | 0.0081 | 0.0101 | 0.0179 |
| Minimum | –0.0691 | –0.1204 | –0.4054 |
| 1st Qu | –0.0040 | –0.0037 | –0.0057 |
| Mean | 0.0002 | 0.0012 | 0.0008 |
| Median | 0.0003 | 0.0006 | 0.0005 |
| 3rd Qu. | 0.0045 | 0.0058 | 0.0083 |
| Maximum | 0.0490 | 0.0709 | 0.1725 |
| Kurtosis | 3.5340 | 12.0282 | 116.7251 |
| Skewness | 0.0035 | –1.0883 | –5.5619 |
| Jarque-Bera test | 3638.847* | 31812.79* | 1538169* |
| Augmented Dickey–Fuller Test | –17.6038* | –15.2872* | –13.0182* |
| Phillips-Perron Unit Root Test | –5585.806* | –4454.225* | –2715.409* |

**Table 2** p-values for a two-sided binomial test of the hypothesis that the model estimated well the extremes of the log-returns at different quantiles against the alternative that this model over or underestimates these quantiles

| models | S&P 500 | NASDAQ | Hang Seng |
|---|---|---|---|
| VaR | 6985 | 5104 | 2673 |
| 0.99 Quantile | 73 (0.67) | 52 (0.88) | 27 (0.92) |
| 0.999 Quantile | 10 (0.25) | 4 (0.82) | 5 (0.20) |
| 0.9999 Quantile | 1 (0.50) | 1 (0.39) | 1 (0.23) |

**Table 3** Backtesting Results: p-values for the theoretically expected number of violations against the violations obtained using the best models fitted for each return. S&P 500 *Model h*, NASDAQ *Model h*, Hang Seng *Model a*

| models | S&P 500 | NASDAQ | Hang Seng |
|---|---|---|---|
| VaR | 253 | 253 | 248 |
| 0.99 Quantile | 3 (0.74) | 1 (0.53) | 4 (0.53) |
| 0.999 Quantile | 1 (0.22) | 0(1) | 0 (1) |
| 0.9999 Quantile | 1(0.02) | 0(1) | 0 (1) |

**Table 4** Results for the S&P 500 returns, *Model a* is a Hawkes model without predictable marks ($\eta = 0$) and without influence of the marks on the estimation of the self-exciting function ($\delta = 0$). *Model b* is a Hawkes model without predictable marks and $\delta > 0$, *Model c* is an ETAS model without predictable marks and without influence of the marks. *Model d* is an ETAS without predictable marks, *Model e* is a Hawkes model without influence of the marks $\delta = 0$ and predictable marks $\eta > 0$, *Model f* is a Hawkes model with $\delta, \eta > 0$, *Model g* is an ETAS model with predictable marks and $\delta = 0$ and *Model h* is the general formulation for an ETAS model with $\delta, \eta > 0$. Standard deviation and 95% confidence intervals are shown in between parentheses. *l.likes* are the results of the maximization of the log-likelihood

| Models | $\tau$ | $\phi$ | $\gamma$ | $\rho$ | $\delta$ | $\xi$ | $\beta$ | $\eta$ | $VaR_{0,01\%}$ | $ES_{0,01\%}$ | *l.like* |
|---|---|---|---|---|---|---|---|---|---|---|---|
| Model a | 0.0105 | 0.0381 | 0.0548 | | | 0.176 | 0.0043 | | 7.6 | 9.4 | 64.3436 |
| | (0.0002) | (0.0071) | (0.0106) | | | (0.0788) | (0.0004) | | (6.73, 8.51) | (8.29, 10.68) | |
| Model b | 0.0109 | 0.0309 | 0.0577 | | 53.84 | 0.1763 | 0.0042 | | 7.92 | 9.81 | 65.5127 |
| | (0.0019) | (0.0078) | (0.0116) | | (46.71) | (0.0787) | (0.0004) | | (7.00, 8.93) | (8.62, 11.1) | |
| Model c | 0.0080 | 0.0445 | 25.86 | 1.48 | | 0.1763 | 0.0042 | | 7.68 | 9.52 | 65.4723 |
| | (0.0024) | (0.0010) | (21.90) | (1.14) | | (0.0787) | (0.0004) | | (6.80, 8.66) | (8.38, 10.8) | |
| Model d | 0.0082 | 0.0360 | 22.61 | 1.35 | 55.86 | 0.1763 | 0.0042 | | 8.01 | 9.92 | 66.7255 |
| | (0.0024) | (0.0105) | (18.49) | (1.01) | (47.34) | (0.0787) | (0.0004) | | (7.08, 9.04) | (8.72, 11.2) | |
| Model e | 0.0107 | 0.0403 | 0.0588 | | | 0.1768 | 0.0026 | 0.0011 | 10.05 | 12.62 | 73.2512 |
| | (0.0018) | (0.0071) | (0.011) | | | (0.0791) | (0.0004) | (0.0003) | (6.96, 13.64 ) | (8.86,16.97) | |
| Model f | 0.0011 | 0.0320 | 0.0615 | | 58.01 | 0.1788 | 0.0026 | 0.0008 | 11.67 | 14.69 | 74.8727 |
| | (0.0018) | (0.0077) | (0.0113) | | (44.13) | (0.0792) | (0.0004) | (0.0003) | (7.98, 15.65 | (10.20, 19.55) | |
| Model g | 0.0085 | 0.0471 | 26.42 | 1.64 | | 0.1733 | 0.0025 | 0.0013 | 10.36 | 12.97 | 74.3078 |
| | (0.0023) | (0.0153) | (22.16) | (1.25) | | (0.0789) | (0.0004) | (0.0004) | (7.19, 13.86) | (9.13, 17.20) | |
| Model h* | 0.0087 | 0.0371 | 21.27 | 1.38 | 64.58 | 0.1749 | 0.0025 | 0.0010 | 12.31 | 15.46 | 76.2481 |
| | (0.0024) | (0.0103) | (16.60) | (0.9846) | (46.63) | (0.0789) | (0.0004) | (0.0004) | (8.42, 16.48) | (10.74, 20.51) | |

*corresponds to the best model fitted in comparison to the other models with the help of a likelihood Ratio test.

**Table 5** Results for the NASDAQ returns, *Model a* is a Hawkes model without predictable marks ($\eta = 0$) and without influence of the marks on the estimation of the self-exciting function ($\delta = 0$). *Model b* is a Hawkes model without predictable marks and $\delta > 0$, *Model c* is an ETAS model without predictable marks and without influence of the marks. *Model d* is an ETAS without predictable marks, *Model e* is a Hawkes model without influence of the marks $\delta = 0$ and predictable marks $\eta > 0$, *Model f* is a Hawkes model with $\delta, \eta > 0$, *Model g* is an ETAS model with predictable marks and $\delta = 0$ and *Model h* is the general formulation for an ETAS model with $\delta, \eta > 0$. Standard deviation and 95% confidence intervals are shown in parentheses. *l.likes* are the results of the maximization of the log-likelihood

| Models | $\tau$ | $\phi$ | $\gamma$ | $\rho$ | $\delta$ | $\xi$ | $\beta$ | $\eta$ | $VaR_{0,01\%}$ | $ES_{0,01\%}$ | *l.like* |
|---|---|---|---|---|---|---|---|---|---|---|---|
| Model a | 0.0118 | 0.0368 | 0.566 | | | 0.2029 | 0.0085 | | 13.51 | 17.55 | –76.07233 |
| | (0.0022) | (0.0089) | (0.0144) | | | (0.0842) | (0.0009) | | (11.61, 15.80) | (14.99, 20.63) | |
| Model b | 0.0123 | 0.0221 | 0.0548 | | 54.15 | 0.2029 | 0.0085 | | 13.62 | 17.70 | –73.23696 |
| | (0.0022) | (0.0089) | (0.0139) | | (41.28) | (0.0842) | (0.0009) | | (11.71, 15.94) | (15.12, 20.82) | |
| Model c | 0.0079 | 0.0449 | 11.95 | 0.6359 | | 0.2029 | 0.0085 | | 14.00 | 18.16 | –74.56616 |
| | (0.0035) | (0.0134) | (10.68) | (0.6471) | | (0.0842) | (0.0009) | | (12.02, 16.39) | (15.51, 21.37) | |
| Model d | 0.0089 | 0.0274 | 12.38 | 0.6788 | 55.78 | 0.2029 | 0.0085 | | 13.98 | 18.14 | –71.73594 |
| | (0.0037) | (0.0120) | (12.22) | (0.7558) | (43.34) | (0.0842) | (0.0009) | | (12.01, 16.37) | (15.48, 21.34) | |
| Model e | 0.0125 | 0.0441 | 0.0704 | | | 0.1428 | 0.0065 | 0.0020 | 10.93 | 13.46 | –72.88907 |
| | (0.0022) | (0.0107) | (0.0184) | | | (0.0832) | (0.0010) | (0.0010) | (10.66, 17.85) | (11.95, 16.20) | |
| Model f | 0.0130 | 0.0266 | 0.0692 | | 55.27 | 0.1175 | 0.0067 | 0.0011 | 10.23 | 12.34 | –68.97337 |
| | (0.0022) | (0.0101) | (0.0177) | | (37.41) | (0.0811) | (0.0009) | (0.0006) | (9.98, 12.38) | (11.12, 15.05) | |
| Model g | 0.0091 | 0.0594 | 8.52 | 0.6409 | | 0.1333 | 0.0062 | 0.0029 | 11.55 | 14.12 | –70.8677 |
| | (0.0033) | (0.0177) | (6.9623) | (0.5606) | | (0.0837) | (0.0010) | (0.0015) | (8.89, 14.28) | (11.04, 17.27) | |
| Model h* | 0.0096 | 0.0359 | 7.63 | 0.5832 | 61.03 | 0.1033 | 0.0064 | 0.0016 | 10.58 | 12.62 | –66.5314 |
| | (0.0033) | (0.0151) | (6.31) | (0.5288) | (41.43) | (0.0805) | (0.0010) | (0.0010) | (8.39, 12.77) | (10.10, 15.65) | |

*corresponds to the best model fitted in comparison to the other models with the help of a likelihood Ratio test.

**Table 6** Results for the Hang Seng returns, *Model a* is a Hawkes model without predictible marks ($\eta = 0$) and without influence of the marks on the estimation of the self-exciting function ($\delta = 0$). *Model b* is a Hawkes model without predictable marks and $\delta > 0$, *Model c* is an ETAS model without predictable marks and without influence of the marks. *Model d* is an ETAS without predictable marks, *Model e* is a Hawkes model without influence of the marks $\delta = 0$ and predictable marks $\eta > 0$, *Model f* is a Hawkes model with $\delta, \eta > 0$, *Model g* is an ETAS model with predictable marks and $\delta = 0$ and *Model h* is the general formulation for an ETAS model with $\delta, \eta > 0$. Standard deviation and 95% confidence intervals are shown in parentheses. *l.likes* are the results of the maximization of the log-likelihood

| Models | $\tau$ | $\phi$ | $\gamma$ | $\rho$ | $\delta$ | $\xi$ | $\beta$ | $\eta$ | $VaR_{0,01\%}$ | $ES_{0,01\%}$ | *l.like* |
|---|---|---|---|---|---|---|---|---|---|---|---|
| Model a* | 0.0132 | 0.0435 | 0.0757 | | | 0.3441 | 0.0118 | | 37.22 | 57.20 | –76.97864 |
| | (0.0030) | (0.0154) | (0.0278) | | | (0.1505) | (0.0022) | | (28.31, 49.42) | (43.16, 76.44) | |
| Model b | 0.0136 | 0.0370 | 0.0838 | | 14.37 | 0.3441 | 0.0118 | | 36.22 | 55.68 | –76.50523 |
| | (0.0031) | (0.0158) | (0.0311) | | (21.22) | (0.1505) | (0.0022) | | (27.57, 48.08) | ( 42.04, 74.39) | |
| Model c | 0.0129 | 0.0459 | 117.61 | 9.29 | | 0.3441 | 0.0118 | | 37.33 | 57.36 | –76.95545 |
| | (0.0031) | (0.0207) | (616.33) | (46.52) | | (0.1505) | (0.0022) | | (28.39, 49.57) | (43.29, 76.67) | |
| Model d | 0.0134 | 0.0388 | 114.47 | 9.86 | 13.95 | 0.3441 | 0.0118 | | 36.37 | 55.91 | –76.49075 |
| | (0.0032) | (0.0200) | (681.03) | (56.71) | (20.70) | (0.1505) | (0.0022) | | (27.68, 48.28) | (42.21, 74.70) | |
| Model e | 0.0135 | 0.0498 | 0.0895 | | | 0.3453 | 0.0094 | 0.0026 | 45.74 | 70.75 | –76.42726 |
| | (0.0030) | (0.0176) | (0.0327) | | | (0.1503) | (0.0026) | (0.0028) | (41.93, 68.66) | (40.89, 82.57) | |
| Model f | 0.0141 | 0.0403 | 0.1011 | | 18.82 | 0.3280 | 0.0093 | 0.0023 | 40.32 | 60.89 | –75.70827 |
| | (0.0031) | (0.0178) | (0.0358) | | (24.27) | (0.1518) | (0.0024) | (0.0022) | (37.20, 62.07) | (36.82, 71.68) | |
| Model g | 0.0132 | 0.0563 | 43.03 | 4.24 | | 0.3463 | 0.0093 | 0.0031 | 46.97 | 72.77 | –76.35493 |
| | (0.0032) | (0.0287) | (141.27) | (12.61) | | (0.1500) | (0.0026) | (0.0035) | (43.24, 70.83) | (41.11, 84.46) | |
| Model h | 0.0138 | 0.0455 | 39.76 | 4.36 | 17.97 | 0.3306 | 0.0092 | 0.0027 | 41.75 | 63.28 | –75.65863 |
| | (0.0033) | (0.0261) | (135.81) | (13.66) | (23.32) | (0.1514) | (0.0025) | (0.0028) | (38.65, 64.48) | (37.37, 74.04) | |

*corresponds to the best model fitted in comparison to the other models with the help of a likelihood Ratio test.

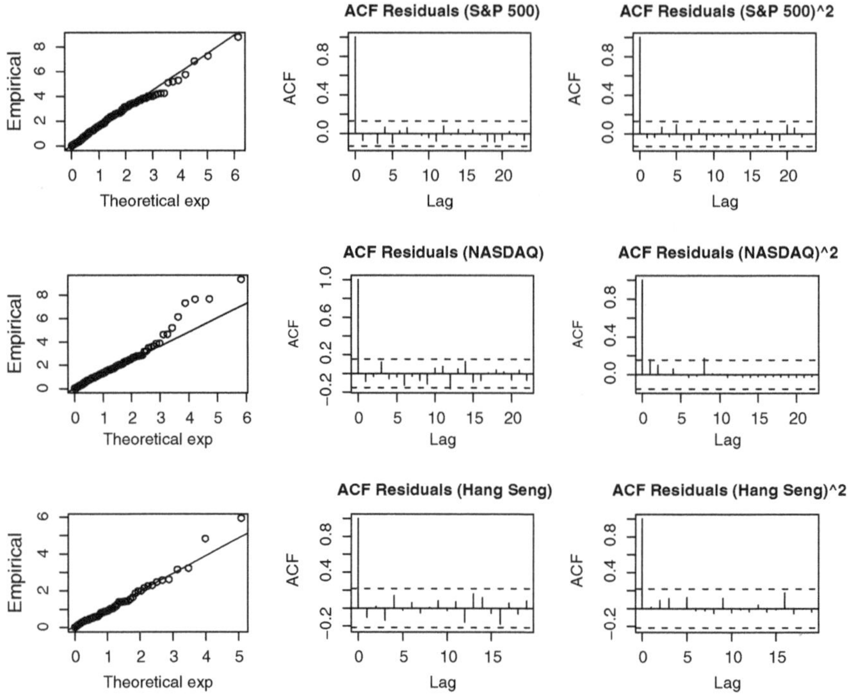


**Fig. 4** qq-plots of the residuals (*left*), autocorrelation function of the residuals (*middle*), autocorrelation function of the square of residuals (*right*), for the returns of the S&P 500 index (*top panel*), the NASDAQ index (*middle panel*) and the Hang Seng index (*bottom panel*)

## References


[1] Beirlant, J., Goegebeu, Y., Segers, J., Teugels, J., Waal, D.D.: Statistics of Extremes: Theory and Applications. Wiley, New York (2004)

[2] Chavez-Demoulin, V., Davison, A., McNeil, A.: A point process approach to value-at-risk estimation. Quant. Fin. **5**, 227–234 (2005)

[3] Chavez-Demoulin, V., Sardy, S.: A Bayesian nonparametric peaks over threshold method to estimate risk measures of a nonstationary financial time series. Research Paper, Department of Mathematics, ETH, Zürich (2004)

[4] Cotter, J., Dowd, K.: Extreme spectral risk measures: An application to futures clearinghouse margin requirements. J. Bank. Fin **30**, 3469–3485 (2006)

[5] Daley, D., Vere-Jones, D.: An Introduction to the Theory of Point Processes. Springer, Berlin (2003)

[6] Embrechts, P., Kluppelberg, C., Mikosch, T.: Modeling Extremal Events. Springer, Berlin (1997)

[7] Falk, M., Hüsler, J., Reiss, R.: Laws of Small Numbers: Extremes and Rare Events. Birkhäuser, Boston (2004)

[8] Ferro, C.A.T., Segers, J.: Inference for clusters of extreme values. J. Roy. Stat. Soc. B **65**, 545–556 (2003)
[9] Hawkes, A.: Spectra of some self-exciting and mutually exciting point processes. Biometrika **58**, 379–402 (1971)
[10] Kagan, Y.Y., Knopoff, L.: Statistical short-term earthquake prediction. Science **235**, 1563–1467 (1987)
[11] Longin, F., Solnik, B.: Extreme correlation of international equity markets. J. Finance **56**, 649–676 (2001)
[12] Matthys, G., Beirlant, J.: Extreme quantile estimation for heavy tailed distributions. Technical Report, Universitair centrum voor Statistiek, Katholieke Universiteit Leuven, Leuven, Belgium (2001)
[13] McNeil, A., Frey, R.: Estimation of tail-related risk measures for heteroscedastic financial time series: an extreme value approach. J. Emp. Finance **7**, 271–300 (2000)
[14] McNeil, A.J.: On extremes and crashes. Risk **11**, 99–104 (1998)
[15] McNeil, A.J., Chavez-Demoulin, V.: Self-exciting processes for extremes in financial time series. Cornell University, Financial Engineering seminar, Talk (2006)
[16] McNeil, A.J., Frey, R., Embrechts, P.: Quantitative Risk Management: Concepts, Techniques, and Tools. Princeton University Press, Princeton (2005)
[17] Novak, S.Y.: Value at risk and the "Black Monday" crash. MUBS Discussion Paper 25, Middlesex University, UK (2004)
[18] Ogata, Y.: Statistical models for earthquake occurrences and residual analysis for point processes. J. Am. Stat. Assoc. **83**, 9–27 (1988)
[19] Smith, R.: Extreme value analysis of environmental time series: An application to trend detection in ground-level ozone. Stat. Sci. **44**, 367–393 (1989)

# Consumption and Income: A Spectral Analysis

D.S.G. Pollock

**Abstract** The relationship between aggregate income and consumption in the United Kingdom is analysed anew. This entails a close examination of the structure of the data, using a variety of spectral methods that depend on the concepts of Fourier analysis. It is found that fluctuations in the rate of growth of consumption tend to precede similar fluctuations in income, which contradicts a common supposition. The difficulty is emphasised of uncovering from the aggregate data a structural equation representing the behaviour of consumers.

## 1 Introduction: The Evolution of the Consumption Function

Over many years, the aggregate consumption function has provided a context in which problems of econometric modelling have been debated and from which significant innovations in methodology have emerged. Whereas such innovations have advanced the subject of econometrics, none of them has been wholly appropriate to the aggregate consumption function itself. This may be one of the reasons why the consumption function has remained a focus of attention.

The vestiges of our misconceptions tend to linger in our minds long after we have consciously amended our beliefs. Our view of the consumption function is particularly prone to the effects of ideas that have not been properly discarded despite their inapplicability. Therefore, in setting a context for our discussion, it is helpful to recount some of the history of the consumption function.

The first difficulties that were encountered in modelling the aggregate consumption function arose from a conflict between Keynesian theory and the empirical findings of Kuznets [12] and others. Whereas the theory of Keynes [11] postulated

D.S.G. Pollock
Department of Economics, University of Leicester, Leicester LE1 7RH
d.s.g.pollock@le.ac.uk

average and marginal propensities to consume that declined with income, it was discovered that income and consumption had maintained a rough proportionality over many years.

At the same time, the econometricians were conscious that there is a double relationship between income and consumption, which follows from the fact that consumption expenditures are a major factor in determining the level of aggregate income. The failure to take account of the second relationship might lead to biases in the estimated coefficients of the consumption function.

Using a static analysis, Haavelmo [7] demonstrated that the estimated marginal propensity to consume was subject to an upward basis that was directly related to the variance of the innovations in consumption and inversely related to the variance of the innovations in income. The latter were attributed to autonomous changes in the rate of investment.

However, Haavelmo also envisaged, in common with other analysts, "that the active dynamic factor in the business cycle is investment, with consumption assuming a passive lagging role." (These are the words of Alvin Hansen [9], as quoted by Haavelmo.) This notion was used by others in reconciling the Keynesian formulation with the empirical findings. The manner in which they did so greatly stimulated the development of dynamic econometric modelling.

Models in which consumption displayed a laggardly response to income were provided by Duesenberry [2], who propounded the relative income hypothesis, by Modigliani and Brumberg [14], who propounded the life-cycle hypothesis—see Modigliani [13], also—and by Friedman [3], who propounded the permanent income hypotheses. According to these models, rapid increases in income will give rise, in the short run, to less-than-proportional increases in consumption, which is in accordance with the Keynesian view. Over longer periods, consumption will gradually regain the long-run relationship with income that was revealed in the empirical findings.

The idea that consumption reacts in a passive and laggardly fashion to the forces impacting upon it also suggested that it might be reasonable to ignore the problem of simultaneous equation bias, to which Haavelmo had drawn attention. The biases would be small if the innovations or disturbances in consumption behaviour were relatively small and if consumers were reacting preponderantly to events of the past.

The two suppositions, upon which the interpretations of the dynamic models largely depended, which were the inertial nature of consumer's behaviour and the relative insignificance of the consumption innovations, have become established preconceptions, despite the lack of evidence to support them. In fact, the evidence that we shall uncover strongly suggests that, in the U.K., the business cycle has been driven by the fluctuations in consumers' expenditure.

For almost two decades, beginning in the mid fifties, successes in modelling the consumption function were seen as grounds for congratulating the econometricians. However, the observations of Granger and Newbold [6] and others on the spurious nature of regression relationships between trended economic variables led many to suspect that the success might be illusory. Whereas such regressions account remarkably well for the level of consumption, they often perform poorly in the far more stringent task of predicting changes in the level of consumption from one period to another. Moreover, as Granger and Newbold [6] emphasised, the standard inferential procedures of linear regression analysis are valid only in application to data that have finite-valued asymptotic moment matrices. The moment matrices of trended variables, such as income and consumption, are unbounded.

An apparent resolution of these difficulties came in the late 1970's with the advent of the error-correction formulation of the consumption function. It was understood that a dynamic regression model in the levels of income and consumption can be expressed, via a linear reparametrisation, as a model that comprises the differences of the variables together with a stationary error term expressing the current disproportion between income and consumption. Such a model, in which all of the variables appear to be from stationary sequences, is amenable to the standard inferential procedures.

The paper of Davidson et al. [1], which adopted an error-correction formulation, succeeded in re-establishing the traditional consumption function within a viable econometric framework. For a model in which the dependent variable was a differenced sequence, it achieved a remarkably high value for the coefficient of determination. It also heralded the incipient notion of a cointegrating relationship between trended variables, which has subsequently proved to be of major importance.

Some doubts have remained concerning the error-correction formulation of the dynamic consumption function. For a start, it is questionable whether the equation is a structural equation that truly represents the behaviour of consumers in the aggregate, as it purports to do. There may be insufficient grounds for ignoring the problems of simultaneous equation bias. There have also been doubts about the statistical significance of the error-correction term, which is included in the equation. We shall raise these doubts anew.

Enough time has elapsed since the publication of the article of Davidson et al. [1] for the data series to have more than doubled in length. In spite of the various economic vicissitudes that are reflected in the extended data set, their model continues to fit remarkably well, with newly estimated coefficients that are not vastly different from the original ones. One of the purposes of the present paper is to examine the basis for this apparent success. The principal purpose is to determine whether the time-honoured presuppositions about the nature of the income-consumption relationship, which were inherited by the consumption function of Davidson et al. [1], have any empirical support.

## 2 The Data and the Four-Period Difference Filter

In evaluating any model, we should begin by inspecting the data. The data series of income and consumption—which is the expenditure on nondurable goods—have two prominent characteristics. The first characteristic is their non-stationarity. Over the extended data, the logarithms of the data, which are plotted in Fig. 1, show upward trends that are essentially linear. The second characteristic of the data series

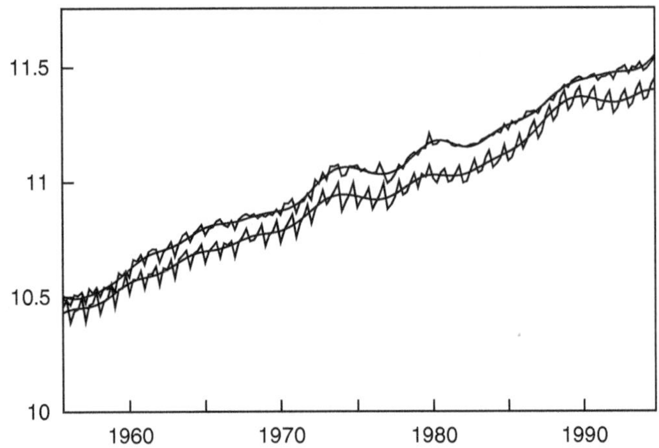


**Fig. 1** The quarterly series of the logarithms of income (upper) and consumption (lower) in the U.K., for the years 1955 to 1994, together with their interpolated trends

is that they both show evident patterns of seasonal variation, which play on the backs of the rising trends.

The seasonal pattern is more evident in the consumption series than it is in the income series. Therefore, we incline to the view that, rather than being transferred from the income stream, the seasonal fluctuations in consumption have their origin in an independent influence that impinges on both income and consumption. This motivates us to look at ways of deseasonalising the data that will remove the effect.

Models like that of Davidson et al. [1] seek to explain an annual growth rate in consumption that is derived from quarterly data. The dependent variable of the model is obtained by passing the logarithms of the consumption series, which we shall denote by $y(t)$, through a four-period difference filter of the form $\nabla_4 = 1 - L^4 = (1-L)(1+L+L^2+L^3)$. Here, $L$ is the lag operator, which has the effect that $Ly(t) = y(t-1)$, where $y(t) = \{y_t; t = 0 \pm 1, \pm 2, \ldots\}$ is a series of observations taken at three-monthly intervals. The filter removes from $y(t)$ both the trend and the seasonal fluctuations; and it removes much else besides.

The squared gain of the filter is depicted in Fig. 2. The operator nullifies the component at zero frequency and it diminishes the power of the elements of the trend whose frequencies are in the neighbourhood of zero. This is the effect of $\nabla = 1 - L$, which is a factor of $\nabla_4$. The filter also removes the elements at the seasonal frequency of $\pi/2$ and at its harmonic frequency of $\pi$, and it attenuates the elements in the neighbourhoods of these frequencies. This is the effect of the four-point summation operator $S_4 = 1+L+L^2+L^3$, which is the other factor of $\nabla_4$. It is also apparent that the filter amplifies the cyclical components of the data that have frequencies in the vicinities of $\pi/4$ and $3\pi/4$; and, as we shall discover later, this is a distortion that can have a marked effect upon some of the estimates that are derived from the filtered data.

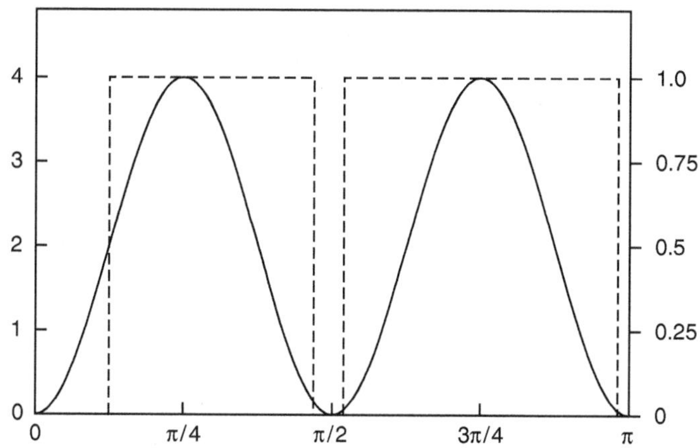


**Fig. 2** The squared gain of the four-period difference filter $\nabla_4 = 1 - L^4$ (continuous line and left scale) and the frequency selection of the deseasonalised detrended data (broken line and right scale)

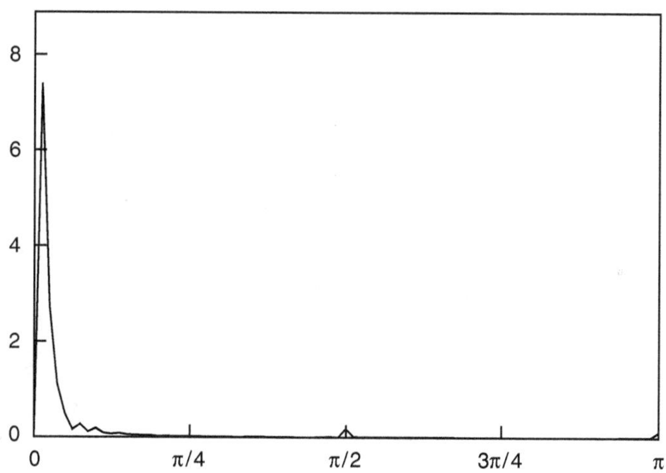


**Fig. 3** The periodogram of the logarithms of consumption in the U.K., for the years 1955 to 1994

The effect of the filter upon the logarithmic consumption series can be see by comparing the periodograms of Figs. 3 and 4. The periodogram of the sample comprised by the vector $y = [y_0, y_1, \ldots, y_{T-1}]'$ is the sequence of the coefficients $\rho_j^2 = \alpha_j^2 + \beta_j^2$, scaled by $T/2$, that come from the Fourier expression

$$\begin{aligned} y_t &= \sum_{j=0}^{[T/2]} \rho_j \cos(\omega_j t - \theta_j) \qquad (1) \\ &= \sum_{j=0}^{[T/2]} \left\{ \alpha_j \cos(\omega_j t) + \beta_j \sin(\omega_j t) \right\}, \end{aligned}$$

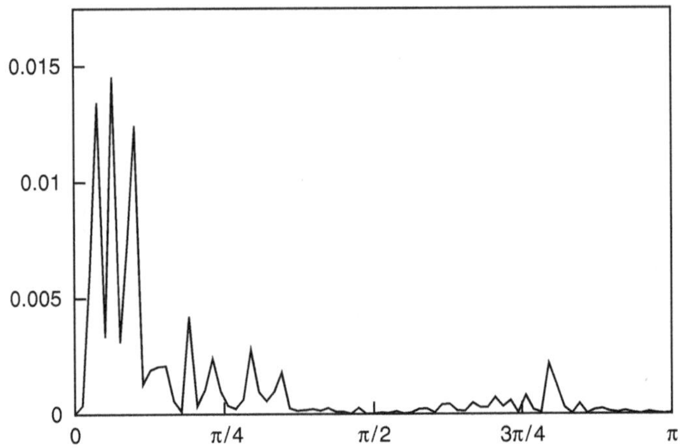


**Fig. 4** The periodogram of the filtered series $\nabla_4 y(t)$ representing the annual growth rate of consumption

where $T$ is the sample size and $[T/2]$ is the integral part of $T/2$. Here, $\omega_j = 2\pi j/T$ is the frequency of a sinusoid that takes $j$ periods to complete a cycle. Its amplitude is $\rho_j$, whilst $\rho_j^2/2$ is its power which is, in other words, its contribution to the variance of the sample.

In the second expression, the parameters are $\alpha_j = \rho_j \cos\theta_j$ and $\beta_j = \rho_j \sin\theta_j$, with $\beta_0 = 0$ and $\beta_{[T/2]} = 0$ if $T$ is an even number. We shall describe $\rho_j \cos(\omega_j t - \theta_j)$ as the $j$th sinusoidal element in the Fourier decomposition of the sample. (For a detailed exposition, see Pollock [17]).

The most striking effect of the filtering is the diminution of the power at the frequencies in the vicinity of zero, which is where the elements of the trend component are to be found, and in the vicinities of $\pi/2$ and $\pi$, where the seasonal elements and their harmonics are to be found. The degree of the amplification of the elements in the vicinities of $\pi/4$ and $3\pi/4$, which is evident in Fig. 4, can be judged in comparison with a periodogram of the detrended data, presented in Fig. 5, which has been obtained by fitting a linear trend.

The methods for detrending and deseasonalising the data that we shall propose are designed to remove the minimum amount of information from the processed series. They avoid the distortions that are induced by the differencing operator.

## 3 The Error-Correction Model and its Implications

The consumption function of Davidson et al. [1] was calculated originally on a data set from the U.K. running from 1958 to 1970, which was a period of relative economic quiescence. When the function is estimated for an extended data period, running from 1956 to 1994, it yields the following results:

$$\nabla_4 y(t) = \underset{(0.40)}{0.70}\nabla_4 x(t) - \underset{(0.60)}{0.156}\nabla\nabla_4 x(t) + \underset{(0.15)}{0.068}\{x(t-4) - y(t-4)\} + e(t) \qquad (2)$$
$$R^2 = 0.77 \qquad \text{D–W} = 0.920.$$

Here $y(t)$ and $x(t)$ represent, respectively, the logarithms of the consumption sequence and the income sequence, without seasonal adjustment. The numbers in parentheses below the estimated coefficients are standard errors. The operators $\nabla = 1 - L$ and $\nabla_4 = 1 - L^4$ are, respectively, the one-period and the four-period difference operator. Therefore, $\nabla_4 y(t)$ and $\nabla_4 x(t)$ represent the annual growth rates of consumption and income, whilst $\nabla_1\nabla_4 x(t)$ represents the acceleration or deceleration in the growth of income.

This specification reflects an awareness of the difficulty of drawing meaningful inferences from a regression equation that incorporates nonstationary variables. The difference operators are effective in reducing the sequences $x(t)$ and $y(t)$ to stationarity. The synthetic sequence $x(t-4) - y(t-4)$ is also presumed to be stationary by virtue of the cointegration of $x(t)$ and $y(t)$; and its role within the equation is to provide an error-correction mechanism, which tends to eliminate any disproportion that might arise between consumption and income.

The specification also bears the impress of some of the earlier experiences in modelling the consumption function that we have described in the introduction. The variable $\nabla_1\nabla_4 x(t)$ with its associated negative-valued coefficient allows the growth of consumption to lag behind the growth of income when the latter is accelerating. This is the sort of response that the analysts of the late 1940's and 1950's, who were intent on reconciling the Keynesian formulations with the empirical findings, were at pains to model.

We can evaluate the roles played by the terms of the RHS of equation (2) by modifying the specification and by observing how the coefficients of the fitted regression are affected and how the goodness of fit is affected.

The first modification is to replace $x(t-4) - y(t-4)$ by a constant dummy variable. The result is a slight change in the estimates of the remaining parameters of the model and a negligible loss in the goodness of fit. This suggests that we can dispense with the error-correction term at little cost:

$$\nabla_4 y(t) = \underset{(0.001)}{0.006} + \underset{(0.53)}{0.682}\nabla_4 x(t) - \underset{(0.66)}{0.160}\nabla\nabla_4 x(t) + e(t) \qquad (3)$$
$$R^2 = 0.76 \qquad \text{D–W} = 0.93.$$

In this connection, we should note that several analysts, including Hylleberg et al. [10], have found that the logarithmic series of consumption and income in the U.K. fail a test for cointegration. This seems to fly is the face of the evident relatedness of the two quantities. However, the finding may be taken as an indication that the relationship is not readily amenable to the linear dynamics of a simple error-correction mechanism.

We should also mention that, in a recent paper, Fenandez-Corugedo et al. [5] have found evidence for an error-correction mechanism within a vector autoregressive system of four equations. Their system has non-durable consumption, labour or non-assets income, the stock of assets and the relative price of durables to non-durables as its variables. However, the factor loadings on the single cointegrating vector indicate that the correction mechanism is present only in the equation of the assets. It is not present in the consumption equation.

The second modification is to eliminate both the error-correction term and the acceleration term $\nabla_1\nabla_4 x(t)$ and to observe how well the annual growth in consumption is explained by the annual growth of income. In this case, we observe that the coefficient of determination of the fitted regression is 0.72, compared with 0.77 for the fully specified model, while the error sum of squares increases to 0.053 from 0.044. We conclude from this that the acceleration term does have some effect:

$$\nabla_4 y(t) = 0.769\nabla_4 x(t) + e(t) \tag{4}$$
$$R^2 = 0.72 \qquad \text{D–W} = 1.15.$$

The fact that the acceleration term enters the consumption function with a negative coefficient seem to suggest that the response of consumption to rapid changes in income is laggardly more often that not. This would fit well with the various hypotheses regarding consumer behaviour that have been mentioned in the introduction. However, the significance of the estimated coefficient is not very great and it is considerably reduced when the coefficient is estimated using only the first third of the data. We shall reconsider the acceleration term at the end of the paper, where we shall discover that its effect is reversed when we analyse the relationship between the trends depicted in Fig. 1.

## 4 A Fourier Method for Detrending the Data

We have seen how the difference operator $1-L$ and the four-point summation operator $S_4 = 1+L+L^2+L^3$ are liable to remove a substantial part of the information that is contained in the data of the consumption series. In this section and the next, we shall propose alternative devices for detrending and for deseasonalising the data that leave much of the information intact. Our basic objective is to remove from the data only those Fourier elements that contribute to the trend or to the seasonality, and to leave the other components of the data unaffected.

A normal requirement for the use of the standard methods of statistical Fourier analysis is that the data in question should be generated by stationary processes, and this requirement is a hardly ever satisfied in econometric analysis. To understand the problems that can arise in applying Fourier methods to trended data, one must recognise that, in analysing a finite data sequence, one is making the implicit assumption that it represents a single cycle of a periodic function that is defined over the entire set of positive and negative integers. This function may be described as the periodic extension of the data sequence.

In the case of a trended sequence, there are bound to be radical disjunctions in the periodic function where one replication of the data sequence ends and another begins. Thus, for example, if the data follow a linear trend, then the function that is the subject of the Fourier analysis will have the appearance of the serrated edge of a saw blade. The saw tooth function has a spectrum that extends across the entire range of frequencies, with ordinates whose absolute values are inversely proportional to the corresponding frequencies—see for example, Hamming [8]. These effects of the trend are liable to be confounded with the spectral effects of the other motions that are present in the data.

The problem is resolved by using an approach that is familiar from the forecasting of ARIMA processes. We begin by differencing the data sequence as many times $d$ as may be necessary to reduce it to a state of stationarity. The income and consumption data need to be differenced twice, giving $d = 2$. We proceed to eliminate the low-frequency sinusoidal elements from the differenced data. Then, by cumulating or 'integrating' the resulting sequence as many times as the original data has been differenced, we will obtain the detrended version of the data. The trend of the data can be obtained, likewise, by cumulating the sum of the low-frequency elements that have been extracted from the differenced data.

To represent these processes, we need to employ the matrix versions of the difference operator and of the summation or cumulation operator, which is its inverse. Let the identity matrix of order $T$ be denoted by

$$I_T = [e_0, e_1, \ldots, e_{T-1}], \tag{5}$$

where $e_j$ represents a column vector that contains a single unit preceded by $j$ zeros and followed by $T - j - 1$ zeros. Then, the finite-sample lag operator is the matrix

$$L_T = [e_1, \ldots, e_{T-1}, 0], \tag{6}$$

which has units on the first subdiagonal and zeros elsewhere. The matrix that takes the $d$-th difference of a vector of order $T$ is given by $\Delta = (I - L_T)^d$.

Taking differences within a vector entails a loss of information. Therefore, if $\Delta = [Q_*, Q]'$, where $Q'_*$ has $d$ rows, then the $d$-th differences of a vector $y = [y_0, \ldots, y_{T-1}]'$ are the elements of the vector $g = [g_d, \ldots, g_{T-1}]'$ that is found in the equation

$$\begin{bmatrix} g_* \\ g \end{bmatrix} = \begin{bmatrix} Q'_* \\ Q' \end{bmatrix} y. \tag{7}$$

The vector $g_* = Q'_* y$ in this equation, which is a transform of the vector $[y_0, \ldots, y_{d-1}]$ of the leading elements of $y$, is liable to be discarded.

The inverse of the difference matrix is the matrix $\Delta^{-1} = \Sigma = [S_*, S]$. This has the effect that

$$S_* g_* + S g = y. \tag{8}$$

The vector $y$ can be recovered from the differenced vector $g$ only if the vector $g_*$ of initial conditions is provided.

The elements of the vector $g = [g_d, \ldots, g_{T-1}]'$ of the differenced data have the following Fourier expression:

$$g_t = \sum_{j=d}^{[T/2]} \{\gamma_j \cos(\omega_j t) + \delta_j \sin(\omega_j t)\}. \tag{9}$$

Let $\omega_C$ be the cut-off frequency that separates the Fourier elements of the trend component from the remainder. Then, by setting $\gamma_j, \delta_j = 0$ when $\omega_j > \omega_C$, we generate the elements of $z = [z_d, \ldots, z_{T-1}]'$, which is the differenced trend component, whereas, by setting $\gamma_j, \delta_j = 0$ when $\omega_j \leq \omega_c$, we generate the elements of $k = [k_d, \ldots, k_{T-1}]'$, which is the remainder.

The vector $z$ requires to be cumulated to form $x = S_* z_* + Sz$, which is the estimate of the tend. The initial conditions in $z_*$ should be chosen so as to ensure that the trend is aligned with the data as closely as possible. The criterion is

$$\text{Minimise} \quad (y - S_* z_* - Sz)'(y - S_* z_* - Sz) \quad \text{with respect to} \quad z_*. \tag{10}$$

The solution for the starting values is

$$z_* = (S_*' S_*)^{-1} S_*' (y - Sz). \tag{11}$$

The cut-off point $\omega_C$ marks the highest frequency amongst the Fourier elements that constitute the trend. The decision of where to place this point should be guided by an appraisal of the spectral structure of the data. Fig. 5 shows the periodogram of the residual sequence obtained by fitting a linear trend through the logarithms of

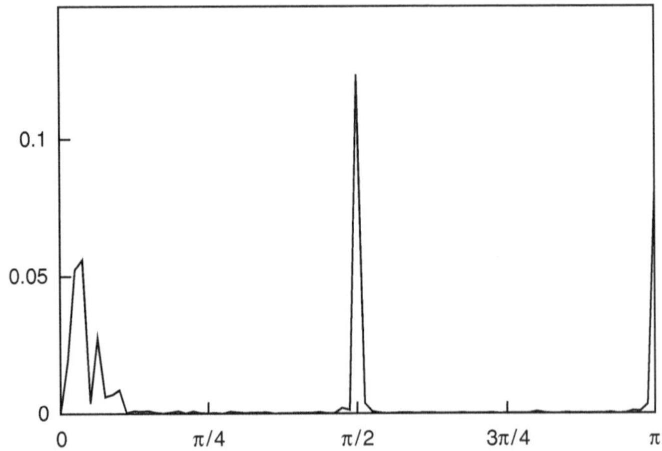


**Fig. 5** The periodogram of the residuals obtained by fitting a linear trend through the logarithmic consumption data of Fig. 1

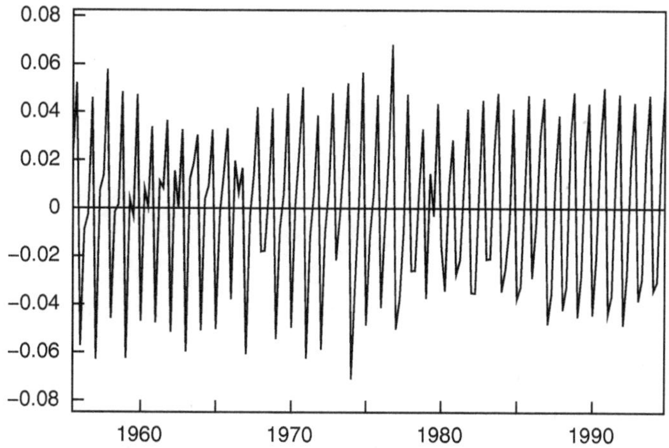


**Fig. 6** The detrended consumption series

the consumption series by least-squares regression. The regression residuals contain exactly the same information as does the twice differenced data sequence; and their periodogram serves to reveal the low-frequency spectral structure of the data. Within the peridogram of the twice-differenced data, this structure is so severely attenuated as to be virtually invisible.

We choose to place the cut-off point at $\omega_C = \pi/8$ radians, which is in a dead space of the periodogram where there are no ordinates of any significant size. Given that the observations are at quarterly intervals, this implies that the trend includes all cycles of four years duration of more. The detrended consumption series is show in Fig. 6. A similar analysis of the income data suggests that the same cut-off point is appropriate. The trends in the consumption and income series that have been calculated on this basis are depicted in Fig. 1.

## 5 A Fourier Method for Deseasonalising the Data

As well as removing the trend from the data, we also wish to remove the seasonal fluctuations. This can be done in much the same way. At its simplest, we can define the differenced seasonal component to consist only of those sinusoidal elements, extracted from the differenced data $\{g_d, \ldots, g_{T-1}\}$, that are at the seasonal frequency and at the harmonically related frequencies. Let $N = T - d$, where $d$ is the degree of differencing. Then, in the case of quarterly data, and on the supposition that $N$ is an even number, the component would be described by the equation

$$u_t = \alpha_{N/4} \cos\left(\frac{\pi t}{2}\right) + \beta_{N/4} \sin\left(\frac{\pi t}{2}\right) + \alpha_{N/2}(-1)^t, \qquad (12)$$

wherein

$$\alpha_{N/4} = \frac{2}{N}\sum_t g_t \cos\left(\frac{\pi t}{2}\right), \tag{13}$$
$$\beta_{N/4} = \frac{2}{N}\sum_t g_t \sin\left(\frac{\pi t}{2}\right),$$
$$\alpha_{N/2} = \frac{1}{N}\sum_t g_t (-1)^t.$$

In fact, this scheme is equivalent to one that uses seasonal dummy variables with the constraint that their associated coefficients must sum to zero. It will generate a pattern of seasonal variation that is the same for every year.

A more complex pattern of seasonality, which will vary gradually over the years, could be obtained by adopting a linear stochastic model with unit roots at the seasonal frequencies or by combining such a model with a "deterministic" trigonometrical or dummy-variable model in the manner suggested by Osborn et al. [16]. However, the desired effect can also be achieved by comprising within the Fourier sum a set of sinusoidal elements whose frequencies are adjacent to the seasonal frequency and to its harmonics.

The combined effect of two elements at adjacent frequencies depends upon whether their sinusoids are in phase, in which case they reinforce each other, or out of phase, in which case they tend to interfere with each other destructively. Two sinusoids whose frequencies are separated by $\theta$ radians will take a total of $\tau = 2\pi/\theta$ periods to move from constructive interference to destructive interference and back again. By this device, a pattern can be generated that evolves over the length of the sample.

It remains to describe how the seasonal elements that have been extracted from the differenced data are to be cumulated to provide an estimate of the seasonal component. It seems reasonable to chose the starting values so as to minimise the sum of squares of the seasonal fluctuations. Let $w = S_*u_* + Su$ be the cumulated seasonal component, where $u_*$ is a vector of $d$ starting values and $u$ is the vector of the seasonal component that has been extracted from the differenced data. Then the criterion is

$$\text{Minimise} \quad (S_*u_* + Su)'(S_*u_* + Su) \quad \text{with respect to} \quad u_*. \tag{14}$$

The solution for the starting values is

$$u_* = -(S_*'S_*)^{-1}S_*'Su. \tag{15}$$

Figure 7 shows the estimated seasonal component of the consumption series. The seasonal series is synthesised from the trigonometric functions at the seasonal frequency of $\pi/2$ and at its harmonic frequency of $\pi$, together with a handful of elements at the adjacent non-seasonal frequencies. It comprises two elements below $\pi/2$ and one above, and it also comprises one element below $\pi$. These choices have

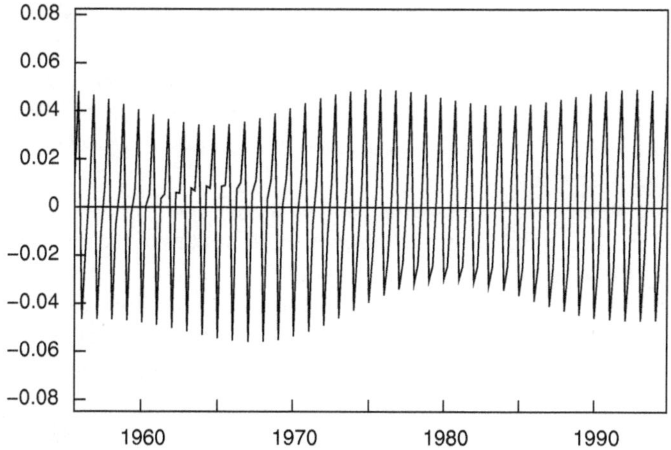


**Fig. 7** The estimated seasonal component of the consumption series

resulted from an analysis of the periodogram of Fig. 5. Figure 2 indicates, via the dotted lines, the frequencies that are present in the detrended and deseasonalised data.

The seasonal component of consumption accounts for the 93 percent of the variation of the detrended consumption series. When the seasonal component is estimated for the income series using the same set of frequencies, it accounts for only 46 percent of the variance of the corresponding detrended series.

## 6 A Re-appraisal of the Income–Consumption Relationship

In the previous section, we have described some new techniques for detrending the data and for extracting the seasonal component. We have discovered that the seasonal fluctuations in consumption are of a greater amplitude than those of the income series. They also appear to be more regular. It is also the case that Hylleberg et al. [10] failed to find cointegration between the two logarithmic series at the seasonal frequencies. These circumstances persuade us to reject the notion that the fluctuations have been transferred from income to consumption. It seems more reasonable to treat the seasonal fluctuations in both series as if they derive from external influences. Therefore, in seeking to establish a relationship between the detrended series, it is best to work with the deseasonalised versions.

When we turn to the deseasonalised and detrended consumption series, we find that its variance amounts to only 7 percent of the variance of the detrended series. It is hardly worthwhile to attempt to model this series. Indeed, the periodogram of Fig. 5 also makes it clear that there is very little information in the data of the consumption sequence that is not attributable either to the trend or to the seasonal component.

If it is accepted that the seasonal component needs no further explanation, then attention may be confined to the trend. The use of ordinary linear statistical methods dictates that any explanation of the consumption trend is bound to be in terms of data elements whose frequencies are bounded by zero and by the cut-off point of $\pi/8$ radians. That is to say, the trend in consumption can only be explained by similar trends in other variables.

Therefore, we turn to the essential parts of the income and the consumption series, which are their trends. We take the annual differences of the logarithmic trends by applying the operator $\nabla_4 = I - L^4$; and the results are a pair of smooth series that represent the annual growth rates of income and consumption. By combining the two series in one graph, which is Fig. 8, we are able to see that, in the main, the fluctuations in the growth in consumption *precede* similar fluctuations in the growth of income.

It may be recalled the income-acceleration term $\nabla\nabla_4 x(t)$ enters the consumption functions of equations (1) and (2) with a negative coefficient. This is in spite of the clear indication of Fig. 8 that the consumption-growth series leads the income-growth series. However, when the smoothed growth series $\nabla_4\hat{y}(t)$ and $\nabla_4\hat{x}(t)$ of Fig. 8 are used in these equations in place of $\nabla_4 x(t)$ and $\nabla_4 y(t)$, the sign on the coefficient of the acceleration term is reversed:

$$\begin{aligned}\nabla_4\hat{y}(t) = \underset{(0.001)}{0.006} + \underset{(0.44)}{0.689}\nabla_4\hat{x}(t) + \underset{(0.170)}{1.055}\nabla\nabla_4\hat{x}(t) + e(t) \\ R^2 = 0.87.\end{aligned} \tag{16}$$

The explanation of this anomaly must lie in the nature of the gain of the four-period difference filter $\nabla_4 = I - L_4$, which is represented in Fig. 2. The effect of

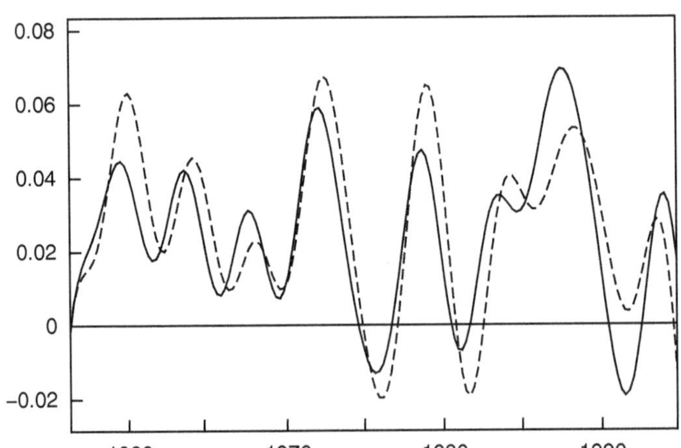


**Fig. 8** The annual differences of the trend of the logarithmic consumption series (solid line) and of the trend of the logarithmic income series (broken line)

the filter is to amplify some of the minor components of the data that lie in the dead spaces of the periodogram of Fig. 5 on either side of the frequencies $\pi/4$ and $3\pi/4$. Thus it can be concluded that, notwithstanding its specious justification, the negative acceleration term is an artefact of the differencing filter. This finding conflicts with the belief that consumption responds in a laggardly fashion to rapid changes in income.

The perception that the series of the annual growth rate in consumption is leading the corresponding series in income can be reaffirmed within the context of a bivariate vector autoregressive model. The model must be applied to the unsmoothed growth rates obtained by taking the four-period differences of the logarithms of the two series. It cannot be applied directly to the smoothed growth-rate series of Fig. 8, which have band-limited spectra. The reason is that an autoregressive model presupposes a spectral density function that is nonzero everywhere in the frequency range except on a set of measure zero.

The bivariate vector autoregressive model takes the form of

$$\nabla_4 y(t) = c_y + \sum_{i=1}^{p} \phi_i \nabla_4 y(t-i) + \sum_{i=1}^{p} \beta_i \nabla_4 x(t-i) + \varepsilon(t), \tag{17}$$

$$\nabla_4 x(t) = c_x + \sum_{i=1}^{p} \psi_i \nabla_4 x(t-i) + \sum_{i=1}^{p} \delta_i \nabla_4 y(t-i) + \eta(t). \tag{18}$$

The terms $c_x$ and $c_y$ stand for small constants, which are eliminated from the model when the differenced series are replaced by deviations about their mean values. The deviations may be denoted by $\tilde{y}(t) = \nabla_4 y(t) - E\{\nabla_4 y(t)\}$ and $\tilde{x}(t) = \nabla_4 x(t) - E\{\nabla_4 x(t)\}$. The expected values can be represented by the corresponding sample means.

In the case of $p = 2$, the estimated equations are

$$\begin{aligned} \tilde{y}(t) = {} & 0.51\tilde{y}(t-1) + 0.34\tilde{y}(t-2) + 0.27\tilde{x}(t-1) - 0.38\tilde{x}(t-2) + e(t), \\ & (0.86) \qquad\quad (0.87) \qquad\quad (0.73) \qquad\quad (0.72) \end{aligned} \tag{19}$$

$$\begin{aligned} \tilde{x}(t) = {} & 0.52\tilde{x}(t-1) - 0.10\tilde{x}(t-2) + 0.16\tilde{y}(t-1) + 0.25\tilde{y}(t-2) + h(t). \\ & (0.93) \qquad\quad (0.92) \qquad\quad (0.11) \qquad\quad (0.11) \end{aligned} \tag{20}$$

To facilitate the analysis of the model, it is helpful to write the equations (17) and (18) in a more summary notation that uses polynomials in the lag operator to represent the various sums. Thus

$$\phi(L)\tilde{y}(t) - \beta(L)\tilde{x}(t) = \varepsilon(t), \tag{21}$$

$$-\delta(L)\tilde{y}(t) + \psi(L)\tilde{x}(t) = \eta(t), \tag{22}$$

where $\phi(L) = 1 - \phi_1 L - \cdots - \phi_p L^p$, $\beta(L) = \beta_1 L + \cdots + \beta_p L^p$, $\psi(L) = 1 - \psi_1 L - \cdots - \psi_p L^p$ and $\delta(L) = \delta_1 L + \cdots + \delta_p L^p$.

The notion that the sequence $\tilde{y}(t)$ is driving the sequence $\tilde{x}(t)$ would be substantiated if the influence of the innovations sequence $\varepsilon(t)$ upon $\tilde{y}(t)$ were found to be stronger than the influence of $\eta(t)$ upon the corresponding sequence $\tilde{x}(t)$. The matter can be investigated via the moving-average forms of the equations, which express $\tilde{x}(t)$ and $\tilde{y}(t)$ as functions only of the innovations sequences $\varepsilon(t)$ and $\eta(t)$. The moving-average equations, which are obtained by inverting equations (21) and (22) jointly, are

$$\tilde{y}(t) = \frac{\psi(L)}{\pi(L)}\varepsilon(t) + \frac{\beta(L)}{\pi(L)}\eta(t), \tag{23}$$

$$\tilde{x}(t) = \frac{\delta(L)}{\pi(L)}\varepsilon(t) + \frac{\phi(L)}{\pi(L)}\eta(t), \tag{24}$$

where $\pi(L) = \phi(L)\psi(L) - \beta(L)\delta(L)$.

Since there is liable to be a degree of contemporaneous correlation between innovations sequences, the variance of the observable sequences $\tilde{y}(t)$ and $\tilde{x}(t)$ will not equal the sum of the variances of the components in $\varepsilon(t)$ and $\eta(t)$ on the RHS. The problem can be overcome by reparametrising the two equations so that each is expressed in terms of a pair of uncorrelated innovations. Such a procedure has been adopted by Geweke [4], for example.

Consider the innovation sequence $\eta(t)$ within the context of equation (23), which is for $\tilde{y}(t)$. We may decompose $\eta(t)$ into a component that lies in the space spanned by $\varepsilon(t)$ and a component $\zeta(t)$ that is in the orthogonal complement of the space. Thus

$$\begin{aligned}\eta(t) &= \frac{\sigma_{\eta\varepsilon}}{\sigma_{\varepsilon}^2}\varepsilon(t) + \left\{\eta(t) - \frac{\sigma_{\eta\varepsilon}}{\sigma_{\varepsilon}^2}\varepsilon(t)\right\} \\ &= \frac{\sigma_{\eta\varepsilon}}{\sigma_{\varepsilon}^2}\varepsilon(t) + \zeta(t),\end{aligned} \tag{25}$$

where $\sigma_{\varepsilon}^2 = V\{\varepsilon(t)\}$ is the variance of the consumption innovations and $\sigma_{\varepsilon\eta}^2 = C\{\varepsilon(t), \eta(t)\}$ is the covariance of the consumption and income innovations. Substituting (25) in equation (23) and combining the terms in $\varepsilon(t)$ gives

$$\tilde{y}(t) = \frac{\alpha(L)}{\pi(L)}\varepsilon(t) + \frac{\beta(L)}{\pi(L)}\zeta(t), \tag{26}$$

where

$$\alpha(L) = \psi(L) + \frac{\sigma_{\eta\varepsilon}}{\sigma_{\varepsilon}^2}\beta(L). \tag{27}$$

We may describe the sequence $\varepsilon(t)$ as the auto-innovations of $\tilde{y}(t)$ and $\zeta(t)$ as the allo-innovations.

By a similar reparametrisation, the equation (24) in $\tilde{x}(t)$ becomes

$$\tilde{x}(t) = \frac{\gamma(L)}{\pi(L)}\eta(t) + \frac{\delta(L)}{\pi(L)}\xi(t), \tag{28}$$

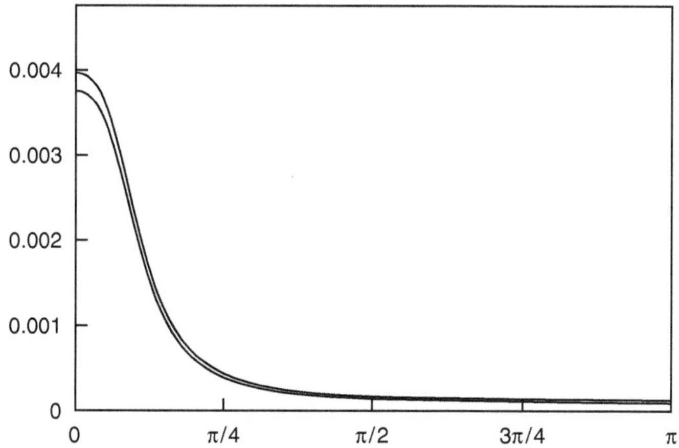


**Fig. 9** The spectrum of the consumption growth sequence $\nabla_4 y(t)$ (the outer envelope) and that of its auto-innovation component $\{\alpha(L)/\pi(L)\}\varepsilon(t)$ (the inner envelope)

where

$$\gamma(L) = \phi(L) + \frac{\sigma_{\eta\varepsilon}}{\sigma_\eta^2}\delta(L), \tag{29}$$

$$\xi(t) = \varepsilon(t) - \frac{\sigma_{\eta\varepsilon}}{\sigma_\eta^2}\eta(t),$$

and where $\eta(t)$ and $\xi(t)$ are mutually uncorrelated. These are, respectively, the auto-innovations and the allo-innovations of $\tilde{x}(t)$.

The relative influences of $\varepsilon(t)$ on $\tilde{y}(t)$ and of $\eta(t)$ on $\tilde{x}(t)$ can now be assessed by an analysis of the corresponding spectral density functions. Figure 9 shows the spectrum of $\tilde{y}(t)$ together with that of its auto-innovation component $\{\alpha(L)/\pi(L)\}\varepsilon(t)$, which is the lower envelope. Figure 10 shows the spectrum of $\tilde{x}(t)$ together with that of its auto-innovation component $\{\gamma(L)/\pi(L)\}\eta(t)$.

From a comparison of the figures, it is clear that the innovation sequence $\varepsilon(t)$ accounts for a much larger proportion of $\tilde{y}(t)$ than $\eta(t)$ does of $\tilde{x}(t)$. Thus, the consumption growth series appears to be driven largely by its auto innovations. These innovations also enter the income growth series to the extent that the latter is not accounted for by its auto innovations. Figure 10 shows that the extent is considerable.

The fact the consumption innovations play a large part in driving the bivariate system implies that the consumption function of Davidson et al. [1], which is equation (2), cannot be properly construed as a structural econometric relationship. For it implies that the estimates are bound to suffer from a simultaneous-equations bias. Nevertheless, in so far as the mechanisms generating the data remain unchanged, the above-mentioned function will retain its status as an excellent predictor of the growth rate of consumption that is based on a parsimonious information set.

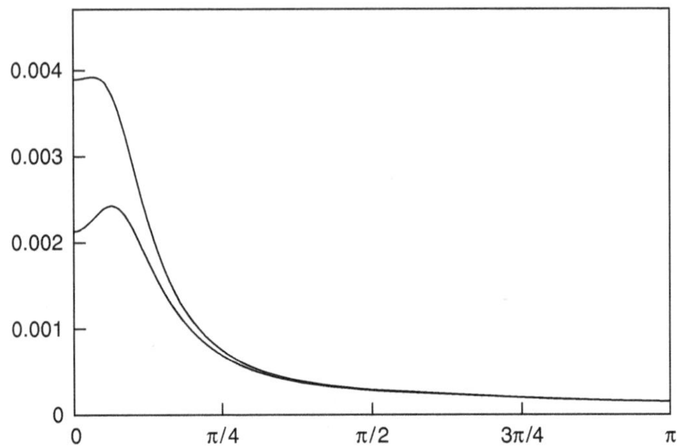


**Fig. 10** The spectrum of the income growth sequence $\nabla_4 x(t)$ (the outer envelope) and that of its auto-innovation component $\{\gamma(L)/\pi(L)\}\eta(t)$ (the inner envelope)

# 7 Conclusions

The traditional macroeconomic consumption function depicts a delayed response of consumption spending to changes in income; and many analysts would expect this relationship to be readily discernible in the macroeconomic data. Instead, the data seem to reflect a delayed response of aggregate income to autonomous changes in consumption. Although the two responses can easily coexist, it is the dominant response that is liable to be discerned in the data at first sight.

A crucial question is whether both responses can be successfully disentangled from the macroeconomics data. The construction of a bivariate autoregressive model is the first step in the process of their disentanglement. However, given the paucity of the information contained in the data, one is inclined to doubt whether the process can be carried much further. Indeed, the efforts that have been devoted to the microeconomic analysis of consumer behaviour in the last twenty years can be construed as a reaction to limited prospects facing macroeconomic investigations.

Much has already been accomplished in the microeconomic analysis of consumer behaviour; and an excellent account of some of the numerous influences that affect consumer behaviour directly has been provided recently by Muellbauer and Latimore [15]. However, what is lacking is a methodology that would enable the consumption behaviour of identifiable social and economic groups to be aggregated into a macroeconomic consumption function.

We have found that, within a bivariate autoregressive system designed to explain the growth rates on income and consumption, the innovations sequence of the consumption equation dominates the corresponding innovations sequence of the income equation. Thus the fluctuations in the growth rate of consumption have been depicted mainly as the result of autonomous influences.

Although the innovations sequences are an artefact of the statistical analysis, they are not entirely devoid of worldly connotations. By a detailed study of the historical circumstances, we should be able to relate the consumption innovations to the fiscal policies of the central governments, the state of the financial markets, the rate of inflation, the political and social climate, and to much else besides. Although some of these influences have been included in macroeconomic consumption functions, it seems that, in the main, there has been a remarkable oversight of the circumstantial details in most attempts at explaining the aggregate level of consumption. The present analysis is, regrettably, no exception.

## References


[1] Davidson, J.E.H., Hendry, D.F., Srba, F., Yeo, S.: Econometric Modelling of the Aggregate Time-Series Relationship between Consumer's Expenditure and Income in the United Kingdom. Econ. J. **88**, 661–692 (1978)

[2] Duesenberry, J.S.: Income, Saving and the Theory of Consumer Behaviour. Princeton University Press, Princeton (1949)

[3] Friedman, M.: A Theory of the Consumption Function. Princeton University Press, Princeton (1957)

[4] Geweke, J.: Measurement of Linear Dependence and Feedback between Multiple Time Series. J. Am. Stat. Assoc. **77**, 304–413 (1982)

[5] Fenandez-Corugedo, E., Price, S., Blake, A.: The Dynamics of Consumers's Expenditure: The U.K. Consumption Function ECM Redux. Discussion paper, Bank of England (2003)

[6] Granger, C.W.J., Newbold, P.: Spurious Regressions in Econometrics. J. Econometrics **2**, 111–120 (1974)

[7] Haavelmo, T.: Methods of Measuring the Marginal Propensity to Consume. J. Am. Stat. Assoc. **42**, 105–122 (1947)

[8] Hamming, R.W.: Digital Filters (3rd ed.). Prentice-Hall, New Jersey (1989)

[9] Hansen, A.H.: Fiscal Policy and Business Cycles. W.W. Norton, New York (1941)

[10] Hylleberg, S., Engle, R.F., Granger, C.W.J., Yoo, B.S.: Seasonal Integration and Cointegration. J. Econometrics **44**, 215–238 (1990)

[11] Keynes, J.M.: The General Theory of Employment, Interest and Money. Macmillan, London (1936)

[12] Kuznets, J.M.: National Product since 1869. National Bureau of Economic Research, New York (1946)

[13] Modigliani F.: The Life-Cycle Hypothesis of Saving Twenty Years Later. In: Parkin, M., Nobay, A.R. (eds.) Contemporary Issues in Economics. Manchester University Press, Manchester (1975)

[14] Modigliani F., Brumberg, R.: Utility Analysis and the Consumption Function: An Interpretation of Cross-Section Data. In: Kurihara, K.K. (ed.) Post-Keynesian Economics. Rutgers University Press, New Brunswick (1954)

[15] Muellbauer, J.N., Latimore, R.: The Consumption Function: A Theoretical and Empirical Overview. In: Pesaran, M., Wickens, M.R. (eds.), Handbook of Applied Econometrics: Macroeconomics, pp. 221–311. Basil Blackwell, Oxford (1995)
[16] Osborn, D.R., Chui, A.P.L., Smith, J.P., Birchenhall, C.R.: Seasonality and the Order of Integration for Consumption. Oxford B. Econ. Stat. **50**, 361–377 (1988)
[17] Pollock, D.S.G.: A Handbook of Time-Series Analysis, Signal Processing and Dynamics. Academic Press, London (1999)

# Part V
# Stochastic Processes

*Financial forecasting appears to be a science that makes astrology look respectable.*

Burton Malkiel

# Improved Estimation Strategy in Multi-Factor Vasicek Model

S. Ejaz Ahmed, Sévérien Nkurunziza, and Shuangzhe Liu

**Abstract** We consider simultaneous estimation of the drift parameters of multivariate Ornstein-Uhlebeck process. In this paper, we develop an improved estimation methodology for the drift parameters when homogeneity of several such parameters may hold. However, it is possible that the information regarding the equality of these parameters may not be accurate. In this context, we consider *Stein-rule* (or shrinkage) estimators to improve upon the performance of the classical maximum likelihood estimator (MLE). The relative dominance picture of the proposed estimators are explored and assessed under an asymptotic distributional quadratic risk criterion. For practical arguments, a simulation study is conducted which illustrates the behavior of the suggested method for small and moderate length of time observation period. More importantly, both analytical and simulation results indicate that estimators based on shrinkage principle not only give an excellent estimation accuracy but outperform the likelihood estimation uniformly.

## 1 Introduction

The Ornstein-Uhlenbeck process has been extensively and successfully used in modelling of different phenomenon such as in biology (Engen and Sæther [8]), in ecology (Engen *et al.* [9], Froda and Nkurunziza [11]), in finance (Schöbel and Zhu [22]). Particulary in field of finance, the Ornstein-Uhlenbeck process is successfully implemented in modelling the term structure of the interest rates. For this reason it is mostly known as Vasicek [27] process in the related literarure. The Vasicek model describes the interaction between equilibrium value of what the interest rate should be and a stochastic movements into the interest rate that results of the unpredictable

Shuangzhe Liu
Faculty of Information Sciences and Engineering, University of Canberra, Canberra ACT 2601, Australia
Shuangzhe.Liu@canberra.edu.au

economic environment (see for e.g. Vasicek [27], Abu-Mostafa [1]). More realistically, the term structure of the interest rates is embedded in a large macroeconomic system (Langetieg [17]). To this end, Langetieg [17] developed a multivariate model of the term structure of interest rate, so-called "multi-factors Vasicek model". Thus, as in Langetieg [17], we consider $p$ instantaneous interest rates, $r_1(t), r_2(t), \ldots, r_p(t)$ which are governed by the stochastic differential equation (SDE) system

$$dr_k(t) = \theta_k(\alpha_k - r_k(t))\,dt + \sigma_k dW_k(t), \quad k = 1,2,\ldots,p, \tag{1}$$

where for each $k = 1,2,\ldots,p$, $\alpha_k > 0$ denotes a steady-state interest rate (or the long-term mean), $\theta_k > 0$ termed as a speed of converging to the steady-state, $\sigma_k > 0$ is a volatility or "randomness level" and $\{W_k(t), t \geqslant 0\}$ is a Wiener process. Consequently, for each $k$, $\theta_k(\alpha_k - r_k(t))$ represents the drift term and thus, $\alpha_k$ and $\theta_k$ are so-called the drift parameters. On the other hand, $\sigma_k dW_k(t)$ is the diffusion term of the process with $\sigma_k$ as the diffusion parameter. In this paper, $\{W_k, t \geqslant 0\}$, $k = 1,2,\ldots,p$ are Wiener processes with possible correlation.

Interestingly, the interest rate Vasicek model has been extensively used in related literature. For example, this model was used to analyze the maturity structure of the public debt both at the Bank of Canada and at the Department of Finance, and in Danish National Bank, see Georges [12]. Abu-Mostafa [1] calibrates the correlated multi-factors Vasicek model of interest rates, and then applied it to Japanese Yen swaps market and U.S. Dollar yield market.

From a statistical perspective, Liptser and Shiryayev [20], Basawa and Prakasa Rao [6] and Kutoyants [16] considered the estimation of drift parameters of the univariate version of model (1) and derived the maximum likelihood estimator (MLE). In this communication, we are interested to form an alternative estimation strategy which performs better than the existing likelihood estimation method of the drift parameters of multivariate Vasicek (or Ornstein-Uhlenbeck) process.

Here the parameter of interest is the speed of converging to the steady-state, i.e., $\theta_k$. Further, we assume that steady-state interest rate $\alpha$ is known. In this case, $X_k(t) = r_k(t) - \alpha_k$. Then, from the multi-factors Vasicek model (1), we get

$$dX_k(t) = -\theta_k X_k(t)dt + \sigma_k dW_k(t), \tag{2}$$

where $\theta_k > 0$, $\sigma_k > 0$. Without loss of generality, we assume that $\{W_k(t), t \geqslant 0, k = 1,2\ldots,p\}$ is a $p$-dimensional Wiener process. Indeed, from $p$-Wiener processes pairwise jointly Gaussian with a non-zero correlation coefficient, one can obtain $p$ independent Wiener process. In practice, the observations are collected at discrete times $0 = t_0 < t_1 < t_2 < \cdots < t_n = T$ and thus, the continuous time modelling is derived through some approximations. Indeed, our statistical procedure is applied by replacing each stochastic integral by its corresponding discrete Riemann-Îto sum. Theoretically, it is well known that the resulting new estimator is asymptotically equivalent to the original estimator obtained under continuous time sampling (see e.g. Le Breton [18]). Of course, as discussed in Dacunha-Castelle and Florens-Zmirou [7] or Florens-Zmirou [10] there is a loss of information due to discretization.

In this paper, we consider the simultaneous estimation problem of the drift parameter vector $\boldsymbol{\theta}$ where $\boldsymbol{\theta} = (\theta_1, \theta_2, \ldots, \theta_p)'$. Further, it is suspected that

$$\theta_1 = \theta_2 = \cdots = \theta_p$$

or nearly equal. Thus, we consider the situations when all the parameters may be equal. This kind of situation may arise in a short time horizon (high-speed factors). Alternatively, the data was collected from various sources under similar conditions. Hence, under these situations it is reasonable to suspect the homogeneity of the several drift parameters. Thus, the main focus of this work is to suggest some estimators of $\boldsymbol{\theta}$ with high estimation accuracy when homogeneity of several drift parameters may hold. In passing, an asymptotic test for the equality of parameters is also suggested. Consequently, we extend the single factor inference problem into a more general form of the Vasicek model, the multi-factors model. In addition, an alternative estimator of variance parameters is also suggested.

The rest of this paper is organized as follows. Section 2 is devoted to testing problem of the equality of several drift parameters. In section 3, we present the shrinkage estimation strategy and outline its supremacy over the MLE. A simulation study is carried out in Section 4. Finally, section 5 offers some concluding remarks. The technical results are presented in the Appendix for a smooth reading of the paper.

## 2 Testing the Homogeneity of the Drift Parameters

Let $\{X_1(t), 0 \leqslant t \leqslant T\}$, $\{X_2(t), 0 \leqslant t \leqslant T\}$,..., $\{X_p(t), 0 \leqslant t \leqslant T\}$ be Ornstein-Uhlenbeck processes (Kutoyants [16], p. 51, Steele [24], p. 140) whose diffusion equations are given by

$$dX_k(t) = -\theta_k X_k(t)dt + \sigma_k dW_k(t) \qquad X_k(0) \text{ fixed}, \tag{3}$$

where $\theta_k > 0$, $\sigma_k > 0$, and $\{W_k(t), 0 \leqslant t \leqslant T\}$ are $p$-independent Wiener processes. Further, let $\boldsymbol{\theta}$ be a $p$-column vector given by $\boldsymbol{\theta} = (\theta_1, \theta_2, \ldots, \theta_p)'$, and let $\Sigma$ be diagonal matrix whose diagonal entrees are $\sigma_1^2, \sigma_2^2, \ldots, \sigma_p^2$, i.e., $\boldsymbol{\Sigma} = \operatorname{diag}\left(\sigma_1^2, \sigma_2^2, \ldots, \sigma_p^2\right)$. Also, let $\boldsymbol{X}(t)$ and $\boldsymbol{W}(t)$ be column vectors given by

$$\boldsymbol{X}(t) = (X_1(t), X_2(t), \ldots, X_p(t))', \boldsymbol{W}(t) = (W_1(t), W_2(t), \ldots, W_p(t))'$$

$$\text{and let } \boldsymbol{V}(t) = \operatorname{diag}(X_1(t), X_2(t), \ldots, X_p(t)),$$

for $0 \leqslant t \leqslant T$. From relation (3), we get

$$d\boldsymbol{X}(t) = -\boldsymbol{V}(t)\boldsymbol{\theta}dt + \Sigma^{\frac{1}{2}} d\boldsymbol{W}(t), \qquad 0 \leqslant t \leqslant T. \tag{4}$$

We consider the following testing problem

$$H_0 : \theta_1 = \theta_2 = \cdots = \theta_p \text{ versus } H_1 : \theta_j \neq \theta_k \quad \text{for some} \quad 1 \leqslant j < k \leqslant p. \tag{5}$$

Noting that the MLE established here corresponds to that given in Lipter and Shiryayev ([20], chapter 17, p. 206-207) or in Kutoyants ([16], p. 63), for the univariate case. Hence, we extend the univariate estimation problem to a multivariate situation. Let

$$\boldsymbol{U}_T = \left( \int_0^T X_1(t)dX_1(t), \int_0^T X_2(t)dX_2(t), \ldots, \int_0^T X_p(t)dX_p(t) \right)',$$

$$\boldsymbol{D}_T = \int_0^T \boldsymbol{V}^2(t)dt, \quad \widehat{\boldsymbol{\theta}}^* = -\boldsymbol{D}_T^{-1}\boldsymbol{U}_T = \left(\widehat{\theta}_1^*, \widehat{\theta}_2^*, \ldots, \widehat{\theta}_p^*\right)'. \tag{6}$$

Also, let $z^+ = \max(z, 0)$. Conditionally to $\boldsymbol{X}_0$, let $\widehat{\boldsymbol{\theta}}$ be maximum likelihood estimator of $\boldsymbol{\theta}$ satisfying the model (4) and $\widetilde{\boldsymbol{\theta}}$ be the restricted MLE (RMLE) of $\boldsymbol{\theta}$ under $H_0$.

**Proposition 1.** *Let $\boldsymbol{e}_p$ be a p-column vector whose all entrees are equal to 1. We have*

$$\widehat{\boldsymbol{\theta}} = \left(\widehat{\theta}_1^{*+}, \widehat{\theta}_2^{*+}, \ldots, \widehat{\theta}_p^{*+}\right)' \text{ and } \widetilde{\boldsymbol{\theta}} = \left(\boldsymbol{e}_p'\Sigma^{-1}\boldsymbol{D}_T\boldsymbol{e}_p\right)^{-1} \boldsymbol{e}_p\boldsymbol{e}_p'\Sigma^{-1}\boldsymbol{D}_T\widehat{\boldsymbol{\theta}}. \tag{7}$$

The proof follows from Proposition 4 which is given in the Appendix.

It is established that $\widehat{\boldsymbol{\theta}}$ and $\widetilde{\boldsymbol{\theta}}$ are strongly consistent for $\boldsymbol{\theta}$ (see Appendix, Proposition 5). Moreover, these estimators are asymptotically normal as $T$ tends to infinity (see, Appendix, Proposition 6). Thus, we suggest the following test for the testing problem (5), when $\Sigma$ is known,

$$\Psi = \begin{cases} 1 \text{ if } T\left(\widehat{\boldsymbol{\theta}} - \widetilde{\boldsymbol{\theta}}\right)' \Sigma^{-1} \left(\widehat{\boldsymbol{\theta}} - \widetilde{\boldsymbol{\theta}}\right) > \chi^2_{p-1;\alpha} \\ 0 \text{ if } T\left(\widehat{\boldsymbol{\theta}} - \widetilde{\boldsymbol{\theta}}\right)' \Sigma^{-1} \left(\widehat{\boldsymbol{\theta}} - \widetilde{\boldsymbol{\theta}}\right) < \chi^2_{p-1;\alpha}, \end{cases} \tag{8}$$

where $\alpha$ is fixed and $0 < \alpha < 1$. Also, we denote by $\chi^2_{p-1}$, the chi-square random variable with $p-1$ degrees of freedom and

$$\Pr\left\{\chi^2_{p-1} > \chi^2_{p-1;\alpha}\right\} = \alpha.$$

When $\Sigma$ is unknown, the test (8) can be modified by replacing $\Sigma$ by a its strongly consistent estimator. Later, we prove that the test $\Psi$ is asymptotically $\alpha$-level. Further, the test obtained for the unknown $\Sigma$ case has asymptotically the same level $\alpha$, and it is as powerful as the test $\Psi$ .

**Corollary 1.** *Under the model (4), the test $\Psi$ in (8) is asymptotically $\alpha$-level test for the testing problem (5).*

**Proof** By Corollary 4 in the Appendix, we have

$$\boldsymbol{\xi}_T'\Sigma^{-1}\boldsymbol{\xi}_T \xrightarrow[T\to\infty]{\mathscr{L}} \chi^2_{p-1},$$

and the rest of proof follows directly.

Furthermore, while for the diffusion process (2), $\Sigma$ is known (equals to the quadratic variation), for the corresponding incomplete sample paths, the covariance matrix $\Sigma$ becomes unknown. Thus, $\Sigma$ needs to be estimated in order to compute $\widehat{\boldsymbol{\theta}}$ and the test statistic given in (8). We replace $\Sigma$ by its corresponding strongly consistent estimator $\widehat{\Sigma}$. Then by Slutsky theorem, the corresponding new estimators are strongly consistent and asymptotically normal.

Now, we suggest an alternative estimator for $(\sigma_1, \sigma_2, \ldots, \sigma_p)'$, that is discrete Riemann-Îto sums corresponding to

$$(\widehat{\sigma}_1, \widehat{\sigma}_2, \ldots, \widehat{\sigma}_p)' \quad \text{where } \widehat{\sigma}_i^2 = \frac{X_i^2(T)}{T} - \frac{2}{T}\int_0^T X_i(t)dX_i(t).$$

Further, if $\sigma_1 = \sigma_2 = \cdots = \sigma_p = \sigma$, then the common value $\sigma^2$ is estimated by

$$\widehat{\sigma}^2 = \frac{1}{p}\sum_{i=1}^{p}\left(\frac{X_i^2(T)}{T} - \frac{2}{T}\int_0^T X_i(t)dX_i(t)\right).$$

Hence

$$\widehat{\Sigma} = \operatorname{diag}\left(\widehat{\sigma}_1^2, \widehat{\sigma}_2^2, \ldots, \widehat{\sigma}_p^2\right) \quad \text{or } \widetilde{\Sigma} = \widehat{\sigma}^2 I_p. \tag{9}$$

**Proposition 2.** *Assume that the model (4) holds. Then,*

*(i)* $\widehat{\Sigma} \xrightarrow[T\to\infty]{a.s.} \Sigma$ *and for any* $0 < \nu < 1$, $T^{\nu}\left(\widehat{\Sigma} - \Sigma\right) \xrightarrow[T\to\infty]{a.s.} 0$;

*(ii) if* $\mathrm{E}\left\{\|\boldsymbol{X}(0)\|^2\right\} < \infty$ *then* $\lim_{T\to\infty}\mathrm{E}\left\{\left\|\widehat{\Sigma} - \Sigma\right\|^2\right\} = 0.$

The proof of Proposition 2 follows from standard stochastic calculus techniques. Also, see for example Nkurunziza and Ahmed [21]. Also, from Proposition 2, we note that the suggested estimator $\widehat{\Sigma}$ converges faster than the classical estimator based on quadratic variation. Since the strongly consistency and normality properties of $\widehat{\boldsymbol{\theta}}$ and $\widetilde{\boldsymbol{\theta}}$ do not change when $\Sigma$ is replaced by $\widehat{\Sigma}$, for the brevity sake, we treat $\Sigma$ as known in the remaining discussions. In the following section, we showcase the main contribution of the paper.

## 3 James-Stein type Shrinkage Estimation

In this section, following Ahmed [4], we consider two James-Stein (J-S) type shrinkage estimators of the drift parameters. The J-S type shrinkage estimator (SE) $\widehat{\theta}^S$ of $\theta$ is defined as

$$\widehat{\theta}^S = \widetilde{\theta} + \{1 - c\psi^{-1}\}(\widehat{\theta} - \widetilde{\theta}), \quad c \in [0, 2(p-2)), \quad p \in [3, \infty), \tag{10}$$

and

$$\psi = T\left(\widehat{\boldsymbol{\theta}} - \widetilde{\boldsymbol{\theta}}\right)' \Sigma^{-1}\left(\widehat{\boldsymbol{\theta}} - \widetilde{\boldsymbol{\theta}}\right),$$

Note that $\psi \geq 0$, and hence, for $\psi < c \Longleftrightarrow 1 - c\psi^{-1} < 0$, that causes a possible inversion of sign or over-shrinkage. The positive-rule shrinkage estimators (PSE) control this drawback satisfactorily. The PSE $\left(\widehat{\theta}^{S+}\right)$ is defined as

$$\widehat{\theta}^{S+} = \widetilde{\theta} + \{1 - c\psi^{-1}\}^{+}\left(\widehat{\theta} - \widetilde{\theta}\right), \tag{11}$$

where $z^{+} = max(0,z)$. Ahmed [4] recommended that the shrinkage estimator should be used as a tool for developing the *PSE* and should not be used as an estimator in its own right. In parametric setups, the SE, PSE, and other related estimators have been extensively studied (Judge and Bock [13] and the references therein). Also, for small sample, Trenkler and Trenkler [26], study some comparison criteria concerning some biased estimators. Large sample properties of these estimators were studied by Sen [23], Ahmed [5] and others. Stigler [25] and Kubokawa [15] provide excellent reviews of (parametric) shrinkage estimators. Jurečková and Sen [14] have an extensive treatise of the asymptotic and interrelationships of robust estimators of location, scale and regression parameters with due emphasis on Stein-rule estimators.

Having all these estimators defined, we need to stress on the regularity conditions under which they have good performance characteristics. Unlike maximum likelihood estimators $\widehat{\theta}$ and $\widetilde{\theta}$, these shrinkage estimators are not linear. Hence, even if the distribution was a standard normal, the finite sample distribution theory of these shrinkage estimators is not simple to obtain, for even normal ones. This difficulty has been largely overcome by asymptotic methods (Ahmed and Saleh [3], Jurečková and Sen [14], and others). These asymptotic methods relate primarily to *convergence in distribution* which may not generally guarantee *convergence in quadratic risk*. This technicality has been taken care of by the introduction of *asymptotic distributional risk* (ADR) (Sen [23]), which, in turn, is based on the concept of a *shrinking neighborhood* of the pivot for which the ADR serves a useful and interpretable role in *asymptotic risk analysis*.

Finally, an interesting feature of this paper is that we consider the shrinkage estimation of drift parameters in the multifactors model, a more general form of the Vasicek model. Based on the reviewed literature, this kind of study is not available for practitioner.

## *3.1 Asymptotic Properties*

Let us consider the following local alternative for the testing problem (5)

$$K_T : \boldsymbol{\theta} = \bar{\theta}\boldsymbol{e}_p + \frac{\boldsymbol{\delta}}{\sqrt{T}} \tag{12}$$

where $\boldsymbol{\delta}$ is a p-column vector with different direction than $\boldsymbol{e}_p$. Also, we assume that $\|\delta\| < \infty$. Further, let

$$\Upsilon = \left(\boldsymbol{e}_p'\Sigma^{-1}\boldsymbol{e}_p\right)^{-1}\boldsymbol{e}_p\boldsymbol{e}_p', \quad \boldsymbol{\delta}^* = \boldsymbol{\delta} - \Upsilon\Sigma^{-1}\boldsymbol{\delta}, \quad \text{and let} \quad \Sigma^* = \Sigma - \Upsilon.$$

Let

$$\boldsymbol{\rho}_T = \sqrt{T}\left(\widehat{\boldsymbol{\theta}} - \bar{\theta}\boldsymbol{e}_p\right), \quad \boldsymbol{\xi}_T = \sqrt{T}\left(\widehat{\boldsymbol{\theta}} - \widetilde{\boldsymbol{\theta}}\right), \quad \boldsymbol{\zeta}_T = \sqrt{T}\left(\widetilde{\boldsymbol{\theta}} - \bar{\theta}\boldsymbol{e}_p\right).$$

**Proposition 3.** *Assume that the model (4) holds. Under the local alternative (12),*

$$\boldsymbol{\rho}_T \xrightarrow[T\to\infty]{\mathscr{L}} \mathscr{N}_p(\boldsymbol{\delta}, \Sigma) \quad \text{and} \quad \begin{pmatrix} \boldsymbol{\zeta}_T \\ \boldsymbol{\xi}_T \end{pmatrix} \xrightarrow[T\to\infty]{\mathscr{L}} \mathscr{N}_{2p}\left(\begin{pmatrix} \boldsymbol{\delta} - \boldsymbol{\delta}^* \\ \boldsymbol{\delta}^* \end{pmatrix}, \begin{pmatrix} \Upsilon & \mathbf{0} \\ \mathbf{0} & \Sigma^* \end{pmatrix}\right).$$

The proof is given in the Appendix.

**Corollary 2.** *Let* $\Xi = \Sigma^{-1} - \Sigma^{-1}\left(\boldsymbol{e}_p'\Sigma^{-1}\boldsymbol{e}_p\right)^{-1}\boldsymbol{e}_p\boldsymbol{e}_p'\Sigma^{-1}$. *If (12) holds, then*

$$\boldsymbol{\xi}_T'\Sigma^{-1}\boldsymbol{\xi}_T \xrightarrow[T\to\infty]{\mathscr{L}} \chi^2_{p-1}\left(\boldsymbol{\delta}^{*'}\Xi\boldsymbol{\delta}^*\right).$$

The proof follows from Proposition 3.

Finally, we establish Corollary 3 which gives the asymptotic power for the test $\Psi$. To this end, let $\Pi_\Psi$ denote the power function of the test $\Psi$.

**Corollary 3.** *Under the conditions of Corollary 2, we have*

$$\lim_{T\to\infty} \Pi_\Psi\left(\boldsymbol{\theta} + \frac{\boldsymbol{\delta}}{\sqrt{T}}\right) = \Pr\left\{\chi^2_{p-1}\left(\boldsymbol{\delta}^{*'}\Xi\boldsymbol{\delta}^*\right) > \chi^2_{p-1;\alpha}\left(\boldsymbol{\delta}^{*'}\Xi\boldsymbol{\delta}^*\right)\right\}.$$

The proof is obtained using Corollary 2.

*Remark*: The effective domain of risk dominance of PSE or SE over MLE is a small neighborhood of the chosen pivot (viz., $\theta = \overline{\theta}\mathbf{e}_p'$); and as we make the observation period $T$ larger and larger, this domain becomes narrower. The corollary 2 shows that for any fixed $\theta \neq \overline{\theta}\boldsymbol{e}_p'$,

$$\psi \xrightarrow[T\to\infty]{\mathscr{L}} \chi^2_{p-1}\left(\boldsymbol{\delta}^{*'}\Xi\boldsymbol{\delta}^*\right) \quad \text{and} \quad T^{-1}\psi \xrightarrow[T\to\infty]{\mathscr{L}} 0 \tag{13}$$

as such, the shrinkage factor $c\psi^{-1} = O_p\left(T^{-1}\right)$, as $T \to \infty$, so that asymptotically there is no shrinkage effect. This justifies the choice of the usual Pitman type of alternatives given in (12).

For an estimator $\widehat{\theta}^{\star}$ of $\theta$, we confine ourselves to a *quadratic loss function* of the form

$$L\left(\widehat{\theta}^{\star}, \theta; \mathbf{W}\right) = \left[\sqrt{T}\left(\widehat{\theta}^{\star} - \theta\right)\right]'\mathbf{W}\left[\sqrt{T}\left(\widehat{\theta}^{\star} - \theta\right)\right], \tag{14}$$

where $\mathbf{W}$ is positive semi-definite (p.s.d). Using the distribution of $\sqrt{T}\left(\widehat{\theta}^{\star} - \theta\right)$ and taking the expected value both sides of (14), we get the expected loss that would be called the *quadratic risk* $\mathrm{R}_T^o\left(\widehat{\theta}^\star,\theta;\mathbf{W}\right)=\mathrm{trace}\left(\mathbf{W}\widehat{\Sigma}_T\right)$, where $\widehat{\Sigma}_T$ is the dispersion matrix of $\sqrt{T}\left(\widehat{\theta}^\star-\theta\right)$. Whenever

$$\lim_{T\to\infty}\widehat{\Sigma}_T=\Sigma$$

exists, $\mathrm{R}_T^o\left(\widehat{\theta}^\star,\theta;\mathbf{W}\right)\to\mathrm{R}^o\left(\widehat{\theta}^\star,\theta;\mathbf{W}\right)=\mathrm{trace}(\mathbf{W}\Sigma)$, which is termed the *asymptotic risk*. In our setup, we denote the distribution of $\sqrt{T}\left(\widehat{\theta}^\star-\theta\right)$ by $\tilde{G}_T(\mathbf{u})$, $\mathbf{u}\in\mathbb{R}^p$. Suppose that $\tilde{G}_T\to\tilde{G}$ (at all points of continuity), as $T\to\infty$, and let $\Sigma_{\tilde{G}}$ be the dispersion matrix of $\tilde{G}$. Then the ADR of $\widehat{\theta}^\star$ is defined as

$$\mathrm{R}^o\left(\widehat{\theta}^\star,\theta;\mathbf{W}\right)=\mathrm{trace}\left(\mathbf{W}\Sigma_{\tilde{G}}\right). \tag{15}$$

The asymptotic bias is defined as

$$B_T^0\left(\widehat{\theta}^\star,\theta\right)=\mathrm{E}\left[\sqrt{T}\left(\widehat{\theta}^\star-\theta\right)\right], \tag{16}$$

Similarly, the *asymptotic distributional bias* (ADB) is

$$B_T^0\left(\widehat{\theta}^\star,\theta\right)=\int\cdots\int\mathbf{x}d\tilde{G}_T(\mathbf{x})\xrightarrow[T\to\infty]{}\left(B\left(\widehat{\theta}^\star,\theta\right)=\int\cdots\int\mathbf{x}d\tilde{G}(\mathbf{x})\right). \tag{17}$$

We present (without derivation) the results on SE and PSE. The proofs are similar to that given in Ahmed [2]. Let $\Delta=\boldsymbol{\delta}^{*'}\Xi\boldsymbol{\delta}^*$ and let $H_\nu(x\,;\Delta)=P\{\chi_\nu^2(\Delta)\le x\}$, $x\in\mathbb{R}^+$.

**Theorem 1.** *Assume that Proposition 3 holds. Then, the ADB functions of the estimators are given as follows:*

$$\begin{aligned}&B\left(\widehat{\theta},\theta\right)=\mathbf{0},\quad B\left(\widetilde{\theta},\theta\right)=-\delta,\quad B\left(\widehat{\theta}^S,\theta\right)=-\delta(p-3)\mathrm{E}\left\{\chi_{p+1}^{-2}(\Delta)\right\}\\&B\left(\widehat{\theta}^{S+},\beta_1\right)=-\delta\left[H_{p+1}(p-3;\Delta)+\mathrm{E}\{\chi_{p+1}^{-2}(\Delta)I\left(\chi_{p+1}^2(\Delta)>(p-3)\right)\right].\end{aligned} \tag{18}$$

Since for the ADB of $\widetilde{\theta}$, $\widehat{\theta}^S$ and $\widehat{\theta}^{S+}$, the component $\delta$ is common and they differ only by scalar factors, it suffices to compare the scalar factors $\Delta$ only. The bias of the $\widetilde{\theta}$ is an unbounded function of $\Delta$. However, the bias function of both $\widehat{\theta}^S$ and $\widehat{\theta}^{S+}$ are bounded in $\Delta$. the ADB of $\widehat{\theta}^S$ and $\widehat{\theta}^{S+}$ starts from the origin at $\Delta=0$, increases to a maximum, and then decreases towards 0. However, the bias curve of $\widehat{\theta}^{S+}$ remains below the curve of SE for all values of $\Delta$.

**Theorem 2.** *Assume that Proposition 3 holds. Then, the ADR functions of the estimators are given as follows:*

$$
\begin{aligned}
&R\left(\widehat{\theta},\theta;\mathbf{W}\right) = trace(\mathbf{W}\Sigma),\\
&R\left(\widetilde{\theta},\theta;\mathbf{W}\right) = trace(\mathbf{W}\Sigma) - \mathrm{trace}(\mathbf{W}\Sigma^{*}) + \delta^{*\prime}\mathbf{W}\delta^{*},\\
&R\left(\widehat{\theta}^{S},\theta;\mathbf{W}\right) = ADR(\widehat{\theta}) + \delta^{*\prime}\mathbf{W}\delta^{*}(p^{2}-3)E(\chi_{p+3}^{-4}(\Delta))\\
&-(p-3)trace(\mathbf{W}\Sigma^{-})\{2\mathrm{E}(\chi_{p+1}^{-2}(\Delta)) - (p-3)\mathrm{E}(\chi_{p+1}^{-4}(\Delta))\},
\end{aligned} \tag{19}
$$

$$
\begin{aligned}
R\left(\widehat{\theta}^{S+},\theta;\mathbf{W}\right) &= ADR(\widehat{\theta}^{S}) + (p-3)trace(\mathbf{W}\Sigma^{*})\\
&\times\Big[2\mathrm{E}\left\{\chi_{p+1}^{-2}(\Delta)I(\chi_{p+1}^{2}(\Delta)\le p-3)\right\}\\
&-(p-3)\mathrm{E}\left\{\chi_{p+1}^{-4}(\Delta)I(\chi_{p+1}^{2}(\Delta)\le p-3)\right\}\Big]\\
&-trace(\mathbf{W}\Sigma)H_{p+1}(p-2;\Delta)\\
&+\delta^{*\prime}\mathbf{W}\delta^{*}\left\{2H_{p+1}(p-3;\Delta) - H_{p+3}(p-3;\Delta)\right\}\\
&-(p-3)\delta^{*\prime}\mathbf{W}\delta^{*}\big[2\mathrm{E}\{\chi_{p+1}^{-2}(\Delta))I(\chi_{p+1}^{2}(\Delta)\le p-3)\}\\
&-2\mathrm{E}\{\chi_{p+3}^{-2}(\Delta)I(\chi_{p+3}^{2}(\Delta)\le p-3)\}\\
&+(p-2)\mathrm{E}\left\{\chi_{p+3}^{-4}(\Delta)I(\chi_{p+3}^{2}(\Delta)\le p-2)\right\}\big].
\end{aligned} \tag{20}
$$

The proof is similar to that given in Ahmed [2]. For a suitable choice of the matrix **W**, risk dominance of the estimators are similar to those under normal theory and can be summarized as follows:

(a) Indeed,

$$R(\widehat{\theta}^{S},\theta;\mathbf{W}) < trace(\mathbf{W}\Sigma) \quad \text{for all} \quad \Delta \in [0,\infty),$$

hence providing greater estimation accuracy than MLE, beating the gold standard. The ADR function of SE is monotone in $\Delta$, the smallest value of the risk is achieved at $\Delta = 0$ and the largest is $trace(\mathbf{W}\Sigma)$. Hence, $\widehat{\theta}^{S}$ outshines the MLE, hence is an admissible estimator with respect to MLE.

(b) $\widehat{\theta}^{S+}$ is superior to $\widehat{\theta}^{S}$ in the entire parameter space induced by $\Delta$. Hence, it is also superior to $\widehat{\theta}$. Most importantly, $\widehat{\theta}^{S+}$ it prevents over-shrinking problem.

# 4 Simulation Study

An extensive Monte Carlo simulation study is conducted to assess the relative risk performance of the all estimators to MLE. In this section, we only report detailed results for $p = 3$ and $p = 5$ in an effort to save the space. We consider the null hypothesis $H_0 : \theta = \overline{\theta}\boldsymbol{e}_p'$ and the length of the time period of observation $T = 30$, $T = 50$. The relative efficiency of the estimators with respect to $\widehat{\theta}$ is defined by

$$RMSE = \frac{\text{risk}\left(\widehat{\theta}\right)}{\text{risk}\left(\text{proposed estimator}\right)}.$$

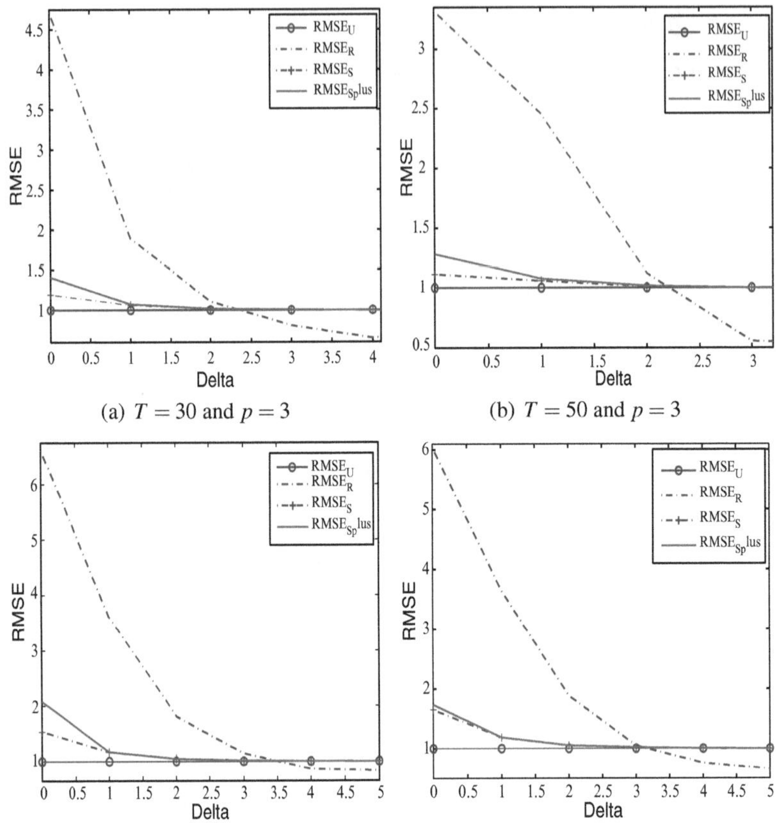


(a) $T = 30$ and $p = 3$

(b) $T = 50$ and $p = 3$

(c) $T = 30$ and $p = 5$

(d) $T = 30$ and $p = 5$

**Fig. 1** Relative efficiency vs $\Delta$

The results are graphically reported in Fig. 1(a)-1(b) for $p = 3$ and in Fig. 1(c)-1(d) for $p = 5$. Graphically, Fig. 1 indicates that a reduction of 50% or more in the risk seems quite realistic depending on the values of $\Delta$ and $p$.

We draw following conclusions:

(a) The behavior of the J-S type shrinkage estimator is robust and have risk-dominance over the MLE.
(b) Comparing the above Figure 1(a)-1(b) to the following Figure 1(c)-1(d), it is observed that for larger vales of $p$ the risk reduction is substantial for shrinkage estimators.
(c) The efficiency of $\widetilde{\theta}$ converges to 0 as $\Delta \to \infty$.

## 5 Concluding Remarks and Outlook for the Future

The Ornstein-Uhlenbeck has paramount application in finance, specifically in modelling the term structure of interest rates in a macro-economic context. We consider the estimation problem for the drift parameter vector of a multivariate Ornstein-Uhlenbeck process. In this context, we suggest shrinkage estimation strategy along with the maximum likelihood estimator. Our suggested shrinkage estimators dominate the MLE. Our simulation studies have provided strong evidence that corroborates with the developed asymptotic theory in this paper. The simulation study indicates that a reduction of 50% or more in the risk seem quite realistic depending on the values of $\Delta$ and $p$ (see Figure 1).

Finally, we stress here that like the statistical models underlying the statistical inferences to be made, the homogeneity of the drift parameters will be susceptible to uncertainty and the practitioners may be reluctant to impose the additional information regarding parameters in the estimation process. However, suggested shrinkage methodology is robust in the sense that it still works better than MLE when such constraint may not hold. Research on statistical implications of these and other estimators for a range of statistical models is ongoing.

**Acknowledgement** The research of Professor Ahmed and Dr. Nkurunziza is supported by grants from Natural Sciences and Engineering Research Council of Canada. Further, the authors would like to thank Drs. R. Ferland and R.S. Mamon for helpful comments and suggestions.

## Appendix

Let $\mu_{\boldsymbol{W}}$ be the measure induced by the Wiener process $\{\boldsymbol{W}(t), t \geq 0\}$. Also, let us denote by $\mu_U$ the probability measure induced by process $\{U(t), t \geq 0\}$.

$$\mu_U(B) = \mathrm{P}\{\omega : U_t(\omega) \in B\}, \text{ where } B \text{ is a Borel set.}$$

The following result plays a central role in establishing the MLE of $\boldsymbol{\theta}$.

**Proposition 4.** *Conditionally to $\boldsymbol{X}_0$, the Radon-Nicodym derivative of $\mu_{\boldsymbol{X}}$ with respect to $\mu_{\boldsymbol{W}}$ is given by*

$$\frac{d\mu_{\boldsymbol{X}}}{d\mu_{\boldsymbol{W}}}(\boldsymbol{X}) = \exp\left\{-\boldsymbol{\theta}'\Sigma^{-1}\boldsymbol{U}_T - \frac{1}{2}\boldsymbol{\theta}'\Sigma^{-1}\boldsymbol{D}_T\boldsymbol{\theta}\right\}. \tag{21}$$

The proof is obtained by applying Theorem (7.7) in Liptser and Shiryayev [19]. It should be noted that, the relation (21) has the same form as the equation 17.24 of Liptser and Shiryayev [19] for the univariate case with the non-random initial value.

In univariate case, Lipter and Shiryayev ([20], Theorem 17.3, Lemma 17.3 and Theorem 17.4) and Kutoyants [16] give some asymptotic results for the maximum likelihood estimator of the drift parameter of an Ornstein-Uhlenbeck process. The ideas of proof are the same as given in these references. Now, we outline the proof of the strong consistency.

**Proposition 5.** ***(Strong consistency)*** *Assume that the model (4) holds.*

*(i) Then,*

$$\Pr\left\{\lim_{T\to\infty}\widehat{\boldsymbol{\theta}}=\boldsymbol{\theta}\right\}=1.$$

*(ii)If $\theta_1=\theta_2=\cdots=\theta_p$, then*

$$\Pr\left\{\lim_{T\to\infty}\widetilde{\boldsymbol{\theta}}=\bar{\theta}\boldsymbol{e}_p\right\}=1.$$

*where $\bar{\theta}$ as the common drift parameter.*

**Proof** For any matrix $A$, we denote by

$$||A||^2=\mathrm{trace}(AA').$$

(i) Let

$$M(T)=\left(\int_0^T X_1(t)dW_1(t),\int_0^T X_2(t)dW_2(t),\ldots,\int_0^T X_p(t)dW_p(t)\right)'.$$

By some computations, we have

$$\boldsymbol{U}_T=-\boldsymbol{D}_T\boldsymbol{\theta}-M(T)\Sigma^{\frac{1}{2}}.$$

Therefore,

$$\widehat{\boldsymbol{\theta}}^*-\boldsymbol{\theta}=\boldsymbol{D}_T^{-1}M(T)\Sigma^{\frac{1}{2}}. \tag{22}$$

Obviously, $M(T)$ is a martingale whose quadratic variation is $\boldsymbol{D}_T$. Moreover, one can verify that

$$\Pr\left\{\lim_{T\to\infty}||\boldsymbol{D}_T||=\infty\right\}=1,$$

and, for any column vector $\boldsymbol{a}$, the process $\boldsymbol{a}'\boldsymbol{D}_T\boldsymbol{a}$ is nondecreasing in $T$. Hence, by strong law of large number for martingale, we get

$$\widehat{\boldsymbol{\theta}}^*-\boldsymbol{\theta}\xrightarrow[T\to\infty]{a.s.}\mathbf{0},$$

that implies that $\widehat{\boldsymbol{\theta}}$ is strongly consistent for $\boldsymbol{\theta}$ and this completes the proof of *(i)*.
(ii)In the similar way, we prove that $\widetilde{\boldsymbol{\theta}}-\boldsymbol{\theta}\xrightarrow[T\to\infty]{a.s.}\mathbf{0}$, which completes the proof.

□

**Proposition 6.** ***(Asymptotic normality)*** *Assume that the model (4) holds and suppose that $\boldsymbol{X}_0$ has the same moment as the invariant distribution.*

*(i) Then,*

$$\sqrt{T}\left(\widehat{\boldsymbol{\theta}}-\boldsymbol{\theta}\right) \xrightarrow[T\to\infty]{\mathscr{L}} \mathscr{N}_p(\mathbf{0},\Sigma), \text{ and } T\left(\widehat{\boldsymbol{\theta}}-\boldsymbol{\theta}\right)'\Sigma^{-1}\left(\widehat{\boldsymbol{\theta}}-\boldsymbol{\theta}\right) \xrightarrow[T\to\infty]{\mathscr{L}} \chi^2_p.$$

*(ii)If* $\theta_1=\theta_2=\cdots=\theta_p$*, then*

$$\sqrt{T}\left(\widetilde{\boldsymbol{\theta}}-\bar{\theta}\boldsymbol{e}_p\right) \xrightarrow[T\to\infty]{\mathscr{L}} \mathscr{N}_p\left(\mathbf{0},\boldsymbol{e}_p\boldsymbol{e}_p'\left(\boldsymbol{e}_p'\Sigma^{-1}\boldsymbol{e}_p\right)^{-1}\right).$$

The proof follows by applying Proposition 1.34 or Theorem 2.8 of Kutoyants ([16], p. 61 and p. 121). Let

$$\boldsymbol{\rho}_T=\sqrt{T}\left(\widehat{\boldsymbol{\theta}}-\bar{\theta}\boldsymbol{e}_p\right),\quad \boldsymbol{\xi}_T=\sqrt{T}\left(\widehat{\boldsymbol{\theta}}-\widetilde{\boldsymbol{\theta}}\right),\quad \zeta_T=\sqrt{T}\left(\widetilde{\boldsymbol{\theta}}-\bar{\theta}\boldsymbol{e}_p\right),$$

$$\Upsilon=\left(\boldsymbol{e}_p'\Sigma^{-1}\boldsymbol{e}_p\right)^{-1}\boldsymbol{e}_p\boldsymbol{e}_p' \quad \text{and} \quad \Sigma^*=\Sigma-\Upsilon.$$

**Proposition 7.** *Assume that Proposition 6 holds. Under* $H_0$ *given in (5), we have*

$$\boldsymbol{\rho}_T \xrightarrow[T\to\infty]{\mathscr{L}} \mathscr{N}_p(\mathbf{0},\Sigma) \quad \text{and} \quad \begin{pmatrix}\boldsymbol{\zeta}_T\\ \boldsymbol{\xi}_T\end{pmatrix} \xrightarrow[T\to\infty]{\mathscr{L}} \mathscr{N}_{2p}\left(\begin{pmatrix}\mathbf{0}\\ \mathbf{0}\end{pmatrix},\begin{pmatrix}\Upsilon & \mathbf{0}\\ \mathbf{0} & \Sigma^*\end{pmatrix}\right).$$

**Proof** Under the null hypothesis $\boldsymbol{\rho}_T=\sqrt{T}\left(\widehat{\boldsymbol{\theta}}-\boldsymbol{\theta}\right)$ and then, by combining Proposition 3, we get the first statement of Proposition. Further, by some computations, we get under the null hypothesis,

$$\begin{pmatrix}\boldsymbol{\zeta}_T\\ \boldsymbol{\xi}_T\end{pmatrix}=\begin{pmatrix}\left(\boldsymbol{e}_p'\Sigma^{-1}\boldsymbol{D}_T\boldsymbol{e}_p\right)^{-1}\boldsymbol{e}_p\boldsymbol{e}_p'\Sigma^{-1}\boldsymbol{D}_T\\ I_p-\left(\boldsymbol{e}_p'\Sigma^{-1}\boldsymbol{D}_T\boldsymbol{e}_p\right)^{-1}\boldsymbol{e}_p\boldsymbol{e}_p'\Sigma^{-1}\boldsymbol{D}_T\end{pmatrix}\sqrt{T}\left(\widehat{\boldsymbol{\theta}}-\boldsymbol{\theta}\right).$$

Again, by combining Proposition 3 and the Slutsky theorem we get the second statement of Proposition, which completes the proof. □

**Corollary 4.** *Under the conditions of Proposition 7 and under* $H_0$*,*

$$\boldsymbol{\xi}_T'\Sigma^{-1}\boldsymbol{\xi}_T \xrightarrow[T\to\infty]{\mathscr{L}} \chi^2_{p-1}.$$

**Proof** The proof is similar to that given in Nkurunziza and Ahmed [21]. For completeness, we outline the proof.

$$\boldsymbol{\xi}_T'\Sigma^{-1}\boldsymbol{\xi}_T=\boldsymbol{\rho}_T'\left(\Sigma^{-1}-\Sigma^{-1}\left(\boldsymbol{e}_p'\Sigma^{-1}\boldsymbol{e}_p\right)^{-1}\boldsymbol{e}_p\boldsymbol{e}_p'\Sigma^{-1}\right)\boldsymbol{\rho}_T+\boldsymbol{\rho}_T'\left(\Xi_T-\Xi\right)\boldsymbol{\rho}_T, \tag{23}$$

where

$$\Xi_T=\left(I_p-\boldsymbol{D}_T\Sigma^{-1}\boldsymbol{e}_p\left(\boldsymbol{e}_p'\Sigma^{-1}\boldsymbol{D}_T\boldsymbol{e}_p\right)^{-1}\right)\Sigma^{-1}\left(I_p-\left(\boldsymbol{e}_p'\Sigma^{-1}\boldsymbol{D}_T\boldsymbol{e}_p\right)^{-1}\boldsymbol{e}_p'\Sigma^{-1}\boldsymbol{D}_T\right),$$

and

$$\Xi = \Sigma^{-1} - \Sigma^{-1}\left(\boldsymbol{e}_p'\Sigma^{-1}\boldsymbol{e}_p\right)^{-1}\boldsymbol{e}_p\boldsymbol{e}_p'\Sigma^{-1}.$$

Combining Proposition 7 and the Slutsky theorem, we deduce that,

$$\boldsymbol{\rho}_T'\left[\Xi_T - \Xi\right]\boldsymbol{\rho}_T \xrightarrow[T\to\infty]{\mathscr{L}} \mathbf{0} \quad \text{and} \quad \boldsymbol{\rho}_T \xrightarrow[T\to\infty]{\mathscr{L}} \mathbf{Z} \sim \mathscr{N}_p\left(\mathbf{0},\Sigma\right).$$

Moreover,

$$\Sigma\left(\Sigma^{-1} - \Sigma^{-1}\left(\boldsymbol{e}_p'\Sigma^{-1}\boldsymbol{e}_p\right)^{-1}\boldsymbol{e}_p\boldsymbol{e}_p'\Sigma^{-1}\right) = \Sigma\Xi$$

is an idempotent matrix, we get

$$\begin{aligned}\boldsymbol{\rho}_T'\left(\Sigma^{-1} - \Sigma^{-1}\left(\boldsymbol{e}_p'\Sigma^{-1}\boldsymbol{e}_p\right)^{-1}\boldsymbol{e}_p\boldsymbol{e}_p'\Sigma^{-1}\right)\boldsymbol{\rho}_T &\xrightarrow[T\to\infty]{\mathscr{L}} \mathbf{Z}'\left(\Sigma^{-1} - \Sigma^{-1}\left(\boldsymbol{e}_p'\Sigma^{-1}\boldsymbol{e}_p\right)^{-1}\boldsymbol{e}_p\boldsymbol{e}_p'\Sigma^{-1}\right)\mathbf{Z}\\ &= \mathbf{Z}'\Xi\mathbf{Z} \sim \chi_r^2,\end{aligned}$$

where

$$r = \text{rank}\left(\Xi\right) = \text{trace}\left(\Sigma\left(\Sigma^{-1} - \Sigma^{-1}\left(\boldsymbol{e}_p'\Sigma^{-1}\boldsymbol{e}_p\right)^{-1}\boldsymbol{e}_p\boldsymbol{e}_p'\Sigma^{-1}\right)\right) = p-1.$$

□

**Proof of Proposition 3** Obviously, $\boldsymbol{\rho}_T = \sqrt{T}\left(\widehat{\boldsymbol{\theta}} - \boldsymbol{\theta}\right) + \boldsymbol{\delta}$, and then, by combining Proposition 6, we get the first statement of Proposition. Further, one can verify that

$$\begin{aligned}\begin{pmatrix}\boldsymbol{\zeta}_T\\ \boldsymbol{\xi}_T\end{pmatrix} &= \begin{pmatrix}\left(\boldsymbol{e}_p'\Sigma^{-1}\boldsymbol{D}_T\boldsymbol{e}_p\right)^{-1}\boldsymbol{e}_p\boldsymbol{e}_p'\Sigma^{-1}\boldsymbol{D}_T\\ I_p - \left(\boldsymbol{e}_p'\Sigma^{-1}\boldsymbol{D}_T\boldsymbol{e}_p\right)^{-1}\boldsymbol{e}_p\boldsymbol{e}_p'\Sigma^{-1}\boldsymbol{D}_T\end{pmatrix}\sqrt{T}\left(\widehat{\boldsymbol{\theta}} - \boldsymbol{\theta}\right)\\ &\quad + \begin{pmatrix}\left(\boldsymbol{e}_p'\Sigma^{-1}\boldsymbol{D}_T\boldsymbol{e}_p\right)^{-1}\boldsymbol{e}_p\boldsymbol{e}_p'\Sigma^{-1}\boldsymbol{D}_T\\ I_p - \left(\boldsymbol{e}_p'\Sigma^{-1}\boldsymbol{D}_T\boldsymbol{e}_p\right)^{-1}\boldsymbol{e}_p\boldsymbol{e}_p'\Sigma^{-1}\boldsymbol{D}_T\end{pmatrix}\boldsymbol{\delta}.\end{aligned}$$

Then, by combining Proposition 6 and the Slutsky theorem we get the second statement of Proposition and that completes the proof. □

## References


[1] Abu-Mostafa, Y.S.: Financial Model Calibration Using Consistency Hints. IEEE Trans. Neural Netw. **12**, 791–808 (2001)

[2] Ahmed, S.E.: Large-sample Pooling Procedure for Correlation. Statist. **41**, 425–438 (1992)

[3] Ahmed, S.E., Saleh, A.K.Md.E.: Improved Nonparametric Estimation of Location Vector in a Multivariate Regression Model. Nonpar. Stat. **11**, 51–78 (1999)

[4] Ahmed, S.E.: Shrinkage Estimation of Regression Coefficients from Censored Data with Multiple Observations. In: Ahmed, S.E., Reid, N. (eds.) Empirical Bayes and and Likelihood inference, pp. 103–120. Springer, New York (2001)
[5] Ahmed, S.E.: . Assessing Process Capability Index for Nonnormal Processes. J. Stat. Plan. Infer. **129**, 195–206 (2005)
[6] Basawa, V.I.; Prakasa Rao, B.L.S.: Statistical Inference for Stochastic Processes. Academic, London (1980)
[7] Dacunha-Castelle, D., Florens-Zmirou, D.: Estimation of the Coefficients of a Diffusion from Discrete Observations. Stochastics **19**, 263–284 (1986)
[8] Engen, S., Sæther, B.E.: Generalizations of the Moran Effect Explaining Spatial Synchrony in Population Fluctuations. Am. Nat. **166**, 603–612 (2005)
[9] Engen, S., Lande, R., Wall, T., DeVries, J.P.: Analyzing Spatial Structure of Communities Using the Two-Dimensional Poisson Lognormal Species Abundance Model. Am. Nat. **160**, 60–73 (2002)
[10] Florens-Zmirou, D.: Approximation Discrete-time Schemes for Statistics of Diffusion Processes. Statistics **20**, 547–557 (1989)
[11] Froda, S., Nkurunziza, S.: Prediction of Predator-prey Populations Modelled by Perturbed ODE. J. Math. Biol. **54**, 407–451 (2007)
[12] Georges, P.: The Vasicek and CIR Models and the Expectation Hypothesis of the Interest Rate Term Structure. The Bank of Canada and Department of Finance, Canada (2003)
[13] Judge, G.G., Bock, M.E.: The Statistical Implications of Pre-test and Stein-rule Estimators in Econometrics. North Holland, Amsterdam (1978)
[14] Jurečková, J., Sen, P.K.: Robust Statistical Procedures. Wiley, New York (1996)
[15] Kubokawa, T.: The Stein Phenomenon in Simultaneous Estimation: A review. In: Ahmed S.E. et al (eds.) Nonparametric Statistics and Related Topics, pp. 143–173. Nova Science, New York (1998)
[16] Kutoyants, A.Y.: Statistical Inference for Ergodic Diffusion Processes. Springer, New York (2004)
[17] Langetieg, T.C.: A Multivariate Model of the Term Structure. J. Fin. **35**, 71–97 (1980)
[18] Le Breton, A.: Stochastic Systems: Modeling, Identification and Optimization I. Mathematical Programming Studies **5**, 124–144 (1976)
[19] Liptser, R.S., Shiryayev, A.N.: Statistics of Random Processes: General Theory, Vol. I. Springer, New York (1977)
[20] Liptser, R.S., Shiryayev, A.N.: Statistics of Random Processes: Applications II. Springer, New York (1978)
[21] Nkurunziza, S., Ahmed, S.E.: Shrinkage Drift Parameter Estimation for Multi-factor Ornstein-Uhlenbeck Processes. University of Windsor, Technical report, WMSR **07-06** (2007)
[22] Schöbel, R., Zhu, J.: Stochastic Volatility With an Ornstein-Uhlenbeck Process: An Extension. Eur. Fin. Rev. **3**, 23–46 (1999)

[23] Sen, P.K.: On the Asymptotic Distributional Risks of Shrinkage and Preliminary Test Versions of Maximum Likelihood Estimators. Sankhya A **48**, 354–371 (1986)
[24] Steele, J.M.: Stochastic Calculus and Financial Applications. Springer, New York (2001)
[25] Stigler, S.M.: The 1988 Neyman Memorial Lecture: A Galtonian Perspective on Shrinkage Estimators. Stat. Sci. **5**, 147–155 (1990)
[26] Trenkler, G., Trenkler, D.: A Note on Superiority Comparisons of Homogeneous Linear Estimators. Commun. Stat. Theor. Meth. **12**, 799–808 (1983)
[27] Vasicek, O.: An Equilibrium Characterization of the Term Structure. J. Fin. Econ. **5**, 177–188 (1977)

# Bounds on Expected Coupling Times in a Markov Chain

Jeffrey J. Hunter

**Abstract** In the author's paper "Coupling and Mixing Times in Markov Chains" (Res. Lett. Inf. Math. Sci, 11, 1–22, 2007) it was shown that it is very difficult to find explicit expressions for the expected time to coupling in a general Markov chain. In this paper simple upper and lower bounds are given for the expected time to coupling in a discrete time finite Markov chain. Extensions to the bounds under additional restrictive conditions are also given with detailed comparisons provided for two and three state chains.

## 1 Introduction

In [5] the derivation of the expected time to coupling in a Markov chain and its relation to the expected time to mixing (as introduced in [4], see also [1], [6]) was explored and the two-state cases and three-state cases were examined in detail.

Considerable difficulty was experienced in attempting to obtain closed form expressions for the expected coupling times. The main thrust of this paper is to explore the derivation of easily computable upper and lower bounds on these expectations.

In Section 2 we summarise the main results on coupling. In Section 3 we derive some new bounds and in Section 4 we compare these bounds with special cases considered in [5].

## 2 Coupling times

Let $P = [p_{ij}]$ be the transition matrix of a finite irreducible, discrete time Markov chain $\{X_n\}$, $(n \geq 0)$, with state space $S = \{1, 2, \ldots, m\}$. Such Markov chains have a unique stationary distribution $\{\pi_j\}$, $(1 \leq j \leq m)$, that, in the case of a regular (finite,

---

Jeffrey J. Hunter
Institute of Information & Mathematical Sciences, Massey University, Private Bag 102-904, North Shore Mail Centre, Auckland 0745, New Zealand,
j.hunter@massey.ac.nz

irreducible and aperiodic) chain, is also the limiting distribution of the Markov chain ([3], Theorem 7.1.2). Let $\boldsymbol{\pi}^T = (\pi_1, \pi_2, \ldots, \pi_m)$ be the stationary probability vector of the Markov chain.

Coupling of Markov chains can be described as follows. Start a Markov chain $\{Y_n\}$, with the same transition matrix $P$ and state space $S$ as for $\{X_n\}$, operating under stationary conditions, so that the initial probability distribution for $Y_0$ is the stationary distribution $\{\pi_j\}$. Start the Markov chain $\{X_n\}$ in an initial state $i$ and allow each Markov chain to evolve, independently, until time $T = n$ when both chains $\{X_n\}$ and $\{Y_n\}$ reach the same state for the first time at this *n-th* trial. We call this the "coupling time" since after time $T$ each chain is coupled and evolves identically as the $\{Y_n\}$ Markov, with each chain having the same distribution at each subsequent trial, the stationary distribution $\{\pi_j\}$.

$\mathbf{Z}_n = (X_n, Y_n), (n \geq 0)$, is a (two-dimensional) Markov chain with state space $S \times S$. The chain is an absorbing chain with absorbing (coupling) states $C = \{(i,i), 1 \leq i \leq m\}$ and transient states $\mathscr{T} = \{(i,j), i \neq j, 1 \leq i \leq m, 1 \leq j \leq m\}$. The transition probabilities, prior to coupling, are given by $P\{\mathbf{Z}_{n+1} = (k,l) | \mathbf{Z}_n = (i,j)\} = p_{ik}p_{jl}$, (see [5]). Once coupling occurs at time $T = n, X_{n+k} = Y_{n+k}$ for all $k \geq 0$.

If $\mathbf{Z}_0 \in C$, coupling of the two Markov chains is instantaneous and the coupling time $T = 0$. Define $T_{ij,kl}$ to be the first passage time from state $(i,j)$ to state $(k,l)$. The time to coupling in state $k$, starting in state $(i,j)$, $(i \neq j)$, is the first passage time $T_{ij,kk}$ to the absorbing state $(k,k)$. Let $T_{ij,C}$ be the first passage time from $(i,j)$, $(i \neq j)$ to the absorbing (coupling) states $C$. Define $T_{ii,C} = 0$, $(1 \leq i \leq m)$, consistent with the coupling occurring instantaneously if $X_0 = Y_0$ (in state $i$).

Under the assumption that the embedded Markov chains, $X_n$ and $Y_n$, are irreducible and aperiodic (i.e regular) the transition matrix for the two dimensional Markov chain can be represented in the canonical form for an absorbing Markov chain, as

$$\tilde{P} = \begin{bmatrix} I & 0 \\ R & Q \end{bmatrix},$$

where $I$ is an $m \times m$ identity matrix, $Q$ is an $m(m-1) \times m(m-1)$ matrix governing the transition probabilities within the transient states $\mathscr{T}$, and $R$ is an $m(m-1) \times m$ matrix governing the transition probabilities from the transient states $\mathscr{T}$ to the absorbing (coupling) states $C$.

Note that if the Markov chains, $X_n$ and $Y_n$ are periodic (period $m$) then coupling either occurs initially or never occurs! We restrict attention to embedded regular chains.

In [5] it was shown that, with probability one, starting in state $(i,j)$ coupling will occur in finite time. Let $\kappa_{ij}^{(C)} = E[T_{ij,C}]$ be the expected time to coupling starting in state $X_0 = i, Y_0 = j$, and let $\boldsymbol{\kappa}^{(C)} \equiv (\kappa_{ij}^{(C)})$ be the column vector (of dimension $m(m-1) \times 1$) of the expected times to coupling. Then all the expected values are finite and, [5],

$$\boldsymbol{\kappa}^{(C)} = (I - Q)^{-1}\boldsymbol{e}. \tag{1}$$

Since the states of the Markov chain $\{Y_n\}$ have at each trial the stationary distribution, and since coupling occurs initially if $i = j$ with $T_{ii,C} = 0$, the expected time to coupling with $X_0$ starting in state $i, (1 \le i \le m)$ is

$$\tau_{C,i} = \sum_{j=1}^{m} \pi_j E[T_{ij,C}] = \sum_{j \neq i} \pi_j \kappa_{ij}^{(C)}. \quad (2)$$

Let $\boldsymbol{\kappa}_1^T = (\kappa_{12}^{(C)}, \ldots, \kappa_{1j}^{(C)}, \ldots, \kappa_{1m}^{C}), \ldots,$
$\boldsymbol{\kappa}_i^T = (\kappa_{i1}^{(C)}, \ldots, \kappa_{i,i-1}^{(C)}, \kappa_{i,i+1}^{(C)}, \ldots, \kappa_{im}^{(C)}), \ldots, \quad \boldsymbol{\kappa}_m^T = (\kappa_{m1}^{(C)}, \ldots, \kappa_{m,m-1}^{(C)}),$
and re-express $\boldsymbol{\kappa}$ as $\boldsymbol{\kappa}^T = (\boldsymbol{\kappa}_1^T, \ldots, \boldsymbol{\kappa}_i^T, \ldots, \boldsymbol{\kappa}_m^T)$.

Define $\boldsymbol{\rho}_i^T = \boldsymbol{\pi}^T[\boldsymbol{e}_1, \boldsymbol{e}_2, \ldots, \boldsymbol{e}_{i-1}, \boldsymbol{e}_{i+1}, \ldots, \boldsymbol{e}_m] = (\pi_1, \ldots, \pi_{i-1}, \pi_{i+1}, \ldots, \pi_m)$, a modification of $\boldsymbol{\pi}^T$ to yield a vector of dimension $1 \times (m-1)$ (with $\pi_i$ removed at the $i-th$ position from $\boldsymbol{\pi}^T$). For $1 \le i \le m$,

$$\tau_{C,i} = \boldsymbol{\rho}_i^T \boldsymbol{\kappa}_i.$$

From (1) observe that $\boldsymbol{\kappa}$ can be obtained by solving the set of linear equations

$$(I - Q)\boldsymbol{\kappa}^{(C)} = \boldsymbol{e}. \quad (3)$$

The $Q$-matrix is of dimension $m(m-1) \times m(m-1)$ and governs the transitions within the $m(m-1)$ transient states. This matrix contains some symmetry. The submatrix of one-step transition probabilities governing transitions between the states $(i, j)$ and $(j, i)$ $(i \neq j)$ has the structure

$$\begin{array}{c} \\ (i,j) \\ (j,i) \end{array} \begin{array}{c} \begin{array}{cc} (i,j) & (j,i) \end{array} \\ \begin{bmatrix} p_{ii}p_{jj} & p_{ij}p_{ji} \\ p_{ji}p_{ij} & p_{jj}p_{ii} \end{bmatrix} \end{array}.$$

The transition probabilities from $(i, j)$ to the other transient states have some symmetrical reciprocity, i.e. for $i \neq j$ and $r \neq s$,

$$P[(X_{n+1}, Y_{n+1}) = (r,s) | (X_n, Y_n) = (i, j)] = p_{ir}p_{js} = P[(X_{n+1}, Y_{n+1}) = (s,r) | (X_n, Y_n) = (j,i)].$$

The one step transition to any coupling state $(k, k)$ has the same probability from either $(i, j)$ or $(j, i)$ i.e.

$$P[(X_{n+1}, Y_{n+1}) = (k,k) | (X_n, Y_n) = (i, j)] = p_{ik}p_{jk} = P[(X_{n+1}, Y_{n+1}) = (k,k) | (X_n, Y_n) = (j,i)].$$

Thus by labelling the states in successive symmetrical pairs, each even numbered row of $Q$ has the same probabilities, but interchanged in pairs, as the previous odd numbered row. Furthermore these pairs of rows have identical probabilities in the same place in the $R$ matrix.

The net effect is that instead of solving the $m(m-1)$ linear equations present in (3), we need only solve a reduced number of $m(m-1)/2$ linear equations. This is effected by observing that $\kappa_{ij}^{(C)} = \kappa_{ji}^{(C)}$ so that only these $m(m-1)/2$ quantities (with $i < j$, say) actually need to be solved. We elaborate further on this later.

We introduce some notation.

$$\text{Let } \mu_{ij} = \sum_{r=1}^{m} p_{ir}p_{jr} = \sum_{r=1}^{m} P\{(X_{n+1},Y_{n+1}) = (r,r)|(X_n,Y_n) = (i,j)\}$$
$$= P\{(X_{n+1},Y_{n+1}) \in C|(X_n,Y_n) = (i,j)\}$$
$$= P\{\text{Coupling occurs at the next trial} \mid \text{The 2-dim MC is in state}(i,j)\}.$$

Observe that $\mu_{ij} = \boldsymbol{p}_i^{(r)T}\boldsymbol{p}_j^{(r)} = \mu_{ji}$ where $\boldsymbol{p}_i^{(r)T} = (p_{i1}, p_{i2}, \ldots, p_{im})$, the $i$-th row of the transition matrix $P$.

## 3 Bounds

In a general Markov chain setting, elemental expressions of the key equations, Eqn. (3), lead, for all $i \neq j$, to

$$\kappa_{ij}^{(C)} - 1 = \sum\sum_{r\neq s} p_{ir}p_{js}\kappa_{rs}^{(C)}. \tag{4}$$

We deduce upper and lower bounds for $\kappa_{ij}^{(C)}$ from Eqns. (4).

**Theorem 1.** *If $\mu_{ij} > 0$ for all $i \neq j$, then, for all $i \neq j$,*

$$\kappa_{\min} \leq \kappa_{ij}^{(C)} \leq \kappa_{\max}, \tag{5}$$

*where* $\kappa_{\min} = \dfrac{1}{\max_{i\neq j}\mu_{ij}}$ *and* $\kappa_{\max} = \dfrac{1}{\min_{i\neq j}\mu_{ij}}$.

*Proof.* Assume that for all

$$r\neq s, \kappa_{rs}^{(C)} \leq \kappa_{\max}. \tag{6}$$

Observe that

$$1 = \left(\sum_{r=1}^{m} p_{ir}\right)\left(\sum_{s=1}^{m} p_{js}\right) = \sum_{r=s} p_{ir}p_{js} + \sum\sum_{r\neq s} p_{ir}p_{js} = \mu_{ij} + \sum\sum_{r\neq s} p_{ir}p_{js}. \tag{7}$$

From Eqn. (4) and Eqn. (7) it follows that

$$\kappa_{ij}^{(C)} \leq 1 + \left(\sum\sum_{r\neq s} p_{ir}p_{js}\right)\kappa_{\max} = 1 + (1-\mu_{ij})\kappa_{\max}. \tag{8}$$

Assumption (6) implies, using inequality (8), that it is sufficient to take $1+(1-\mu_{ij})\kappa_{\max} \leq \kappa_{\max}$ and hence that $\mu_{ij}\kappa_{\max} \geq 1$, i.e. $\kappa_{\max} \geq \frac{1}{\mu_{ij}}$ for all $i \neq j$. This is satisfied by taking $\kappa_{\max} = \max_{i\neq j}\dfrac{1}{\mu_{ij}} = \dfrac{1}{\min_{i\neq j}\mu_{ij}} = \dfrac{1}{\mu_{\min}}$.

Similarly let us assume that for all

$$r \neq s, \kappa_{\min} \leq \kappa_{rs}^{(C)}. \tag{9}$$

From Eqn. (4) and Eqn. (7) we have that

$$\kappa_{ij}^{(C)} \geq 1 + \left( \sum_{r \neq s} \sum p_{ir} p_{js} \right) \kappa_{\min} = 1 + (1 - \mu_{ij}) \kappa_{\min}. \tag{10}$$

Similar to the argument used above, using assumption (9) and inequality (10), we require $1 + (1 - \mu_{ij})\kappa_{\min} \geq \kappa_{\min}$ and hence that $\mu_{ij}\kappa_{\min} \leq 1$. Thus, $\kappa_{\min} \leq \dfrac{1}{\mu_{ij}}$ for all $i \neq j$, which is satisfied by taking $\kappa_{\min} = \min\limits_{i \neq j} \dfrac{1}{\mu_{ij}} = \dfrac{1}{\max_{i \neq j} \mu_{ij}} = \dfrac{1}{\mu_{\max}}$. □

**Corollary 1.** *Provided $\mu_{ij} > 0$ for all $i \neq j$,*

$$\frac{(1 - \pi_i)}{\mu_{\max}} = (1 - \pi_i)\kappa_{\min} \leq \tau_{C,i} \leq (1 - \pi_i)\kappa_{\max} = \frac{(1 - \pi_i)}{\mu_{\min}}. \tag{11}$$

*Proof.* Inequalities (11) follow directly from Eqn. (2) and Eqn.( 5). □

If the stationary distribution $\{\pi_i\}$ of the underlying Markov chain is unknown then a simpler, but slightly larger, upper bound for $\tau_{C,i}$ valid for all $i$ follows from (11):

$$\tau_{C,i} < \kappa_{\max} = \frac{1}{\min_{i \neq j} \mu_{ij}} = \frac{1}{\mu_{\min}} = \frac{1}{\min_{i \neq j} \sum_{r=1}^{m} p_{ir} p_{jr}}. \tag{12}$$

**Corollary 2.** *If the underlying Markov chain consists of independent trials, i.e. the transition probabilities $p_{ij} = p_j$, then for all $i, j$,*

$$\kappa_{ij}^{(C)} = \frac{1}{\sum_{r=1}^{m} p_r^2}. \tag{13}$$

*Proof.* Observe that $\mu_{ij} = \sum_{r=1}^{m} p_r^2 \equiv \mu$. Thus $\min_{i \neq j} \mu_{ij} = \max_{i \neq j} \mu_{ij} = \mu$ and from (5) we deduce $\dfrac{1}{\mu} = \kappa_{\min} \leq \kappa_{ij}^{(C)} \leq \kappa_{\max} = \dfrac{1}{\mu}$ leading to Eqn. (13).

Expression (13) can also be derived directly in this special case by solving Equations (4) (see also Eqn. (5.7) of [5]). □

In [5] it was shown that, under the condition of independent trials,

$$\tau_{C,i} = \frac{\sum_{j \neq i} p_j}{\sum_{k=1}^{m} p_k^2} = \frac{1 - p_i}{\sum_{k=1}^{m} p_k^2}.$$

Since $1 - 2\sum_{r<s} p_r p_s = 1 - \left[ \left(\sum_{k=1}^{m} p_k\right)^2 - \left(\sum_{k=1}^{m} p_k^2\right) \right] = \sum_{k=1}^{m} p_k^2$,

$$\tau_{C,i} = \frac{1-p_i}{\sum_{k=1}^m p_k^2} = \frac{1-\pi_i}{1-2\sum_{r<s} p_r p_s}.$$

Thus the bounds given by Corollary 1 are tight under independence assumptions. The interval $(\kappa_{\min}, \kappa_{\max})$, or its width $\kappa_{\max} - \kappa_{\min}$, could be used as a measure of the departure of the underlying MC from independence.

If expression (12) is used when the conditions of Corollary 1 are violated, the upper bound grossly overestimates the maximum value of $\tau_{C,i}$. In those chains, if at least one $\mu_{ij} = 0$, the upper bound will be $\infty$. This will occur in those examples where $p_{ij} = 1$ for some pair $(i, j)$, with $i \neq j$, and $p_{rj} = 0$ for some $r \neq i$ .

Since there are instances when some of the $\mu_{ij}$ could be zero, it is necessary to explore these cases in more detail. We consider the reduced number of linear equations alluded to in Section 2 above.

Define, for all $i \neq j$ and $r \neq s$,

$$\begin{aligned}\alpha_{i,j}^{(r,s)} &= P\{(X_{n+1}, Y_{n+1} = (r,s)|(X_n, Y_n) \in \{(i,j),(j,i)\}\}\\ &= P\{\text{One step transition to state } (r,s) \text{ from either } (i,j) \text{ or } (j,i)\}\\ &= p_{ir}p_{js} + p_{jr}p_{is}.\end{aligned}$$

Observe that $\alpha_{i,j}^{(r,s)} = \alpha_{j,i}^{(r,s)} = \alpha_{i,j}^{(s,r)} = \alpha_{j,i}^{(s,r)}$. In each of these situations we shall write the expression in the form $\alpha_{i,j}^{(r,s)}$ with $i < j$ and $r < s$.

Further since for $i \neq j$, $\kappa_{ij}^{(C)} = \kappa_{ji}^{(C)}$ we write the common value as simply $\kappa_{ij}$ with $i < j$.

Thus from (4) above,

$$\begin{aligned}\kappa_{ij}^{(C)} &= \sum\sum_{r \neq s} p_{ir}p_{js}\kappa_{rs}^{(C)} = \sum\sum_{r<s} p_{ir}p_{js}\kappa_{rs}^{(C)} + \sum\sum_{r>s} p_{ir}p_{js}\kappa_{rs}^{(C)}\\ &= \sum\sum_{r<s} p_{ir}p_{js}\kappa_{rs}^{(C)} + \sum\sum_{s<r} p_{js}p_{im}\kappa_{sr}^{(C)} = \sum\sum_{r<s}(p_{ir}p_{js} + p_{jr}p_{is})\kappa_{rs}\\ &= \sum\sum_{r<s} \alpha_{i,j}^{(r,s)}\kappa_{rs}.\end{aligned}$$

Thus for all $i < j$,

$$\kappa_{ij} - 1 = \sum\sum_{r<s} \alpha_{i,j}^{(r,s)}\kappa_{rs}. \tag{14}$$

Equation (14) is the reduced variant of the linear equations (4).

Note that, using Equation (7), the parameters $\alpha_{i,j}^{(r,s)}$ have the property that for all $i < j$,

$$\begin{aligned}\sum_{r<s} \alpha_{i,j}^{(r,s)} &= \sum_{r<s}(p_{ir}p_{js} + p_{jr}p_{is}) = \sum_{r<s} p_{ir}p_{js} + \sum_{s>r} p_{is}p_{jr}\\ &= \sum_{r<s} p_{ir}p_{js} + \sum_{r>s} p_{ir}p_{js} = \sum_{r \neq s} p_{ir}p_{js} = 1 - \sum_{r} p_{ir}p_{jr}\\ &= 1 - \mu_{ij}.\end{aligned} \tag{15}$$

**Theorem 2.** *Without loss of generality, assume $a < b$ and $i < j$. If $\mu_{ab} = 0$ and $\mu_{ij} > 0$ for all $(i,j) \neq (a,b)$, then for all $(i,j) \neq (a,b)$*

$$\kappa_{\min} \leq \kappa_{ij} \leq \kappa_{\max}, \tag{16}$$

*with*

$$\frac{1}{1-\alpha_{a,b}^{(a,b)}} + \kappa_{\min} \leq \kappa_{ab} \leq \frac{1}{1-\alpha_{a,b}^{(a,b)}} + \kappa_{\max}, \tag{17}$$

*where*

$$\kappa_{\min} = \min_{i<j,(i,j)\neq(a,b)} \left[\frac{\lambda_{i,j}^{(a,b)}}{\mu_{ij}}\right] \text{ and } \kappa_{\max} = \max_{i<j,(i,j)\neq(a,b)} \left[\frac{\lambda_{i,j}^{(a,b)}}{\mu_{ij}}\right], \tag{18}$$

*with*

$$\lambda_{i,j}^{(a,b)} = 1 + \frac{\alpha_{i,j}^{(a,b)}}{1-\alpha_{a,b}^{(a,b)}}. \tag{19}$$

*Proof.* From the reduced equations (14), with $a < b$ and $i < j$,
$\kappa_{ab} - 1 = \alpha_{a,b}^{(a,b)}\kappa_{ab} + \sum\sum_{r<s,(r,s)\neq(a,b)} \alpha_{a,b}^{(r,s)}\kappa_{rs}$
implying $\kappa_{ab}(1-\alpha_{a,b}^{(a,b)}) = 1 + \sum\sum_{r<s,(r,s)\neq(a,b)} \alpha_{a,b}^{(r,s)}\kappa_{rs}$.

From (15) $\sum_{r<s} \alpha_{a,b}^{(r,s)} = 1 - \mu_{ab} = 1$, so that $\sum_{r<s,(r,s)\neq(a,b)} \alpha_{a,b}^{(r,s)} = 1 - \alpha_{a,b}^{(a,b)}$.
Assuming (16), i.e. $\kappa_{\min} \leq \kappa_{ij} \leq \kappa_{\max}$,
$1 + \{1-\alpha_{a,b}^{(a,b)}\}\kappa_{\min} \leq 1 + \sum\sum_{r<s,(r,s)\neq(a,b)} \alpha_{a,b}^{(r,s)}\kappa_{rs} = \kappa_{ab}(1-\alpha_{a,b}^{(a,b)} \leq$
$1 + \{1-\alpha_{a,b}^{(a,b)}\}\kappa_{\max}$ and result (17) follows.
The Theorem will follow once we establish the values for the bounds (18).

For $i < j$, from equations (14),

$$\kappa_{ij} - 1 - \alpha_{i,j}^{(a,b)}\kappa_{ab} = \sum \sum_{r<s,(r,s)\neq(a,b)} \alpha_{i,j}^{(r,s)}\kappa_{rs}. \tag{20}$$

Now from (15), $\sum\sum_{r<s,(r,s)\neq(a,b)} \alpha_{i,j}^{(r,s)} = 1 - \mu_{ij} - \alpha_{i,j}^{(a,b)}$, so that from Eqn. (20),

$$(1-\mu_{ij}-\alpha_{i,j}^{(a,b)})\kappa_{\min} \leq \kappa_{ij} - 1 - \alpha_{i,j}^{(a,b)}\kappa_{ab} \leq \left(1-\mu_{ij}-\alpha_{i,j}^{(a,b)}\right)\kappa_{\max},$$

or that

$$(1-\mu_{ij}-\alpha_{i,j}^{(a,b)})\kappa_{\min} + 1 + \alpha_{i,j}^{(a,b)}\kappa_{ab} \leq \kappa_{ij} \leq$$
$$(1-\mu_{ij}-\alpha_{i,j}^{(a,b)})\kappa_{\max} + 1 + \alpha_{i,j}^{(a,b)}\kappa_{ab}.$$

Using (17), the above expression is bounded above and below as

$$(1-\mu_{ij}-\alpha_{i,j}^{(a,b)})\kappa_{\min}+1+\alpha_{i,j}^{(a,b)}\left\{\frac{1}{1-\alpha_{a,b}^{(a,b)}}+\kappa_{\min}\right\}\leq\kappa_{ij}\leq$$

$$(1-\mu_{ij}-\alpha_{i,j}^{(a,b)})\kappa_{\max}+1+\alpha_{i,j}^{(a,b)}\left\{\frac{1}{1-\alpha_{a,b}^{(a,b)}}+\kappa_{\max}\right\}$$

which simplifies, using Eqn. (19), to $(1-\mu_{ij})\kappa_{\min}+\lambda_{i,j}^{(a,b)}\leq\kappa_{ij}\leq(1-\mu_{ij})\kappa_{\max}+\lambda_{i,j}^{(a,b)}$.

Since we require the lower and upper quantities of the above expression to be bounded below by $\kappa_{\min}$ and above by $\kappa_{\max}$, respectively, we further require, for all $i<j$, $\kappa_{\min}\leq(1-\mu_{ij})\kappa_{\min}+\lambda_{i,j}^{(a,b)}$ and $(1-\mu_{ij})\kappa_{\max}+\lambda_{i,j}^{(a,b)}\leq\kappa_{\max}$ implying $\mu_{ij}\kappa_{\min}\leq\lambda_{i,j}^{(a,b)}$ and $\lambda_{i,j}^{(a,b)}\leq\mu_{ij}\kappa_{\max}$ leading to expressions (18). □

Theorem 2 requires $\mu_{ab}=\sum_{a=1}^{m}p_{ar}p_{br}=0$. This implies that $p_{ar}p_{br}=0$ for all $r$. In particular $p_{aa}p_{ba}=0$ and $p_{ab}p_{bb}=0$. Thus there are four possible cases:
$(i)$ $p_{aa}=0$ and $p_{bb}=0$, $(ii)$ $p_{aa}=0$ and $p_{ab}=0$, $(iii)$ $p_{ba}=0$ and $p_{bb}=0$. $(iv)$ $p_{ba}=0$ and $p_{ab}=0$.
These conditions will place restrictions, in particular, on $\alpha_{a,b}^{(a,b)}=p_{aa}p_{bb}+p_{ba}p_{ab}$. For the respective cases: $(i)$ $\alpha_{a,b}^{(a,b)}=p_{ba}p_{ab}$, $(ii)$ $\alpha_{a,b}^{(a,b)}=0$, $(iii)$ $\alpha_{a,b}^{(a,b)}=0$, $(iv)$ $\alpha_{a,b}^{(a,b)}=p_{aa}p_{bb}$.
A simplification of Eqn. (18) and (19) for each of these special cases can now be displayed.

Let us extend Theorem 2 to the situation where we have two distinct pairs of states $(a,b)$ and $(c,d)$, where $\mu_{ab}=0$ and $\mu_{cd}=0$.

**Theorem 3.** *Without loss of generality, assume $a<b$, $c<d$ (with $(a,b)\neq(c,d)$) and $i<j$. If $\mu_{ab}=0$, $\mu_{cd}=0$ and $\mu_{ij}>0$ for all $(i,j)\neq(a,b)$ and $(c,d)$, then for all $(i,j)\neq(a,b),(c,d)$*

$$\kappa_{\min}\leq\kappa_{ij}\leq\kappa_{\max},\tag{21}$$

*with*

$$\frac{1+\alpha_{a,b}^{(c,d)}-\alpha_{c,d}^{(c,d)}}{\tau_2}+\kappa_{\min}\leq\kappa_{ab}\leq\frac{1+\alpha_{a,b}^{(c,d)}-\alpha_{c,d}^{(c,d)}}{\tau_2}+\kappa_{\max},\tag{22}$$

$$\frac{1+\alpha_{c,d}^{(a,b)}-\alpha_{a,b}^{(a,b)}}{\tau_2}+\kappa_{\min}\leq\kappa_{cd}\leq\frac{1+\alpha_{c,d}^{(a,b)}-\alpha_{a,b}^{(a,b)}}{\tau_2}+\kappa_{\max},\tag{23}$$

*where*

$$\kappa_{\min}=\min_{i<j,(i,j)\neq(a,b),(c,d)}\left[\frac{\lambda_{i,j}^{(a,b;c,d)}}{\mu_{ij}}\right],\text{ and }\kappa_{\max}=\max_{i<j,(i,j)\neq(a,b),(c,d)}\left[\frac{\lambda_{i,j}^{(a,b;c,d)}}{\mu_{ij}}\right],\tag{24}$$

*with*

$$\lambda_{i,j}^{(a,b;c,d)} = 1 + \frac{\alpha_{i,j}^{(a,b)}(1+\alpha_{a,b}^{(c,d)} - \alpha_{c,d}^{(c,d)}) + \alpha_{i,j}^{(c,d)}(1+\alpha_{c,d}^{(a,b)} - \alpha_{a,b}^{(a,b)})}{\tau_2}, \tag{25}$$

*where*

$$\tau_2 = (1-\alpha_{a,b}^{(a,b)})(1-\alpha_{c,d}^{(c,d)}) - \alpha_{a,b}^{(c,d)}\alpha_{c,d}^{(a,b)}. \tag{26}$$

*Proof.* From the reduced equations (14), for distinct pairs $(a,b)$, $(c,d)$ and $(i,j)$ with $a<b$, $c<d$ and $i<j$,

$$\kappa_{ab} = 1 + \alpha_{a,b}^{(a,b)}\kappa_{ab} + \alpha_{a,b}^{(c,d)}\kappa_{cd} + \Delta_{ab}, \tag{27}$$

$$\kappa_{cd} = 1 + \alpha_{c,d}^{(a,b)}\kappa_{ab} + \alpha_{c,d}^{(c,d)}\kappa_{cd} + \Delta_{cd}, \tag{28}$$

$$\kappa_{ij} = 1 + \alpha_{i,j}^{(a,b)}\kappa_{ab} + \alpha_{i,j}^{(c,d)}\kappa_{cd} + \Delta_{ij}, \tag{29}$$

where for all $(i,j)$, $\Delta_{ij} = \sum\sum_{r<s,(r,s)\neq(a,b),(c,d)} \alpha_{i,j}^{(r,s)}\kappa_{rs}$.
From Eqns. (27) and (28),

$$B\begin{bmatrix}\kappa_{ab}\\ \kappa_{cd}\end{bmatrix} \equiv \begin{bmatrix}1-\alpha_{a,b}^{(a,b)} & -\alpha_{a,b}^{(c,d)}\\ -\alpha_{c,d}^{(a,b)} & 1-\alpha_{c,d}^{(c,d)}\end{bmatrix}\begin{bmatrix}\kappa_{ab}\\ \kappa_{cd}\end{bmatrix} = \begin{bmatrix}1+\Delta_{ab}\\ 1+\Delta_{cd}\end{bmatrix}.$$

Since $\det(B) = \tau_2$, as given by (26), taking the inverse of $B$ yields

$$\begin{bmatrix}\kappa_{ab}\\ \kappa_{cd}\end{bmatrix} = B^{-1}\begin{bmatrix}1+\Delta_{ab}\\ 1+\Delta_{cd}\end{bmatrix} = \frac{1}{\tau_2}\begin{bmatrix}1-\alpha_{c,d}^{(c,d)} & \alpha_{a,b}^{(c,d)}\\ \alpha_{c,d}^{(a,b)} & 1-\alpha_{a,b}^{(a,b)}\end{bmatrix}\begin{bmatrix}1+\Delta_{ab}\\ 1+\Delta_{cd}\end{bmatrix},$$

so that

$$\begin{bmatrix}\kappa_{ab}\\ \kappa_{cd}\end{bmatrix} = \frac{1}{\tau_2}\begin{bmatrix}1+\alpha_{a,b}^{(c,d)} - \alpha_{c,d}^{(c,d)} + (1-\alpha_{c,d}^{(c,d)})\Delta_{ab} + \alpha_{a,b}^{(c,d)}\Delta_{cd}\\ 1+\alpha_{c,d}^{(a,b)} - \alpha_{a,b}^{(a,b)} + \alpha_{c,d}^{(a,b)}\Delta_{ab} + (1-\alpha_{a,b}^{(a,b)})\Delta_{cd}\end{bmatrix}. \tag{30}$$

Since, for all $(i,j)$,

$$\sum\sum_{r<s,(r,s)\neq(a,b),(c,d)} \alpha_{i,j}^{(r,s)} = 1 - \alpha_{i,j}^{(a,b)} - \alpha_{i,j}^{(c,d)} - \mu_{ij}, \tag{31}$$

$\sum\sum_{r<s,(r,s)\neq(a,b),(c,d)} \alpha_{a,b}^{(r,s)} = 1 - \alpha_{a,b}^{(a,b)} - \alpha_{a,b}^{(c,d)}$, and $\sum\sum_{r<s,(r,s)\neq(a,b),(c,d)} \alpha_{c,d}^{(r,s)} = 1 - \alpha_{c,d}^{(a,b)} - \alpha_{c,d}^{(c,d)}$.

Assuming (21) i.e. for $(i,j)\neq(a,b),(c,d)$, $\kappa_{\min} \leq \kappa_{ij}^{(C)} \leq \kappa_{\max}$, from the definition of $\Delta_{ij}$,

$$(1-\alpha_{a,b}^{(a,b)} - \alpha_{a,b}^{(c,d)})\kappa_{\min} \leq \Delta_{ab} \leq (1-\alpha_{a,b}^{(a,b)} - \alpha_{a,b}^{(c,d)})\kappa_{\max}$$

and

$$(1-\alpha_{c,d}^{(a,b)} - \alpha_{c,d}^{(c,d)})\kappa_{\min} \leq \Delta_{cd} \leq (1-\alpha_{c,d}^{(a,b)} - \alpha_{c,d}^{(c,d)})\kappa_{\max}.$$

From these above two bounds, the bounds given by Eqns. (22) and (23) now follow upon simplification from Eqns. (30).

Now, from Eqn. (29), using the upper and lower bounds given by Eqns. (22) and (23) together with (31), it is easily shown, using definition (25), that

$$\lambda_{i,j}^{(a,b;c,d)} + (1-\mu_{ij})\kappa_{\min} \leq \kappa_{ij} \leq \lambda_{i,j}^{(a,b;c,d)} + (1-\mu_{ij})\kappa_{\max}.$$

Since the left hand side and the right hand side of the above equation must be bounded below by $\kappa_{\min}$ and bounded above by $\kappa_{\max}$ respectively, the expressions given by (24) now follow. □

Theorem 3 can be extended further to incorporate the situation of multiple pairs of states each with zero probability of a one step to coupling. Note that there must be at least one pair of states where a single step takes the chain to a coupling state, since coupling occurs with probability one. (Otherwise, the chain is either an absorbing chain or consists of periodic states.)

**Theorem 4.** *Suppose* $\mu_{a_i,b_i} = 0, (a_i < b_i)$ *for* $i = 1,2,\ldots,n$ *and* $\mu_{ij} > 0$ *otherwise (with* $n < m(m-1)/2$*). Then, for* $(i,j) \notin \{(a_1,b_1),\ldots,(a_n,b_n)\}$,

$$\kappa_{\min} \leq \kappa_{ij} \leq \kappa_{\max}, \tag{32}$$

*with, for* $i = 1,2,\ldots,n$,

$$\sum_{j=1}^{n} A_{ij} + \kappa_{min} \leq \kappa_{a_i b_i} \leq \sum_{j=1}^{n} A_{ij} + \kappa_{\max}, \tag{33}$$

*where*

$$\kappa_{\min} = \min_{i<j,(i,j)\neq(a_1,b_1),\ldots,(a_n,b_n)} \left[\frac{\lambda_{ij}}{\mu_{ij}}\right] \text{ and } \kappa_{\max} = \max_{i<j,(i,j)\neq(a_1,b_1),\ldots,(a_n,b_n)} \left[\frac{\lambda_{ij}}{\mu_{ij}}\right], \tag{34}$$

*with*

$$\lambda_{ij} = 1 + \sum_{r=1}^{n}\sum_{s=1}^{n} A_{rs}\alpha_{i,j}^{(a_r,b_r)}, \tag{35}$$

*and* $[A_{rs}] = (I-A)^{-1}$ *and A is the* $n \times n$ *matrix* $A = [a_{rs}] = \left[\alpha_{a_r,b_r}^{(a_s,b_s)}\right]$.

*Proof.* From the reduced equations (14), for distinct pairs $(a_i,b_i)$, $(i = 1,2,\ldots,n)$ with $a_i < b_i$,

$$\kappa_{a_i b_i} = 1 + \sum_{k=1}^{n} \alpha_{a_i,b_i}^{(a_k,b_k)} \kappa_{a_k b_k} + \Delta_{a_i b_i}, \tag{36}$$

and for $i < j$, with $(i,j) \neq (a_i,b_i)$,

$$\kappa_{ij} = 1 + \sum_{k=1}^{n} \alpha_{i,j}^{(a_k,b_k)} \kappa_{a_k b_k} + \Delta_{ij}, \tag{37}$$

where, for all $(i,j)$, $\Delta_{ij} = \sum\sum_{r<s,(r,s)\neq(a_1,b_1),\ldots,(a_n,b_n)} \alpha_{i,j}^{(r,s)} \kappa_{rs}$.

Let $\boldsymbol{\kappa}^T = (\kappa_{a_1 b_1}, \kappa_{a_2 b_2}, \ldots, \kappa_{a_n b_n})$ and $\boldsymbol{\Delta}^T = (\Delta_{a_1 b_1}, \Delta_{a_2 b_2}, \ldots, \Delta_{a_n b_n})$. From Eqn. (36) $\boldsymbol{\kappa} = \boldsymbol{e} + A\boldsymbol{\kappa} + \boldsymbol{\Delta}$, i.e. $(I-A)\boldsymbol{\kappa} = \boldsymbol{e} + \boldsymbol{\Delta}$ implying $\boldsymbol{\kappa} = (I-A)^{-1}(\boldsymbol{e} + \boldsymbol{\Delta})$.

Now for $i = 1, 2, \ldots, n$, using Eqn. (32),

$$\left(\sum \sum_{r<s,(r,s)\neq(a_1 b_1),\ldots,(a_n b_n)} \alpha_{a_i,b_i}^{(r,s)}\right) \kappa_{\min} \leq \Delta_{a_i b_i} \leq$$

$$\left(\sum \sum_{r<s,(r,s)\neq(a_1 b_1),\ldots,(a_n b_n)} \alpha_{a_i,b_i}^{(r,s)}\right) \kappa_{\max}.$$

Since, from (15), $\sum\sum_{r<s} \alpha_{i,j}^{(r,s)} = 1 - \mu_{ij}$, it follows, under the conditions of the theorem for $i = 1, 2, .., n$, that

$$\left(1 - \sum_{k=1}^{n} \alpha_{a_i,b_i}^{(a_k,b_k)}\right) \kappa_{\min} \leq \Delta_{a_i b_i} \leq \left(1 - \sum_{k=1}^{n} \alpha_{a_i,b_i}^{a_k,b_k}\right) \kappa_{\max},$$

i.e.

$$\left(1 - \sum_{k=1}^{n} a_{ik}\right) \kappa_{\min} \leq \Delta_{a_i b_i} \leq \left(1 - \sum_{k=1}^{n} a_{ik}\right) \kappa_{\max}.$$

Expressing these element-wise inequalities in matrix form yields,

$$\kappa_{\min}(\boldsymbol{e} - A\boldsymbol{e}) \leq \boldsymbol{\Delta} \leq (\boldsymbol{e} - A\boldsymbol{e})\kappa_{\max},$$

or

$$\kappa_{\min}(I-A)\boldsymbol{e} \leq (I-A)\boldsymbol{\kappa} - \boldsymbol{e} \leq (I-A)\boldsymbol{e}\kappa_{\max}.$$

Now if $\boldsymbol{x}$ is a non-negative vector ($\boldsymbol{x} \geq \boldsymbol{0}$) and $B$ is a nonnegative matrix then $B\boldsymbol{x} \geq \boldsymbol{0}$. Note that $A$ is a sub-stochastic matrix (since there is at least one pair of states $(c,d) \notin \{(a_1,b_1),\ldots,(a_n,b_n)\}$ with $\alpha_{a_i,b_i}^{(c,d)} > 0$ for at least one $i$, so that there is at least one row of $A$ with a row-sum less than 1). Consequently $A$ has a maximal eigenvalue less than 1. This implies that $\sum_{k=0}^{\infty} A^k = (I-A)^{-1}$ with $(I-A)^{-1}$ non-singular. Consequently $(I-A)^{-1} \geq 0$, (see [2], Theorem 4.6.6), leading to

$$(I-A)^{-1}\boldsymbol{e} + \kappa_{\min}\boldsymbol{e} \leq \boldsymbol{\kappa} \leq (I-A)^{-1}\boldsymbol{e} + \kappa_{\max}\boldsymbol{e},$$

which leads in element form to Eqn. (33).

Now $\sum\sum_{r<s,(r,s)\neq(a_1,b_1),\ldots,(a_n,b_n)} \alpha_{i,j}^{(r,s)} = 1 - \sum_{r=1}^{n} \alpha_{i,j}^{(a_r,b_r)} - \mu_{ij}$ so that for $(i,j) \notin \{(a_1,b_1),\ldots,(a_n,b_n)\}$

$$\left(1 - \sum_{r=1}^{n} \alpha_{i,j}^{(a_r,b_r)} - \mu_{ij}\right) \kappa_{\min} \leq \Delta_{ij} \leq \left(1 - \sum_{r=1}^{n} \alpha_{i,j}^{(a_r,b_r)} - \mu_{ij}\right) \kappa_{\max}.$$

From Eqn. (37), for $(i,j) \notin \{(a_1,b_1),\ldots,(a_n,b_n)\}$,

$$1+\sum_{r=1}^{n} \alpha_{i,j}^{(a_r,b_r)} \kappa_{a_r b_r} + \left(1-\sum_{r=1}^{n} \alpha_{i,j}^{(a_r,b_r)} - \mu_{ij}\right) \kappa_{\min} \leq \kappa_{ij} \leq$$

$$1+\sum_{r=1}^{n} \alpha_{i,j}^{(a_r,b_r)} \kappa_{a_r b_r} + \left(1-\sum_{r=1}^{n} \alpha_{i,j}^{(a_r,b_r)} - \mu_{ij}\right) \kappa_{\max}.$$

Now, from Eqn. (33),

$$1+\sum_{r=1}^{n} \alpha_{i,j}^{(a_r,b_r)} \left(\sum_{s=1}^{n} A_{rs} + \kappa_{\min}\right) + \left(1-\sum_{r=1}^{n} \alpha_{i,j}^{(a_r,b_r)} - \mu_{ij}\right) \kappa_{\min} \leq \kappa_{ij},$$

and

$$\kappa_{ij} \leq 1+\sum_{r=1}^{n} \alpha_{i,j}^{(a_r,b_r)} \left(\sum_{s=1}^{n} A_{rs} + \kappa_{\max}\right) + \left(1-\sum_{r=1}^{n} \alpha_{i,j}^{(a_r,b_r)} - \mu_{ij}\right) \kappa_{\max}.$$

From Eqn. (32) we require, for the lower bound,

$$\kappa_{\min} \leq 1+\left(\sum_{r=1}^{n} \alpha_{i,j}^{(a_r,b_r)}\right) \left(\sum_{s=1}^{n} A_{rs} + \kappa_{\min}\right) + \left(1-\sum_{r=1}^{n} \alpha_{i,j}^{(a_r,b_r)} - \mu_{ij}\right) \kappa_{\min},$$

implying, for all $(i,j) \notin \{(a_1,b_1),\ldots,(a_n,b_n)\}$, that $\mu_{ij}\kappa_{\min} \leq 1+\sum_{r=1}^{n}\sum_{s=1}^{n} A_{rs}\alpha_{i,j}^{(a_r,b_r)} \equiv \lambda_{ij}$, leading to the first bound in (34) and expression (25).

Similarly for the upper bound we require, for all $(i,j)\notin\{(a_1,b_1),\ldots,(a_n,b_n)\}$, $\lambda_{ij} = 1+\sum_{r=1}^{n}\sum_{s=1}^{n} A_{rs}\alpha_{i,j}^{(a_r,b_r)} \leq \mu_{ij}\kappa_{\max}$ leading to the second bound in (34). □

Note that Theorem 2 follows from Theorem 4 when $n=1$ with $(a_1,b_1)=(a,b)$ where $A=[a_{11}]=\left[\alpha_{a_1,b_1}^{(a_1,b_1)}\right]=\left[\alpha_{a,b}^{(a,b)}\right]$, $[A_{11}]=(I-A)^{-1}=(1-\alpha_{a,b}^{(a,b)})^{-1}$.

Similarly, Theorem 3 follows from Theorem 4 when $n=2$ with $(a_1,b_1)=(a,b)$, $(a_2,b_2)=(c,d)$ where $A=\begin{bmatrix}\alpha_{a_1,b_1}^{(a_1,b_1)} & \alpha_{a_1,b_1}^{(a_2,b_2)}\\ \alpha_{a_2,b_2}^{(a_1,b_1)} & \alpha_{a_2,b_2}^{(a_2,b_2)}\end{bmatrix}=\begin{bmatrix}\alpha_{a,b}^{(a,b)} & \alpha_{a,b}^{(c,d)}\\ \alpha_{c,d}^{(a,b)} & \alpha_{c,d}^{(c,d)}\end{bmatrix}$ and $[A_{rs}]=(I-A)^{-1}=\begin{bmatrix}A_{11} & A_{12}\\ A_{21} & A_{22}\end{bmatrix}=\frac{1}{\tau_2}\begin{bmatrix}1-\alpha_{c,d}^{(c,d)} & \alpha_{a,b}^{(c,d)}\\ \alpha_{c,d}^{(a,b)} & 1-\alpha_{a,b}^{(a,b)}\end{bmatrix}$ with $\tau_2=\det(I-A)=(1-\alpha_{a,b}^{(a,b)})(1-\alpha_{c,d}^{(c,d)})-\alpha_{a,b}^{(c,d)}\alpha_{c,d}^{(a,b)}$.

# 4 Special cases

*Example 1 (Two-state Markov chains).*

Let $P = \begin{bmatrix} p_{11} & p_{12} \\ p_{21} & p_{22} \end{bmatrix} = \begin{bmatrix} 1-a & a \\ b & 1-b \end{bmatrix}, (0 < a \leq 1, 0 < b \leq 1)$, be the transition matrix of a two-state Markov chain with state space $S = \{1,2\}$. Let $d = 1-a-b$.

If $-1 < d < 1$, the Markov chain is regular with a unique stationary distribution given by

$$\pi_1 = \frac{b}{a+b}, \pi_2 = \frac{a}{a+b}$$

Note that $\mu_{12} = \mu_{21} = p_{11}p_{21} + p_{12}p_{22} = (1-a)b + a(1-b) = a+b-2ab \equiv \mu$.

Note that $\mu \neq 0$, since if $\mu = 0$ then $a(1-b) + b(1-a) = 0$, i.e. $a(1-b) = 0$ and $(1-a)b = 0$. Thus either *(i)* $a = 0$ and $b = 0$ or *(ii)* $a = 1$ and $b = 1$. Case *(i)* is impossible since this implies both states are absorbing, while case 2 implies the chain is periodic period 2. In both cases coupling never occurs.

In this special case, expressions for the expected number of trials to coupling can be found explicitly since the solution of equations (3) for, $(\boldsymbol{I} - \boldsymbol{Q})\boldsymbol{\kappa}^{(C)} = \boldsymbol{e}$ is easily effected with

$$\kappa_{12}^{(C)} = \kappa_{21}^{(C)} = \frac{1}{(a+b-2ab)} = \frac{1}{\mu} = \kappa_{\min} = \kappa_{\max}.$$

Further it was shown in [5] that

$\tau_{C,1} = \dfrac{a}{(a+b)(a+b-2ab)}$, implying $\tau_{C,1} = \dfrac{\pi_2}{\mu} = (1-\pi_1)\kappa_{\min} = (1-\pi_1)\kappa_{\max}$

and $\tau_{C,2} = \dfrac{b}{(a+b)(a+b-2ab)} = \dfrac{\pi_1}{\mu} = (1-\pi_2)\kappa_{\min} = (1-\pi_2)\kappa_{\max}$.

Thus the inequalities (5) and (11) are in fact equalities, with $(1-\pi_i)\kappa_{\min} = \tau_{C,i} = (1-\pi_i)\kappa_{\max}$.

*Example 2 (Three-state Markov chains (Explicit solutions of the $\kappa_{ij}$)). .*

Let $P = \begin{bmatrix} p_{11} & p_{12} & p_{13} \\ p_{21} & p_{22} & p_{23} \\ p_{31} & p_{32} & p_{33} \end{bmatrix} = \begin{bmatrix} 1-b-c & b & c \\ d & 1-d-f & f \\ g & h & 1-g-h \end{bmatrix}$ be the transition matrix of a Markov chain with state space $S = \{1,2,3\}$.

Note that $0 < b+c \leq 1$, $0 < d+f \leq 1$ and $0 < g+h \leq 1$. Let

$$\begin{aligned}
\Delta_1 &= p_{23}p_{31} + p_{21}p_{32} + p_{21}p_{31} = fg + dh + dg, \\
\Delta_2 &= p_{31}p_{12} + p_{32}p_{13} + p_{32}p_{12} = gb + hc + hb, \\
\Delta_3 &= p_{12}p_{23} + p_{13}p_{21} + p_{13}p_{23} = bf + cd + cf, \\
\Delta &= \Delta_1 + \Delta_2 + \Delta_3 = fg + dh + dg + gb + hc + hb + bf + cd + cf.
\end{aligned}$$

The Markov chain, with the above transition matrix, is irreducible (and hence a stationary distribution exists) if and only if $\Delta_1 > 0$, $\Delta_2 > 0$, $\Delta_3 > 0$, with stationary probability vector

$$(\pi_1, \pi_2, \pi_3) = \frac{1}{\Delta}(\Delta_1, \Delta_2, \Delta_3). \tag{38}$$

Observe that

$$\begin{aligned}
\mu_{12} = \mu_{21} &= p_{11}p_{21} + p_{12}p_{22} + p_{13}p_{23} = (1-b-c)d + b(1-d-f) + cf \\
&= b + d - 2bd - cd - bf + cf, \\
\mu_{23} = \mu_{32} &= p_{21}p_{31} + p_{22}p_{32} + p_{23}p_{33} = dg + (1-d-f)h + f(1-g-h) \\
&= h + f - 2fh - dh - fg + dg, \\
\mu_{13} = \mu_{31} &= p_{31}p_{11} + p_{32}p_{12} + p_{33}p_{13} = g(1-b-c) + hb + (1-g-h)c \\
&= c + g - 2cg - bg - ch + bh.
\end{aligned}$$

Using the reduced equations (14) with just three parameters $\kappa_{12}, \kappa_{13}$ and $\kappa_{23}$ yields

$$\begin{aligned}
\kappa_{12} &= 1 + \alpha_{1,2}^{(1,2)}\kappa_{12} + \alpha_{1,2}^{(1,3)}\kappa_{13} + \alpha_{1,2}^{(2,3)}\kappa_{23} \\
\kappa_{13} &= 1 + \alpha_{1,3}^{(1,2)}\kappa_{12} + \alpha_{1,3}^{(1,3)}\kappa_{13} + \alpha_{1,3}^{(2,3)}\kappa_{23} \\
\kappa_{23} &= 1 + \alpha_{2,3}^{(1,2)}\kappa_{12} + \alpha_{2,3}^{(1,3)}\kappa_{13} + \alpha_{2,3}^{(2,3)}\kappa_{23}
\end{aligned}$$

where $\alpha_{a,j}^{(r,s)} = p_{ir}p_{js} + p_{jr}p_{is}$. In matrix form,

$$\begin{bmatrix} 1-\alpha_{1,2}^{(1,2)} & -\alpha_{1,2}^{(1,3)} & -\alpha_{1,2}^{(2,3)} \\ -\alpha_{1,3}^{(1,2)} & 1-\alpha_{1,3}^{(1,3)} & -\alpha_{1,3}^{(2,3)} \\ -\alpha_{2,3}^{(1,2)} & -\alpha_{2,3}^{(1,3)} & 1-\alpha_{2,3}^{(2,3)} \end{bmatrix} \begin{bmatrix} \kappa_{12} \\ \kappa_{13} \\ \kappa_{23} \end{bmatrix} = B\boldsymbol{\kappa} = \boldsymbol{e}. \tag{39}$$

In [5] we were unable to find compact expressions for the solutions of (39) in all cases and special cases were considered. However, the structure exhibited by Eqn. (39) now permits a simple solution:
First note that

$$\boldsymbol{\kappa} = B^{-1}\boldsymbol{e} = \frac{1}{\tau_3}\begin{bmatrix} \tau_{11} & \tau_{12} & \tau_{13} \\ \tau_{21} & \tau_{22} & \tau_{23} \\ \tau_{31} & \tau_{32} & \tau_{33} \end{bmatrix}\boldsymbol{e} = \frac{1}{\tau_3}\begin{bmatrix} \tau_{11} + \tau_{12} + \tau_{13} \\ \tau_{21} + \tau_{22} + \tau_{23} \\ \tau_{31} + \tau_{32} + \tau_{33} \end{bmatrix} \tag{40}$$

where
$\tau_{11} = (1-\alpha_{1,3}^{(1,3)})(1-\alpha_{2,3}^{(2,3)}) - \alpha_{1,3}^{(2,3)}\alpha_{2,3}^{(1,3)}$, $\tau_{12} = \alpha_{1,2}^{(1,3)}(1-\alpha_{2,3}^{(2,3)}) + \alpha_{1,2}^{(2,3)}\alpha_{2,3}^{(1,3)}$,
$\tau_{13} = \alpha_{1,2}^{(1,3)}\alpha_{1,3}^{(2,3)} + \alpha_{1,2}^{(2,3)}(1-\alpha_{1,3}^{(1,3)})$, $\tau_{21} = \alpha_{1,3}^{(1,2)}(1-\alpha_{2,3}^{(2,3)}) + \alpha_{1,3}^{(2,3)}\alpha_{2,3}^{(1,2)}$,
$\tau_{22} = (1-\alpha_{1,2}^{(1,2)})(1-\alpha_{2,3}^{(2,3)}) - \alpha_{2,3}^{(1,2)}\alpha_{1,2}^{(2,3)}$, $\tau_{23} = (1-\alpha_{1,2}^{(1,2)})\alpha_{1,3}^{(2,3)} + \alpha_{1,2}^{(2,3)}\alpha_{1,3}^{(1,2)}$,
$\tau_{31} = \alpha_{1,3}^{(1,2)}\alpha_{2,3}^{(1,3)} + (1-\alpha_{1,3}^{(1,3)})\alpha_{2,3}^{(1,2)}$, $\tau_{32} = (1-\alpha_{1,2}^{(1,2)})\alpha_{2,3}^{(1,3)} + \alpha_{1,2}^{(1,3)}\alpha_{2,3}^{(1,2)}$,
$\tau_{33} = (1-\alpha_{1,2}^{(1,2)})(1-\alpha_{1,3}^{(1,3)}) - \alpha_{1,2}^{(1,3)}\alpha_{1,3}^{(1,2)}$,
and $\det(B) = \tau_3$ with the following equivalent forms:

$$\begin{aligned}\tau_3 &= (1-\alpha_{1,2}^{(1,2)})\tau_{11} - \alpha_{1,2}^{(1,3)}\tau_{21} - \alpha_{1,2}^{(2,3)}\tau_{31} \\ &= -\alpha_{1,3}^{(1,2)}\tau_{12} + (1-\alpha_{1,3}^{(1,3)})\tau_{22} - \alpha_{1,3}^{(2,3)}\tau_{32} \\ &= -\alpha_{2,3}^{(1,2)}\tau_{13} - \alpha_{2,3}^{(1,3)}\tau_{23} + (1-\alpha_{2,3}^{(2,3)})\tau_{33}.\end{aligned}$$

Using the observations, from Eqns. (15), that

$$\begin{aligned}\alpha_{1,2}^{(1,2)} + \alpha_{1,2}^{(1,3)} + \alpha_{1,2}^{(2,3)} + \mu_{12} &= 1, \\ \alpha_{1,3}^{(1,2)} + \alpha_{1,3}^{(1,3)} + \alpha_{1,3}^{(2,3)} + \mu_{13} &= 1, \\ \alpha_{2,3}^{(1,2)} + \alpha_{2,3}^{(1,3)} + \alpha_{2,3}^{(2,3)} + \mu_{23} &= 1,\end{aligned} \tag{41}$$

it can be shown that $\tau_3$ can be re-expressed as one of the following equivalent forms

$$\begin{aligned}\tau_3 &= \mu_{12}\tau_{11} + \mu_{13}\tau_{12} + \mu_{23}\tau_{13} = \mu_{12}\tau_{21} + \mu_{13}\tau_{22} + \mu_{23}\tau_{23} \\ &= \mu_{12}\tau_{31} + \mu_{13}\tau_{32} + \mu_{23}\tau_{33}.\end{aligned}$$

Thus from Eqn. (40),

$$\kappa_{12} = \frac{\tau_{11}+\tau_{12}+\tau_{13}}{\tau_3}, \kappa_{13} = \frac{\tau_{21}+\tau_{22}+\tau_{23}}{\tau_3}, \kappa_{23} = \frac{\tau_{31}+\tau_{32}+\tau_{33}}{\tau_3}. \tag{42}$$

Further $\tau_{C,1} = \pi_2\kappa_{12} + \pi_3\kappa_{13}$, $\tau_{C,2} = \pi_1\kappa_{12} + \pi_3\kappa_{23}$, $\tau_{C,3} = \pi_1\kappa_{13} + \pi_2\kappa_{23}$ so that

$$\tau_{C,1} = \frac{\Delta_2\kappa_{12} + \Delta_3\kappa_{13}}{\Delta}, \tau_{C,2} = \frac{\Delta_1\kappa_{12} + \Delta_3\kappa_{23}}{\Delta}, \tau_{C,3} = \frac{\Delta_1\kappa_{13} + \Delta_2\kappa_{23}}{\Delta}$$

implying

$$\begin{aligned}\tau_{C,1} &= \frac{\Delta_2(\tau_{11}+\tau_{12}+\tau_{13}) + \Delta_3(\tau_{21}+\tau_{22}+\tau_{23})}{\Delta\tau_3}, \\ \tau_{C,2} &= \frac{\Delta_1(\tau_{11}+\tau_{12}+\tau_{13}) + \Delta_3(\tau_{31}+\tau_{32}+\tau_{33})}{\Delta\tau_3}, \\ \tau_{C,3} &= \frac{\Delta_1(\tau_{21}+\tau_{22}+\tau_{23}) + \Delta_2(\tau_{31}+\tau_{32}+\tau_{33})}{\Delta\tau_3}. \square\end{aligned}$$

We now explore the derivation of simple bounds for $\kappa_{ij}$ utilising Theorems 1, 2 and 3 for the special cases considered in [5] where coupling occurred. We initially restrict attention to the cases where all the $\mu_{ij}$ are positive (Example 3). Other cases when $\mu_{12} = 0, \mu_{13} > 0, \mu_{23} > 0$ (Example 4) and $\mu_{12} = 0, \mu_{13} = 0, \mu_{23} > 0$, (Example 5) follow after Example 3.

*Example 3 (Three-state Markov chains (with all $\mu_{ij}$ positive.)).*
First observe that in Example 2, *Case* 1 (when $p_{12} = p_{23} = p_{31} = 1$) and *Case* 2 (when $p_{12} = p_{32} = 1, p_{21} + p_{23} = 1$) each involve a periodic Markov chain (period 3 for Case 1 and period 2 for Case 2). In Case 1 coupling either occurs initially

or never occurs. In Case 2 coupling either occurs initially, after one step, or never occurs. For coupling to occur with probability one we need to restrict attention to regular (irreducible, aperiodic, finite) Markov chains. Thus we omit further consideration of these two cases.

*Case 3: "Constant movement" with* $p_{11} = p_{22} = p_{33} = 0$

The transition matrix $P = \begin{bmatrix} 0 & b & 1-b \\ 1-f & 0 & f \\ g & 1-g & 0 \end{bmatrix} = \begin{bmatrix} 0 & p_{12} & p_{13} \\ p_{21} & 0 & p_{23} \\ p_{31} & p_{32} & 0 \end{bmatrix}$, with $0 < b < 1$, $0 < f < 1, 0 < g < 1$. It is easily seen that $\mu_{12} = p_{13}p_{23} = (1-b)f$, $\mu_{23} = p_{21}p_{31} = (1-f)g$ and $\mu_{13} = p_{32}p_{12} = b(1-g)$. Under the stated conditions, all of these parameters are positive so that the conditions of Theorem 1 are satisfied.

With $\mu_{\min} = \min\{(1-b)f, (1-f)g, b(1-g)\}$,
and $\mu_{\max} = \max\{(1-b)f, (1-f)g, b(1-g)\}$, Theorem 1 leads to
$\kappa_{\min} = \dfrac{1}{\mu_{\max}} \leq \kappa_{ij} \leq \kappa_{\max} = \dfrac{1}{\mu_{\min}}$.

Since

$$\begin{aligned}
\Delta_1 &\equiv p_{23}p_{31} + p_{21}p_{32} + p_{21}p_{31} = fg + 1 - f = 1 - f(1-g), \\
\Delta_2 &\equiv p_{31}p_{12} + p_{32}p_{13} + p_{32}p_{12} = gb + 1 - g = 1 - g(1-b), \\
\Delta_3 &\equiv p_{12}p_{23} + p_{13}p_{21} + p_{13}p_{23} = bf + 1 - b = 1 - b(1-f), \\
\Delta &\equiv \Delta_1 + \Delta_2 + \Delta_3 = 3 - f(1-g) - g(1-b) - b(1-f).
\end{aligned}$$

Using (38), the stationary probabilities can be derived. Bounds on the expected coupling times follow from application of Eqn. (11) yielding

$$\begin{aligned}
\frac{2-g(1-b)-b(1-f)}{[3-f(1-g)-g(1-b)-b(1-f)]\mu_{\max}} &\leq \tau_{C,1} \leq \frac{2-g(1-b)-b(1-f)}{[3-f(1-g)-g(1-b)-b(1-f)]\mu_{\min}}, \\
\frac{2-f(1-g)-b(1-f)}{[3-f(1-g)-g(1-b)-b(1-f)]\mu_{\max}} &\leq \tau_{C,2} \leq \frac{2-f(1-g)-b(1-f)}{[3-f(1-g)-g(1-b)-b(1-f)]\mu_{\min}}, \\
\frac{2-f(1-g)-g(1-b)}{[3-f(1-g)-g(1-b)-b(1-f)]\mu_{\max}} &\leq \tau_{C,3} \leq \frac{2-f(1-g)-g(1-b)}{[3-f(1-g)-g(1-b)-b(1-f)]\mu_{\min}}.
\end{aligned}$$

Computation of $\tau_{C,i}$, for all values of the parameters in [5] showed that

$$2.6667 \leq \min_{1 \leq i \leq 3} \tau_{C,i} < \infty.$$

For all combinations of $b = f = g$, the ratios $r_{L,i} = \dfrac{\text{lower bound of } \tau_{C,i}}{\tau_{C,i}}$ and $r_{U,i} = \dfrac{\text{upper bound of } \tau_{C,i}}{\tau_{C,i}}$ are both equal to 1, leading to the result that lower bound of $\tau_{C,i} = \tau_{C,i} =$ upper bound of $\tau_{C,i}$. This is not equivalent to the independence condition implied under Corollary 2 but arises due to the symmetry of the transition matrix in each situation, with the stationary probabilities all equal to 1/3.

Taking all combinations of $b, f$, and $g$ in steps of 0.1 between 0.1 and 0.9 we achieve considerable variability between the ratios.

In particular, $0.097 \le r_{L,i} \le 1$ with the minimal ratio being achieved at $(b,c,f) = (0.1,0.1,0.9)$ and $(0.9,0.1,0.9)$ for $r_{L,1}$, at $(0.9,0.1,0.1)$ and $(0.9,0.9,0.1)$ for $r_{L,2}$, and at $(0.1,0.9,0.1)$ and $(0.1,0.9,0.9)$ for $r_{L,3}$.

Further, $1 \le r_{U,i} \le 14.063$ with the maximal ratio being achieved at $(b,c,f) = (0.5,0.9,0.1)$ for $r_{U,1}$, at $(0.1,0.5,0.9)$ for $r_{U,2}$, and at $(0.9,0.1,0.5)$ for $r_{U,3}$.

*Case 4: "Independent trials"*

For this case $P = \begin{bmatrix} p_1 & p_2 & p_3 \\ p_1 & p_2 & p_3 \\ p_1 & p_2 & p_3 \end{bmatrix}$, so that $p_{ij} = p_j$ for all $i,j$ implying that the Markov chain is equivalent to independent trials on the state space S = {1, 2, 3}.
For all $i \neq j, \mu_{ij} = p_1^2 + p_2^2 + p_3^2 = 1 - 2p_1p_2 - 2p_2p_3 - 2p_3p_1$.
Now $\Delta_1 = p_1$, $\Delta_2 = p_2$, $\Delta_3 = p_3$, $\Delta = p_1 + p_2 + p_3 = 1$, implying $\pi_1 = p_1$, $\pi_2 = p_2$, $\pi_3 = p_3$.

For all $i$, it was shown in [5] that $\tau_{C,i} = \dfrac{1 - p_i}{1 - 2p_1p_2 - 2p_1p_3 - 2p_2p_3} = \dfrac{1-\pi_i}{\mu_{\min}} = \dfrac{1-\pi_i}{\mu_{\max}}$. Thus each inequality in (11) is in fact an equality, with the upper and lower bounds coinciding, as observed in Corollary 2.

*Case 5: "Cyclic drift " $p_{13} = p_{21} = p_{32} = 0$ with*

$$P = \begin{bmatrix} p_{11} & p_{12} & 0 \\ 0 & p_{22} & p_{23} \\ p_{31} & 0 & p_{33} \end{bmatrix} = \begin{bmatrix} 1-b & b & 0 \\ 0 & 1-f & f \\ g & 0 & 1-g \end{bmatrix}.$$

For this case $\mu_{12} = p_{12}p_{22} = b(1-f), \mu_{23} = p_{23}p_{33} = f(1-g), \mu_{13} = p_{31}p_{11} = g(1-b)$, with $\mu_{\min} = \min\{b(1-f), f(1-g), g(-b)\}$ and
$\mu_{\max} = \max\{b(1-f), f(1-g), g(1-b)\}$.
Thus for $0 < b < 1, 0 < f < 1, 0 < g < 1$, all the $\mu_{ij}$ parameters are positive and the results of Theorem 1 can be applied.

Further $\Delta_1 = fg$, $\Delta_2 = gb$, $\Delta_3 = bf$, $\Delta = fg + gb + bf$ so that expressions for the stationary probabilities follow from Eqn. (38). Using Eqn. (11) this leads to the following bounds on the expected times to coupling:

$$\frac{b(g+f)}{[fg+gb+bf]\mu_{\max}} \le \tau_{C,1} \le \frac{b(g+f)}{[fg+gb+bf]\mu_{\min}},$$
$$\frac{f(g+b)}{[fg+gb+bf]\mu_{\max}} \le \tau_{C,2} \le \frac{f(g+b)}{[fg+gb+bf]\mu_{\min}},$$
$$\frac{g(f+b)}{[fg+gb+bf]\mu_{\max}} \le \tau_{C,3} \le \frac{g(f+b)}{[fg+gb+bf]\mu_{\min}}.$$

As for Case 3, we explore the ratios $r_{L,i} = \dfrac{\text{lower bound of } \tau_{C,i}}{\tau_{C,i}}$ and $r_{U,i} = \dfrac{\text{upper bound of } \tau_{C,i}}{\tau_{C,i}}$.

When $b = f = g$, both ratios are equal to 1, leading to the lower bound of $\tau_{C,i} = \tau_{C,i} =$ upper bound of $\tau_{C,i}$. As for Case 3, this is not equivalent to the independence condition implied under Corollary 2 but arises due to the symmetry of the transition matrix in each situation with the stationary probabilities all equal to 1/3.

Taking all combinations of $b$, $f$, and $g$ in steps of 0.1 between 0.1 and 0.9 we achieve less variability between the lower ratios $r_{L,i}$, but much more variability between the upper ratios $r_{U,i}$ than was present in Case 3.

In particular, $0.185 \leq r_{L,i} \leq 1$ with the minimal ratio being achieved at $(b, f, g) = (0.1, 0.9, 0.1)$ for $r_{L,1}$, at $(0.1, 0.1, 0.9)$ for $r_{L,2}$, and $(0.9, 0.1, 0.1)$ for $r_{L,3}$.

Further $1 \leq r_{U,i} \leq 67.69$ with the maximal ratio being achieved at $(b, f, g) = (0.9, 0.1, 0.9)$ for $r_{U,1}$, at $(0.9, 0.9, 0.1)$ for $r_{U,2}$, and $(0.1, 0.9, 0.9)$ for $r_{U,3}$.
From Eqn. (12) simple upper bounds, valid for all $i$, can be given as

$$\tau_{C,i} < \frac{1}{\mu_{\min}} = \frac{1}{\min(p_{12}p_{22}, p_{11}p_{31}, p_{23}p_{33})} = \frac{1}{\min(b(1-f), f(1-g), g(1-b))}.$$

*Case 6: "Constant probability state selection"*

In this case, with $P = \begin{bmatrix} 1-a & \frac{a}{2} & \frac{a}{2} \\ \frac{b}{2} & 1-b & \frac{b}{2} \\ \frac{c}{2} & \frac{c}{2} & 1-c \end{bmatrix}$, $(0 < a \leq 1, 0 < b \leq 1, 0 < c \leq 1)$.

Observe that

$$\mu_{12} = \frac{2(a+b)-3ab}{4}, \mu_{13} = \frac{2(a+c)-3ac}{4}, \mu_{23} = \frac{2(b+c)-3bc}{4}$$

with

$$\mu_{\min} = \min\left(\frac{2(a+b)-3ab, 2(b+c)-3bc, 2(a+c)-3ac}{4}\right),$$

$$\mu_{\max} = \max\left(\frac{2(a+b)-3ab, 2(b+c)-3bc, 2(a+c)-3ac}{4}\right).$$

Further $\Delta_1 = \frac{3bc}{4}$, $\Delta_2 = \frac{3ac}{4}$, $\Delta_3 = \frac{3ab}{4}$ and thus $\Delta = \frac{3(bc+ac+ab)}{4}$. This leads to expressions for the stationary probabilities and hence to the following bounds for the $\tau_{C,i}$:

$$\frac{a(b+c)}{[bc+ac+ab]\mu_{\max}} \leq \tau_{C,1} \leq \frac{a(b+c)}{[bc+ac+ab]\mu_{\min}},$$
$$\frac{b(a+c)}{[bc+ac+ab]\mu_{\max}} \leq \tau_{C,2} \leq \frac{b(a+c)}{[bc+ac+ab]\mu_{\min}},$$
$$\frac{c(a+b)}{[bc+ac+ab]\mu_{\max}} \leq \tau_{C,3} \leq \frac{c(a+b)}{[bc+ac+ab]\mu_{\min}}.$$

Paralleling the procedures of cases 3 and 5 we obtain the following observations for the ratios $r_{L,i}$ and $r_{U,i}$. Firstly both $r_{L,i} = r_{U,i} = 1$, implying equality of the lower and upper bounds of $\tau_{C,i}$, and equal to $\tau_{C,i}$ occur at all cases when $a = b = c$, with the stationary probabilities all the same. In this case there is much less variability between the actual values of the expected times to coupling and the associated lower and upper bounds.

In particular it can be shown that for all values of $(a,b,c)$ in the ranges $0.1(0.1)1.0$, $0.277 \leq r_{L,i} \leq 1$ and $1 \leq r_{U,i} \leq 2.17$. The lower ratio $r_{L,i} = 0.277$ occurs at the following sets of values of $(a,b,c)$: $(0.1,0.1,1)$ and $(0.1,1,0.1)$ for $r_{L,1}$, $(0.1,0.1,1)$ and $(1,0.1,0.1)$ for $r_{L,2}$, and $(0.1,1,0.1)$ and $(1,0.1,0.1)$ for $r_{L,3}$. The upper ratio $r_{U,i} = 2.17$ occurs at $(a,b,c) = (1,0.1,0.1)$ for $r_{U,1}$, $(0.1,1,0.1)$ for $r_{U,2}$, and $(0.1,0.1,1)$ for $r_{U,3}$.

These bounds, especially the upper bounds, are much tighter than those exhibited in Cases 3 and 5, highlighting the efficacy of the procedure of Theorem 1 when the transition matrix is a positive matrix.

*Example 4 (Three-state Markov chains ($\mu_{12} = 0, \mu_{13} > 0, \mu_{23} > 0$)).*

Let $P = \begin{bmatrix} p_{11} & p_{12} & p_{13} \\ p_{21} & p_{22} & p_{23} \\ p_{31} & p_{32} & p_{33} \end{bmatrix} = \begin{bmatrix} 1-b-c & b & c \\ d & 1-d-f & f \\ g & h & 1-g-h \end{bmatrix}$ be the transition matrix of a Markov chain with state space $S = \{1,2,3\}$.

Note that $0 < b+c \leq 1, 0 < d+f \leq 1$ and $0 < g+h \leq 1$.

Observe that $\mu_{12} = p_{11}p_{21} + p_{12}p_{22} + p_{13}p_{23} = 0$ implies $p_{11}p_{21} = 0$, $p_{12}p_{22} = 0$ and $p_{13}p_{23} = 0$.

Thus eight cases need to be considered:

$$\begin{array}{llll} (i) & p_{11} = 0, p_{12} = 0, \text{ and } p_{13} = 0, & (ii) & p_{11} = 0, p_{12} = 0, \text{ and } p_{23} = 0, \\ (iii) & p_{11} = 0, p_{22} = 0, \text{ and } p_{13} = 0, & (iv) & p_{11} = 0, p_{22} = 0, \text{ and } p_{23} = 0, \\ (v) & p_{21} = 0, p_{12} = 0, \text{ and } p_{13} = 0, & (vi) & p_{21} = 0, p_{12} = 0, \text{ and } p_{23} = 0, \\ (vii) & p_{21} = 0, p_{22} = 0, \text{ and } p_{13} = 0, & (viii) & p_{21} = 0, p_{22} = 0, \text{ and } p_{23} = 0. \end{array}$$

Of these cases (*i*) and (*viii*) are impossible since $p_{11} + p_{12} + p_{13}$ and $p_{21} + p_{22} + p_{23}$ must be 1. Also cases (*v*) and (*vi*) are impossible (since the above restrictions would imply, respectively, that $p_{11} = 1$ and $p_{22} = 1$ and hence, respectively, that states 1 and 2, are absorbing.)

This leads to four remaining possibilities (with (*ii*), (*iii*), (*iv*), (*vii*) relabelled as (*a*), (*b*), (*c*) (*d*))

$(a) p_{11} = 0, p_{12} = 0$, and $p_{23} = 0$, with $p_{13} = 1$,
$(b) p_{11} = 0, p_{22} = 0$, and $p_{13} = 0$, with $p_{12} = 1$,
$(c) p_{11} = 0, p_{22} = 0$, and $p_{23} = 0$, with $p_{21} = 1$,
$(d) p_{21} = 0, p_{22} = 0$, and $p_{13} = 0$, with $p_{23} = 1$.

For case (*a*): $P_a = \begin{bmatrix} 0 & 0 & 1 \\ d & 1-d & 0 \\ g & h & 1-g-h \end{bmatrix}$, with $\mu_{13} = 1-g-h > 0$, $\mu_{23} = dg + (1-d)h > 0$; $\alpha_{1,2}^{(1,2)} = 0$, $\alpha_{1,3}^{(1,2)} = 0$, $\alpha_{2,3}^{(1,2)} = dh + (1-d)g$, and $0 < d \leq 1$, $0 \leq g < 1$, $0 < h < 1$, $0 < g+h < 1$.

For case $(b)$: $P_b = \begin{bmatrix} 0 & 1 & 0 \\ d & 0 & 1-d \\ g & h & 1-g-h \end{bmatrix}$, with $\mu_{13} = h > 0$, $\mu_{23} = dg + (1-d)(1-g-h) > 0$; $\alpha_{1,2}^{(1,2)} = d$, $\alpha_{1,3}^{(1,2)} = g$, $\alpha_{2,3}^{(1,2)} = dh$, and $0 \le d < 1, 0 \le g < 1$, $0 < h < 1$, $0 < g+h \le 1$.

For case $(c)$: $P_c = \begin{bmatrix} 0 & b & 1-b \\ 1 & 0 & 0 \\ g & h & 1-g-h \end{bmatrix}$, with $\mu_{13} = bh + (1-b)(1-g-h) > 0$, $\mu_{23} = g > 0$; $\alpha_{1,2}^{(1,2)} = b$, $\alpha_{1,3}^{(1,2)} = bg$, $\alpha_{2,3}^{(1,2)} = h$, and $0 \le b < 1$, $0 < g < 1$, $0 \le h < 1$, $0 < g+h \le 1$.

For case $(d)$: $P_d = \begin{bmatrix} 1-b & b & 0 \\ 0 & 0 & 1 \\ g & h & 1-g-h \end{bmatrix}$, with $\mu_{13} = (1-b)g + bh > 0$, $\mu_{23} = 1-g-h > 0$; $\alpha_{1,2}^{(1,2)} = 0$, $\alpha_{1,3}^{(1,2)} = (1-b)h$, $\alpha_{2,3}^{(1,2)} = 0$, and $0 < b \le 1$, $0 < g < 1$, $0 \le h < 1$, $0 < g+h < 1$.

Note that there is some symmetry between cases $(a)$ and $(d)$, and between cases $(b)$ and $(c)$.
Case $(d)$ converts to Case $(a)$ by relabelling the states $\{1,2,3\}$ as $\{2,1,3\}$ and changing the parameters $(b,g,h)$ to $(d,h,g)$. This same procedure will also convert Case $(c)$ to Case $(b)$.

From Theorem 2, $\dfrac{1}{1-\alpha_{1,2}^{(1,2)}} + \kappa_{\min} \le \kappa_{12} \le \dfrac{1}{1-\alpha_{1,2}^{(1,2)}}$, $\kappa_{\min} \le \kappa_{13} \le \kappa_{\max}$, $\kappa_{\min} \le \kappa_{23} \le \kappa_{\max}$. $\kappa_{\min} = \min\left[\dfrac{1}{\mu_{13}}\left\{1 + \dfrac{\alpha_{1,3}^{(1,2)}}{1-\alpha_{1,2}^{(1,2)}}\right\}, \dfrac{1}{\mu_{23}}\left\{1 + \dfrac{\alpha_{2,3}^{(1,2)}}{1-\alpha_{1,2}^{(1,2)}}\right\}\right]$, $\kappa_{\max} = \max\left[\dfrac{1}{\mu_{13}}\left\{1 + \dfrac{\alpha_{1,3}^{(1,2)}}{1-\alpha_{1,2}^{(1,2)}}\right\}, \dfrac{1}{\mu_{23}}\left\{1 + \dfrac{\alpha_{2,3}^{(1,2)}}{1-\alpha_{1,2}^{(1,2)}}\right\}\right]$. These expressions, with substitution as above for the special cases, together with explicit calculations for $\kappa_{ij}$ provided by equations (42) lead to the following observations.

For each of the following parameter selections: case $(a)$ with $(d,g,h) = (1,0.3,0.5)$, case $(b)$ with $(d,g,h) = (0,0.5,0.3)$, $(0.6,0.4,0.4)$, case $(c)$ with $(b,g,h) = (0,0.3,0.5)$, $(0.6,0.4,0.4)$, and case $(d)$ with $(b,g,h) = (1,0.5,0.3)$ the lower bound for each $\kappa_{ij}$ = upper bound for $\kappa_{ij}$ = exact value of $\kappa_{ij}$, providing an effective way of evaluating $\kappa_{ij}$. Further, at each of the above parameter selections, for $i = 1,2,3$, the lower bound for each $\tau_{C,i}$ = upper bound for $\tau_{C,i}$ = exact value of $\tau_{C,i}$.

For each $(i,j)$ with $i < j$, let $s_{L,ij} = \dfrac{\text{lower bound of } \kappa_{ij}}{\kappa_{ij}}$ and $s_{U,ij} = \dfrac{\text{upper bound of } \kappa_{ij}}{\kappa_{ij}}$, and for $i = 1,2,3$, let $r_{L,i} = \dfrac{\text{lower bound of } \tau_{C,i}}{\tau_{C,i}}$ and $r_{U,i} = \dfrac{\text{upper bound of } \tau_{C,i}}{\tau_{C,i}}$.

In every case $s_{L,ij} \leq 1, r_{L,i} \leq 1, s_{U,ij} \geq 1$ and $r_{U,i} \geq 1$.

Minimal extreme values, with the parameters taking increments of 0.1 in the restricted ranges for each case, occur at the following parameter selections:

Case $(a)$: $s_{L,12} = 0.305,\ s_{L,23} = 0.173,\ r_{L,2} = r_{L,3} = 0.186,$ at $(d,g,h) = (0.1,0,0.1)$, $s_{L,13} = 0.223$ and $r_{L,1} = 0.234$ at $(d,g,h) = (1,0.8,0.1)$.

Case $(b)$: $s_{L,12} = s_{L,13} = r_{L,1} = r_{L,2} = 0.100,\ s_{L,23} = r_{L,3} = 0.011$ at $(d,g,h) = (0.9,0,0.9)$.

Case $(c)$: $s_{L,12} = s_{L,23} = r_{L,1} = r_{L,3} = 0.100,\ s_{L,13} = r_{L,2} = 0.011$ at $(b,g,h) = (0.9,0.9,0)$.

Case $(d)$: $s_{L,12} = 0.305,\ s_{L,13} = 0.173,\ r_{L,1} = r_{L,3} = 0.186$ at $(b,g,h) = (0.1,0.1,0)$, $s_{L,23} = 0.223$ and $r_{L,2} = 0.234$ at $(b,g,h) = (1,0.1,0.8)$.

Maximal extreme values, with the parameters taking increments of 0.1 in the restricted ranges for each case, occur at the following parameter selections:

Case $(a)$: $s_{U,12} = 46.54,\ s_{U,13} = 87.62,\ s_{U,23} = 44.67,\ r_{U,1} = 80.46,$ $r_{U,2} = 44.84,\ r_{U,3} = 58.18$, at $(d,g,h) = (0.9,0,0.1)$.

Case $(b)$: $s_{U,12} = r_{U,1} = 82.90,\ s_{U,13} = r_{U,2} = 81.99$, at $(d,g,h) = (0.9,0,0.9)$, $s_{U,23} = r_{U,3} = 29.25$ at $(d,g,h) = (0.9,0.9,0.1)$.

Case $(c)$: $s_{U,12} = r_{U,1} = 82.90,\ s_{U,23} = r_{U,3} = 81.99$, at $(b,g,h) = (0.9,0.9,0)$, $s_{U,13} = r_{U,2} = 29.25$ at $(b,g,h) = (0.9,0.1,0.9)$.

Case $(d)$: $s_{U,12} = 46.54,\ s_{U,13} = 44.67,\ s_{U,23} = 87.62,\ r_{U,1} = 44.84,$ $r_{U,2} = 80.46,\ r_{U,3} = 58.18$, at $(b,g,h) = (0.9,0.1,0)$.

These extremal ratios for the lower bound, (resp. the upper bound) are in many instances smaller (resp. larger) that those experienced when the $\mu_{ij}$ are all positive.

*Example 5 (Three-state Markov chains ($\mu_{12} = 0, \mu_{13} = 0, \mu_{23} > 0$)).*

Let $P = \begin{bmatrix} p_{11} & p_{12} & p_{13} \\ p_{21} & p_{22} & p_{23} \\ p_{31} & p_{32} & p_{33} \end{bmatrix} = \begin{bmatrix} 1-b-c & b & c \\ d & 1-d-f & f \\ g & h & 1-g-h \end{bmatrix}$ be the transition matrix of a Markov chain with state space $S = \{1,2,3\}$.

Note that $0 < b+c \leq 1, 0 < d+f \leq 1$ and $0 < g+h \leq 1$.

Consider the four possibilities from $\mu_{12} = 0$ cases:
$(a)$ $p_{11} = 0,\ p_{12} = 0$, and $p_{23} = 0$, with $p_{13} = 1$,
$(b)$ $p_{11} = 0,\ p_{22} = 0$, and $p_{13} = 0$, with $p_{12} = 1$,
$(c)$ $p_{11} = 0,\ p_{22} = 0$, and $p_{23} = 0$, with $p_{21} = 1$,
$(d)$ $p_{21} = 0,\ p_{22} = 0$, and $p_{13} = 0$, with $p_{23} = 1$.

For case $(a)$: $\mu_{13} = 1-g-h = 0 \Rightarrow h = 1-g,\ \mu_{23} = dg + (1-d)(1-g) > 0,$

$P_a = \begin{bmatrix} 0 & 0 & 1 \\ d & 1-d & 0 \\ g & 1-g & 0 \end{bmatrix}$ with $0 < d \leq 1, 0 \leq g < 1, 0 < h = 1-g \leq 1$.

For case $(b)$: with $\mu_{13} = 0 \Rightarrow h = 0$, $\mu_{23} = dg + (1-d)(1-g) > 0$,

$$P_b = \begin{bmatrix} 0 & 1 & 0 \\ d & 0 & 1-d \\ g & 0 & 1-g \end{bmatrix} \text{ with } 0 \le d < 1, 0 < g \le 1, h = 0.$$

For case $(c)$: $\mu_{13} = bh + (1-b)(1-g-h) = 0$, $\mu_{23} = g > 0$, implies $bh = 0$ and $(1-b)(1-g-h) = 0$. This leads to four possibilities: $b = 0$ and $b = 1$ (impossible); $b = 0$ and $g + h = 1$; $h = 0$ and $b = 1$ (which is impossible since state 3 is then transient); $h = 0$ and $g = 1$ (which doesn't lead to coupling since the chain is then periodic with period 2). Thus there is only one possibility:

$$P_c = \begin{bmatrix} 0 & 0 & 1 \\ 1 & 0 & 0 \\ g & 1-g & 0 \end{bmatrix} \text{ with } 0 < g < 1, 0 < h = 1-g < 1,$$ (which is a special case of $(a)$ with $d = 1$).

For case $(d)$: $\mu_{13} = (1-b)g + bh = 0$, $\mu_{23} = 1-g-h > 0$ implies $(1-b)g$ and $bh = 0$. There are four possibilities: $b = 1$ and $g = 0$ (impossible since state 1 is then transient); $b = 1$ and $h = 0$; $g = 0$ and $b = 0$ (impossible since state 1 is then absorbing); $g = 0$ and $h = 1$ (which is impossible since state 1 is then transient). Thus there is only one possibility:

$$P_d = \begin{bmatrix} 0 & 1 & 0 \\ 0 & 0 & 1 \\ g & 0 & 1-g \end{bmatrix} \text{ with } b = 1, 0 < g < 1, h = 0,$$ (which is a special case of $(b)$ with $d = 0$).

Thus effectively there are only two non trivial cases to consider – case $(a)$ with $0 \le d \le 1$, $0 < g \le 1$ and case $(b)$, with $0 \le d \le 1$, $0 < g \le 1$. (The symmetry, as present in Example 4, effectively reduces this to one case.)

In computing the bounds for the special cases above, for $\kappa_{12}$, $\kappa_{13}$ and $\kappa_{23}$, using the procedure of Theorem 3, first observe that $\kappa_{\min} = \dfrac{\lambda_{2,3}^{(1,2;1,3)}}{\mu_{23}} = \kappa_{\max}$, leading to

$$\kappa_{23} = \frac{\lambda_{2,3}^{(1,2;1,3)}}{\mu_{23}}, \quad \kappa_{12} = \frac{1 + \alpha_{1,2}^{(1,3)} - \alpha_{1,3}^{(1,3)}}{\tau_2} + \frac{\lambda_{2,3}^{(1,2;1,3)}}{\mu_{23}},$$

$$\kappa_{13} = \frac{1 + \alpha_{1,3}^{(1,2)} - \alpha_{1,2}^{(1,2)}}{\tau_2} + \frac{\lambda_{2,3}^{(1,2;1,3)}}{\mu_{23}},$$

where

$$\lambda_{2,3}^{(1,2;1,3)} = 1 + \frac{\alpha_{2,3}^{(1,2)}(1 + \alpha_{1,2}^{(1,3)} - \alpha_{1,3}^{(1,3)}) + \alpha_{2,3}^{(1,3)}(1 + \alpha_{1,3}^{(1,2)} - \alpha_{1,2}^{(1,2)})}{\tau_2}$$

with

$$\tau_2 = (1 - \alpha_{1,2}^{(1,2)})(1 - \alpha_{1,3}^{(1,3)}) - \alpha_{1,2}^{(1,3)}\alpha_{1,3}^{(1,2)}.$$

Simplification using the observations from Eqn. (41), that since $\mu_{12} = 0$ and $\mu_{13} = 0$, $\alpha_{1,2}^{(1,2)} + \alpha_{1,2}^{(1,3)} + \alpha_{1,2}^{(2,3)} = 1$, $\alpha_{1,3}^{(1,2)} + \alpha_{1,3}^{(1,3)} + \alpha_{1,3}^{(2,3)} = 1, \alpha_{2,3}^{(1,2)} + \alpha_{2,3}^{(1,3)} + \alpha_{2,3}^{(2,3)} + \mu_{23} = 1$.

Further, in cases $(a)$ and $(c)$:

$$\alpha_{1,2}^{(1,2)} = 0,\ \alpha_{1,2}^{(1,3)} + \alpha_{1,2}^{(2,3)} = 1,\ \alpha_{1,3}^{(1,2)} = 0, \alpha_{1,3}^{(1,3)} + \alpha_{1,3}^{(2,3)} = 1,\ \alpha_{2,3}^{(1,3)} = 0,$$

$$\alpha_{2,3}^{(2,3)} = 0,\ \tau_2 = \alpha_{1,3}^{(2,3)},\ \lambda_{23}^{(1,2;1,3)} = \frac{\alpha_{1,3}^{(2,3)} + \alpha_{2,3}^{(1,2)}\left(1 + \alpha_{1,2}^{(1,3)} - \alpha_{1,3}^{(1,3)}\right)}{\alpha_{1,3}^{(2,3)}},$$

while in cases $(b)$ and $(d)$:

$$\alpha_{12}^{(1,3)} = 0,\ \alpha_{12}^{(1,2)} + \alpha_{12}^{(2,3)} = 1,\ \alpha_{13}^{(1,3)} = 0,\ \alpha_{13}^{(1,2)} + \alpha_{13}^{(2,3)} = 1,\ \alpha_{23}^{(1,2)} = 0,$$

$$\alpha_{23}^{(2,3)} = 0,\ \tau_2 = \alpha_{1,2}^{(2,3)},\ \lambda_{2,3}^{(1,2;1,3)} = \frac{\alpha_{1,2}^{(2,3)} + \alpha_{2,3}^{(1,2)} + \alpha_{2,3}^{(1,3)}\left(1 + \alpha_{1,3}^{(1,2)} - \alpha_{1,2}^{(1,2)}\right)}{\alpha_{1,2}^{(2,3)}}.$$

Thus in this example, all the bounds are exact, with agreement to the explicit solutions of equations (38) being obtained as in Example 3, i.e. $\kappa_{ij}(\text{exact}) = \kappa_{ij}(\text{bound})$ leading to the ratios

$$r_{L,i} = \frac{\text{lower bound of } \tau_{C,i}}{\tau_{C,i}} = \frac{\text{upper bound of } \tau_{C,i}}{\tau_{C,i}} = r_{U,i} = 1, \text{ in each case.}$$

The computation procedure of Theorem 3 is thus an alternative procedure for evaluating the $\kappa_{ij}$ in the case of a three-state chain when any two of the parameters $\mu_{ab}$ and $\mu_{cd}$ are both zero.

**Acknowledgement** The author thanks Götz Trenkler for his friendship and salutes him on the occasion of his retirement, in particular recognising his contributions to matrices and statistics. I am particularly grateful for his interest in my work using generalized inverses in solving a variety of problems involving Markovian kernels.

## References

[1] Aldous, D.J., Fill, J.A.: Reversible Markov Chains and Random Walks on Graphs (Book in preparation) See http://www.stat.Berkeley.EDU/users/aldous/book.html

[2] Hunter, J.J.: Mathematical Techniques of Applied Probability, Volume 1, Discrete Time Models: Basic Theory. Academic, New York (1983)

[3] Hunter, J.J.: Mathematical Techniques of Applied Probability, Volume 2, Discrete Time Models: Techniques and Applications, Academic, New York (1983)

[4] Hunter, J.J.: Mixing Times with Applications to Perturbed Markov Chains, Linear Algebra Appl. **417** 108–123 (2006)

[5] Hunter, J.J.: Coupling and Mixing times in a Markov chain, Res. Lett. Inf. Math. Sci. **11**, 1–22 (2007). (Submitted to Linear Algebra Appl.)
[6] Lovasz, L., Winkler, P.: Mixing Times. In: Aldous, D. Propp, J. (eds.) Microsurveys in Discrete Probability, DIMACS Series in Discrete Math. Theor. Comp. Sci., 85–133. AMS (1998)

# Multiple Self-decomposable Laws on Vector Spaces and on Groups: The Existence of Background Driving Processes

Wilfried Hazod

**Abstract** Following K. Urbanik, we define for simply connected nilpotent Lie groups $G$ multiple self-decomposable laws as follows: For a fixed continuous one-parameter group $(T_t)$ of automorphisms put $L^{(0)} := M^1(\mathbb{G})$ and $L^{(m+1)} := \left\{\mu \in M^1(\mathbb{G}) : \forall t > 0 \ \exists \ \nu(t) \in L^{(m)} : \ \mu = T_t(\mu) * \nu(t)\right\}$ for $m \geq 0$.

Under suitable commutativity assumptions it is shown that also for $m > 0$ there exists a background driving Lévy process with corresponding continuous convolution semigroup $(\nu_s)_{s\geq 0}$ determining $\mu$ and vice versa. Precisely, $\mu$ and $\nu_s$ are related by iterated Lie Trotter formulae.

## 1 Introduction

Self-decomposable laws or class $L$−laws were introduced by P. Lévy within the frame of limit distributions of normalized sums of independent (not necessarily identically distributed) random variables. In the past various types of distributions which are well-known in statistical applications turned out to be self-decomposable. See e.g. Z. Jurek [14], K. Sato [20] for a survey. Recently the self-decomposability property and the related additive processes – one- and multidimensional – turned out to be important for model building in Mathematical Finance. See e.g., [3] for a survey and for references.

K. Urbanik [25] extended the concept of self-decomposability to finite dimensional vector spaces $\mathbb{V}$ with operator normalization. See also [21] or the monograph [12], and the literature mentioned there.

Closely related to self-decomposability are generalized Ornstein–Uhlenbeck processes and Mehler semigroups of transition operators and, on the other hand, stable

Wilfried Hazod
Fakultät für Mathematik, Technische Universität Dortmund, D-44221 Dortmund, Germany
wilfried.hazod@mathematik.tu-dortmund.de

hemigroups and M-semigroups of probabilities. (For details and hints to the literature see e.g., [12, 5, 7, 8, 9, 22, 1, 19].) Let $(X_t)_{t\geq 0}$ be a stochastically continuous additive process taking values in $\mathbb{V}$ then the distributions $\nu(s,t)$ of the increments $X_s^{-1}X_t,\ s\leq t$, form a *continuous convolution hemigroup*, i.e. $(s,t)\mapsto \nu(s,t)$ is continuous and

$$\nu(s,t)\star\nu(t,r)=\nu(s,r)\quad \text{for } s\leq t\leq r. \tag{1}$$

A hemigroup $(\nu(s,t))_{s\leq t}$ is called *stable* w.r.t. a continuous one-parameter group of vector space automorphisms $\mathbb{T}=(T_t)_{t\in\mathbb{R}}\subseteq \mathrm{GL}(\mathbb{V})$ if for all $r$, for all $s\leq t$

$$T_r(\nu(s,t))=\nu(s+r,t+r),\quad \nu(s,t)=T_s(\nu(0,t-s)). \tag{2}$$

Put $\nu(s):=\nu(0,s),\ s\geq 0$. Then, as easily verified,

$$\nu(s+t)=\nu(s)\star\ T_s(\nu(t)),\ 0\leq s\leq t. \tag{3}$$

Continuous families $(\nu(s))_{s\geq 0}$ of probabilities satisfying (3) are called *M-semigroups* or skew semigroups. (The corresponding transition operators are generalized Mehler semigroups.)

$\mu\in M^1(\mathbb{V})$ is called (operator) self-decomposable w.r.t. $\mathbb{T}$ if $\forall t\geq 0$

$$\mu=T_t(\mu)\star\nu(t) \text{ for some } \nu(t)\in M^1(\mathbb{V}) \tag{4}$$

$\nu(t)$ is called *cofactor*. Stable hemigroups and M-semigroups are interesting objects of investigation in their own right. Furthermore, we have: If $\mu$ is self-decomposable (w.r.t. $\mathbb{T}$) then the cofactors $(\nu(t))_{t\geq 0}$ form a M-semigroup and $(T_s(\nu(t-s)))_{s\leq t}$ is a stable hemigroup. Hence in view of the above mentioned connections, self-decomposable laws with contracting $\mathbb{T}$ are limits of (generalized) Ornstein–Uhlenbeck processes (resp. of the corresponding M-semigroups). (Cf. [22], see also [9]).

K. Urbanik [26] introduced *multiple self-decomposability* defining nested classes of self-decomposable laws $L^{(m)}(\mathbb{T})$ inductively: $L^{(0)}(\mathbb{T}):=M^1(\mathbb{V})$, $L^{(1)}(\mathbb{T})$ the set of self-decomposable laws,

$$L^{(m+1)}(\mathbb{T}) := \left\{\mu:\mu=\ T_t(\mu)\star\nu(t) \text{ with } \nu(t)\in L^{(m)}(\mathbb{T}),\ t>0\right\}. \tag{5}$$

See also e.g., [4, 16, 12, 21, 23]. The concepts of self-decomposability, M-semigroups, stable hemigroups generalize to contractible (hence simply connected nilpotent) Lie groups $\mathbb{G}$, where $\mathbb{T}\subseteq \mathrm{Aut}(\mathbb{G})$ denotes a subgroup of automorphisms. The afore mentioned defining equations $(1),(2),(4)$ are used verbatim in this more general situation. See e.g., [5, 7, 9, 10, 19]. For self-decomposable laws on groups in connection with limit theorems see e.g. [24]. In particular, also multiply self-decomposability and the classes $L^{(m)}(\mathbb{T})$ (5) make sense in the group case.

For vector spaces $\mathbb{V}$, self-decomposable laws $\mu$ and their cofactors $\nu(t)$ are infinitely divisible. Hence for any fixed $s\geq 0$ there exists a Lévy process $\left(Z_t^{(s)}\right)_{t\geq 0}$

such that $Z_1^{(s)}$ is distributed according to $\nu(s)$. But the interesting objects are additive processes $(X_t)_{t\geq 0}$ with – in general non-stationary – increments $\left(X_s^{-1}X_t\right)_{s\leq t}$ distributed according to $\nu(s,t)$ with (1) and (2). There exist *hidden* Lévy processes $(Y_t)_{t\geq 0}$ (uniquely determined up to equivalence) *driving* $(X_t)_{t\geq 0}$, i.e., we have a random integral representation

$$X_t = \int_0^t T_u dY_u\,, \quad t \geq 0. \tag{6}$$

$(Y_t)_{t\geq 0}$ is called *background driving Lévy process*. See e.g. [14, 12, 13, 15, 20, 2].

For group-valued processes only weak versions of (6) are known: Let $(\nu(s,t))_{s\leq t}$ be a stable hemigroup (with corresponding M-semigroup $\nu(t) := \nu(0,t) : t \geq 0$) then there exists a uniquely determined continuous convolution semigroup $(\nu_t)_{t\geq 0}$ related to the M-semigroup $(\nu(t))_{t\geq 0}$ by Lie-Trotter formulas

$$\nu(t) = \lim_n \bigstar_{k=0}^{[nt]-1} T_{k/n}(\nu_{1/n}), \quad \nu(s,t) = \lim_n \bigstar_{k=[ns]}^{[nt]-1} T_{k/n}(\nu_{1/n}) \tag{7}$$

and

$$\nu_t = \lim_n \nu(1/n)^{[nt]} = \lim_n \nu(t/n)^n. \tag{8}$$

By (slight) abuse of language we call in the sequel the continuous convolution semigroup $(\nu_t)_{t\geq 0}$ the background driving Lévy process of the M-semigroup $(\nu(t))_{t\geq 0}$ resp. of the stable hemigroup $(\nu(s,t))_{s\leq t}$.

Let $\mathbb{T}$ be contracting. Then $\lim_{t\to\infty} \nu(t) =: \mu$ exists (and is self-decomposable then) iff $\nu(t)$ possesses finite logarithmic moments for some – hence all – $t > 0$. The M-semigroup of cofactors $(\nu(t))_{t\geq 0}$ possesses finite logarithmic moments $(t > 0)$ iff the background driving process $(\nu_t)_{t\geq 0}$ shares this property $(t > 0)$. (For vector spaces see e.g., [12], for groups see e.g., [5, 10]).

The aim of this paper is to prove the existence of background driving processes for *multiple self-decomposable laws* $\mu \in L^{(m)}(\mathbb{T})$ and to investigate the correspondences between multiple-cofactors and background driving processes in this case. In particular, to obtain analogues of the Lie Trotter formulas (7) and (8) for the multiple self-decomposable case.

The main results are new even for vector spaces. They are formulated and proved for the group case (under a commutativity assumption). But the proofs are written in such a way that they can easily extended to other convolution structures, e.g., to matrix cone hypergroups, structures closely related to Wishart distributions. (See [6] and the literature mentioned there.) Wishart distributions are not infinitely divisible w.r.t the usual convolution on the vector space of Hermitean matrices, but w.r.t. the new convolution structures they are even stable, hence completely self-decomposable. So, even when the investigations here are motivated by purely mathematical questions, there might be statistical applications in the future.

## 2 Multiple Self-decomposability

Let $\mathbb{G}$ be a contractible (hence simply connected nilpotent) Lie group, let $\mathbb{T} = (T_t)_{t\in\mathbb{R}}$ be a continuous one-parameter group in $\mathrm{Aut}(\mathbb{G})$, $T_{t+s} = T_t T_s$, $t,s \in \mathbb{R}$. We defined in the introduction the classes $L^{(m)}(\mathbb{T})$ (cf. (5)).

**Proposition 1.** *Let $\mu \in L^{(m)}(\mathbb{T})$ for some $m$. Then for all $t_1,\dots,t_m \in \mathbb{R}_+$ and $1 \le k \le m$ there exist $\nu^{(k)}(t_1,\dots,t_k) \in L^{(m-k+1)}(\mathbb{T}) \subseteq M^1(\mathbb{G})$ such that*

$$\begin{aligned}\mu = {} & T_{t_1}(\mu)\star T_{t_2}\left(\nu^{(1)}(t_1)\right)\star\cdots \\ & \star\; T_{t_m}\left(\nu^{(m-1)}(t_1,\dots,t_{m-1})\right)\star\nu^{(m)}(t_1,\dots,t_m). \end{aligned} \tag{9}$$

*The measures $\nu^{(k)}(t_1,\dots,t_k)$ are called $k-$cofactors.*

[[ $\mu = T_{t_1}(\mu)\star\nu^{(1)}(t_1) = T_{t_1}(\mu)\star T_{t_2}\left(\nu^{(1)}(t_1)\right)\star\nu^{(2)}(t_1,t_2)$, since $\nu^{(1)}(t_1)$ is $T_{t_2}$-decompsable. Per iteration we obtain finally

$$\begin{aligned}\mu = {} & T_{t_1}(\mu)\star(T_{t_2}\left(\nu^{(1)}(t_1)\right))\star T_{t_3}\left(\nu^{(2)}(t_1,t_2)\right)\star\cdots \\ & \cdots\star T_{t_m}\left(\nu^{(m-1)}(t_1,\dots,t_{m-1})\right)\star\nu^{(m)}(t_1,\dots,t_m).\end{aligned}$$
]]

For the main result, the subsequent Theorem 1, we assume additional conditions: The convolution factors in (9), i.e., the probabilities

$$\left\{T_t(\mu),\dots T_{t_k}(\nu^{(k-1)}(t_1,\dots,t_k)),\ \nu^{(m)}(t_1,\dots,t_m)\right\} \qquad 1\le k\le m,\ t_i\in\mathbb{R}_+ \tag{10}$$

commute.
For all $\nu\in L^{(1)}(\mathbb{T})$ the convolution operator is injective, i.e.,

$$\nu\star\rho = \nu\star\rho' \Rightarrow \rho = \rho' \tag{11}$$

(hence in particular, the cofactors are uniquely determined)
and $\mathbb{T} = (T_t)$ is contracting, i.e.,

$$\forall x\in X\in\mathbb{G}\ \lim_{t\to\infty} T_t(x) = e. \tag{12}$$

Note that (10) and (11) are obviously true for vector spaces: In this case, the convolution semigroup is commutative, therefore (10) is trivial, and $\nu$ is infinitely divisible, hence the Fourier transform has no zeros. Whence (11) follows. For injectivity in the group case see e.g., the discussion in [18].

**Theorem 1.** *Let $\mu \in L^{(m)}(\mathbb{T})$. Assume (10), (11) and (12). Then there exists a uniquely determined continuous convolution semigroup $\left(\nu_t^{(m)}\right)_{t\ge0}$, the $m^{th}$-background driving Lévy process, such that the k-cofactors, $1\le k\le m$, and $\mu$ are* *uniquely determined by* $\left(\nu_t^{(m)}\right)_{t\geq 0}$*. Furthermore,* $\nu_t^{(m)}$ *possesses finite* $\log_+^m(\cdot)$*–moments for all* $t > 0$.

*Hence – under (10), (11) and (12) – there exists a bijective mapping between Lévy processes with finite* $\log_+^m(\cdot)$*–moments and m-self-decomposable laws. (To simplify notations we write* $\nu_t := \nu_t^{(m)}$*.)*

*For* $m = 2$ *the result can be formulated in the following way: For a continuous convolution semigroup* $\left(\nu_t = \nu_t^{(2)}\right)_{t\geq 0}$ *with finite* $\log_+^2(\cdot)$*–moments there exists a uniquely determined 2-self-decomposable law* $\mu$ *with cofactors* $\nu^{(1)}(s)$, $\nu^{(2)}(s,t), s, t \geq 0$*, such that*

$$\nu^{(2)}(s,t) = \lim_N \lim_M \mathop{\bigstar}_{j=0}^{[Nt]-1} \mathop{\bigstar}_{k=0}^{[Ms]-1} T_{\frac{k}{M}+\frac{j}{N}}\left(\nu_{\frac{1}{N}\cdot\frac{1}{M}}\right)$$

*and*

$$\nu(s) = \nu^{(1)}(s) = \lim_{t\to\infty} \nu^{(2)}(s,t), \quad \mu = \lim_{s\to\infty} \nu(s).$$

*Conversely, let* $\mu$ *be 2-self-decomposable, let* $\left(\nu^{(2)}(s,t)\right)_{s,t\in\mathbb{R}_+}$ *be corresponding 2-cofactors, then there exists a continuous convolution semigroup* $\left(\nu_r = \nu_r^{(2)}\right)_{r\geq 0}$*, uniquely determined by* $\mu$*, such that for* $r = s\cdot t$*,* $r,s,t \geq 0$

$$\nu_r = \nu_r^{(2)} = \nu_{s\cdot t}^{(2)} = \lim_N \lim_M \left(\nu^{(2)}(s/M,\, t/N)\right)^{N\cdot M}.$$

The proof will be carried out only for $m = 2$, the general case follows along the same lines by induction. It relies on a space-time enlargement $\Gamma = \mathbb{G} \rtimes \mathbb{R}$, a semidirect extension of $\mathbb{G}$ by the real line via the automorphism group $\mathbb{T}$. The construction provides the means to investigate multi-parameter-analogues of M-semigroups $\left(\nu^{(m)}(t_1,\dots,t_m)\right)_{t_i\geq 0}$, the $m$-cofactors of $\mu \in L^{(m)}(\mathbb{T})$. Multi-parameter M-semigroups are – via space-time continuous convolution semigroups and Lie-Trotter formulas – related to multi-parameter continuous convolution semigroups, *the* $m^{th}$*-background driving Lévy processes.*

## 3 The Toolbox

We consider as afore mentioned the *space-time* group $\Gamma = \mathbb{G} \rtimes \mathbb{R}$, a semidirect product with group operation $(x,s)(y,t) = (xT_s(y), s+t)$, $x,y \in \mathbb{G}, s,t \in \mathbb{R}$. Let $M_*^1(\Gamma) := \{\rho \otimes \varepsilon_r : \rho \in M^1(\mathbb{G}), r \in \mathbb{R}\}$, a closed subsemigroup of $M^1(\Gamma)$. For probabilities in $M_*^1(\Gamma)$, convolution has a considerably simple form:

$$(\rho \otimes \varepsilon_s) * (\rho' \otimes \varepsilon_{s'}) = (\rho \star T_s(\rho')) \otimes \varepsilon_{s+s'},$$

where $*$ denotes convolution on $\Gamma$ and $\star$ denotes convolution on $\mathbb{G}$.

If $(\mu(t))_{t\geq 0}$ is a M-semigroup on $\mathbb{G}$ then $(\lambda_t := \mu(t) \otimes \varepsilon_t)_{t\geq 0}$ is a continuous convolution semigroup, the *space-time semigroup*, and conversely, for a continuous convolution semigroup $(\lambda_t := \mu(t) \otimes \varepsilon_t)_{t\geq 0}$ of probabilities on $\Gamma$ the space-component $(\mu(t))_{t\geq 0}$ is a M-semigroup. A continuous convolution semigroup $(\lambda_t)_{t\geq 0}$ is characterized by the *generating functional* $\mathscr{A} := \frac{d^+}{dt}\lambda_t|_{t=0}$ which has for $(\lambda_t)_{t\geq 0} \subseteq M^1_*(\Gamma)$ a pleasant form:

$$\mathscr{A} = B \oplus \varepsilon_0 + \varepsilon_e \oplus P, \tag{13}$$

where $B := \frac{d^+}{dt}\mu(t)|_{t=0}$ and $P$ is a differential operator of $1^{st}$ order. In particular, $B := \frac{d^+}{dt}\mu(t)|_{t=0}$ exists for any M-semigroup, and $B$ is the generating functional of a continuous convolution semigroup, $(\mu_t)_{t\geq 0}$ say. This semigroup is called background driving Lévy process, as afore mentioned. Applying the Lie-Trotter formula for generating functionals to the decomposition $(13)$ we obtain as announced

$$\mu(t) = \lim_{n\to\infty} \mathop{\bigstar}_{k=0}^{n-1} T_{\frac{t}{n}k}\left(\mu_{\frac{t}{n}}\right) = \lim_{n\to\infty} \mathop{\bigstar}_{k=0}^{[nt]-1} T_{\frac{k}{n}}\left(\mu_{\frac{1}{n}}\right) \tag{7}$$

and conversely,

$$\mu_t = \lim_{n\to\infty} \mu\,(t/n)^n = \lim_{n\to\infty} \mu\,(1/n)^{[nt]}. \tag{8}$$

Convergence is uniform on compact subsets of $\mathbb{R}_+$.

For the background of probabilities on groups the reader is referred to, e.g., [11, 5], for details concerning the decomposition $(13)$, see e.g. [5, 9].

Putting things together we obtain

**Proposition 2.** *(a) Let* $(\mu(t))_{t\geq 0} \subseteq M^1(\mathbb{G})$ *be a continuous M-semigroup. Then* $(8)$ *defines a (uniquely determined) continuous convolution semigroup* $(\mu_t)_{t\geq 0} \subseteq M^1(\mathbb{G})$.

*(b) Conversely, let* $(\mu_t)_{t\geq 0}$ *be a continuous convolution semigroup. Then* $(7)$ *defines a (uniquely determined) continuous M-semigroup* $(\mu(t))_{t\geq 0}$.

In the sequel we shall tacitly make use of the following well-known result. (We formulate a version which is adapted to our situation):

**Lemma 1.** *(a) Let* $\mathbb{G}$ *be a second countable locally compact group and let* $\mathbb{R}_+ \ni t \mapsto \alpha_t^{(n)} \in M^1(\mathbb{G})$ *be a sequence of functions which are assumed (i) to be weakly continuous, (ii)* $\forall t \geq 0$ *there exists* $\lim_{n\to\infty} \alpha_t^{(n)} =: \alpha_t \in M^1(\mathbb{G})$, *where (iii)* $(\alpha_t)_{t\geq 0}$ *satisfies the semigroup condition* $\alpha_{s+t} = \alpha_s \star \alpha_t$, $s,t \geq 0$.

*Then* $(\alpha_t)_{t\geq 0}$ *is a continuous convolution semigroup.*

*(b) As a corollary we obtain: Let* $\mathbb{G}$ *be a contractible Lie group, let* $\mathbb{T} = (T_t) \subseteq \mathrm{Aut}(\mathbb{G})$ *be contracting as before. Let (1)* $t \mapsto \alpha^{(n)}(t) \in M^1(\mathbb{G})$ *be continuous and (2) assume* $\lim_{n\to\infty} \alpha^{(n)}(t) =: \alpha(t) \in M^1(\mathbb{G})$ *to exist. Assume further (3)* $(\alpha(t))_{t\geq 0}$ *to satisfy the M-semigroup condition* $\alpha(s+t) = \alpha(s) \star T_s(\alpha(t))$, $s,t \geq 0$.

*Then* $(\alpha(t))_{t\geq 0}$ *is a continuous M-semigroup.*

⟦ To prove a) consider the convolution operators acting on $C_c(\mathbb{G}) \subseteq C_0(\mathbb{G}) \cap L^2(\mathbb{G})$: $R_\mu f(x) := \int f(xy)d\mu(y), L_\mu f(x) := \int f(yx)d\mu(y)$. Let $f,g \in C_c(\mathbb{G})$. Then

$$\begin{aligned}\langle R_\mu f, g\rangle &= \int R_\mu f(x)\overline{g(x)}d\omega_{\mathbb{G}}(x) = \int\int f(xy)d\mu(y)\overline{g(x)}d\omega_{\mathbb{G}}(x)\\ &= \int\int f(xy)\overline{g(x)}d\omega_{\mathbb{G}}(x)d\mu(y) =: \langle L_\nu f, \mu\rangle,\end{aligned}$$

where $\omega_{\mathbb{G}}$ denotes a Haar measure and $\nu := \overline{g}\cdot\omega_{\mathbb{G}}$ denotes the measure with density $\overline{g}$. Applying this formula to $\mu = \alpha_t^{(n)}$ and to $\alpha_t$, we obtain that $t \mapsto \langle R_{\alpha_t} f, g\rangle$ is measurable for all $f,g \in C_c(\mathbb{G})$. A density argument shows that $(R_{\alpha_t})_{t\geq 0}$ is a $C_0$ contraction semigroup on $L^2(\mathbb{G})$, measurable w.r.t. the weak operator topology. Since $L^2(\mathbb{G})$ is separable by assumption, continuity (in the strong operator topology) follows. Then, as well known and easily verified, weak continuity of $t \mapsto \alpha_t$ follows.

To prove (b) we notice that $(\beta_t := \alpha(t)\otimes\varepsilon_t)_{t\geq 0} \subseteq M^1(\Gamma)$ satisfies the assumptions of a). Hence continuity of $t \mapsto \beta_t$ follows, and therefore $t \mapsto \alpha(t)$ is continuous. ⟧

**Definition 1.** (a) A family $(\nu(s,t))_{s,t\geq 0} \subseteq M^1(\mathbb{G})$ is called 2-M-semigroup if for fixed $s \geq 0$ resp. $t \geq 0$, $t \mapsto \nu(s,t)$ resp. $s \mapsto \nu(s,t)$ are continuous M-semigroups. (Analogously, m-M-semigroups are defined for $m \geq 2$.).

(b) A family $(\nu_{s,t})_{s,t\geq 0} \subseteq M^1(\mathbb{G})$ is called continuous 2-semigroup if for fixed $s \geq 0$ resp. $t \geq 0$, $t \mapsto \nu_{s,t}$ resp. $s \mapsto \nu_{s,t}$ are continuous convolution semigroups.

In the following we assume throughout (in view of (10)) that

$$\{T_r(\nu(s,t)),\ r,s,t \geq 0\} \tag{14}$$

commute.

Applying Proposition 2 for fixed $s$ resp. for fixed $t$ we obtain

**Proposition 3.** *Let* $(\nu(s,t))_{s,t\geq 0}$ *be a 2-M-semigroup. Then for fixed* $s \geq 0$ *resp.* $t \geq 0$, *there exist continuous convolution semigroups* $\left(\rho_t^{(s)}\right)_{t\geq 0}$ *resp.* $\left(\sigma_s^{(t)}\right)_{s\geq 0}$ *such that for fixed* $t \geq 0$ *resp.* $s \geq 0$ $s \mapsto \rho_t^{(s)}$ *and* $t \mapsto \sigma_s^{(t)}$ *are continuous M-semigroups. The correspondence is given by the Lie-Trotter formulas* (7) *and* (8)*:*

$$\rho_t^{(s)} = \lim_n \nu(s,t/n)^n, \quad \sigma_s^{(t)} = \lim_m \nu(s/m,t)^m$$

*and conversely* (15)

$$\nu(s,t) = \lim_n \mathop{\bigstar}_{k=0}^{[nt]-1} T_{k/n}\left(\rho_{1/n}^{(s)}\right) = \lim_m \mathop{\bigstar}_{j=0}^{[ms]-1} T_{j/m}\left(\sigma_{1/m}^{(t)}\right).$$

⟦ Continuity follows since convergence in (7) and (8) is uniform on compact subsets. Alternatively, this follows by Lemma 1. To prove the M-semigroup property of e.g., $\left(\rho_t^{(s)}\right)_{s\geq 0}$, note that for $t \geq 0, s_1, s_2 \geq 0$ we have

$$\begin{aligned}\rho_t^{(s_1+s_2)} &= \lim_n \nu(s_1+s_2,t/n)^n = \lim_n \left(\nu(s_1,t/n) \star T_{s_1}\left(\nu(s_2,t/n)\right)\right)^n \\ &\overset{(14)}{=} \lim_n \nu(s_1,t/n)^n \star T_{s_1}\left(\lim_n \nu(s_2,t/n)^n\right) = \rho_t^{(s_1)} \star T_{s_1}\left(\rho_t^{(s_2)}\right).\end{aligned}$$

The other assertions are proved analogously. ⟧

**Proposition 4.** *Let, as in Proposition 3, $(\nu(s,t))_{s,t\geq 0}$ be a 2-M-semigroup. Define for $s,t \geq 0$:*

$$\begin{aligned}\nu_{s,t} &:= \lim_n \left(\sigma_s^{(t/n)}\right)^n = \lim_n \lim_m \left(\nu\left(s/m,t/n\right)\right)^{m\cdot n} \\ &\textit{and} \qquad\qquad\qquad\qquad\qquad\qquad\qquad\qquad (16)\\ \overline{\nu}_{s,t} &:= \lim_n \left(\rho_s^{(t/n)}\right)^n = \lim_m \lim_n \left(\nu\left(s/m,t/n\right)\right)^{n\cdot m}.\end{aligned}$$

*Then we have:*
*$(s,t) \mapsto \nu_{s,t}$ and $(s,t) \mapsto \overline{\nu}_{s,t}$ are continuous 2-semigroups (cf. Definition 1).*

⟦ Continuity follows by Lemma 1. We have to show the 2-semigroup property: $s \mapsto \sigma_s^{(u)}$ is a continuous convolution semigroup for all $u$, therefore also $s \mapsto \nu_{s,t}$ is a continuous convolution semigroup for all fixed $t$. (Recall that we assumed that all convolution factors commute (14)).

For fixed $s \geq 0$, $t \mapsto \sigma_s^{(t)}$ is a M-semigroup. Hence by (7) and (8), $t \mapsto \nu_{s,t}$ is a continuous convolution semigroup. The other assertions are proved analogously. ⟧

Conversely, we obtain with a similar proof:

**Proposition 5.** *Let $(\nu_{s,t})_{s,t\geq 0}$ be a continuous 2-semigroup. Define*

$$\begin{aligned}\nu(s,t) &:= \lim_n \lim_m \mathop{\bigstar}_{k=0}^{[nt]-1} \mathop{\bigstar}_{j=0}^{[ms]-1} T_{\frac{k}{n}+\frac{j}{m}}\left(\nu_{1/m,1/n}\right) \\ &= \lim_n \lim_m \mathop{\bigstar}_{k=0}^{n-1} \mathop{\bigstar}_{j=0}^{m-1} T_{\frac{t}{n}k+\frac{s}{m}j}\left(\nu_{s/m,\, t/n}\right)\end{aligned}$$

*for $s,t \geq 0$. Then $(\nu_{s,t})_{s,t\geq 0}$ is a continuous 2-M-semigroup.*

⟦ Continuity follows by Lemma 1. Furthermore, for fixed $s \geq 0$, $t \mapsto \sigma_s^{(t)} = \lim_m \mathop{\bigstar}_{j=0}^{[ms]-1} T_{j/m}\left(\nu_{1/m,\, t}\right)$ is a M-semigroup, and for fixed $t \geq 0$, $s \mapsto \sigma_s^{(t)}$ is a continuous convolution semigroup. Moreover, $\left(\nu(s,t) = \lim_n \bigstar_{k=0}^{[nt]-1} T_{k/n}\left(\sigma_s^{(1/n)}\right)\right)_{s,t\geq 0}$ is a 2-M-semigroup. $]\!]$

Finally, for continuous 2-semigroups we obtain the following representation:

**Proposition 6.** *Let* $(\mu_{s,t})_{s,t\geq 0}$ *be a continuous 2-semigroup. Then there exists a uniquely determined continuous convolution semigroup* $(\alpha_r)_{r\geq 0} \subseteq M^1(\mathbb{G})$ *such that* $\mu_{s,t} = \alpha_{s\cdot t}$, $s,t \geq 0$. *In fact,* $\alpha_r = \mu_{r,1} = \mu_{1,r}$, $r \geq 0$.

*Conversely, to any continuous convolution semigroup* $(\alpha_r)_{r\geq 0}$ *the mapping* $(s,t) \mapsto \mu_{s,t} := \alpha_{s\cdot t}$ *defines a continuous 2-semigroup.*

$[\![$ For fixed $t \geq 0$, $s \mapsto \mu_{s,t}$ is a continuous convolution semigroup. Let $B(t) := \frac{d^+}{ds}\mu_{s,t}|_{s=0}$ denote the generating functional. Hence for all test functions $f \in \mathscr{D}(\mathbb{G})$, $\mathbb{R}_+ \ni t \mapsto \langle B(t), f\rangle$ is measurable. Furthermore, the semigroup property $\mu_{s,t_1+t_2} = \mu_{s,t_1} \star \mu_{s,t_2}$ yields $\langle B(t_1+t_2), f\rangle = \langle B(t_1), f\rangle + \langle B(t_2), f\rangle$. Whence $\langle B(t), f\rangle = t \cdot \langle B(1), f\rangle$ follows. This holds for any $f$, whence, with $B := B(1)$ we obtain: $B(t) = t \cdot B$.

Put $\beta_s^{(t)} := \mu_{s,t}$ and $\alpha_s := \beta_s^{(1)} = \mu_{s,1}$. The continuous convolution semigroup $\left(\beta_s^{(t)}\right)_{s\geq 0}$ is generated by $B(t) = t \cdot B$. Whence $\beta_s^{(t)} = \beta_{s\cdot t}^{(1)} = \alpha_{s\cdot t}$ follows. Hence, $\mu_{s,t} = \alpha_{s\cdot t}$ as asserted.

The converse is obvious. $]\!]$

# 4 Proof of Theorem 1

As afore announced, to simplify notations we shall prove Theorem 1 for $m = 2$ only. Let $\mathbb{T}$ be a contracting group of automorphisms, let $\mu \in L^{(2)}(\mathbb{T})$. For $s,t \geq 0$ we have $\mu = T_t(\mu) \star \nu^{(1)}(t) = T_s\left(T_t(\mu) \star \nu^{(1)}(t)\right) \star \nu(s) = T_{t+s}(\mu) \star T_s\left(\nu^{(1)}(t)\right) \star \nu^{(1)}(s)$. On the other hand, $\mu = T_{t+s}(\mu) \star \nu^{(1)}(s+t)$. By the injectivity assumption (11) and commutativity (10), we obtain $\nu^{(1)}(s+t) = \nu(s) \star T_s(\nu^{(1)}(t))$, i.e., the 1-cofactors form a M-semigroup. (Note that independently from the injectivity assumption, 1-cofactors $\left(\nu^{(1)}(s)\right)_{s\geq 0}$ may be chosen in such a way. Cf. [7]).

Applying these considerations to the 1-cofactors $\nu^{(1)}(s)$ instead of $\mu$ we obtain for fixed $s$: $\nu^{(1)}(s) = T_t(\nu^{(1)}(s)) \star \nu^{(2)}(s,t), \forall t \geq 0$, and $t \mapsto \nu^{(2)}(s,t)$ is a continuous M-semigroup.

**Claim:** For fixed $t \geq 0$, $s \mapsto \nu^{(2)}(s,t)$ is a M-semigroup. Hence the 2-cofactors $\left(\nu^{(2)}(s,t)\right)_{s,t\geq 0}$ form a 2-M-semigroup (cf. Definition 1).

$[\![$ Let $s_1, s_2, r \geq 0$. The injectivity assumption (11) yields uniqueness of the cofactors, hence

$$\nu^{(1)}(s_1+s_2) = T_r\left(\nu^{(1)}(s_1+s_2)\right) \star \nu^{(2)}(s_1+s_2,r)$$
$$\overset{(14)}{=} \nu^{(2)}(s_1+s_2,r) \star T_r\left(\nu^{(1)}(s_1+s_2)\right).$$

On the other hand, 1-cofactors being M-semigroups,

$$\nu^{(1)}(s_1+s_2) = \nu^{(1)}(s_1) \star T_{s_1}\left(\nu^{(1)}(s_2)\right)$$
$$= \text{(by self-decomposability of 1-cofactors)}$$
$$\overset{\forall r\geq 0}{=} T_r\left(\nu^{(1)}(s_1)\right) \star \nu^{(2)}(s_1,r) \star T_{s_1}\left(T_r\left(\nu^{(1)}(s_2)\right) \star \nu^{(2)}(s_2,r)\right)$$
$$\overset{(14)}{=} \left(\nu^{(2)}(s_1,r) \star T_{s_1}\left(\nu^{(2)}(s_2,r)\right)\right) \star T_r\left(\nu^{(1)}(s_1) \star T_{s_1}\left(\nu^{(1)}(s_2)\right)\right)$$
$$= \nu^{(2)}(s_1,r) \star T_{s_1}\left(\nu^{(2)}(s_2,r)\right) \star T_r\left(\nu^{(1)}(s_1+s_2)\right).$$

Again by the injectivity assumption (11) we may identify the cofactors to obtain $\nu^{(2)}(s_1+s_2,r) = \nu^{(2)}(s_1,r) \star T_{s_1}\left(\nu^{(2)}(s_2,r)\right)$, $r,s_1,s_2 \geq 0$.

The claim is proved. ]]

Applying the tools in Sect. 3 (Propositions 4 – 6) we obtain :
There exists a uniquely determined continuous convolution semigroup $(\nu_r)_{r\geq 0} \doteq \left(\nu_r^{(2)}\right)_{r\geq 0}$ such that for all $r,s,t \geq 0$, $r = s\cdot t$

$$\nu_r = \nu_{s\cdot t} = \lim_N \lim_M \left(\nu^{(2)}\left(s/M, t/N\right)\right)^{N\cdot M}$$

and conversely (cf. Proposition 5 )

$$\nu^{(2)}(s,t) = \lim_N \lim_M \mathop{\bigstar}_{k=0}^{[Nt]-1} \mathop{\bigstar}_{j=0}^{[Ms]-1} T_{\frac{k}{N}+\frac{j}{M}}\left(\nu_{\frac{1}{N}\cdot\frac{1}{M}}\right).$$

By assumption, $\mathbb{T}$ is contracting. Hence $\nu^{(2)}(s,t) \overset{t\to\infty}{\longrightarrow} \nu^{(1)}(s)$, $\forall s \geq 0$, furthermore, $\nu^{(1)}(s) \overset{s\to\infty}{\longrightarrow} \mu$ and thus $\lim_{s\to\infty}\lim_{t\to\infty} \nu^{(2)}(s,t) = \mu$.

Note that in view of the 2-M-semigroup property this yields

$$\nu^{(2)}(M\cdot s, N\cdot t) = \mathop{\bigstar}_{k=0}^{[Nt]-1} \mathop{\bigstar}_{j=0}^{[Ms]-1} T_{kt+js}\left(\nu^{(2)}(s,t)\right) \overset{M,N\to\infty}{\longrightarrow} \mu.$$

These convolution products converge iff $\nu^{(2)}(s,t)$ has finite $\log_+^2(\cdot)$–moments, i.e., iff $\int_{\mathbb{G}} \left(\log_+(\|x\|)\right)^2 d\nu^{(2)}(s,t)(x) < \infty$. (For vector spaces see e.g., [4, 12, 21], for groups see [17]).

**Claim:** $\int_{\mathbb{G}} \left(\log_{+}(||x||)\right)^2 d\nu^{(2)}(s,t)(x)$ is finite iff the $2^{nd}$-background driving Lévy process shares this property, i.e., iff $\int_{\mathbb{G}} \left(\log_{+}(||x||)\right)^2 d\nu_r^{(2)}(x)$ is finite for $r > 0$.

We sketch a proof in complete analogy to [5, 10] (for the case $m = 1$):

⟦ Let $\varphi : \mathbb{G} \to \mathbb{R}_+$ be a continuous sub-multiplicative function equivalent with $\log_+^2(||\cdot||)$ and let $\psi : \Gamma \to \mathbb{R}_+$ be an analogous function on the space-time group.

For fixed $t > 0$ let $\left(\lambda_s^{(t)} := \nu^{(2)}(s,t) \otimes \varepsilon_s\right)_{s\geq 0}$ be the space-time continuous convolution semigroup. Since $\lambda_s^{(t)} \in M_*^1(\Gamma)$, $\int_{\mathbb{G}} \varphi d\nu(s,t) < \infty$ iff $\int_{\Gamma} \psi d\lambda_s^{(t)} < \infty$. This is the case iff the Lévy measure $\gamma^{(t)}$ of $\left(\lambda_s^{(t)}\right)_{s\geq 0}$ fulfills $\int_{\complement U} \psi d\gamma^{(t)} < \infty$ for all neighbourhoods $U$ of the unit in $\Gamma$.

Since $\lambda_t^{(s)} \in M_*^1(\Gamma)$ it follows easily that this is again equivalent with $\int_{\complement V} \varphi d\eta^{(t)} < \infty$ for all neighbourhoods $V$ of the unit in $\mathbb{G}$, where $\eta^{(t)}$ denotes the Lévy measure of $B(t) := \frac{\partial^+}{\partial s} \nu^{(2)}(s,t)\,|_{s=0}$.

But $B(t)$ is the generating functional of the continuous convolution semigroup $\left(\sigma_s^{(t)}\right)_{s\geq 0}$. Hence the above integrals are finite iff $\int_{\mathbb{G}} \varphi d\sigma_s^{(t)} < \infty$, $s > 0$, hence iff $\int_{\mathbb{G}} \left(\log_{+}(||x||)\right)^2 d\sigma_s^{(t)} < \infty$ for all $t > 0$.

Repeating these arguments and replacing $t \mapsto \nu(s,t)$ by $t \mapsto \sigma_s^{(t)}$ we obtain finally:

$\int \left(\log_{+}(||x||)\right)^2 d\nu^{(2)}(s,t) < \infty$ iff $\int \left(\log_{+}(||x||)\right)^2 d\nu_{s\cdot t}^{(2)} < \infty$ $(\forall s,t > 0)$, as asserted. ⟧

Theorem 1 is proved. □

## 4.1 *Concluding Remark*

At a first glance the foregoing construction appears asymmetric: The Lie Trotter formula is applied first to $s$ then to $t$, consequently the $2^{nd}$ background process was constructed via the family of continuous convolution semigroups $\sigma_s^{(t)}$. Switching to the space-time semigroups we obtained differentiability of $\left(\nu^{(2)}(s,t)\right)_{s,t\in\mathbb{R}_+}$ (evaluated at test functions). In particular, for fixed $t \geq 0$ and for $s = 0$, $B(t) := \frac{\partial^+}{\partial s} \nu^{(2)}(s,t)\,|_{s=0}$, $t \geq 0$, is the generating functional of the continuous convolution semigroup $\left(\sigma_s^{(t)}\right)_{s\geq 0}$, i.e., $\frac{\partial^+}{\partial s} \sigma_s^{(t)}\,|_{s=0} = B(t)$. Adopting the notation $\left(\sigma_s^{(t)} =: \mathrm{Exp}(s \cdot B(t))\right)_{s\geq 0}$, for $t \geq 0$, this yields $\frac{\partial^+}{\partial t} \sigma_s^{(t)}|_{t=0} = \frac{\partial^+}{\partial t} \mathrm{Exp}(sB(t))|_{t=0} =: s \cdot C$ where $C$ is the generating functional of the background driving process $(\nu_r)_{r\geq 0}$, i.e., $\nu_r = \mathrm{Exp}(r \cdot C)$. In other words, – explaining the afore mentioned asymmetry – we obtain

$$C = \frac{\partial^+}{\partial t} \mathrm{Exp}\left(\frac{\partial^+}{\partial s} \nu^{(2)}(s,t)\Big|_{s=0}\right)\Big|_{t=0}.$$

Interchanging the role of $s$ and $t$, $\left(\sigma_s^{(t)}\right)_{s,t\geq 0}$ and $\left(\rho_t^{(s)}\right)_{t,s\geq 0}$ and $M$ and $N$, we obtain analogously $\frac{\partial^+}{\partial s}\rho_t^{(s)}|_{s=0} = t\cdot\overline{C}$, the generating functional of a Lévy process $(\overline{\nu}_r)_{r\geq 0}$, and moreover

$$\nu^{(2)}(s,t) = \lim_M \lim_N \mathop{\bigstar}_{j=0}^{[Ms]-1} \mathop{\bigstar}_{k=0}^{[Nt]-1} T_{\frac{k}{N}+\frac{j}{M}}\left(\overline{\nu}_{\frac{1}{N}\cdot\frac{1}{M}}\right).$$

## References


[1] Becker-Kern, P: Stable and semistable hemigroups: domains of attraction and selfdecomposability. J. Theor. Prob. **16**, 573–598 (2001)

[2] Becker-Kern, P.: Random integral representation of operator-semi-self-similar processes with independent increments. Stoch. Proc. Appl. **109**, 327–344 (2004)

[3] Bingham, N.H.: Lévy processes and selfdecomposability in finance. Prob. Math. Stat. **26**, 367–378 (2006)

[4] Bunge, J.B.: Nested classes of $C$-decomposable laws. Ann. Prob. **25**, 215–229 (1997)

[5] Hazod, W., Siebert, E.: Stable Probability Measures on Euclidean Spaces and on Locally Compact Groups. Structural Properties and Limit Theorems. Mathematics and its Applications vol. **531**. Kluwer, Dordrecht (2001)

[6] Hazod, W.: Probability on matrix-cone hypergroups: Limit theorems and structural properties (2008) (to be published)

[7] Hazod, W.: On some convolution semi-and hemigroups appearing as limit distributions of normalized products of group-valued random variables. In: Heyer, H., Marion, J. (eds.) Analysis on Infinite-Dimensional Lie Groups, pp. 104–121. Marseille (1997), World Scientific, Singapore (1998)

[8] Hazod, W.: Stable hemigroups and mixing of generating functionals. J. Math. Sci. **111**, 3830–3840 (2002)

[9] Hazod, W.: On Mehler semigroups, stable hemigroups and selfdecomposability. In: Heyer, H., Hirai, T., Kawazoe, T., Saito, K. (eds.) Infinite Dimensional Harmonic Analysis III. Proceedings 2003, pp. 83–98. World Scientific, Singapore (2005)

[10] Hazod, W., Scheffler, H-P.: Strongly $\tau$-decomposable and selfdecomposable laws on simply connected nilpotent Lie groups. Mh. Math. **128**, 269–282 (1999)

[11] Heyer, H.: Probability Measures on Locally Compact Groups. Springer, Berlin, Heidelberg, New York (1977)

[12] Jurek, Z.I., Mason, D.: Operator Limit Distributions in Probability Theory. J. Wiley, New York (1993)

[13] Jurek, Z.I., Vervaat, W.: An integral representation for self-decomposable Banach space valued random variables. Z. Wahrsch. **62**, 247–262 (1983)

[14] Jurek, Z.I.: Selfdecomposability: an exception or a rule? Annales Univ. M. Curie-Sklodowska, Lublin. Sectio A, 174–182 (1997)

[15] Jurek, Z.I.: An integral representation of operator-selfdecomposable random variables. Bull. Acad. Polon. Sci. **30**, 385–393 (1982)

[16] Jurek, Z.I.: The classes $L_m(Q)$ of probability measures on Banach spaces. Bull. Acad. Polon. Sci. **31**, 51–62 (1983)

[17] Kosfeld, K.: Dissertation, Technische Universität Dortmund (forthcoming)

[18] Kunita H.: Analyticity and injectivity of convolution semigroups on Lie groups. J. Funct. Anal. **165**, 80–100 (1999)

[19] Kunita H.: Stochastic processes with independent increments in a Lie group and their self similar properties. In: Stochastic differential and difference equations. Prog. Syst. Control Theor. **23**, 183–201 (1997)

[20] Sato, K.: Lévy Processes and Infinitely Divisible Distributions. Cambridge University Press, Cambridge (1999)

[21] Sato, K.:Class $L$ of multivariate distributions and its subclasses. J. Multivariate Anal. **10**, 207–232 (1980)

[22] Sato, K., Yamasato, M.: Operator-selfdecomposable distributions as limit distributions of processes of Ornstein–Uhlenbeck type. Nagoya Math. J. **97**, 71–94 (1984)

[23] Sato, K., Yamasato, M.: Completely operator-selfdecomposable distributions and operator stable distributions. J. Multivariate Anal. **10**, 207–232 (1980)

[24] Shah, R.: Selfdecomposable measures on simply connected nilpotent groups. J. Theoret. Prob. **13**, 65–83 (2000)

[25] Urbanik, K.: Lévy's probability measures on Euclidean spaces. Studia Math. **44**, 119–148 (1972)

[26] Urbanik, K.: Limit laws for sequences of normed sums satisfying some stability conditions. J. Multivariate Anal. **3**, 225–237 (1973)

# Part VI
# Matrix Algebra and Matrix Computations

*Math is like love – a simple idea but it can get complicated.*

Anonymous

# Further Results on Samuelson's Inequality

Richard William Farebrother

**Abstract** In this paper we show that Samuelson's [11] inequality is essentially due to Gauss [6] whilst a more general result of the same type is due to Aitken [1, 2]. We also show that the adding-up constraint on the deviations from sample means implicit in Trenkler and Puntanen's [14] multivariate generalisation of Samuelson's Inequality can be regarded as a special case of a more general formulation involving a set of linear constraints on the deviations.

## 1 Gauss's Variant of Samuelson's Inequality

Given $n$ observations $y_1, y_2, ..., y_n$ on a single variable $Y$, we define their sample mean $\bar{y} = \frac{1}{n}\sum_{i=1}^{n} y_i$ and sample variance $s^2 = \frac{1}{n}\sum_{i=1}^{n} d_i^2$ (assumed positive) where $d_i = y_i - \bar{y}$ for $i = 1, 2, ..., n$.

In this context, a result that is variously attributed to Samuelson [11], Thompson [13] and Laguerre [8] asserts that $d_j^2 \leq (n-1)s^2$ for all $j = 1, 2, ..., n$, see Olkin [10], Jensen [7] or Trenkler and Puntanen [14].

This basic result may readily be established by applying Lagrange's method to the optimisation problem:

$$\text{Maximise} \qquad d_j^2/s^2$$

subject to

$$\sum_{i=1}^{n} d_i = 0$$

$$\sum_{i=1}^{n} d_i^2 = ns^2$$

or to:

$$\text{Maximise} \qquad d_j^2/s^2 = (\sum_{i \neq j} d_i/s)^2$$

Richard William Farebrother
11 Castle Road, Bayston Hill, Shrewsbury SY3 0NF, UK
R.W.Farebrother@Manchester.ac.uk

subject to

$$\sum_{i\neq j}(d_i/s)^2 + [\sum_{i\neq j} d_i/s]^2 = n$$

However, we prefer to offer an alternative approach and to apply it in a more general context:

$$\text{Maximise} \qquad d_j^2/s^2$$

subject to

$$\mathbf{X}'\mathbf{d} = \sum_{i=1}^{n} \mathbf{x}_i d_i = \mathbf{0}$$

$$\sum_{i=1}^{n} d_i^2 = ns^2$$

where $\mathbf{X}$ is a given $n\times q$ matrix of rank $q$.

Assuming that the optimal value of $d_j^2/s^2$ is nonzero, we define $a_i = -d_i/d_j$ and reformulate this problem as:

$$\text{Minimise} \qquad ns^2/d_j^2 = 1 + \mathbf{a}'_{[j]}\mathbf{a}_{[j]}$$

subject to

$$\mathbf{X}'_{[j]}\mathbf{a}_{[j]} = \mathbf{x}_j$$

where $\mathbf{X}_{[j]}$ represents the $n\times q$ matrix $\mathbf{X}$ with its $j$th row $\mathbf{x}'_j$ deleted and $\mathbf{a}_{[j]}$ represents the $n\times 1$ matrix $\mathbf{a}$ with its $j$th element $a_j$ deleted.

Now, this reformulation of the problem is familiar to statisticians as one defining the best (minimum variance) linear unbiased estimators of the $q$ slope parameters of the standard linear statistical model. Further, the so-called Gauss–Markov Theorem asserts that the optimal (BLUE) solution to this problem is given by setting $\mathbf{a}_{[j]} = \mathbf{X}_{[j]}(\mathbf{X}'_{[j]}\mathbf{X}_{[j]})^{-1}\mathbf{x}_j$ whence $ns^2/d_j^2 = 1+\mathbf{x}'_j(\mathbf{X}'_{[j]}\mathbf{X}_{[j]})^{-1}\mathbf{x}_j$ and thus the maximal value of $d_j^2/s^2 = n/[1+\mathbf{x}'_j(\mathbf{X}'_{[j]}\mathbf{X}_{[j]})^{-1}\mathbf{x}_j]$, so that $\alpha s^2 - d_j^2$ is nonnegative for all $\alpha \geq n/[1+\mathbf{x}'_j(\mathbf{X}'_{[j]}\mathbf{X}_{[j]})^{-1}\mathbf{x}_j] = n[1-\mathbf{x}'_j(\mathbf{X}'\mathbf{X})^{-1}\mathbf{x}_j]$.

In particular, when $\mathbf{X}$ is an $n\times 1$ column of ones, then we have to set $a_i = 1/(n-1)$ for all $i\neq j$ whence $ns^2/d_j^2 = 1+\sum_{i\neq j} a_i^2 = n/(n-1)$ and thus the maximal value of $d_j^2/s^2 = n-1$, so that $\alpha s^2 - d_j^2$ is nonnegative for all $\alpha \geq n-1$. And we deduce that Samuelson's Inequality is essentially a corollary of the Gauss–Markov Theorem first stated by Gauss in 1823, see [3, 4].

## 2 First Multivariate Result

The univariate problem outlined in Sect. 1 may readily be generalised to the case of a $n\times p$ matrix $\mathbf{D}$ of rank $p$ satisfying the conditions $\mathbf{X}'\mathbf{D} = \mathbf{0}$ for some $n\times q$ matrix $\mathbf{X}$ of rank $Q \leq p$.

Let $\mathbf{P} = [\mathbf{J}\ \mathbf{K}]$ be an $n \times n$ permutation matrix partitioned by its first $q$ columns and the remaining $n-q$ columns, then the expression $\mathbf{X}'\mathbf{D} = \mathbf{X}'\mathbf{J}\mathbf{J}'\mathbf{D} + \mathbf{X}'\mathbf{K}\mathbf{K}'\mathbf{D} = \mathbf{0}$ implies that the $q \times p$ matrix $\mathbf{J}'\mathbf{D}$ satisfies the condition $\mathbf{J}'\mathbf{D} = -\mathbf{Z}'\mathbf{K}'\mathbf{D}$ where $\mathbf{Z} = \mathbf{K}'\mathbf{X}(\mathbf{J}'\mathbf{X})^{-1}$.

In particular, if $p = q$ then we may be interested in solving the following generalisation of the problem outlined in Sect. 1:

$$\text{Maximise} \qquad [det(\mathbf{J}'\mathbf{D})]^2/det(n\mathbf{S})$$

subject to

$$\mathbf{X}'\mathbf{J}\mathbf{J}'\mathbf{D} + \mathbf{X}'\mathbf{K}\mathbf{K}'\mathbf{D} = \mathbf{0}$$

$$\mathbf{D}'\mathbf{J}\mathbf{J}'\mathbf{D} + \mathbf{D}'\mathbf{K}\mathbf{K}'\mathbf{D} = n\mathbf{S}$$

where the $p \times p$ positive definite matrix $\mathbf{S} = \frac{1}{n}\mathbf{D}'\mathbf{D}$.

Assuming that the optimal solution to this problem is strictly positive, so that $\mathbf{J}'\mathbf{D}$ is nonsingular, then we may define $\mathbf{A} = -\mathbf{K}'\mathbf{D}(\mathbf{J}'\mathbf{D})^{-1}$ and rewrite this problem as:

$$\text{Minimise} \qquad det[n(\mathbf{D}'\mathbf{J})^{-1}\mathbf{S}(\mathbf{J}'\mathbf{D})^{-1}] = det[\mathbf{I}_q + \mathbf{A}'\mathbf{A}]$$

$$\text{subject to} \qquad \mathbf{Z}'\mathbf{A} = \mathbf{I}_q$$

Now, Aitken [1, 1] has shown that $det(\mathbf{A}'\mathbf{A} + \mathbf{B}'\mathbf{B})$ is minimised subject to $\mathbf{Z}'\mathbf{A} = \mathbf{I}_q$ and $\mathbf{B} = \mathbf{I}_q$ by setting $\mathbf{A} = \mathbf{Z}(\mathbf{Z}'\mathbf{Z})^{-1}$, so that $det[n(\mathbf{D}'\mathbf{J})^{-1}\mathbf{S}(\mathbf{J}'\mathbf{D})^{-1}] \geq det[\mathbf{I}_q + (\mathbf{Z}'\mathbf{Z})^{-1}]$ whence $[det(\mathbf{J}'\mathbf{D})]^2 \leq det(n\mathbf{S})/det[\mathbf{I}_q + (\mathbf{Z}'\mathbf{Z})^{-1}]$.

In particular, if $p = q = 1$, the $n \times 1$ matrix $\mathbf{J}$ is the $j$th column of $\mathbf{I}_n$ and $\mathbf{S} = s^2$, then, as in Sect. 1, we have $d_j^2 \leq ns^2/[1 + x_j^2/\sum_{i \neq j} x_i^2]$.

## 3 Second Multivariate Result

Aitken's [1, 1] generalisation of his analysis to the trace and other spurs of lower order than the determinant are of little interest in the present context. We therefore consider a different multivariate generalisation of the problem of Sect. 1 that seeks to determine the smallest value of $\alpha$ for which the $p \times p$ matrix $\alpha\mathbf{S} - \mathbf{d}_j\mathbf{d}_j'$ is nonnegative definite when the $n \times p$ matrix $\mathbf{D}$ satisfies the conditions $\mathbf{X}'\mathbf{D} = \mathbf{0}$ where $\mathbf{X}$ is an $n \times q$ matrix of rank $q, \mathbf{d}_j$ represents the $j$th row of $\mathbf{D}$, and $\mathbf{S} = \frac{1}{n}\mathbf{D}'\mathbf{D}$.

Again supposing that $\mathbf{J}'\mathbf{X}$ is nonsingular, we find that the $q \times p$ matrix $\mathbf{J}'\mathbf{D}$ satisfies $\mathbf{J}'\mathbf{D} = -\mathbf{Z}'\mathbf{K}'\mathbf{D}$ where $\mathbf{Z} = \mathbf{K}'\mathbf{X}(\mathbf{J}'\mathbf{X})^{-1}$. In this context, our problem is to determine the smallest value of $\alpha = n\gamma$ for which the $p \times p$ matrix

$$\mathbf{T}(\gamma) = n\gamma\mathbf{S} - \mathbf{D}'\mathbf{J}\mathbf{J}'\mathbf{D}$$

$$= \gamma\mathbf{D}'\mathbf{K}\mathbf{K}'\mathbf{D} - (1-\gamma)\mathbf{D}'\mathbf{J}\mathbf{J}'\mathbf{D}$$

$$= \mathbf{D}'\mathbf{K}[\gamma\mathbf{I}_{n-q} - (1-\gamma)\mathbf{Z}\mathbf{Z}']K'D$$

is nonnegative definite. Now, this condition is clearly satisfied if (but not only if) $\gamma \geq (1-\gamma)\lambda_*$ or $\alpha/n = \gamma \geq \mu_*$ where $\lambda_*$ is the largest eigenvalue of the $(n-q)\times(n-q)$ matrix $\mathbf{ZZ}'$, $\mathbf{v}_*$ is the corresponding $(n-q)\times 1$ eigenvector and $\mu_* = \lambda_*/(1+\lambda_*)$ is the largest eigenvalue of the $q\times q$ matrix $[\mathbf{I}_q + (\mathbf{Z}'\mathbf{Z})^{-1}]^{-1} = \mathbf{J}'\mathbf{X}(\mathbf{X}'\mathbf{X})^{-1}X'J$.

Conversely, if $\mathbf{T}(\gamma)$ is nonnegative definite for all choices of $\mathbf{D}$ including those for which the first column of $\mathbf{K}'\mathbf{D}$ is a multiple of $\mathbf{v}_*$, then the upper left element of $\mathbf{T}(\gamma)$ must be nonnegative and $\alpha$ must satisfy $\alpha/n \geq \mu_*$.

## 4 Main Results

Summarising the results established in Sect. 3, we have:

**Theorem 1.** *Let* $\mathbf{X}$ *be an* $n\times q$ *matrix of rank* $q$*, and let* $\mathbf{Y}$ *be an* $n\times p$ *matrix such that* $\mathbf{D} = \mathbf{MY}$ *have rank* $p$ *and* $\mathbf{S} = \frac{1}{n}\mathbf{D}'\mathbf{D}$ *is positive definite, where* $\mathbf{M} = \mathbf{I}_n - \mathbf{X}(\mathbf{X}'\mathbf{X})^{-1}\mathbf{X}'$*. Further, let* $\mathbf{P} = [\mathbf{j}\ \mathbf{K}]$ *be an* $n\times n$ *permutation matrix partitioned by its first* $q$ *columns and the remaining* $n-q$ *columns in such a way that* $\mathbf{J}'\mathbf{X}$ *is nonsingular, then the* $p\times p$ *matrix* $\alpha\mathbf{S} - \mathbf{D}'\mathbf{JJ}'\mathbf{D}$ *is nonnegative definite for all choices of* $\mathbf{Y}$ *if and only if* $\alpha/n \geq 1 - 1/(1+\lambda_*)$ *where* $\lambda_*$ *is the largest eigenvalue of the* $(n-q)\times(n-q)$ *matrix* $\mathbf{ZZ}' = \mathbf{K}'\mathbf{X}(\mathbf{X}'\mathbf{JJ}'\mathbf{X})^{-1}\mathbf{X}'\mathbf{K}$*. Moreover, this condition holds as an equality if the eigenvector of* $\mathbf{ZZ}'$ *corresponding to its largest eigenvalue* $\lambda_*$ *may be expressed as a linear combination of the columns of* $\mathbf{K}'\mathbf{D}$*.*

Further, on specialising this general result to the case $q = 1$, we have:

**Theorem 2.** *Let* $\mathbf{x}$ *be an* $n\times 1$ *nonzero matrix, and let* $\mathbf{Y}$ *be an* $n\times p$ *matrix such that* $\mathbf{D} = \mathbf{MY}$ *have rank* $p$ *and* $\mathbf{S} = \frac{1}{n}\mathbf{D}'\mathbf{D}$ *is positive definite, where* $\mathbf{M} = \mathbf{I}_n - \mathbf{x}(\mathbf{x}'\mathbf{x})^{-1}\mathbf{x}'$*. Further, suppose that the* $j$*th element of* $\mathbf{x}$ *is nonzero and that* $\mathbf{D}$ *has* $j$*th row* $\mathbf{d}'_j$*, then the* $p\times p$ *matrix* $\alpha\mathbf{S} - \mathbf{d}_j\mathbf{d}'_j$ *is nonnegative definite for all choices of* $\mathbf{Y}$ *if and only if* $\alpha/n \geq 1 - x_j^2/(\Sigma_{i=1}^n x_i^2)$*. Moreover, this condition holds as an equality if the elements of* $\mathbf{x}$ *other than the* $j$*th may be expressed as a linear combination of the columns of* $\mathbf{D}$*.*

*Remark 1.* Suppose that $x_j \neq 0$ and there is a $p\times 1$ matrix $\mathbf{c}$ such that $\mathbf{y}'_i\mathbf{c} = x_i$ for all $i \neq j$ and $\mathbf{y}'_j\mathbf{c} \neq x_j$ (or $\mathbf{D}$ will have less than full column rank), then $\mathbf{d}'_i\mathbf{c} = gx_i$ for all $i \neq j$ where $g = x_j(x_j - \mathbf{y}'_j\mathbf{c})/(\Sigma_{h=1}^n x_h^2) \neq 0$ as required by the final part of Theorem 2.

*Remark 2.* Trenkler and Puntanen's [14] result may be obtained from our Theorem 2 by setting $\mathbf{x}$ equal to an $n\times 1$ column of ones. That due to Samuelson [11] then follows by setting $p = 1$.

*Remark 3.* If the $n \times p$ matrix $\mathbf{D}$ satisfies $\mathbf{X}'\mathbf{d} = \mathbf{0}$ where $\mathbf{X}$ is an $n \times q$ nonzero matrix, but we only require a bound on the value of $\alpha$ such that $\alpha\mathbf{S} - \mathbf{d}_j\mathbf{d}'_j$ is nonnegative definite for all $\mathbf{Y}$, then we should apply Theorem 2 for all linear combinations $\mathbf{x} = \mathbf{Xc}$ of the columns of the $n \times q$ matrix $\mathbf{X}$ satisfying $\mathbf{x}'_j\mathbf{c} = 1$.

In this way we obtain $\alpha\mathbf{S} - \mathbf{d}_j\mathbf{d}'_j$ is nonnegative definite for all $\mathbf{Y}$ if and only if $\alpha/n \geq 1 - 1/(max_{\mathbf{x}'_j\mathbf{c}=1}\mathbf{c}'\mathbf{X}'\mathbf{Xc})$.

## 5 Historical Remark

Let $\mathbf{x}$, $\mathbf{y}$ and $\mathbf{w}$ be $n \times 1$ nonzero matrices and let $\mathbf{X}_* = [\mathbf{x}\ \mathbf{w}]$ then we may define $\mathbf{M} = \mathbf{I}_n - \mathbf{x}(\mathbf{x}'\mathbf{x})^{-1}\mathbf{x}'$ and $\mathbf{M}_* = \mathbf{I}_n - \mathbf{X}_*(\mathbf{X}'_*\mathbf{X}_*)^{-1}\mathbf{X}'_*$ and deduce that:

$$\mathbf{y}'\mathbf{M}_*\mathbf{y} = \mathbf{y}'\mathbf{My} - \mathbf{y}'\mathbf{Mw}(\mathbf{w}'\mathbf{Mw})^{-1}\mathbf{w}'\mathbf{My}.$$

Using a variant of Gauss's notation, Farebrother ([4], pp.101 and 195–196) has shown that this fundamental result appears in Gauss [5] as:

$$[ll,2] = [ll,1] - [lb,1]^2/[bb,1]$$

and in Laplace [9] as:

$$[p^{(2)}p^{(2)}] = [p^{(1)}p^{(1)}] - [p^{(1)}b^{(1)}]^2/[b^{(1)}b^{(1)}]$$

and again in Laplace [9] as:

$$[p^{(2)}p^{(2)}] = \frac{[pp][qq][rr] - [pp][qr]^2 - [qq][pr]^2 - [rr][pq]^2 + 2[pq][pr][qr]}{[qq][rr] - [qr]^2}.$$

The result now known as Samuelson's Inequality follows immediately from these expressions by setting $\mathbf{a} = \mathbf{r} = \mathbf{x}$ equal to an $n \times 1$ column of ones, $\mathbf{b} = \mathbf{q} = \mathbf{w}$ equal to the $j$th column of $\mathbf{I}_n$ and $\mathbf{l} = \mathbf{p} = \mathbf{y}$ before noting that $[ll,2] = [p^{(2)}p^{(2)}]$ is non-negative. But this hypothetical connection with the work of Gauss and Laplace falls to the ground as neither author seems to have contemplated setting $\mathbf{b} = \mathbf{q}$ equal to the $j$th column of $\mathbf{I}_n$.

On the other hand, augmenting $\mathbf{x}$ by the $j$th column of $\mathbf{I}_n$ has exactly the same effect on the sum of squared deviations function as deleting the $j$th row from $[\mathbf{x}\ \mathbf{y}]$. So that Gauss [6] would have come close to a variant of Samuelson's Inequality if he had also explicitly stated the relation between the adjusted and unadjusted sum of squared deviations when a single row is deleted from the data set in parallel with his earlier statement of the corresponding relationship when a single row is added, see Farebrother ([4], pp. 146–147).

## References

[1] Aitken, A.C.: On least squares and linear combination of observations. Proc. Roy. Soc. Edinb. A **55**, 42–47 (1935)
[2] Aitken, A.C.: On the estimation of many statistical parameters. Proc. Roy. Soc. Edinb. A **62**, 33–37 (1948)
[3] Farebrother, R.W.: Some early statistical contributions to the theory and practice of linear algebra. Linear Algebra Appl. **237**, 205–224 (1996)
[4] Farebrother, R.W.: Fitting Linear Relationships: A History of the Calculus of Observations 1750–1900. Springer, New York (1999)
[5] Gauss, C.F.: Disquisitio de elementis ellipticis Palladis ex oppositionibus annorum 1803, 1804, 1805, 1807, 1808, 1809. In: Commentationes Societatis Regiae Scientiarum Gottingensis Recentiores **1**, pp. 2–26 (1811). Reprinted in his *Werke*, vol. 6, pp. 1–24, Göttingen (1874)
[6] Gauss, C.F.: Theoria combinationis observationum erroribus minimis obnoxiae: pars posterior. In: Commentationes Societatis Regiae Scientiarum Gottingensis Recentiores **5**, pp. 63–90 (1823). Reprinted in his *Werke*, vol. 4, pp. 57–93, Göttingen (1880). Reprinted with an English translation by Stewart (1995, pp. 50–97). [For details of translations into other languages, see Farebrother (1999).]
[7] Jensen, S.T.: The Laguerre-Samuelson inequality with extensions and applications in statistics and matrix theory. M.Sc. Thesis, McGill University, Montréal, Canada (1999)
[8] Laguerre, E.N.: Sur une méthode pour obtenir par approximation les racines d'une équation algébrique qui a toutes ses racines réelles. Nouvelles Annales de Mathématiques (Paris) 2$^e$ Série **19**, 161–171 and 193–202 (1880)
[9] Laplace, P.S.: Premier Supplément to *Théorie Analytique des Probabilités* (1816), Mme. Courcier, Paris, 1820. Reprinted in his *Oeuvres Complètes*, vol. 7, Imprimerie Royale, Paris (1847) and Gauthier-Villars et Fils, Paris (1886)
[10] Olkin, I.: A matrix formulation on how deviant an observation can be. Am. Stat. **46**, 205–209 (1992)
[11] Samuelson, P.A.: How deviant can you be? J. Am. Stat. Assoc. **63**, 1522–1525 (1968)
[12] Stewart, G.W.: Theory of the Combination of Observations Least Subject to Errors. SIAM Publ., Philadelphia, Pennsylvania (1995)
[13] Thompson, W.R.: On a criterion for the rejection of observations and the distribution of the ratio of deviation to sample standard deviation. Ann. Math. Stat. **6**, 214–219 (1935)
[14] Trenkler, G., Puntanen, S.J.: A multivariate version of Samuelson's Inequality. Linear Algebra Appl. **410**, 143–149 (2006)

# Revisitation of Generalized and Hypergeneralized Projectors

Oskar Maria Baksalary

**Abstract** The notions of generalized and hypergeneralized projectors, introduced by Groß and Trenkler [Generalized and hypergeneralized projectors, Linear Algebra Appl. 264 (1997) 463–474], attracted recently considerable attention. The list of publications devoted to them comprises now over ten positions, and the present paper briefly discusses some of the results available in the literature. Furthermore, several new characteristics of generalized and hypergeneralized projectors are established with the use of Corollary 6 in Hartwig and Spindelböck [Matrices for which $A^*$ and $A^\dagger$ commute. Linear Multilinear Algebra 14 (1984) 241–256].

## 1 Introduction

Let $\mathbb{C}_{m,n}$ be the set of $m \times n$ complex matrices. The symbols $\mathbf{A}^*$, $\mathscr{R}(\mathbf{A})$, and $\text{rk}(\mathbf{A})$ will denote the conjugate transpose, range (column space), and rank, respectively, of $\mathbf{A} \in \mathbb{C}_{m,n}$. Additionally, $\mathbf{I}_n$ will mean the identity matrix of order $n$.

Two matrix generalized inverses will be of interest in the present paper. Namely, $\mathbf{A}^\dagger \in \mathbb{C}_{n,m}$ will stand for the Moore–Penrose inverse of $\mathbf{A} \in \mathbb{C}_{m,n}$, i.e., for the unique matrix satisfying the equations

$$\mathbf{A}\mathbf{A}^\dagger\mathbf{A} = \mathbf{A},\ \mathbf{A}^\dagger\mathbf{A}\mathbf{A}^\dagger = \mathbf{A}^\dagger,\ \mathbf{A}\mathbf{A}^\dagger = (\mathbf{A}\mathbf{A}^\dagger)^*,\ \mathbf{A}^\dagger\mathbf{A} = (\mathbf{A}^\dagger\mathbf{A})^*,$$

and $\mathbf{A}^\# \in \mathbb{C}_{n,n}$ will be the group inverse of $\mathbf{A} \in \mathbb{C}_{n,n}$, i.e., the unique matrix satisfying the equations

$$\mathbf{A}\mathbf{A}^\#\mathbf{A} = \mathbf{A},\ \mathbf{A}^\#\mathbf{A}\mathbf{A}^\# = \mathbf{A}^\#,\ \mathbf{A}\mathbf{A}^\# = \mathbf{A}^\#\mathbf{A}.$$

Oskar Maria Baksalary
Faculty of Physics, Adam Mickiewicz University, ul. Umultowska 85, PL 61-614 Poznań, Poland
baxx@amu.edu.pl

Recall that the existence of the group inverse is restricted to the matrices of index one, i.e., satisfying $\text{rk}(\mathbf{A}^2) = \text{rk}(\mathbf{A})$. Henceforth, whenever the group inverse occurs, it is assumed to exist.

Several known classes of matrices will be recalled in what follows. The symbols $\mathbb{C}_{m,n}^{\mathsf{PI}}$ and $\mathbb{C}_{m,n}^{\mathsf{CA}}$ will denote the subsets of $\mathbb{C}_{m,n}$ comprising partial isometries and contractions, i.e.,

$$\mathbb{C}_{m,n}^{\mathsf{PI}} = \{\mathbf{A} \in \mathbb{C}_{m,n} \colon \ \mathbf{A}\mathbf{A}^*\mathbf{A} = \mathbf{A}\} = \{\mathbf{A} \in \mathbb{C}_{m,n} \colon \ \mathbf{A}^\dagger = \mathbf{A}^*\}, \tag{1}$$

$$\mathbb{C}_{m,n}^{\mathsf{CA}} = \{\mathbf{A} \in \mathbb{C}_{m,n} \colon \ \|\mathbf{A}\mathbf{x}\| \leqslant \|\mathbf{x}\| \text{ for all } \mathbf{x} \in \mathbb{C}_{n,1}\}, \tag{2}$$

where $\|.\|$ is a vector norm. Moreover, $\mathbb{C}_n^{\mathsf{P}}$, $\mathbb{C}_n^{\mathsf{OP}}$, $\mathbb{C}_n^{\mathsf{QP}}$, $\mathbb{C}_n^{\mathsf{N}}$, $\mathbb{C}_n^{\mathsf{SD}}$, $\mathbb{C}_n^{\mathsf{EP}}$, and $\mathbb{C}_n^{\mathsf{WEP}}$ will stand for the sets consisting of oblique projectors (idempotent matrices), orthogonal projectors (Hermitian idempotent matrices), and quadripotent, normal, star-dagger, EP (range-Hermitian), and weak-EP matrices, respectively, i.e.,

$$\mathbb{C}_n^{\mathsf{P}} = \{\mathbf{A} \in \mathbb{C}_{n,n} \colon \ \mathbf{A}^2 = \mathbf{A}\}, \tag{3}$$

$$\mathbb{C}_n^{\mathsf{OP}} = \{\mathbf{A} \in \mathbb{C}_{n,n} \colon \ \mathbf{A}^2 = \mathbf{A} = \mathbf{A}^*\} = \{\mathbf{A} \in \mathbb{C}_{n,n} \colon \ \mathbf{A}^2 = \mathbf{A} = \mathbf{A}^\dagger\}, \tag{4}$$

$$\mathbb{C}_n^{\mathsf{QP}} = \{\mathbf{A} \in \mathbb{C}_{n,n} \colon \ \mathbf{A}^4 = \mathbf{A}\}, \tag{5}$$

$$\mathbb{C}_n^{\mathsf{N}} = \{\mathbf{A} \in \mathbb{C}_{n,n} \colon \ \mathbf{A}\mathbf{A}^* = \mathbf{A}^*\mathbf{A}\}, \tag{6}$$

$$\mathbb{C}_n^{\mathsf{SD}} = \{\mathbf{A} \in \mathbb{C}_{n,n} \colon \ \mathbf{A}^*\mathbf{A}^\dagger = \mathbf{A}^\dagger\mathbf{A}^*\}, \tag{7}$$

$$\mathbb{C}_n^{\mathsf{EP}} = \{\mathbf{A} \in \mathbb{C}_{n,n} \colon \ \mathbf{A}\mathbf{A}^\dagger = \mathbf{A}^\dagger\mathbf{A}\} = \{\mathbf{A} \in \mathbb{C}_{n,n} \colon \ \mathscr{R}(\mathbf{A}) = \mathscr{R}(\mathbf{A}^*)\}, \tag{8}$$

$$\mathbb{C}_n^{\mathsf{WEP}} = \{\mathbf{A} \in \mathbb{C}_{n,n} \colon \ \mathbf{A}\mathbf{A}^\dagger\mathbf{A}^\dagger\mathbf{A} = \mathbf{A}^\dagger\mathbf{A}\mathbf{A}\mathbf{A}^\dagger\}. \tag{9}$$

The concepts of generalized and hypergeneralized projectors were introduced by Groß and Trenkler [13, pp. 465, 466]. They may be viewed as weakened versions of the two characterizations of orthogonal projectors in (4), obtained by deleting in each of them the idempotency requirement. Their explicit specifications are restated in the following.

**Definition.** A matrix $\mathbf{A} \in \mathbb{C}_{n,n}$ is called:

(a) Generalized projector whenever $\mathbf{A}^2 = \mathbf{A}^*$,
(b) Hypergeneralized projector whenever $\mathbf{A}^2 = \mathbf{A}^\dagger$.

The two sets of matrices specified in Definition will henceforth be denoted by $\mathbb{C}_n^{\mathsf{GP}}$ and $\mathbb{C}_n^{\mathsf{HGP}}$, respectively.

According to our knowledge, 12 papers (including two in print) and one problem (i.e., [8]) were devoted so far to generalized and/or hypergeneralized projectors (additionally, problem [9] was submitted for publication). In the cornerstone paper [13], Groß and Trenkler established several characterizations of the sets $\mathbb{C}_n^{\mathsf{GP}}$ and $\mathbb{C}_n^{\mathsf{HGP}}$, observing, for instance, that

$$\mathbb{C}_n^{\mathsf{GP}} = \mathbb{C}_n^{\mathsf{QP}} \cap \mathbb{C}_n^{\mathsf{N}} \cap \mathbb{C}_{n,n}^{\mathsf{PI}} = \mathbb{C}_n^{\mathsf{QP}} \cap \mathbb{C}_n^{\mathsf{N}} = \mathbb{C}_n^{\mathsf{HGP}} \cap \mathbb{C}_{n,n}^{\mathsf{PI}} \tag{10}$$

and

$$\mathbb{C}_n^{\mathsf{HGP}} = \mathbb{C}_n^{\mathsf{QP}} \cap \mathbb{C}_n^{\mathsf{EP}}; \tag{11}$$

see Theorems 1, 2, and Corollary in [13]. Characterization (10) was later supplemented, on the one hand, by

$$\mathbb{C}_n^{\mathsf{GP}} = \mathbb{C}_n^{\mathsf{QP}} \cap \mathbb{C}_{n,n}^{\mathsf{PI}} = \mathbb{C}_n^{\mathsf{QP}} \cap \mathbb{C}_n^{\mathsf{EP}} \cap \mathbb{C}_{n,n}^{\mathsf{PI}},$$

with the first part given in [5, Theorem] and the second in [3, p. 301], and, on the other hand, by

$$\mathbb{C}_n^{\mathsf{GP}} = \mathbb{C}_n^{\mathsf{SD}} \cap \mathbb{C}_n^{\mathsf{QP}} \cap \mathbb{C}_n^{\mathsf{WEP}} = \mathbb{C}_{n,n}^{\mathsf{CA}} \cap \mathbb{C}_n^{\mathsf{QP}} \cap \mathbb{C}_n^{\mathsf{WEP}},$$

provided in [4, Theorems 3 and 4]. Similarly, characterization (11) was in [3, Theorem 3] supplemented by

$$\mathbb{C}_n^{\mathsf{HGP}} = \mathbb{C}_n^{\mathsf{QP}} \cap \mathbb{C}_n^{\mathsf{WEP}}.$$

Further relevant observations pointed out in [13, Theorems 1 and 2] were that the nonzero eigenvalues of any generalized and hypergeneralized projector are cubic roots of unity, the fact being a consequence of the quadripotency property. This result was later explored by Stewart [17], who characterized classes $\mathbb{C}_n^{\mathsf{GP}}$ and $\mathbb{C}_n^{\mathsf{HGP}}$ by their spectral decompositions. Similar considerations to the ones in [17] were, with respect to generalized projectors, carried out by Du and Li [11] in the more general settings of the infinite dimensional Hilbert space.

Other inspiring considerations, involving pairs of either generalized or hypergeneralized projectors, were given in Sects. 3 and 4 in [13]. In particular, the observations originating from Theorems 5 and 6 therein, respectively, that for $\mathbf{G}_1, \mathbf{G}_2 \in \mathbb{C}_n^{\mathsf{GP}}$,

$$\mathbf{G}_1\mathbf{G}_2 = \mathbf{0} = \mathbf{G}_2\mathbf{G}_1 \;\Leftrightarrow\; \mathbf{G}_1 + \mathbf{G}_2 \in \mathbb{C}_n^{\mathsf{GP}},$$
$$\mathbf{G}_1\mathbf{G}_2 = \mathbf{G}_2^* = \mathbf{G}_2\mathbf{G}_1 \;\Leftrightarrow\; \mathbf{G}_1 - \mathbf{G}_2 \in \mathbb{C}_n^{\mathsf{GP}},$$

were followed by the complete solution to the problem of when a linear combination of two generalized projectors is also a generalized projector established in [1]. Furthermore, Remarks on pp. 271, 272 in [13], respectively, according to which for $\mathbf{H}_1, \mathbf{H}_2 \in \mathbb{C}_n^{\mathsf{HGP}}$,

$$\mathbf{H}_1\mathbf{H}_2 = \mathbf{0} = \mathbf{H}_2\mathbf{H}_1 \;\Rightarrow\; \mathbf{H}_1 + \mathbf{H}_2 \in \mathbb{C}_n^{\mathsf{HGP}},$$

$$\mathbf{H}_2^*\mathbf{H}_2 = \mathbf{H}_2^*\mathbf{H}_1,\; \mathbf{H}_2\mathbf{H}_2^* = \mathbf{H}_1\mathbf{H}_2^* \;\Rightarrow\; \mathbf{H}_1 - \mathbf{H}_2 \in \mathbb{C}_n^{\mathsf{HGP}},$$

inspired considerations in [2] and [6] leading to a still partial answer to the question of when a linear combination of two hypergeneralized projectors also belongs to the set $\mathbb{C}_n^{\mathsf{HGP}}$.

An interesting direction of research – clearly inspired by [13] – was proposed by Benítez and Thome [10], who considered the set of so called $k$-generalized projectors composed of matrices $\mathbf{K} \in \mathbb{C}_{n,n}$ satisfying $\mathbf{K}^k = \mathbf{K}^*$ for given integer $k > 1$. Several characterizations of the set were established in [10], including the ones which refer to the question of when a linear combination of two commuting $k$-generalized projectors is also a $k$-generalized projector. The notion of a $k$-generalized projector was later considered in [12] and [15] in the settings of the infinite dimensional Hilbert spaces.

In the recent paper [4], several results dealing with generalized and hypergeneralized projectors were derived by utilizing a useful representation of square matrices established by Hartwig and Spindelböck in Corollary 6 in [14]. This result is recalled below.

**Lemma 1.** *Let* $\mathbf{A} \in \mathbb{C}_{n,n}$ *be of rank r. Then there exists unitary* $\mathbf{U} \in \mathbb{C}_{n,n}$ *such that*

$$\mathbf{A} = \mathbf{U}\begin{pmatrix} \mathbf{\Sigma K} & \mathbf{\Sigma L} \\ \mathbf{0} & \mathbf{0} \end{pmatrix}\mathbf{U}^*, \tag{12}$$

*where* $\mathbf{\Sigma} = \mathrm{diag}(\sigma_1 \mathbf{I}_{r_1}, \ldots, \sigma_t \mathbf{I}_{r_t})$ *is the diagonal matrix of singular values of* $\mathbf{A}$, $\sigma_1 > \sigma_2 > \ldots > \sigma_t > 0$, $r_1 + r_2 + \ldots + r_t = r$, *and* $\mathbf{K} \in \mathbb{C}_{r,r}$, $\mathbf{L} \in \mathbb{C}_{r,n-r}$ *satisfy*

$$\mathbf{KK}^* + \mathbf{LL}^* = \mathbf{I}_r. \tag{13}$$

From (12) it follows that

$$\mathbf{A}^\dagger = \mathbf{U}\begin{pmatrix} \mathbf{K}^*\mathbf{\Sigma}^{-1} & \mathbf{0} \\ \mathbf{L}^*\mathbf{\Sigma}^{-1} & \mathbf{0} \end{pmatrix}\mathbf{U}^*, \tag{14}$$

and, provided that $\mathbf{A}$ is of index one,

$$\mathbf{A}^\# = \mathbf{U}\begin{pmatrix} \mathbf{K}^{-1}\mathbf{\Sigma}^{-1} & \mathbf{K}^{-1}\mathbf{\Sigma}^{-1}\mathbf{K}^{-1}\mathbf{L} \\ \mathbf{0} & \mathbf{0} \end{pmatrix}\mathbf{U}^*. \tag{15}$$

The lemma below, which will be useful in the subsequent considerations, is obtained straightforwardly by combining representations (12) and (14) with specifications given in (1), (3)–(6), (8), and Definition. It is recalled here (in the reduced form) after [4, Lemma 1].

**Lemma 2.** *Let* $\mathbf{A} \in \mathbb{C}_{n,n}$ *be of rank r and have representation* (12). *Then:*

(i) $\mathbf{A} \in \mathbb{C}_{n,n}^{\mathsf{PI}}$ *if and only if* $\mathbf{\Sigma} = \mathbf{I}_r$,
(ii) $\mathbf{A} \in \mathbb{C}_n^{\mathsf{P}}$ *if and only if* $\mathbf{\Sigma K} = \mathbf{I}_r$,
(iii) $\mathbf{A} \in \mathbb{C}_n^{\mathsf{OP}}$ *if and only if* $\mathbf{L} = \mathbf{0}$, $\mathbf{\Sigma} = \mathbf{I}_r$, $\mathbf{K} = \mathbf{I}_r$,
(iv) $\mathbf{A} \in \mathbb{C}_n^{\mathsf{QP}}$ *if and only if* $(\mathbf{\Sigma K})^3 = \mathbf{I}_r$,
(v) $\mathbf{A} \in \mathbb{C}_n^{\mathsf{N}}$ *if and only if* $\mathbf{L} = \mathbf{0}$, $\mathbf{K\Sigma} = \mathbf{\Sigma K}$,

(vi) $\mathbf{A} \in \mathbb{C}_n^{\mathsf{EP}}$ *if and only if* $\mathbf{L} = \mathbf{0}$,

(vii) $\mathbf{A} \in \mathbb{C}_n^{\mathsf{GP}}$ *if and only if* $\mathbf{L} = \mathbf{0}$, $\mathbf{\Sigma} = \mathbf{I}_r$, $\mathbf{K}^3 = \mathbf{I}_r$,

(viii) $\mathbf{A} \in \mathbb{C}_n^{\mathsf{HGP}}$ *if and only if* $\mathbf{L} = \mathbf{0}$, $(\mathbf{\Sigma K})^3 = \mathbf{I}_r$.

The usefulness of the representation provided in Lemma 1 to explore various classes of matrices, such as EP, normal, and Hermitian, as well as oblique and orthogonal projectors, was demonstrated in [7] and [18], respectively, whereas its applicability to deal with generalized and hypergeneralized projectors was shown in [4]. In the next section, the considerations in [4] are extended and further characterizations of the sets $\mathbb{C}_n^{\mathsf{GP}}$ and $\mathbb{C}_n^{\mathsf{HGP}}$ with the use of Lemma 1 are obtained.

## 2 Results

We begin with proving the theorem below, which is quoted here after Baksalary and Liu [5], where it constitutes the main result. As already mentioned, equivalences (a) $\Leftrightarrow$ (b) $\Leftrightarrow$ (c) given therein were originally obtained by Groß and Trenkler [13, Theorem 1], so the crucial part established in [5] concerns the equivalence (a) $\Leftrightarrow$ (d). The proof of this result in [5] is relatively extensive and involved, which is not the case in the proof given below based on Lemma 1.

**Theorem 1.** *For any* $\mathbf{A} \in \mathbb{C}_{n,n}$, *the statements* (a)–(d) *below are mutually equivalent:*

(a) $\mathbf{A} \in \mathbb{C}_n^{\mathsf{GP}}$,

(b) $\mathbf{A} \in \mathbb{C}_n^{\mathsf{QP}} \cap \mathbb{C}_n^{\mathsf{N}} \cap \mathbb{C}_{n,n}^{\mathsf{PI}}$,

(c) $\mathbf{A} \in \mathbb{C}_n^{\mathsf{QP}} \cap \mathbb{C}_n^{\mathsf{N}}$,

(d) $\mathbf{A} \in \mathbb{C}_n^{\mathsf{QP}} \cap \mathbb{C}_{n,n}^{\mathsf{PI}}$.

*Proof.* First observe that the validity of the implication (a) $\Rightarrow$ (b) is clearly seen from Lemma 2, whereas the implication (b) $\Rightarrow$ (c) is satisfied trivially.

To show part (c) $\Rightarrow$ (d) notice that from points (iv) and (v) of Lemma 2 it follows that $\mathbf{A} \in \mathbb{C}_n^{\mathsf{QP}} \cap \mathbb{C}_n^{\mathsf{N}}$ is equivalent to $\mathbf{L} = \mathbf{0}$, $\mathbf{K\Sigma} = \mathbf{\Sigma K}$, $(\mathbf{\Sigma K})^3 = \mathbf{I}_r$. From the last two relationships we get $\mathbf{\Sigma}^3\mathbf{K}^3 = \mathbf{I}_r$. Hence, taking into account that combining $\mathbf{L} = \mathbf{0}$ with (13) implies $\mathbf{K}^* = \mathbf{K}^{-1}$, we obtain $\mathbf{\Sigma}^3 = (\mathbf{K}^{-1})^3 = (\mathbf{K}^*)^3 = \mathbf{K}^3$. Thus, it is seen that $\mathbf{K}^3 = \mathbf{I}_r$, and, in consequence, $\mathbf{\Sigma} = \mathbf{I}_r$, what completes the present step of the proof.

It remains to show that (d) $\Rightarrow$ (a), what will be accomplished by demonstrating that $\mathbf{K}^3 = \mathbf{I}_r$ implies $\mathbf{L} = \mathbf{0}$. First notice that $\mathbf{K}^3 = \mathbf{I}_r$ entails $\mathbf{K} \in \mathbb{C}_r^{\mathsf{QP}}$, i.e., in view of the nonsingularity of $\mathbf{K}$, the eigenvalues of $\mathbf{K}$ are cubic roots of unity. Secondly, in the light of the Schur's triangularization theorem [16, p. 508], there exists unitary $\mathbf{W} \in \mathbb{C}_{r,r}$ and upper-triangular matrix $\mathbf{T} \in \mathbb{C}_{r,r}$ such that $\mathbf{K} = \mathbf{WTW}^*$. Thus, $\mathbf{KK}^* = \mathbf{WTT}^*\mathbf{W}^*$, and taking traces on both sides of (13) gives

$$\mathrm{trace}(\mathbf{TT}^*) + \mathrm{trace}(\mathbf{LL}^*) = r. \tag{16}$$

Direct calculations show that $\mathrm{trace}(\mathbf{TT}^*)$ is equal to the sum of squared moduli of the (nonzero) entries of $\mathbf{T}$. Since $\mathbf{T}$ has on its diagonal $r$ eigenvalues whose moduli equal one, we obtain

$$\mathrm{trace}(\mathbf{TT}^*) = r + \sum_{i=1}^{r} \sum_{\substack{j=1 \\ i<j}}^{r} |t_{ij}|^2, \tag{17}$$

where $t_{ij}$, $i,j = 1,...,r$, are entries of $\mathbf{T}$. Combining (16) with (17) and the fact that $\mathrm{trace}(\mathbf{LL}^*) \geqslant 0$, leads to the conclusion that (16) is satisfied merely when $\mathbf{T}$ is a diagonal matrix and $\mathbf{L} = \mathbf{0}$. The proof is thus complete. □

The next theorem proves characterizations of the set $\mathbb{C}_n^{\mathsf{GP}}$.

**Theorem 2.** *Let* $\mathbf{A} \in \mathbb{C}_{n,n}$. *Then the following conditions are equivalent:*

(i) $\mathbf{A} \in \mathbb{C}_n^{\mathsf{GP}}$, (ii) $\mathbf{A} = \mathbf{A}^*\mathbf{A}^\dagger$,

(iii) $\mathbf{A} = \mathbf{A}^\dagger\mathbf{A}^*$, (iv) $\mathbf{A} = \mathbf{A}^*\mathbf{A}^\#$,

(v) $\mathbf{A} = \mathbf{A}^\#\mathbf{A}^*$, (vi) $\mathbf{A} = \mathbf{A}^*\mathbf{A}^*$.

*Proof.* We establish part (i) $\Leftrightarrow$ (ii) only, for the remaining equivalences can be shown similarly. From (12) and (14) it follows that

$$\mathbf{A}^*\mathbf{A}^\dagger = \mathbf{U}\begin{pmatrix} \mathbf{K}^*\boldsymbol{\Sigma}\mathbf{K}^*\boldsymbol{\Sigma}^{-1} & \mathbf{0} \\ \mathbf{L}^*\boldsymbol{\Sigma}\mathbf{K}^*\boldsymbol{\Sigma}^{-1} & \mathbf{0} \end{pmatrix}\mathbf{U}^*.$$

Thus, condition (ii) of the theorem is satisfied if and only if $\mathbf{L} = \mathbf{0}$, $\boldsymbol{\Sigma}\mathbf{K} = \mathbf{K}^*\boldsymbol{\Sigma}\mathbf{K}^*\boldsymbol{\Sigma}^{-1}$, or, in view of the nonsingularity of $\boldsymbol{\Sigma}$, equivalently,

$$\mathbf{L} = \mathbf{0}, \quad \boldsymbol{\Sigma}\mathbf{K}\boldsymbol{\Sigma} = \mathbf{K}^*\boldsymbol{\Sigma}\mathbf{K}^*. \tag{18}$$

Taking the conjugate transposes on both sides of the latter condition in (18) gives $\boldsymbol{\Sigma}\mathbf{K}^*\boldsymbol{\Sigma} = \mathbf{K}\boldsymbol{\Sigma}\mathbf{K}$. In view of $\mathbf{K}^* = \mathbf{K}^{-1}$, by postmultiplying $\boldsymbol{\Sigma}\mathbf{K}^*\boldsymbol{\Sigma} = \mathbf{K}\boldsymbol{\Sigma}\mathbf{K}$ by $\mathbf{K}^*$ we obtain $\boldsymbol{\Sigma}\mathbf{K}^*\boldsymbol{\Sigma}\mathbf{K}^* = \mathbf{K}\boldsymbol{\Sigma}$. Hence, by using the latter condition in (18), we arrive at $\boldsymbol{\Sigma}^2 = \mathbf{I}_r$, or, in other words, $\boldsymbol{\Sigma} = \mathbf{I}_r$. Substituting this relationship into the latter condition in (18) shows that $\mathbf{K}^3 = \mathbf{I}_r$. Thus, referring to point (vii) of Lemma 2 leads to the conclusion that equivalence (i) $\Leftrightarrow$ (ii) indeed holds. □

Equivalence (i) $\Leftrightarrow$ (vi) of Theorem 2 expresses the known fact that $\mathbf{A} \in \mathbb{C}_n^{\mathsf{GP}} \Leftrightarrow \mathbf{A}^* \in \mathbb{C}_n^{\mathsf{GP}}$, being a part of characterization (2.18) in [3]. Further observations of this kind are that $\mathbf{A} \in \mathbb{C}_n^{\mathsf{GP}}$ can be equivalently expressed as $(\mathbf{A}^\dagger)^2 = (\mathbf{A}^\dagger)^*$, obtained on account of Theorem 5 in [3], or as $(\mathbf{A}^\#)^2 = (\mathbf{A}^\#)^*$, following form Theorem 2 in [4].

The theorem below lists necessary and sufficient conditions for $\mathbf{A} \in \mathbb{C}_n^{\mathsf{HGP}}$.

**Theorem 3.** *Let* $\mathbf{A} \in \mathbb{C}_{n,n}$. *Then the following conditions are equivalent:*

(i) $\mathbf{A} \in \mathbb{C}_n^{\mathsf{HGP}}$, (ii) $\mathbf{A} = \mathbf{A}^\dagger\mathbf{A}^\#$,

(iii) $\mathbf{A} = \mathbf{A}^\#\mathbf{A}^\dagger$, (iv) $\mathbf{A} = \mathbf{A}^\dagger\mathbf{A}^\dagger$.

*Proof.* Also this time the proof will be limited to part (i) $\Leftrightarrow$ (ii) only. From (14) and (15) it follows that

$$\mathbf{A}^\dagger\mathbf{A}^\# = \mathbf{U}\begin{pmatrix}\mathbf{K}^*\mathbf{\Sigma}^{-1}\mathbf{K}^{-1}\mathbf{\Sigma}^{-1} & \mathbf{K}^*\mathbf{\Sigma}^{-1}\mathbf{K}^{-1}\mathbf{\Sigma}^{-1}\mathbf{K}^{-1}\mathbf{L} \\ \mathbf{L}^*\mathbf{\Sigma}^{-1}\mathbf{K}^{-1}\mathbf{\Sigma}^{-1} & \mathbf{L}^*\mathbf{\Sigma}^{-1}\mathbf{K}^{-1}\mathbf{\Sigma}^{-1}\mathbf{K}^{-1}\mathbf{L}\end{pmatrix}\mathbf{U}^*.$$

Hence, condition (ii) of the theorem is equivalent to the conjunction $\mathbf{L} = \mathbf{0}$, $\mathbf{\Sigma K} = \mathbf{K}^*\mathbf{\Sigma}^{-1}\mathbf{K}^{-1}\mathbf{\Sigma}^{-1}$. In view of $\mathbf{L} = \mathbf{0} \Rightarrow \mathbf{K}^* = \mathbf{K}^{-1}$, the conjunction can be expressed as $\mathbf{L} = \mathbf{0}$, $(\mathbf{\Sigma K})^3 = \mathbf{I}_r$, being necessary and sufficient conditions for $\mathbf{A} \in \mathbb{C}_n^{\mathsf{HGP}}$ given in point (viii) of Lemma 2. □

Equivalence (i) $\Leftrightarrow$ (iv) of Theorem 3 expresses the known fact that $\mathbf{A} \in \mathbb{C}_n^{\mathsf{HGP}} \Leftrightarrow \mathbf{A}^\dagger \in \mathbb{C}_n^{\mathsf{HGP}}$ established in [3, Theorem 5]. Additional observations are that $\mathbf{A} \in \mathbb{C}_n^{\mathsf{HGP}}$ if and only if either $(\mathbf{A}^*)^2 = (\mathbf{A}^*)^\dagger$, derived from (2.18) in [3], or $(\mathbf{A}^\#)^2 = (\mathbf{A}^\#)^\dagger$, being consequence of Theorem 2 in [4].

As was pointed out in [10, p. 152], if $\mathbf{A}$ is a $k$-generalized projector, i.e., satisfies $\mathbf{A}^k = \mathbf{A}^*$ for given integer $k > 1$, then $\mathbf{A}^k = \mathbf{A}^\dagger$. Extending this implication to the equivalence leads to what follows.

**Theorem 4.** *Let* $\mathbf{A} \in \mathbb{C}_{n,n}$ *and* $k \in \mathbb{N}$, $k \geqslant 1$. *Then* $\mathbf{A}^k = \mathbf{A}^\dagger$ *if and only if* $\mathbf{A} \in \mathbb{C}_n^{\mathsf{EP}}$ *and* $\mathbf{A}^{k+2} = \mathbf{A}$.

*Proof.* It is seen from (12) that for integer $k \geqslant 1$

$$\mathbf{A}^k = \mathbf{U}\begin{pmatrix}(\mathbf{\Sigma K})^k & (\mathbf{\Sigma K})^{k-1}\mathbf{\Sigma L} \\ \mathbf{0} & \mathbf{0}\end{pmatrix}\mathbf{U}^*,$$

with $(\mathbf{\Sigma K})^0 = \mathbf{I}_r$. Hence, $\mathbf{A}^k = \mathbf{A}^\dagger$ is equivalent to $\mathbf{L} = \mathbf{0}$, $(\mathbf{\Sigma K})^{k+1} = \mathbf{I}_r$, whereas $\mathbf{A}^{k+2} = \mathbf{A}$ is satisfied if and only if

$$(\mathbf{\Sigma K})^{k+2} = \mathbf{\Sigma K}, \quad (\mathbf{\Sigma K})^{k+1}\mathbf{\Sigma L} = \mathbf{\Sigma L}. \tag{19}$$

In view of (13), combining the former condition in (19) postmultiplied by $\mathbf{K}^*$ with the latter condition in (19) postmultiplied by $\mathbf{L}^*$ shows that (19) is equivalent to $(\mathbf{\Sigma K})^{k+1} = \mathbf{I}_r$. Hence, that assertion follows. □

**Acknowledgement** The author is very grateful to Professor Dr. Götz Trenkler for introducing him to Corollary 6 in Hartwig and Spindelböck [14]. Warm words of thanks are directed towards the Alexander von Humboldt Foundation for its financial support.

## References

[1] Baksalary, J.K., Baksalary, O.M.: On linear combinations of generalized projectors. Linear Algebra Appl. **388**, 17–24 (2004)

[2] Baksalary, J.K., Baksalary, O.M., Groß, J.: On some linear combinations of hypergeneralized projectors. Linear Algebra Appl. **413**, 264–273 (2006)
[3] Baksalary, J.K., Baksalary, O.M., Liu, X.: Further properties of generalized and hypergeneralized projectors. Linear Algebra Appl. **389**, 295–303 (2004)
[4] Baksalary, J.K., Baksalary, O.M., Liu, X., Trenkler, G.: Further results on generalized and hypergeneralized projectors. Linear Algebra Appl. **429**, 1038–1050 (2008)
[5] Baksalary, J.K., Liu, X.: An alternative characterization of generalized projectors. Linear Algebra Appl. **388**, 61–65 (2004)
[6] Baksalary, O.M., Benítez, J.: On linear combinations of two commuting hypergeneralized projectors. Computers & Mathematics with Applications **56**, 2481–2489 (2008) (Dedicated to Professor Götz Trenkler on the occasion of his 65th birthday.)
[7] Baksalary, O.M., Trenkler, G.: Characterizations of EP, normal, and Hermitian matrices. Linear Multilinear Algebra **56**, 299–304 (2008)
[8] Baksalary, O.M., Trenkler, G.: Problem 37-2 "Rank of a generalized projector". IMAGE **37**, 32 (2006)
[9] Baksalary, O.M., Trenkler, G.: Problem "A Property of the Range of Generalized and Hypergeneralized Projectors". IMAGE. Submitted.
[10] Benítez, J., Thome, N.: Characterizations and linear combinations of $k$-generalized projectors. Linear Algebra Appl. **410**, 150–159 (2005)
[11] Du, H.-K., Li, Y.: The spectral characterization of generalized projections. Linear Algebra Appl. **400**, 313–318 (2005)
[12] Du, H.-K., Wang, W.-F., Duan, Y.-T.: Path connectivity of $k$-generalized projectors. Linear Algebra Appl. **422**, 712–720 (2007)
[13] Groß, J., Trenkler, G.: Generalized and hypergeneralized projectors. Linear Algebra Appl. **264**, 463–474 (1997)
[14] Hartwig, R.E., Spindelböck, K.: Matrices for which $A^*$ and $A^\dagger$ commute. Linear Multilinear Algebra **14**, 241–256 (1984)
[15] Lebtahi, L., Thome, N.: A note on $k$-generalized projectors, Linear Algebra Appl. **420**, 572–575 (2007)
[16] Meyer, C.D.: Matrix Analysis and Applied Linear Algebra. SIAM, Philadelphia (2000)
[17] Stewart, G.W.: A note on generalized and hypergeneralized projectors. Linear Algebra Appl. **412**, 408–411 (2006)
[18] Trenkler, G.: On oblique and orthogonal projectors. In: Brown, P., Liu, S., Sharma, D. (eds.) Contributions to Probability and Statistics: Applications and Challenges, Proceedings of the International Statistics Workshop, pp. 178–191. World Scientific, Singapore (2006)

# On Singular Periodic Matrices

Jürgen Groß

**Abstract** In this note we recall the concept of a singular periodic square matrix, admitting a positive integer power greater than one which is identical to the matrix itself. Characterizations involving the group inverse of a matrix are given and relationships with normal and EP matrices are investigated.

## 1 Introduction

Let $\mathbb{C}_{m,n}$ denote the set of complex $m \times n$ matrices. The symbols $A^*$, $\mathscr{R}(A)$, and $\mathrm{rk}(A)$ will stand for the conjugate transpose, the range, and the rank of a given matrix $A \in \mathbb{C}_{m,n}$.

For the following definitions we refer to [3]. A generalized inverse $A^-$ of $A \in \mathbb{C}_{m,n}$ is any solution to the matrix equation $AXA = A$ with respect to $X$. The unique solution $X$ to the four equations

$$AXA = A, XAX = X, AX = (AX)^*, XA = (XA)^*$$

is called the Moore–Penrose inverse of $A$ denoted by $A^\dagger$. A matrix $A \in \mathbb{C}_{n,n}$ is said to be normal if $AA^* = A^*A$. A matrix $A \in \mathbb{C}_{n,n}$ is called EP (or range-Hermitian) if $\mathscr{R}(A) = \mathscr{R}(A^*)$, or equivalently $AA^\dagger = A^\dagger A$. A matrix $A \in \mathbb{C}_{n,n}$ is a group matrix if there is a solution to the three equations

$$AXA = A, XAX = X, AX = XA \,.$$

with respect to $X$. The unique solution $X$ is called the group inverse of $A$ and is denoted by $A^\#$. It is well known that $A^\#$ exists if and only if $\mathrm{rk}(A) = \mathrm{rk}(A^2)$. Moreover, $A \in \mathbb{C}_{n,n}$ is EP if and only if $A^\dagger = A^\#$.

Jürgen Groß
Carl von Ossietzky Universität Oldenburg, Fakultät V, Institut für Mathematik, D-26111 Oldenburg, Germany
j.gross@uni-oldenburg.de

Mirsky [7, p. 298] calls a matrix $A \in \mathbb{C}_{n,n}$ periodic if $A^k = I_n$ for some integer $k \geq 1$, where $I_n$ denotes the $n \times n$ identity matrix. A periodic matrix is this sense is necessarily nonsingular. A slight modification extends the definition to singular matrices.

**Definition 1.** A matrix $A \in \mathbb{C}_{n,n}$ is called $(k+1)$-potent if $A^{k+1} = A$. A matrix $A \in \mathbb{C}_{n,n}$ is called periodic if it is $(k+1)$-potent for some integer $k \geq 1$.

Sometimes a matrix $A \in \mathbb{C}_{n,n}$ is called $(k+1)$-potent if $k$ is the smallest integer $k \geq 1$ such that $A^{k+1} = A$, thereby making the number $k$ unique. However, the above notion appears to be suitable for the following derivations.

In the following section we collect some observations concerning periodic matrices and connect them to recent results in the literature.

## 2 Results

Two well known examples of periodic matrices are the cases $k = 1$ (idempotent matrices) and $k = 2$ (tripotent matrices) in Definition 1. Groß [4] investigate quadripotent matrices ($k = 3$ in Definition 1) being in addition normal or EP. They call the corresponding two classes of matrices (quadripotent normal and quadripotent EP) the sets of generalized and hypergeneralized projectors, respectively. Referring to such projectors and pointing out the possibility of generalizations, quite recently [10] has demonstrated that quadripotent EP matrices are necessarily diagonalizable. It can be shown, however, that quadripotency alone makes a matrix diagonalizable.

**Theorem 1.** *A matrix from $\mathbb{C}_{n,n}$ is periodic if and only if it is similar to a diagonal matrix and has all its nonzero roots equal to roots of unity.*

*Proof.* The proof follows along the same lines as the proof of Theorem 10.2.6 in [7], being identical to the assertion except for the omission of term 'nonzero' in order to characterize nonsingular periodic matrices.

Consider the polynomial $p(t) = t^{k+1} - t$ for some integer $k \geq 1$. It is well known that it has $k+1$ distinct roots, namely 0 and the $k$ distinct $k$th roots of unity. Hence, $p(t)$ can be written as the product of $k+1$ distinct linear factors.

Now, if $A$ is periodic, i.e. $A^{k+1} = A$ for some integer $k \geq 1$, then $p(t)$ annihilates $A$, i.e. $p(A) = 0$. Since $p(t)$ can be written as the product of distinct linear factors, this shows that $A$ is diagonalizable, where clearly the nonzero eigenvalues of $A$ are necessarily $k$th roots of unity.

Conversely, if $A$ is similar to a diagonal matrix and has all its nonzero roots equal to roots of unity, then $A$ can be written as $A = S\Lambda S^{-1}$ for some nonsingular matrix $S$ and some diagonal matrix $\Lambda$ containing only zeros and roots of unity on its main diagonal. But then, there exists $k \geq 1$ such that $\Lambda^k$ is identical to $\Lambda$ but with each root of unity replaced by 1. Hence $\Lambda^{k+1} = \Lambda$ and therefore $A^{k+1} = A$, showing that $A$ is periodic. $\square$

Similar to Theorem 1 the following refers to matrices satisfying the identity $A^{k+1} = A$ for some fixed integer $k \geq 1$.

**Theorem 2.** *A matrix $A \in \mathbb{C}_{n,n}$ is $(k+1)$-potent for some integer $k \geq 1$ if and only if $A$ is similar to a diagonal matrix and has all its nonzero roots equal to kth roots of unity.*

The proof of Theorem 2 is almost identical to the proof of Theorem 1 with an obvious alteration in the 'if' part. Note that Theorem 2 has already been acknowledged in the literature, see, e.g. [9, p. 60].

It is easily seen that any diagonalizable matrix satisfies $\mathrm{rk}(A) = \mathrm{rk}(A^2)$ and is therefore a group matrix. Moreover, the following is obvious.

**Theorem 3.** *A matrix $A \in \mathbb{C}_{n,n}$ is $(k+1)$-potent for some integer $k \geq 2$ if and only if $A^\# = A^{k-1}$.*

It might be of interest to recall that for an arbitrary group matrix $A \in \mathbb{C}_{n,n}$, its group inverse is necessarily a polynomial of finite degree in $A$, and although being necessarily unique, it can be written as $A^\# = A(A^3)^- A$, where $(A^3)^-$ is an arbitrary generalized inverse of $A^3$, cf. the results in [9, Chapt. 4] and [3, Sect. 4.4].

Note that when $k = 1$, then $A$ is $(k+1)$-potent if and only if $A^{k-1} = I_n$ is a generalized inverse of $A$, but in that case $I_n$ is not the group inverse of $A$, the latter being $A$ itself. If $A$ is $(k+1)$-potent, $k \geq 1$, then for $A^- = A^{k-1}$ we have

$$\mathrm{rk}(A) = \mathrm{rk}(AA^-) = \mathrm{tr}(AA^-) = \mathrm{tr}(A^k),$$

where $\mathrm{tr}(\cdot)$ denotes the trace of a square matrix.

A further obvious observations is that if $A^\dagger = A^{k-1}$ for some $k \geq 1$, then $AA^\dagger = A^\dagger A$, showing that necessarily $A^\dagger = A^\#$. Hence we may state the following.

**Theorem 4.** *For a matrix $A \in \mathbb{C}_{n,n}$ and an integer $k \geq 2$, the following statements are equivalent:*

*1. $A^\dagger = A^{k-1}$;*
*2. $A^\# = A^{k-1}$ and $A^\# = A^\dagger$;*
*3. $A$ is $(k+1)$-potent and EP.*

When investigating relationships between several classes of matrices, [6] define a matrix $A$ to be bi-EP, bi-normal, or bi-dagger if $AA^\dagger A^\dagger A = A^\dagger AAA^\dagger$, $AA^*A^*A = A^*AAA^*$, or $(A^\dagger)^2 = (A^2)^\dagger$, respectively. Among other results, the authors show the following, see also the scheme in [6, p. 247].

**Lemma 1.** *For a matrix $A \in \mathbb{C}_{n,n}$ the following statements are equivalent*

*1. $A$ is EP (and thus a group matrix);*
*2. $A$ is a group matrix and bi-EP;*
*3. $A$ is a group matrix and bi-normal;*
*4. $A$ is a group matrix and bi-dagger;*

Since a $(k+1)$-potent matrix is necessarily a group matrix, it is therefore clear that the term 'EP' in statement (3) of Theorem 4 may also be replaced by one of the terms 'bi-EP', 'bi-normal', or 'bi-dagger', thus yielding three further equivalent statements in Theorem 4.

Obviously unaware of the results by [6], recently [2] redefine the concepts of bi-EP and bi-normal matrices, calling them weak-EP and weak-normal, respectively, and give some results in connection with quadripotent matrices. Their Lemma 3, however, is a rederivation of the equivalence between statements (1) and (2) in Lemma 1, and their Theorem 3 is then a special case of Theorem 4. Moreover, in view of Theorem 4 and the equivalence between (1) and (3) in Lemma 1, the implication in their Theorem 4 can be strengthened to an equivalence statement.

Hartwig [6, p. 246] also show that the class of EP matrices is contained in the class of bi-normal matrices, the class of bi-normal matrices is contained in the class of bi-dagger matrices, and the class of bi-dagger matrices is contained in the class of bi-EP matrices. As noted above, considering $(k+1)$-potent matrices within the class of bi-EP matrices, being the widest among the aforementioned ones, does not yield a different characterization. On the other hand, one may also consider $(k+1)$-potent matrices within narrower classes than the class of EP matrices, an obvious one being the class of normal matrices. Before characterizing $(k+1)$-potent normal matrices we state the following Lemma.

**Lemma 2.** *Let $z \in \mathbb{C}$ be nonzero, and let $n \neq 2$ be a positive integer. Then $z^n = z\bar{z}$ if and only if $z^n = 1$.*

*Proof.* The case $n = 1$ is trivial, so that we assume $n \geq 3$ in the following. If $z^n = 1$, then $|z| = 1$ and hence $|z|^2 = z\bar{z} = 1$, showing that $z^n = z\bar{z}$. Conversely suppose $z^n = z\bar{z} = |z|^2$ and write $z$ in its polar decomposition as $z = |z|\,\mathrm{e}^{\mathrm{i}\theta}$. Then $|z|^n\,\mathrm{e}^{\mathrm{i}n\theta} = |z|^2$, showing that $|z|^n = |z|^2$ and $n\theta = 2\pi k$ for some integer $k$. Hence, $|z| = \sqrt[n-2]{1}$ and thus $z = \mathrm{e}^{\mathrm{i}2\pi k/n}$. But there are exactly $n$ different numbers $\mathrm{e}^{\mathrm{i}2\pi k/n}$, corresponding to the choices $k = 0, \ldots, n-1$, being well-known as the $n$ roots of $z^n = 1$. □

Just as our Theorem 4 generalizes Theorem 2 in [4], see also the correction in [2, Lemma 2], the following result generalizes Theorem 1 from [4].

**Theorem 5.** *For a matrix $A \in \mathbb{C}_{n,n}$ and an integer $k \geq 3$, the following statements are equivalent:*

*1. $A^* = A^{k-1}$;*
*2. $A^\# = A^{k-1}$ and $A^\# = A^*$;*
*3. $A^\# = A^{k-1}$ and $A^\dagger = A^*$;*
*4. $A$ is $(k+1)$-potent and normal.*

*Proof.* It is clear that (2) implies (1). If (1) is satisfied, then obviously $AA^* = A^*A$. Hence, $A$ is normal and can therefore be written in the form $A = U\Lambda U^*$ for some unitary matrix $U$ and some diagonal matrix $\Lambda$. Then the identity $A^* = A^{k-1}$ is satisfied if and only if $\Lambda^* = \Lambda^{k-1}$. This means that every diagonal element $\lambda$ of $\Lambda$ must satisfy $\overline{\lambda} = \lambda^{k-1}$. If $\lambda$ is nonzero, this is equivalent to $\lambda\overline{\lambda} = \lambda^k$, implying $\lambda^k = 1$

from Lemma 2. This shows that $A$ is diagonalizable and has all its nonzero roots equal to $k$th roots of unity. Then, from Theorems 2 and 3 it follows $A^{\#} = A^{k-1}$ and hence (2).

For the equivalence between (2) and (3) note that

$$A^{\#} = A^{*} \quad \Leftrightarrow \quad A^{\#} = A^{\dagger} \text{ and } A^{\dagger} = A^{*} ,$$

where $A^{\dagger} = A^{*}$ means that $A$ is a partial isometry. Hence, it is clear that (2) implies (3). If (3) is satisfied, then from Corollary 6 in [6] it is easily deduced that $A$ can be decomposed as

$$A = U \begin{pmatrix} K & L \\ 0 & 0 \end{pmatrix} U^{*} ,$$

where $U$ is unitary, $K^k = I_a$, $a = \mathrm{rk}(A)$, and $KK^* + LL^* = I_a$. Then $K$ is a contraction ($I_a - KK^*$ is Hermitian nonnegative definite), $K$ is a group matrix, and the eigenvalues of $K$ have absolute values 1. These properties show that $K$ is necessarily unitary, see, e.g. Theorem 2.8 and the subsequent remarks in [5], and compare also [3, Ex. 6.4.50, p. 225]. Hence, $KK^* = I_a$ and thus $LL^* = 0$, i.e. $L = 0$. But then $A$ is obviously EP, showing $A^{\dagger} = A^{\#}$ and hence $A^{\#} = A^{*}$.

For the equivalence between (2) and (4) it is clear that (2) implies (4). If (4) is satisfied, then $A^{\#} = A^{k-1}$ and $A = U\Lambda U^*$ for some unitary matrix $U$ and some diagonal matrix $\Lambda$. Each nonzero diagonal element of $\Lambda$ is a $k$th root of unity, showing that $\Lambda\Lambda^*$ has only 0 and 1 entries on its main diagonal and is thus idempotent. Then necessarily $AA^*$ is idempotent, being equivalent to $AA^*A = A$. But then, $A^* = A^{\#}$, showing (2). □

*Remark 1.* The equivalence between statements (2), (3), and (4) in Theorem 5 also holds for $k = 2$.

The equivalence between (3) and (4) in Theorem 5 shows that for $k \geq 2$ a square matrix is $(k+1)$-potent and normal if and only if it is $(k+1)$-potent and a partial isometry, an equivalence statement which is also seen to be true for $k = 1$. For the case $k = 3$, this result has originally been established by [1]. Our proof is considerably shorter though, since it depends on the cited references.

If $A, B \in \mathbb{C}_{n,n}$ are both idempotent, then $AB = BA$ is sufficient for $AB$ to be idempotent. Moreover, $AB = 0 = BA$ is necessary and sufficient for $A + B$ to be idempotent, and $AB = A = BA$ is necessary and sufficient for $B - A$ to be idempotent, see [9, Sect. 5.1]. In general, for $(k+1)$-potent matrices we may state the following.

**Theorem 6.** *Let $A, B \in \mathbb{C}_{n,n}$ be two $(k+1)$-potent matrices for an integer $k \geq 1$ such that $AB = BA$, and let $H = AB$. Then:*

1. *$H$ is $(k+1)$-potent;*
2. *If $H = 0$, then $A + B$ is $(k+1)$-potent;*
3. *If $H = A^2$, then $B - A$ is $(k+1)$-potent.*

*Proof.* We consider $k \geq 2$. Since $A$ and $B$ are diagonalizable and commute, we can write $A = S\Lambda_1 S^{-1}$ and $B = S\Lambda_2 S^{-1}$ for a nonsingular matrix $S$ and diagonal matrices $\Lambda_1$ and $\Lambda_2$, containing the eigenvalues of $A$ and $B$ on their main diagonal, respectively.

Let $\lambda$ denote any fixed (the $i$th, say) diagonal element of $\Lambda_1$ and let $\mu$ denote the corresponding ($i$th) diagonal element of $\Lambda_2$.

Then $(\lambda\mu)^k = 1$ if $\lambda$ and $\mu$ are both nonzero and $\lambda\mu = 0$ otherwise, thus confirming statement (1). For statement (2) note that $H = 0$ implies $\lambda\mu = 0$, so that clearly $(\lambda+\mu)^k = 1$ or $\lambda+\mu = 0$. For statement (3) note that when $A$ is a group matrix, the identities $AB = BA = A^2$ are equivalent to $A^\# B = BA^\# = AA^\#$. In view of $A^\# = A^{k-1}$, the latter means that $\lambda^{k-1}\mu = 0$ if $\lambda = 0$ and $\lambda^{k-1}\mu = 1$ if $\lambda \neq 0$. Then $\mu - \lambda = \mu$ if $\lambda = 0$ and $\mu - \lambda = 0$ if $\lambda \neq 0$. Thus $(\mu-\lambda)^k = 1$ or $\mu - \lambda = 0$. □

In general, statements (2) and (3) of Theorem 6 cannot be reversed. Indeed, when $A$ and $B$ are two commuting $(k+1)$-potent matrices, then $A+B$ can be $(k+1)$-potent but $AB \neq 0$. Consider, e.g. the two $1\times 1$ matrices

$$A = (1) \quad \text{and} \quad B = (-\frac{1}{2} + \mathrm{i}\frac{\sqrt{3}}{2}),$$

being both 7-potent. Then $A+B = \frac{1}{2} + \mathrm{i}\frac{\sqrt{3}}{2}$ is also 7-potent, but $AB \neq 0$. Similarly, when $A$ and $B$ are two commuting $(k+1)$-potent matrices, then $B-A$ can be $(k+1)$-potent but $AB \neq A^2$. Consider, e.g. the two $1\times 1$ matrices

$$A = (-1) \quad \text{and} \quad B = (-\frac{1}{2} + \mathrm{i}\frac{\sqrt{3}}{2}),$$

both of which are 7-potent. Then $B - A = \frac{1}{2} + \mathrm{i}\frac{\sqrt{3}}{2}$ is also 7-potent, but $AB \neq A^2$.

In the proof of Theorem 6 we have noted that $AB = BA = A^2$ is equivalent to $A^\# B = BA^\# = AA^\#$ when $A$ is a group matrix, see also Lemma 2.2 in [8]. When $A$ and $B$ are both group matrices, the binary relation defined by

$$A \overset{\#}{\leqslant} B \quad :\Leftrightarrow \quad A^\# B = BA^\# = AA^\#$$

specifies a partial order, the so-called sharp partial order introduced by [8]. The following result states that $(k+1)$-potency is inherited downwards by the sharp partial order.

**Theorem 7.** *Let $A, B \in \mathbb{C}_{n,n}$ be two group matrices. If $B$ is $(k+1)$-potent and $A \overset{\#}{\leqslant} B$, then $A$ is $(k+1)$-potent.*

*Proof.* If $B$ is $(k+1)$-potent and $A \overset{\#}{\leqslant} B$, then $BA^\# = B^{k+1}A^\# = AA^\#$. Then also $B^{k+1}A^\# = B^k AA^\# = B^{k-1}AAA^\#$, the last identity holding in view of $BA = AA$. Since $AAA^\# = A$ it follows that $B^{k-1}A = AA^\#$. But since $B^{k-1}A = A^{k-1}A$ in view of $BA = AA$, it follows $A^k = AA^\#$, implying $A^{k+1} = A$. □

As noted by [8, p. 21], if $A$ and $B$ are group matrices and $A$ is below $B$ with respect to the sharp partial order, then $B - A$ is a group matrix and $B - A$ is below

$B$ with respect to the sharp partial order. Hence, if $A$ and $B$ are $(k+1)$-potent and $A$ is below $B$ with respect to the sharp partial order, then $B-A$ is necessarily $(k+1)$-potent from Theorem 7, showing an alternative derivation of statement (3) from Theorem 6.

Another binary relation, the so-called minus partial order, is defined in the set $\mathbb{C}_{m,n}$ by

$$A \overset{-}{\leqslant} B \quad :\Leftrightarrow \quad A_1^- A = A_1^- B \text{ and } AA_2^- = BA_2^- \text{ for some } A_1^-, A_2^- \in A\{1\} ,$$

where $A\{1\}$ stands for the set of all generalized inverses of $A$. It is well known that $A \overset{-}{\leqslant} B$ if and only if

$$\text{rk}(B-A) = \text{rk}(B) - \text{rk}(A) ,$$

explaining why the minus partial order is also called the rank-subtractivity partial order.

Clearly, if $A \overset{\#}{\leqslant} B$ for group matrices $A$ and $B$, then also $A \overset{-}{\leqslant} B$. However, the inheritance property from Theorem 7 does not hold in general when the sharp partial order is replaced by the minus partial order. To see this, consider the $2 \times 2$ matrices

$$A = \begin{pmatrix} \frac{1}{2} & \lambda \\ \frac{1}{2} & \lambda \end{pmatrix} \quad \text{and} \quad B = \begin{pmatrix} 1 & \lambda \\ 0 & \lambda \end{pmatrix} ,$$

where $\lambda = -\frac{1}{2} + \mathrm{i}\frac{\sqrt{3}}{2}$, so that $\lambda^3 = 1$. Then $A \overset{-}{\leqslant} B$, $B$ is 4-potent, and $A$ is diagonalizable but not periodic.

## References


[1] Baksalary, J.K., Liu, X.: An alternative characterization of generalized projectors. Linear Algebra Appl. **388**, 61–65 (2004)

[2] Baksalary, J.K., Baksalary, O.M., Liu, X.: Further properties of generalized and hypergeneraliued projectors. Linear Algebra Appl. **389**, 295–303 (2004)

[3] Ben-Irael, A., Greville, T.E.: Generalized Inverses. Theory and Applications (2nd edn.). Springer, New York (2003)

[4] Groß, J., Trenkler, G.: Generalized and hypergeneralized projectors. Linear Algebra Appl. **264**, 463–474 (1997)

[5] Hartwig, R.E., Spindelböck, K.: Partial isometries, contractions and EP matrices. Linear Multilinear Algebra **13**, 295–310 (1983)

[6] Hartwig, R.E., Spindelböck, K.: Matrices for which $A^*$ and $A^\dagger$ commute, Linear and Multilinear Algebra **14**, 241–256 (1984)

[7] Mirsky, L.: An Introduction to Linear Algebra. Dover Publication, New York (1990) [First published at the Clarendon Press, Oxford (1955)]

[8] Mitra, S.K.: On group inverses and the sharp order. Linear Algebra Appl. **92**, 17–37 (1987)

[9] Rao, C.R., Mitra, S.K.: Generalized Inverse of Matrices and its Applications. Wiley, New York (1971).

[10] Stewart, G.W.: A note on generalized and hypergeneralized projectors. University of Maryland, Institute for Advanced Computer Studies, Technical Report 2004-57 [ftp://thales.cs.umd.edu/pub/reports/Contents.html] (2004)

# Testing Numerical Methods Solving the Linear Least Squares Problem

Claus Weihs

**Abstract** The paper derives a general method for testing algorithms solving the Least-Squares-Problem (LS-Problem) of a linear equation system. This test method includes the generation of singular test matrices with arbitrary condition, full column rank and exactly representable generalized inverses, as well as a method for choosing general right hand sides. The method is applied to three LS-Problem solvers in order to assess under what conditions the error in the least squares solution is only linearly dependent on the condition number.

## 1 The Linear Model: Testing of Algorithms

The linear model is probably the most used model in statistics. For its popularity alone it should be of particular importance to analyze the numerical problems in the estimation of unknown coefficients. However, for this purpose some systematic approach is necessary, i.e. some sort of an experimental design of testing. Testing sporadic examples can possibly show a completely distorted image. In particular, the well favored practice to test new algorithms based on standard problems from literature does not assess the superiority of the algorithms in the general case. Instead, it is necessary to cover the entire space of possible inputs, namely of the coefficients matrices and the right hand side vectors, with as little examples as possible. Additionally it would be helpful to construct representative test matrices for which the accuracy of the estimates of their generalized inverse can be calculated easily, e.g. for which, in the ideal case, the generalized inverse can be exactly computed. In Weihs [6] singular test matrices with full column rank that satisfy such a property were successfully constructed. Moreover, representative right hand side vectors were suggested. In this paper the most important principles of testing computer

Claus Weihs
Fakultät Statistik, Technische Universität Dortmund, D-44221 Dortmund
weihs@statistik.tu-dortmund.de

algorithms from this diploma thesis and their application to important LS-Problem solvers are discussed. After motivating the usage of condition numbers (in Sect. 2), the numerical methods used for solving the least squares problem are introduced (in Sect. 3). The construction of numerically favorable representative test matrices and representative right hand sides follows in Sects. 4 and 5.1. Here, the focus is on the exact determination of the least-squares solution and on the full coverage of the diversity of possible problems, avoiding too specific test examples. Finally, the testing method is explained (Sect. 5.2) and test results (Sect. 5.3) for the introduced numerical methods are discussed.

## 2 Condition of the Least-Squares Problem

**Definition 1 (Least-Squares Problem).** Let $A \in L(m,n)$, $b \in \mathbb{R}^m$.
Then, $x_0 \in \mathbb{R}^n$ with $\|b - Ax_0\|_2 = min_{x\in\mathbb{R}^n} \|b - Ax\|_2$ is called Least-Squares-Solution (LS-Solution) of the linear equation system $Ax = b$.

Practically, the coefficients matrix $A \in L(m,n)$, i.e. the real-valued matrix A with m rows and n columns, and the right hand side $b \in \mathbb{R}^m$ of the LS-Problem, are not exactly known in the general case (e.g. as the accuracy of measuring and representation is finite). So it becomes desirable to find a measure for the sensitivity of an LS-Problem to "disturbances" in the data $A$, $b$.

**Definition 2 (Condition numbers for the LS-Problem).** If the data $A$, $b$ contain relative errors of size $\delta$, then let the relative error in the LS-Solution $x_0$ be constrained by $f(\kappa)\delta$, where $f(\kappa)$ is a function of $\kappa$. Such measures $\kappa$ are called condition numbers for the LS-Problem.

The spectral condition number $K(A) := s_1(A)/s_r(A)$ is such a condition number, as motivated below. Here $s_j(A) := \sqrt{\lambda_j(AA^T)}$, $j = 1,\ldots,r$, r := rank(A), are the singular values of A, where $\lambda_1(AA^T) \geq \lambda_2(AA^T) \geq \ldots \geq \lambda_r(AA^T) > \lambda_{r+1}(AA^T) = \ldots = \lambda_m(AA^T) = 0$ are the eigenvalues of $AA^T$. One can show that $K(A) = \|A\|_2\|A^+\|_2$, where $\|A\|_2 := \max_{\|x\|_2=1}\|Ax\|_2 = \max\{\sqrt{\lambda_i}|\lambda_i \text{ eigenvalue of } AA^T,\ 1 \leq i \leq m\}$. Note that the positive eigenvalues of $AA^T$ and $A^TA$ are the same.

The so-called F-condition number $K_F(A) := \|A\|_F\|A^+\|_F$, which corresponds to the so-called Frobenius norm $\|A\|_F := \left(\sum_{i=1}^m\sum_{j=1}^n |a_{ij}|^2\right)^{0.5}$ of $A \in L(m,n)$, is another condition number of the LS-Problem, as $K(A) \leq K_F(A) \leq \min(m,n)K(A)$.

Notice that the LS-Problem can generally have more than one solution. A unique LS-Solution exists, if and only if $A$ has full column rank, since only then $A^TA$ can be inverted, and the normal equations $A^TAx = A^Tb$ can be uniquely solved.

From now on all results presented apply solely to LS-Problems with coefficient matrices with full column rank. Thus, let in the following:

$$A \in L(m,n), \quad m \geq n, \quad rank(A) = n.$$

Condition bounds: v.d. Sluis [5] introduced an upper as well as a lower bound for errors in the LS-Solution of a disturbed linear equation system that can occur in the worst-case (without specific information about the coefficients matrix, the selected method, etc. the worst-case has to be assumed realistic!).

**Theorem 1 (v.d. Sluis [5], p. 245/6, Theorem 4.3).** *Let $A, dA \in L(m,n)$, $m \geq n$, $rank(A) = n$ and $b, db \in \mathbb{R}^m$, as well as*

$$\|dA\|_2 \leq \delta \|A\|_2, \quad \|db\|_2 \leq \delta \|b\|_2 \text{ and } \mu := \delta \frac{s_1(A)}{s_n(A)} < 1.$$

*Furthermore let $x_0$ be the LS-Solution of the system $Ax = b$, let $r_0 := b - Ax_0$ be the corresponding residual and $dx_0$ chosen so that $(x_0 + dx_0)$ is the LS-Solution of the disturbed system $(A + dA)(x_0 + dx) = b + db$. Then:*

*1. For every pair $(A, b)$ and any kind of "disturbance" $(dA, db)$ it is valid that*

$$\|dx_0\|_2 \leq \frac{\delta}{s_n(A)} \left[ \frac{s_1(A)\|r_0\|_2}{s_n(A)(1-\mu^2)} + \frac{s_1(A)\|x_0\|_2}{1-\mu} + \frac{\|b\|_2}{1-\mu} \right].$$

*2. For every pair $(A, b)$ there is a "disturbance" $(dA, db)$ so that*

$$\|dx_0\|_2 \geq \frac{\delta}{s_n(A)} \left[ \frac{s_1(A)\|r_0\|_2}{s_n(A)(1-\mu^2)} + \frac{\|b\|_2}{1-\mu^2} \right].$$

*3. For every pair $(A, b)$ there is a "disturbance" $(dA, db)$ so that*

$$\|dx_0\|_2 \geq \frac{\delta}{s_n(A)} \left[ s_1(A)\|x_0\|_2 + \|b\|_2 \right].$$

*Proof.* see v.d. Sluis [5], pp. 246–248 □

Hence, an upper bound for the relative error is:

$$\frac{\|dx_0\|_2}{\|x_0\|_2} \leq \frac{\delta}{s_n(A)} \left[ \frac{s_1(A)\|r_0\|_2}{s_n(A)\|x_0\|_2(1-\mu^2)} + \frac{s_1(A)}{1-\mu} + \frac{\|b\|_2}{\|x_0\|_2(1-\mu)} \right].$$

With respect to v.d. Sluis (1.), (2.) it appears to be realistic to add an "amplifying factor" $\delta K^2(A) = \delta \frac{s_1^2(A)}{s_n^2(A)}$ for $\|r\|_2$ in the case of $1 = \|A\|_2 = s_1(A)$. However, for $\|A\|_2 \neq 1$ this amplifying factor would be $\delta \frac{s_1(A)}{s_n^2(A)} = \delta \frac{K(A)}{s_n(A)}$.

Thus, at least in the case of coefficient matrices $A$ with full column rank, K(A) is a condition number for the LS-Problem. Unfortunately, the lower bounds reveal the importance of the terms of the upper bound. Nevertheless v.d. Sluis does not indicate how realistic the upper bounds are in practice. The intension of the cited diploma thesis was to analyze by means of long test series, how realistic it is to assume a dependence of the error in the LS-Solution on $K^2(A)$ for different types of matrices and LS-Problem solvers.

# 3 Calculation of the Least-Squares-Solution

The most common method to calculate the Least-Squares-Solution is by the normal equations:

$$A^T Ax = A^T b.$$

It is known, however, that this method is numerically problematic already for not very badly conditioned LS-Problems, because the condition number of $A^T A$ is equal to the square of the condition number of $A$. This resulted in the development of many alternative methods trying to avoid this problem.

In the following only two of these methods are introduced: at first the Gram–Schmidt process, an orthogonalization process, secondly the method of Greville that turns out to be particularly useful in the construction of test matrices. There are certainly far more methods to construct the LS-Solution, cp. Lawson and Hanson [2], which are not discussed here.

## 3.1 Gram–Schmidt Method

The Gram–Schmidt (GS-) orthogonalization process produces a so called full-rank decomposition (frd)

$$A = BC \quad \text{with} \quad B \in L(m,k), C \in L(k,n), rank(B) = rank(C) = k \tag{1}$$

of a matrix A, that substantially simplifies the calculation of the generalized inverse $A^+$ of A; since:

**Theorem 2.** *If $A = BC$ is an frd of A, then:*

$$A^+ = C^T(CC^T)^{-1}(B^T B)^{-1} B^T \tag{2}$$

In fact, the Gram–Schmidt orthogonalization process produces a special frd of a matrix A, namely a so-called triangular decomposition since the matrix C is upper triangular:

GS-triangular decomposition (cp. Peters and Wilkinson [3], p. 313): The following algorithm for triangular decomposition of a matrix $A \in L(m,n)$ consists of $n := rank(A)$ "central steps", where $Q_1 := A$ is successively transformed into $Q_2, \ldots, Q_{n+1}$, where, if $(q_1\ q_2\ \cdots\ q_{s-1}\ a_s^{(s)}\ a_{s+1}^{(s)}\ \cdots\ a_n^{(s)})$ is a column representation of $Q_s$ and $(q_1, \ldots, q_{s-1})$ forms an orthogonal system with $q_i^T q_i = d_i$, $1 \leq i \leq s-1$, the s-th central step looks as follows:

- *Pivot strategy*: Let $\|a_{i_0}^{(s)}\|_2$ be the maximum of $\|a_i^{(s)}\|_2$, $i = s, \ldots, n$. (If more than one column with maximum norm exists, e.g. choose the one with the smallest index.) Then interchange the columns $s$ and $i_0$ in $Q_s$. (Even after interchanging, the i-th column of $Q_s$ is still denoted as $a_i^{(s)}$, $i = s, \ldots, n$.)

- Conduct the s-th step of the MGS (Modified GS) algorithm and save the values $u_{si} := r_{si}$ for $i = s+1, \ldots, k := n$ and $d_s := q_s^T q_s$:
  *MGS algorithm:*
  1. Set $q_s := a_s^{(s)}$, $d_s := q_s^T q_s$.
  2. For $i = s+1, \ldots, k$ calculate $r_{si} := (a_i^{(s)})^T \frac{q_s}{d_s}$ and $a_i^{(s+1)} := a_i^{(s)} - r_{si} q_s$.

In the Gram–Schmidt orthogonalization process linearly independent column vectors of A are orthogonalized. Here we use the MGS variant of the GS process which is much more stable numerically than the classical GS process (cp. Rice [4]).

Apparently $Q_{n+1}$ has the form: $Q = Q_{n+1} = (q_1\ q_2\ \ldots\ q_n)$, and hence: $\tilde{A} = QU$, where $\tilde{A}$ is the matrix A possibly after some column interchanges, and $U \in L(n,n)$ is an upper triangular matrix, defined by $u_{si}$, $1 \leq s \leq n$, $s+1 \leq i \leq n$, and $u_{ss} := 1$, $1 \leq s \leq n$. So an frd of $\tilde{A}$ is found, for which the generalized inverse can be generated according to Theorem 2. Some row interchanges in the generalized inverse may be necessary to generate $A^+$ because of pivoting.

## 3.2 Method of Greville

Now we proceed with the method of Greville [1] for the calculation of the generalized inverse that, thus, implicitly generates the LS-Solution. With this method the calculation of the generalized inverse $A^+$ of $A \in L(m,n)$ can be carried out in n resp. m steps. In the j-th step the generalized inverse of $A_j$ (resp. $A_{(j)}$) := (first j columns (resp. rows) of A) is calculated.

**Theorem 3.** • *Let $A = [a_1\ \ldots\ a_n]$ be a column representation of $A \in L(m,n)$ with full column rank, and*

$$A_1 := [a_1], \quad A_j := [A_{j-1}\ a_j], \quad j = 2, \ldots, n, \text{ as well as}$$
$$d_j := A_{j-1}^+ a_j, \quad c_j := a_j - A_{j-1} d_j \text{ and } b_j^T := c_j^+ = (c_j^T c_j)^{-1} c_j^T.$$

*Then:*

$$A_j^+ = [A_{j-1}\ a_j]^+ = \begin{bmatrix} A_{j-1}^+ - d_j b_j^T \\ b_j^T \end{bmatrix}. \tag{3}$$

• *Let $A = \begin{bmatrix} a_1^T \\ \vdots \\ a_m^T \end{bmatrix}$ be a row representation of $A \in L(m,n)$ with full row rank, and*

$$A_{(1)} := [a_1^T], \quad A_{(j)} := \begin{bmatrix} A_{(j-1)} \\ a_j^T \end{bmatrix}, \quad j = 2, \ldots, m, \text{ as well as}$$
$$d_j^T := a_j^T A_{(j-1)}^+, \quad c_j^T := a_j^T - d_j^T A_{(j-1)} \text{ and } b_j := c_j^{T+} = (c_j^T c_j)^{-1} c_j.$$

*Then:*

$$A^+_{(j)} = \begin{bmatrix} A_{(j-1)} \\ a_j^T \end{bmatrix}^+ = \begin{bmatrix} A^+_{j-1} - b_j d_j^T & b_j \end{bmatrix} \tag{4}$$

Theorem 3 apparently provides two construction rules for the calculation of $A^+$, recursively by $A_j^+$ resp. $A^+_{(j)}$. Obviously, one important condition is $c_j \neq 0 \Leftrightarrow a_j$ is not linear combination of the columns of $A_{j-1}$. Analogously, this is true for the row version. As we either want to assume maximum column rank or maximum row rank, $c_j$ resp. $c_j^T$ can never be zero. The variant of Greville's theorem dealing with the case $c_j = 0$ is not relevant here.

## 4 Test Matrices

An exact investigation of the accuracy of the methods for the calculation of LS-Solutions introduced in Sect. 3 is only possible, if the LS-Solutions that should be calculated by the methods are known exactly. For this reason some types of test matrices are described in the following, the generalized inverses of which are not only known, but can also be computed without any roundoff errors.

### *4.1 Non-singular Test Matrices*

Zielke [7] specified different types of non-singular test matrices, the elements of which can be chosen in such a way that not only the test matrix itself, but also its generalized inverse is integer and of relatively simple structure. In the following we restrict ourselves to one special type of these test matrices that only contains one freely selectable parameter matrix Z:

$$A_Z(Z,n,p) = \begin{bmatrix} Z+I_p & Z+2I_p & \dots & Z+(m-1)I_p & Z \\ Z+I_p & Z+2I_p & \dots & Z+(m-2)I_p & Z \\ \vdots & \vdots & & \vdots & \vdots \\ Z+I_p & Z+2I_p & \dots & Z+(m-2)I_p & Z \\ Z+I_p & Z+I_p & \dots & Z+(m-2)I_p & Z \\ Z & Z+I_p & \dots & Z+(m-2)I_p & Z-I_p \end{bmatrix},$$

$$A_Z(Z,n,p)^{-1} = \begin{bmatrix} -Z-(m-2)I_p & I_p & \dots & I_p & 2I_p & Z \\ & & & I_p & -I_p & \\ & (0) & & \iddots & -I_p & \\ & & I_p & \iddots & & (0) \\ I_p & & -I_p & & & \\ Z+(m-2)I_p & -I_p & \dots & -I_p & -I_p & -Z-I_p \end{bmatrix}$$

with n = overall number of columns, $I_p$ = identity matrix with p columns, Z $\in L(p,p)$, $m := n/p \geq 3$.

*Example 1.* Let $Z = 998$. Then:

$$A_Z(998,3,1) = \begin{bmatrix} 999 & 1000 & 998 \\ 999 & 999 & 998 \\ 998 & 999 & 997 \end{bmatrix}, \ A_Z(998,3,1)^{-1} = \begin{bmatrix} -999 & 2 & 998 \\ 1 & -1 & 0 \\ 999 & -1 & -999 \end{bmatrix}.$$

The inverse can be constructed as follows: Select the first and the last "row". Then continue selecting "rows" from below, until $m$ "rows" are obtained. In the last but one row set all entries to zero except for the "first" entry, set to $I_p$, and the "second", set to $-I_p$. Continue for the last but two rows setting all entries to zero except for the "second" entry, set to $I_p$, and the "third", set to $-I_p$, etc.

'Unless they are not only used to expose gross failures in the algorithm, test matrices should have the worst possible condition, i.e. a high condition number. Then it is possible to test the quality of a method with respect to error propagation avoiding enormous computational costs' (see Zielke [7], p. 34).

This raises the question how test matrices with high condition can be constructed. This is generally neither possible nor sensible using random numbers for all entries! In contrast, we can control the condition of $A_Z(Z,n,p)$ by means of free parameters. It can be shown (see Zielke [7], p. 47) that

$$K_F(A_Z(Z,n,1)) \cong 2nZ^2,$$

if Z is integer. So the F-condition number increases with the square of the free parameter and linearly with the rank $n$. Thus even for small ranks $n$ one gets test matrices with very high condition numbers without having to set the free parameters to high values. So for instance: $K_F(A_Z(10^3,3,1)) \cong 6 \cdot 10^6$, $K_F(A_Z(10^3,5,1)) \cong 10 \cdot 10^6 \cong 10^7$ and $K_F(A_Z(10^5,5,1)) \cong 10^{11}$.

## 4.2 Singular Test Matrices

Zielke [8] gave an overview of singular test matrices. He proofed (in Theorem 4) that the generalized inverse of a singular integer matrix A has to be non-integer if A does not result from a non-singular matrix by adding or inserting zero rows and/or columns and A is not the zero matrix. Therefore, if one is interested in exact least-squares solutions then, obviously, the best one could reach are generalized inverses only containing non-integers exactly representable on the computer with a rather short mantissa. This has been realized in Weihs [6] in a very general manner.

By means of row canceling in matrices with full row rank and by using the method of Greville to build the generalized inverse based on Zielke's nonsingular matrices singular test matrices can be successfully constructed with properties similarly good as for Zielke's matrices. Then test matrices with full column rank can be created by transposing. Unfortunately, directly dealing with columns leads to identical rows and hence to identical observations.

For the derivation of such singular test matrices from Zielke matrices a 'converse' of the row version of Greville's Theorem (see Theorem 3) is needed. Thereto let $A_{(j_1,\ldots,j_p;i_1,\ldots,i_q)}$ be the matrix that was created by successively canceling the columns $j_1 \neq \ldots \neq j_p$ and the rows $i_1 \neq \ldots \neq i_q$ of A.

**Theorem 4 (Converse of the row version of Greville's Theorem).** *Let $A_j \in L(j,n)$ be of maximum row rank, $a_j^T \in L(1,n)$, and*
*$A_1 := [a_1^T]$, $A_j := \begin{bmatrix} A_{j-1} \\ a_j^T \end{bmatrix}$, $b_j :=$ (j-th column of $A_j^+ \in L(n,j)$). Then:*

$$A_{j-1}^+ = \left( I_n - \frac{b_j b_j^T}{b_j^T b_j} \right) A_{j\ (j)}^+, \tag{5}$$

*where $I_n$ is the identity matrix with n columns.*

*Proof.* see Weihs [6] □

**Corollary 1.** *Let $A_j \in L(j,n)$ be of maximum row rank, $A_{j-1} := A_{j(;i_1,\ldots,i_k)}$, $1 \leq i_1,\ldots,i_k \leq j$, $1 \leq k < j$, and $b_i^{(k)}$ := (i-th column of $A_j^+$ after k cancelations), $1 \leq i \leq j-1$, $0 \leq k < j$. So $b_i^{(0)}$ is just the i-th column of $A_j^+$ itself. Then:*

$$A_{j-1}^+ = \left( I_n - \frac{b_{i_k}^{(k-1)} b_{i_k}^{(k-1)^T}}{b_{i_k}^{(k-1)^T} b_{i_k}^{(k-1)}} - \cdots - \frac{b_{i_1}^{(0)} b_{i_1}^{(0)^T}}{b_{i_1}^{(0)^T} b_{i_1}^{(0)}} \right) A_{j\ (i_1,\ldots,i_k)}^+. \tag{6}$$

Then from Corollary 1 it follows:

**Corollary 2.** *Let $A_n := A_Z(Z,n,1)$ and $A_{n-1} := A_{n(;i)}$, $1 < i < n-1$. Then:*

$$A_{n-1}^+ = \begin{bmatrix} 0.75 & 0 \ldots & 0 & -0.25 & 0.25 & 0 & \ldots 0 & 0.25 \\ & 1 & & & & & & \\ (0) & & \ddots & & & (0) & & \\ & & 1 & & & & & \\ -0.25 & 0 \ldots & 0 & 0.75 & 0.25 & 0 & \ldots 0 & 0.25 \\ 0.25 & 0 \ldots & 0 & 0.25 & 0.75 & 0 & \ldots 0 & -0.25 \\ & & & & & 1 & & \\ & & (0) & & & & \ddots & (0) \\ & & & & & & 1 & \\ 0.25 & 0 \ldots & 0 & 0.25 & -0.25 & 0 & \ldots 0 & 0.75 \\ & & & & \uparrow & & & \end{bmatrix} A_{n\ (i)}^+$$

*(n-i+1)-th column*

Obviously, $A_{n-1}^{+}$ is exactly representable if $A_n^{+}$ is. Unfortunately this result cannot easily be generalized for the case of several cancelations, because cancelations of successive rows of $A_Z(Z,n,1)$ lead to generalized inverses not exactly representable. So at most $\lfloor (n-2)/2 \rfloor$ rows of $A_Z(Z,n,1)$ can be canceled. In the case of canceling non-successive rows of $A_Z(Z,n,1)$ one can show:

**Corollary 3.** *Let* $A_n := A_Z(Z,n,1)$, $A_{n-1} := A_{n(;i_1,i_2,\ldots,i_k)}$, $1 < i_1 < \ldots < i_k < n-1$, $|i_p - i_q| > 1$ *for* $p \neq q$. *Then:*

$$A_{n-1}^{+} = \begin{bmatrix} 1-\tilde{m} & 0 \ldots 0 & -1/m & 1/m & \ldots & -1/m & 1/m & 0 \ldots 0 & \tilde{m} \\ & 1 & & & & & & & \\ (0) & \ddots & & & & (0) & & & \\ & 1 & & & & & & & \\ -1/m & 0 \ldots 0 & 1-\tilde{m} & \tilde{m} & \ldots & 1/m & -1/m & 0 \ldots 0 & 1/m \\ 1/m & 0 \ldots 0 & \tilde{m} & 1-\tilde{m} & \ldots & -1/m & 1/m & 0 \ldots 0 & -1/m \\ & & \ldots & \ldots & \ldots & \ldots & \ldots & & \\ -1/m & 0 \ldots 0 & 1/m & -1/m & \ldots & 1-\tilde{m} & \tilde{m} & 0 \ldots 0 & 1/m \\ 1/m & 0 \ldots 0 & -1/m & 1/m & \ldots & \tilde{m} & 1-\tilde{m} & 0 \ldots 0 & -1/m \\ & & & & & & & 1 & \\ & & & (0) & & & & \ddots & (0) \\ & & & & & & & 1 & \\ \tilde{m} & 0 \ldots 0 & 1/m & -1/m & \ldots & 1/m & -1/m & 0 \ldots 0 & 1-\tilde{m} \end{bmatrix} A_n^{+}\,(i_1,\ldots,i_k)$$

$\uparrow$ $(n-i_k+1)$-*th* $\quad\uparrow$ $(n-i_1+1)$-*th column*

with $m := 2(k+1)$, $\tilde{m} := \frac{k}{m}$. (Notice that there are $k$ double columns with $(-1/m \;\; 1/m)$ in the first row.)

Obviously, it is important here that $\frac{1}{m} = \frac{1}{(2k+2)}$ can be computed exactly. But this is true only for $k = 2^i - 1$, $i \in \mathbb{N}$, since then $m = 2^{i+1}$. From this another restriction follows for the canceling of rows of $A_Z(Z,n,1)$:

If $n = 3$ no row must be canceled as $1 < i < n-1 = 2$ would be required. If $n = 4$ only the second row is possible to cancel. And also for $n = 5$ only one row can be canceled, namely row 2 or 3. One should keep in mind that from the condition $k = 2^i - 1$, $i \in \mathbb{N}$, only $k = 1,3,7,15$, resp. not successive rows may be canceled. For these matrices of the type $A_Z(Z,n,1)$ with $n = 4,8,16,32$, resp. rows are needed.

For $n = 4$ consider the following example: Let $Z = 998$:

$$A_Z(998,4,1) = \begin{bmatrix} 999 & 1000 & 1001 & 998 \\ 999 & 1000 & 1000 & 998 \\ 999 & 999 & 1000 & 998 \\ 998 & 999 & 1000 & 997 \end{bmatrix},$$

$$A_Z(998,4,1)^{-1} = \begin{bmatrix} -1000 & 1 & 2 & 998 \\ 0 & 1 & -1 & 0 \\ 1 & -1 & 0 & 0 \\ 1000 & -1 & -1 & -999 \end{bmatrix},$$

and

$$A_Z(998,4,1)_{(;2)} = \begin{bmatrix} 999 & 1000 & 1001 & 998 \\ 999 & 999 & 1000 & 998 \\ 998 & 999 & 1000 & 997 \end{bmatrix},$$

$$A_Z(998,4,1)^{+}_{(;2)} = \begin{bmatrix} 0.75 & -0.25 & 0.25 & 0.25 \\ -0.25 & 0.75 & 0.25 & 0.25 \\ 0.25 & 0.25 & 0.75 & -0.25 \\ 0.25 & 0.25 & -0.25 & 0.75 \end{bmatrix} \begin{bmatrix} -1000 & 2 & 998 \\ 0 & -1 & 0 \\ 1 & 0 & 0 \\ 1000 & -1 & -999 \end{bmatrix}$$
$$= \begin{bmatrix} -499.75 & 1.5 & 498.75 \\ 500.25 & -1.5 & -499.25 \\ -499.25 & 0.5 & 499.25 \\ 499.75 & -0.5 & -499.75 \end{bmatrix}$$

Up to this point only the case $p = 1$ was covered. For $p > 1$ there are further possibilities for canceling rows leading to exactly representable generalized inverses (see Weihs [6], pp. 98). Note that transposing results in test matrices with full column rank with 1, 3, 7, 15 degrees of freedom and $n = 4, 8, 16, 32$ observations, respectively. So with the above instructions sensible degrees of freedom for applications in statistics are automatically attained. Even adding linearly dependent rows to the former created test matrices with full column rank could be realized in such a way that the corresponding generalized inverses are exactly representable (see Weihs [6], pp. 101).

Overall, we have succeeded in constructing (at least with respect to condition) general exactly representable test matrices with exactly representable generalized inverses. With these matrices it appears possible to search the space $L(m,n)$ of coefficients matrices in a way adequate for the LS-Problem, namely by choosing matrices with condition numbers covering a whole range.

## 5 Test Method

In addition to test matrices, the test algorithm also requires the right hand sides for which the methods for solving the LS-Problem are to be tested. The next subsection introduces one possible generation method. Then the test algorithm is explained and test results are discussed.

### *5.1 Selection of the Right Hand Sides*

Right hand sides should be preferably constructed to be general and as exactly representable as possible. Let us assume that the generalized inverse $A^+$ of a test matrix is known exactly, then the LS-Solution $x_0 := A^+ b_0$ of the system $Ax = b_0$ with an arbitrary vector $b_0$ can be determined and with it $y_0 := Ax_0$ as well as $r_0 := b_0 - y_0$. Then $r_0$ is orthogonal to $y_0 (r_0 \perp y_0)$ as $r_0^T y_0 = 0$.

So we have obtained an orthogonal decomposition $b_0 = y_0 + r_0$ with $y_0 \in im(A)$ and $y_0 \perp r_0$. $y_0$ is the projection of $b_0$ on im(A) (image of A) and $r_0$ is the corresponding residual. Now let $r_0 \neq 0$. By compressing or stretching $r_0$ we can construct right hand sides $b$ for arbitrary design matrices with the LS-Solution $x_0$ and a projection $y_0$ which intersects im(A) at an arbitrary angle. For that purpose let $b := y_0 + r$ with $r \perp y_0$. Then for the angle $\phi(b)$ between $b$ and $y_0$ it is true :

$$\cos\phi(b) := \frac{y_0^T b}{\|y_0\|_2 \|b\|_2} = \frac{y_0^T (y_0 + r)}{\|y_0\|_2 \|b\|_2} \overset{r \perp y_0}{=} \frac{\|y_0\|_2^2}{\|y_0\|_2 \|b\|_2} = \frac{\|y_0\|_2}{\|b\|_2}, \tag{7}$$

and we can show:

$$\tan\phi(b) = \frac{\sin\phi(b)}{\cos\phi(b)} = \frac{\|r\|_2}{\|y_0\|_2} \quad \text{and} \quad \phi(b) = \tan^{-1}\left(\frac{\|r\|_2}{\|y_0\|_2}\right). \tag{8}$$

For the evaluation of the accuracy of the calculated LS-Solution in dependence of the angle between right hand side $b$ and im(A) we now use a method that generally allows constructing right hand sides intersecting im(A) at "very small" angles as well as at "very large" and "medium sized" angles (around 45°): Let
$\tan\phi_j := \mathbf{2^{-21}}, 2^{-19}, \mathbf{2^{-3}}, 2^{-1}, \mathbf{2}, 2^3, \mathbf{2^{19}}, 2^{21},\ j = 1, \ldots, 8$. Then:
$\phi_j \cong 5 \cdot 10^{-7}, 2 \cdot 10^{-6}, 0.12, 0.46, 1.1, 1.45, 1.570794, 1.570796$, resp.
$\phi_j \triangleq (\mathbf{3 \cdot 10^{-5}})^\circ, (10^{-4})^\circ, \mathbf{1.8}^\circ, 26.6^\circ, \mathbf{63.4}^\circ, 82.9^\circ, \mathbf{89.99989}^\circ, 89.99997^\circ$.

ALG.RS: The following algorithm selects right hand sides $b_k$, $k = 1, \ldots, 4$, for given $y_0$, $r_0$ of "not too different 2-norm", so that: $\phi(b_k) \in [\phi_{2k-l}, \phi_{2k})$:

- Determine the biggest $i \in \{19, 17, \ldots, 1, -1, \ldots, -19\}$ so that $\|r_0\|_2 \geq 2^i \|y_0\|_2$. If such an $i$ does not exist, set $i := -21$.
- Set $b_k := y_0 + c_k r_0 =: y_0 + r_k$ with $c_k := 2^{-21-i}, 2^{-3-i}, 2^{1-i}, 2^{19-i}$, $k = 1, \ldots, 4$. If $y_0$ and $r_0$ are of not too different 2-norm, then: $c_k 2^i \leq c_k \frac{\|r_0\|_2}{\|y_0\|_2} = \tan\phi(b_k) < c_k 2^{i+2}$, i.e. $\phi(b_k) \in [\phi_{2k-l}, \phi_{2k})$.

Now let $P$ be the number of right hand sides to be constructed for LS-Problems with a coefficients matrix A as in Sects. 4.1, 4.2 (the generalized inverse $A^+$ of which, thus, is exactly available in general). Then the following steps are run through P times:

- Choose an arbitrary vector $b_0 \in \mathbb{R}^m$, e.g. at random.
- Determine the LS-Solution $x_0 := A^+ b_0$ of the system $Ax = b_0$ by using the exact generalized inverse $A^+$, the projection $y_0 := Ax_0$ and the residual $r_0 := b_0 - y_0$.
- If $r_0 = 0$, then choose $y_0 = b_0$ as right hand side, else choose $y_0$, $b_1, \ldots, b_4$ as right hand sides, where $b_1, \ldots, b_4$ are constructed by means of ALG.RS.

Notice that one should restrict the size of $b_0$, e.g. by $\|b_0\|_\infty \leq 1000$. As $x_0$, $y_0$, $r_0$, $b_1, \ldots, b_4$ should be determined as exactly as possible, one should always work with the maximum potential accuracy in the calculation of the right hand sides.

## 5.2 Test Algorithm

Based on Sect. 4 and on the creation of right hand sides in Sect. 5.1 the following test method was identified to assess the LS-Problem solvers in Sect. 3.

1. Specify invertible test matrices of the type $A_Z(Z,n,1)$ with $n = 4,8,16,32,64$, where the integer number $Z$ can be chosen freely. Use test matrices with specified size orders for the condition numbers, if the machine accuracy is $2.2 \times 10^{-16}$:
   class 1 (K1): $10^4 < K_F < 10^6$ : critical for simple accuracy,
   class 2 (K2): $10^{10} < K_F < 10^{12}$ : still uncritical for double accuracy,
   class 3 (K3): $10^{14} < K_F < 10^{16}$ : critical for double accuracy,
   class 4 (K4): $10^{18} < K_F$ : more than critical for double accuracy.
2. Cancel $k = 1,3,7,15,31$ rows from these test matrices and use the exactly representable inverses of $A_Z(Z,n,1)$ to construct exact generalized inverses of the matrices. Transpose these matrices, in order to obtain matrices with full column rank and their exact generalized inverses.
3. Generate different right hand sides (b1–b4) for the chosen test matrices by means of the algorithm ALG.RS. Calculate the exact LS-Solutions by multiplying the exact generalized inverses by their exact right hand sides.
4. Apply the MGS algorithm for Gram–Schmidt orthogonalization, the column version of the Greville-method and the normal equations method to the generated test problems. Record the accuracy of the results.
5. Repeat the entire procedure at least 100 times.
6. Compare the results. Which ranking of the methods does arise? What changes happen for different angles of the right hand sides?

The accuracy of the results can be characterized, e.g. by the mean value and the standard deviation of the relative errors $\|dx_0\|/\|x_0\|$ of the LS-Solutions over the repetitions.

To assess the obtained accuracy let us consider the theoretical error bounds. Let $x_b$ be the LS-Solution of the system $Ax = b$, let $r_b := b - Ax_b$ be the corresponding residual and $dx_b$ be chosen in such a way that $(x_b + dx_b)$ is the LS-Solution of the disturbed system $(A+dA)(x+dx) = b+db$. If only small disturbances in $A$ and $b$ are allowed, one can generally set $\mu = 0$ in Theorem 1, so one gets:

$$\begin{aligned}
\frac{\|dx_b\|_2}{\|x_b\|_2} &\leq \delta K(A)\left[\|A^+\|\frac{\|r_b\|}{\|x_b\|}+1\right]+\delta\|A^+\|\frac{\|b\|}{\|x_b\|} \\
&\leq \delta K(A)\left[K(A)\frac{\|r_b\|}{\|y_0\|}+1+\frac{\|b\|}{\|y_0\|}\right], \\
&\quad \text{as } \|y_0\| := \|Ax_b\| \leq \|A\|\,\|x_b\| \\
&= \delta K(A)\left[K(A)\tan\phi(b)+1+1/\cos\phi(b)\right].
\end{aligned}$$

From this bound it has to be suspected (as mentioned before) that $\delta K(A)^2$ appears as an amplification factor of $\|r_b\|/\|y_0\|$ $(= \tan\phi(b))$ in the relative error in the LS-Solution. However, alternatives to the normal equations were mainly developed to avoid such a dependence. Thus, for the comparison of the results of the different

LS-Problem solvers we will compare the percentages of the results in the so-called solution class L (meaning linear) defined by

$$\frac{\|dx_b\|}{\|x_b\|} \leq \min(\delta K_F(A)\,[1 + \tan\phi(b) + 1/\cos\phi(b)]\,, 10^{-2}). \tag{9}$$

Note that this requires an accuracy of at least 2 decimal places. We set $\delta = 10^{-16}$ = machine accuracy. The percentage of the LS-Solutions in solution class L provides information about the probability that the reviewed methods for LS-Solution depend only linearly on K(A).

## 5.3 Test Results

The simulation results are summarized in Figs. 1–3. The $x$-axis shows the tupel $(Z, n)$, the $y$-axis the percentage in solution class L. The illustrations point out that with increasing condition of the test matrices (classes K1-K4) a linear dependence on K(A) becomes more and more unrealistic. A general statement about the dependence of the size of the angle between b and $y_0$ cannot be made.

The MGS-process performs best. For not too badly conditioned matrices (classes K1-K2) a linear dependence on K(A) can be assumed, except for almost right angles between b and $y_0$ (b4). Besides, it appears also to be realistic to assume a linear dependence on K(A) in the case of small angles (b1,b2) in combination with K3-matrices. When increasing the size of the angle, the linear dependence on K(A) tends to become more unrealistic.

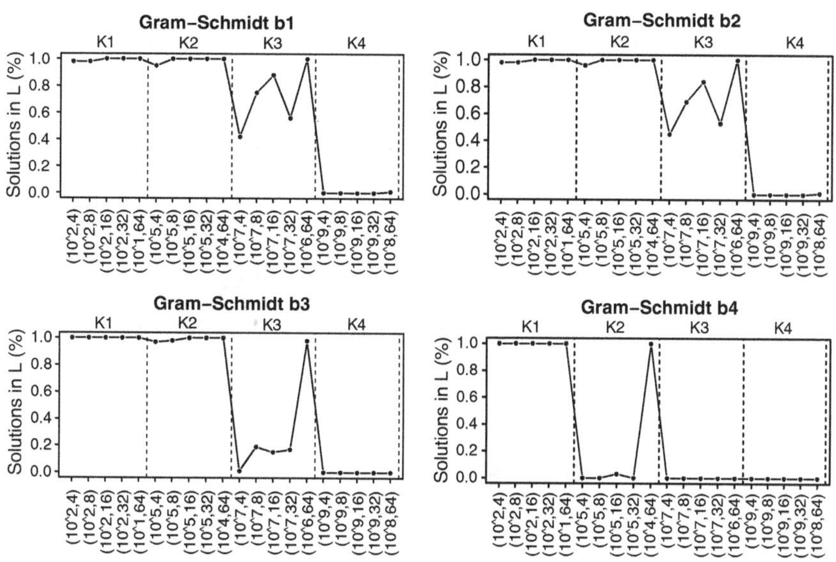


**Fig. 1** Class-L-percentages for MGS-process

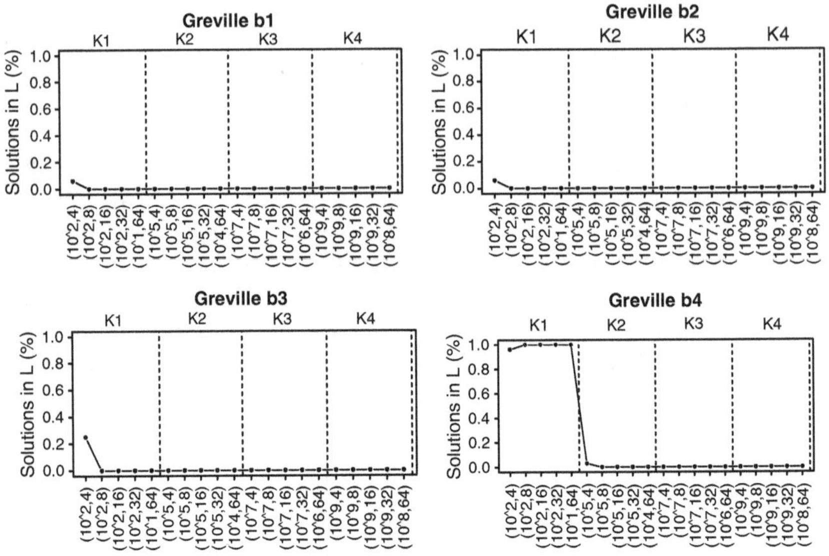


**Fig. 2** Class-L-percentages for Greville's method

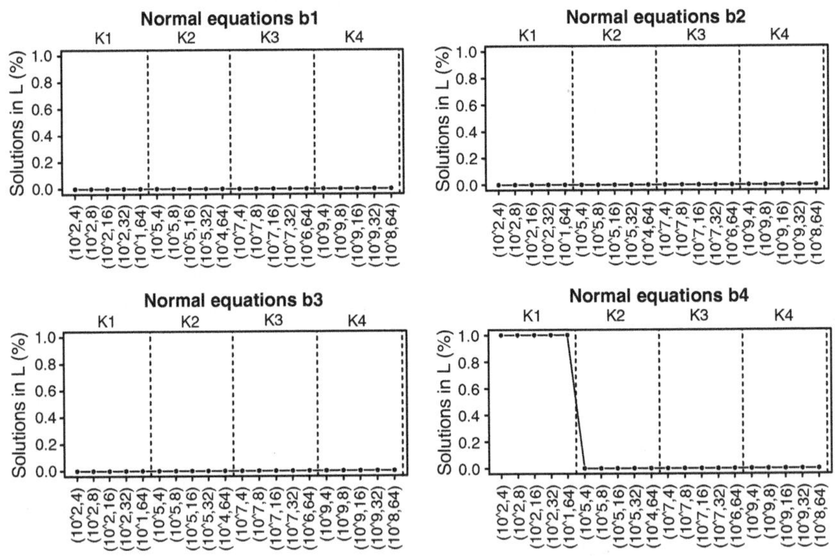


**Fig. 3** Class-L-percentages for normal equations

For the method of Greville and the normal equations there is no identifiable linear dependence of K(A), except when the test matrix has a good condition (class K1) and the angle between b and $y_0$ is big (b4). For such a large angle (near 90 degrees) and a small condition number the term with 1/cos is dominant in the upper bound, because cos is next to 1, tan is next to 0 and K(A) is small enough. As seen from the examples here even $K(A) = 10^5$ is too large already.

# 6 Conclusion

The paper derived a general test method for testing LS-Problem solvers. This test method includes the generation of singular test matrices with arbitrary condition, full column rank and exactly representable generalized inverses, and a method for choosing general right hand sides. Applying the test method shows that for the MGS-orthogonalization process the dependence of the error in the least squares solution is in general not dependent on the square of the condition number if the condition is not too bad and the angle between the right hand side and the projection is not too large. For Greville's method and the normal equations squared dependence on K(A) has to be expected.

**Acknowledgement** I would like to thank Dr. Heike Trautmann for running the simulations and producing the graphics illustrating the results, as well as Gerd Kopp and Matthias Redecker for supporting the production of the text.

# References

[1] Greville, T.N.E.: Some Applications of the Pseudoinverse of a Matrix. SIAM Rev. **2**, 485–494 (1943)
[2] Lawson, C.L., Hanson, R.J.: Solving Least Squares Problems. Prentice Hall, New Jersey (1974)
[3] Peters, G., Wilkinson, J.H.: The Least-Squares Problem and Pseudoinverses. Comput. J. **13**, 309–316 (1970)
[4] Rice, J.R.: Experiments on Gram-Schmidt Orthogonalization. Math. Comput. **20**, 325–328 (1966)
[5] v.d. Sluis, A.: Stability of the Solutions of Linear Least Squares Problems. Num. Mathematik **23**, 241–254 (1975)
[6] Weihs, C.: Kondition des linearen Ausgleichsverfahrens, Testmatrizen, Vergleich von Lösungsverfahren. Diploma thesis, Universität Bonn (1977)
[7] Zielke, G.: Testmatrizen mit maximaler Konditionszahl. Computing **13**, 33–54 (1974)
[8] Zielke, G.: Report on Test Matrices for Generalized Inverses. Computing **36**, 105–162 (1986)

# On the Computation of the Moore–Penrose Inverse of Matrices with Symbolic Elements

Karsten Schmidt

**Abstract** In this paper potential difficulties in using Greville's method for the computation of the Moore–Penrose inverse of a matrix that also contains symbolic elements are discussed. For the actual computation of the Moore–Penrose inverse of matrices whose elements are not numeric only, a Computer Algebra System has to be used. Initially, the computation of the Moore–Penrose inverse of a vector is considered which is a simple task if it only has numeric elements. If it contains symbolic elements, it might also be straightforward, but might turn out to be difficult. As Greville's method – an iterative algorithm that needs n steps for the computation of the Moore–Penrose inverse of an m by n matrix – requires the computation of the Moore–Penrose inverse of a vector in each step, the difficulty just mentioned might prevent the actual computation of the Moore–Penrose inverse of a matrix with symbolic elements.

## 1 Introduction

For any matrix $\boldsymbol{A} \in \mathbb{R}^{m \times n}$ a unique matrix $\boldsymbol{A}^+ \in \mathbb{R}^{n \times m}$ exists, which satisfies the following four conditions

$$\boldsymbol{A}\boldsymbol{A}^+\boldsymbol{A} = \boldsymbol{A} \tag{1}$$
$$\boldsymbol{A}^+\boldsymbol{A}\boldsymbol{A}^+ = \boldsymbol{A}^+ \tag{2}$$
$$(\boldsymbol{A}^+\boldsymbol{A})' = \boldsymbol{A}^+\boldsymbol{A} \tag{3}$$
$$(\boldsymbol{A}\boldsymbol{A}^+)' = \boldsymbol{A}\boldsymbol{A}^+ \tag{4}$$

$\boldsymbol{A}^+$ is the Moore–Penrose inverse (or pseudoinverse) of $\boldsymbol{A}$. The concepts of the Moore–Penrose inverse and, more generally, the so-called generalized inverses (any

Karsten Schmidt
Fakultät Wirtschaftswissenschaften, Fachhochschule Schmalkalden, Germany
kschmidt@fh-sm.de

matrix $\boldsymbol{A}^- \in \mathbb{R}^{n\times m}$ satisfying the condition $\boldsymbol{AA}^-\boldsymbol{A} = \boldsymbol{A}$) go back to Moore [4] and Penrose [5]. Some standard references related to generalized inverses and to the Moore–Penrose inverse are Greville [3], Rao [6], Rao/Mitra [7], and Ben-Israel/Greville [1]. Note that $\boldsymbol{A}^-$ is not unique in general.

If $\boldsymbol{A}$ is square and nonsingular, its inverse $\boldsymbol{A}^{-1}$ exists. Obviously, in this case the above conditions (1) to (4) are satisfied when $\boldsymbol{A}^{-1}$ is substituted for $\boldsymbol{A}^+$. Hence, if $\boldsymbol{A}$ is a nonsingular matrix, we have $\boldsymbol{A}^+ = \boldsymbol{A}^{-1}$ (and $\boldsymbol{A}^{-1}$ is the only generalized inverse).

In this paper we will discuss problems that might occur if we use a computer to find the Moore–Penrose inverse of a matrix that also contains symbolic elements. For the actual computation of the Moore–Penrose inverse of matrices whose elements are not numeric only, we have to use a Computer Algebra System. We will be using *Derive*, a very popular Computer Algebra System, which nevertheless was discontinued by Texas Instruments in 2007.

In the next section we will consider the special case of a matrix $\boldsymbol{A}$ having only one column, i.e. we will consider the computation of the Moore–Penrose inverse of a vector in *Derive*. We will see that this is a straightforward task if the vector has only numeric elements. Computation of the Moore–Penrose inverse of a vector which contains symbolic elements might also be uncomplicated, but might turn out to be a problem.

We will then proceed to the computation of the Moore–Penrose inverse of a matrix in *Derive*. We will apply Greville's method which is an iterative algorithm that needs $n$ steps for the computation of the Moore–Penrose inverse of an $m$ by $n$ matrix. We will start again with matrices containing only numeric elements, and then consider the case of symbolic elements. As Greville's method requires the computation of the Moore–Penrose inverse of a vector in each step, the potential problem described in Sect. 2 might avoid the actual computation of the Moore–Penrose inverse of a matrix with symbolic elements.

In the last section we will consider a way out of the potential problem described in Sect. 2.

## 2 Computation of the Moore–Penrose Inverse of a Vector

The Moore–Penrose inverse of a (column) vector $\boldsymbol{a} \in \mathbb{R}^n$ is given by

$$\boldsymbol{a}^+ = \begin{cases} \frac{1}{\boldsymbol{a}'\boldsymbol{a}}\boldsymbol{a}' & \text{if } \boldsymbol{a} \neq \boldsymbol{o} \\ \boldsymbol{o}' & \text{if } \boldsymbol{a} = \boldsymbol{o} \end{cases} \tag{5}$$

where $\boldsymbol{o}$ denotes the ($n$ by 1) zero vector. Apparently, $\boldsymbol{a}^+$ is a row vector.

Since a vector is nothing else but a matrix with only one column, it should be declared in *Derive* as such. The *Derive* function `MPIV` for the computation of the Moore–Penrose inverse of a vector $\boldsymbol{a}$ given below works as follows (cf. Schmidt [8]): the function first checks if the actual parameter which has been passed on is indeed a (column) vector. If not, an error message appears on the screen. If the parameter turns out to be a vector, the function tests if $\boldsymbol{a}$ is a zero vector by computing $\boldsymbol{a}'\boldsymbol{a}$ and checking if this is equal to 0. If so, the Moore–Penrose inverse of $\boldsymbol{a} = \boldsymbol{o}$ is simply $\boldsymbol{a}^+ = \boldsymbol{o}'$. If $\boldsymbol{a}'\boldsymbol{a}$ is greater than 0, $\boldsymbol{a}^+$ equals the transpose of $\boldsymbol{a}$ (a row vector) divided by the scalar $\boldsymbol{a}'\boldsymbol{a}$.

```
         MPIV( a ) :=
           If DIM(a ' ) = 1
              If (a ' .a)↓1 ↓1 = 0
#1:              0.a'
                 a' /(a'.a) ↓1 ↓1
              "This is not a column vector !"
```

If a vector has symbolic elements, the `MPIV` function might not be able to compute its Moore–Penrose inverse. To exemplify this, consider the following set of vectors:

$$\boldsymbol{a} = \begin{pmatrix} 0 \\ 2 \end{pmatrix}; \quad \boldsymbol{b} = \begin{pmatrix} x \\ 2 \end{pmatrix}; \quad \boldsymbol{c} = \begin{pmatrix} 0 \\ x \end{pmatrix}$$

We define these vectors in *Derive* as matrices with one column:

```
#2:   a :=  [ 0 ]
            [ 2 ]

#3:   b :=  [ x ]
            [ 2 ]

#4:   c :=  [ 0 ]
            [ x ]
```

Clearly, as $\boldsymbol{a}'\boldsymbol{a} = 4 \neq 0$, according to (5) we have

$$\boldsymbol{a}^+ = \frac{1}{\boldsymbol{a}'\boldsymbol{a}}\boldsymbol{a}' = \frac{1}{4}(0 \quad 2) = (0 \quad \frac{1}{2})$$

Moving on to $\boldsymbol{b}$ we find that $\boldsymbol{b}'\boldsymbol{b} = x^2 + 4$ which means that $\boldsymbol{b}'\boldsymbol{b} > 0$ for any $x \in \mathbb{R}$. Hence we get

$$\boldsymbol{b}^+ = \frac{1}{\boldsymbol{b}'\boldsymbol{b}}\boldsymbol{b}' = \frac{1}{x^2+4}(x \quad 2) = \begin{pmatrix} \frac{x}{x^2+4} & \frac{2}{x^2+4} \end{pmatrix}$$

Considering now $\boldsymbol{c}$, it turns out that $\boldsymbol{c}'\boldsymbol{c} = x^2$, i.e. we have $\boldsymbol{c}'\boldsymbol{c} = 0$ for $x = 0$, and $\boldsymbol{c}'\boldsymbol{c} > 0$ otherwise. Therefore

$$\boldsymbol{c}^+ = \begin{cases} \frac{1}{\boldsymbol{c}'\boldsymbol{c}}\boldsymbol{c}' = \frac{1}{x^2}(0 \quad x) = (0 \quad \frac{1}{x}) & \text{if } x \neq 0 \\ \boldsymbol{o}' & \text{if } x = 0 \end{cases}$$

Let us now see how the `MPIV` function copes with these vectors:

#5: $$\mathrm{MPIV}(a) = \left[\left[0,\ \frac{1}{2}\right]\right]$$

#6: $$\mathrm{MPIV}(b) = \left[\left[\frac{x}{x^2+4},\ \frac{2}{x^2+4}\right]\right]$$

#7: $$\mathrm{MPIV}(c) = \mathrm{IF}\left(x = 0,\ 0 \cdot \begin{bmatrix} 0 \\ x \end{bmatrix}^{`},\ \begin{bmatrix} 0 \\ x \end{bmatrix}^{`} \cdot \left(\begin{bmatrix} 0 \\ x \end{bmatrix}^{`} \cdot \begin{bmatrix} 0 \\ x \end{bmatrix}\right)_{1,1}^{-1}\right)$$

The `MPIV` function has no problem computing the Moore–Penrose inverse of any vector which contains numbers only, and therefore computes $\boldsymbol{a}^+$ without difficulty. Note that as we define a column vector as a matrix with one column, its Moore–Penrose inverse (a row vector) is a matrix with one row. *Derive* reminds us of this detail by using double brackets.

The function also computes the Moore–Penrose inverse of vector $\boldsymbol{b}$ although it contains the symbol $x$, since for any value of $x$ we have $\boldsymbol{b} \neq \boldsymbol{o}$. Vector $\boldsymbol{c}$, on the other hand, would be a zero vector if the second element, $x$, equalled 0. Therefore, the `MPIV` function would not be able to actually compute $\boldsymbol{c}^+$ and thus shows the two possible results in the form `IF(I,II,III)` where `I` is a condition, `II` the formula if the condition is true ( $\boldsymbol{o}'$ from (5)), and `III` the formula if the condition is false ( $\frac{1}{\boldsymbol{c}'\boldsymbol{c}}\boldsymbol{c}'$ from (5)).

## 3 Computation of the Moore–Penrose Inverse of a Matrix

For the computation of the Moore–Penrose inverse of a matrix we apply Greville's method (Ben-Israel/Greville [1], p. 263; cf. Büning/Naeve/Trenkler/Waldmann [2], pp. 194–196, for an alternative method), which finds the Moore–Penrose inverse in a finite number of steps (the following description of the algorithm is taken from Schmidt/Trenkler [9], pp. 131–132).

We first consider the column notation of the $m$ by $n$ matrix $\boldsymbol{A}$

$$\boldsymbol{A} = [\boldsymbol{a}_1 \quad \boldsymbol{a}_2 \quad \cdots \quad \boldsymbol{a}_n]$$

and denote the $m$ by $k$ submatrix, which comprises the first $k$ columns of $\boldsymbol{A}$, by

$$\boldsymbol{A}_k = [\boldsymbol{a}_1 \quad \boldsymbol{a}_2 \quad \cdots \quad \boldsymbol{a}_k]$$

Hence,

$$\boldsymbol{A}_k = [\boldsymbol{A}_{k-1} \quad \boldsymbol{a}_k]$$

Furthermore, we define the following vectors for $j \geq 2$:

$$d'_j = a'_j A^{+'}_{j-1} A^{+}_{j-1}$$

$$c_j = \left(I - A_{j-1} A^{+}_{j-1}\right) a_j$$

$$b'_j = c^{+}_j + \frac{1 - c^{+}_j c_j}{1 + d'_j a_j} d'_j$$

Note that $d'_j$ is a row vector, $c_j$ a column vector (and hence $c^{+}_j$ a row vector) and $b'_j$ a row vector. Then we have

$$A^{+}_j = \begin{bmatrix} A_{j-1} & a_j \end{bmatrix}^{+} = \begin{bmatrix} A^{+}_{j-1} - A^{+}_{j-1} a_j b'_j \\ b'_j \end{bmatrix} \tag{6}$$

Since $A_1 = a_1$ is a matrix having only one column, its Moore–Penrose inverse can be computed by (5). Using (6) we can then iteratively calculate $A^{+}_2, A^{+}_3, \ldots, A^{+}_n = A^{+}$.

The MPI function given below works as follows (cf. Schmidt [8]): initially, the MPIV function is called with the first column of $A$ as a parameter. The result returned is the first row of $A^{+}$ (which is only an intermediate result). The MPI function then proceeds to the second column of $A$ and computes the second intermediate $A^{+}$ by transforming the previous result and appending another row. This is repeated for all columns of $A$. After as many steps as the number of columns of $A$, $A^{+}$ is computed by the MPI function.

```
        MPI(A, APLUS, aj, dt, c, bt, J) :=
          Prog
            APLUS := MPIV(A COL [1])
            J := 2
            Loop
              If J > DIM(A')
#2:              RETURN APLUS
              aj := A COL [J]
              dt := aj` · APLUS` · APLUS
              c := (IDENTITY_MATRIX(DIM(A)) – A COL [1, ..., J –1]·APLUS)·aj
              bt := MPIV(c) + (1 – MPIV(c)·c)/(1 + dt·aj)·dt
              APLUS := APPEND(APLUS – APLUS·aj·bt, bt)
              J :+ 1
```

Note that in each step the MPIV function is called. Hence, in the case of symbolic elements the MPI function might be unable to compute $A^{+}$. To exemplify this, consider the matrices

$$A = \begin{pmatrix} 1 & 0 \\ 2 & 0 \end{pmatrix}; \quad B = \begin{pmatrix} x & 0 \\ 2 & 0 \end{pmatrix}; \quad C = \begin{pmatrix} 1 & 0 \\ 2 & x \end{pmatrix}$$

which we define in *Derive*:

$$\#3:\quad A := \begin{bmatrix} 1 & 0 \\ 2 & 0 \end{bmatrix}$$

$$\#4:\quad B := \begin{bmatrix} x & 0 \\ 2 & 0 \end{bmatrix}$$

$$\#5:\quad C := \begin{bmatrix} 1 & 0 \\ 2 & x \end{bmatrix}$$

Clearly, as ***A*** has only numeric elements the `MPI` function easily computes its Moore–Penrose inverse. The `MPI` function is also able to compute $\boldsymbol{B}^+$ since the first column of ***B*** is not a zero vector whatever the value of $x$ is, and the second column of ***B*** is a zero vector anyway.

$$\#6:\quad \mathrm{MPI(A)} = \begin{bmatrix} \frac{1}{5} & \frac{2}{5} \\ 0 & 0 \end{bmatrix}$$

$$\#7:\quad \mathrm{MPI(B)} = \begin{bmatrix} \frac{x}{x^2+4} & \frac{2}{x^2+4} \\ 0 & 0 \end{bmatrix}$$

$$\#8:\quad \mathrm{MPI(C)} = \mathrm{APPEND}\left(\left[\left[-\frac{2\cdot x}{5}\right]\right]\cdot\left(\left(\left(\mathrm{IF}\left(x = 0, 0\cdot\begin{bmatrix} -\frac{2\cdot x}{5} \\ \frac{x}{5} \end{bmatrix}, \begin{bmatrix} -\frac{2\cdot x}{5} \\ \frac{x}{5} \end{bmatrix}, \left(\right.\right.\right.\right.\right.\right.$$

However, trying to compute $\boldsymbol{C}^+$ with the `MPI` function is not successful since the second column of ***C*** turns into a zero vector if $x = 0$. #8 in the above screenshot is not the entire output generated by the `MPI` function which is several lines long and contains 4 unsolved `IF` expressions.

## 4 A Way Out

The problem the `MPI` function might have in computing the Moore–Penrose inverse of a matrix with symbolic elements is entirely due to the `MPIV` function which is called in two statements within the `MPI` function. Therefore, in order to find a way out we have to look for a remedy to overcome the problem the `MPIV` function might have in computing the Moore–Penrose inverse of a vector with symbolic elements.

Let us reconsider the vectors

$$\boldsymbol{a} = \begin{pmatrix} 0 \\ 2 \end{pmatrix}; \quad \boldsymbol{b} = \begin{pmatrix} x \\ 2 \end{pmatrix}; \quad \boldsymbol{c} = \begin{pmatrix} 0 \\ x \end{pmatrix}$$

from Sect. 2 and determine their rank. Clearly, $r(\boldsymbol{a}) = 1$, and $r(\boldsymbol{b}) = 1$ for any $x \in \mathbb{R}$. Since the value of $x$ is crucial as to whether $\boldsymbol{c}$ is a zero vector or not, we have

$$r(\boldsymbol{c}) = \begin{cases} 1 & \text{if } x \neq 0 \\ 0 & \text{if } x = 0 \end{cases}$$

But if we compute the rank of the three vectors in *Derive*

```
#5:                                        RANK(a) = 1
#6:                                        RANK(b) = 1
#7:                                        RANK(c) = 1
```

we get $r(\boldsymbol{c}) = 1$, i.e. *Derive* does not make a case differentiation. Apparently, the single $x$ value which turns $\boldsymbol{c}$ into a zero vector, is neglected.

Why not use this viewpoint for an alternative `MPIV` function which does *not* consider the particular case in which a vector becomes a zero vector for a certain value of a symbolic element of this vector, but *only* the infinite number of cases where this vector is not a zero vector? The respective `MPIV0` function is given below:

```
          MPIV0(a) :=
            If DIM (a') = 1
#1:             a ' /(a' ·a) ↓1 ↓1
                "This is not a column vector !"
```

Unsurprisingly, this *Derive* function is not capable of computing the Moore–Penrose inverse of a zero vector, which is not a real problem as long as we are interested in the computation of the Moore–Penrose inverse of vectors only (and not vectors as columns of matrices). Using the `MPIV0` function for the computation of the Moore–Penrose inverse of the three vectors, we get not only the same $\boldsymbol{a}^{+}$ and $\boldsymbol{b}^{+}$ as in Sect. 2, but now also $\boldsymbol{c}^{+}$.

| | |
|---|---|
| #8: | $\text{MPIV0(a)} = \left[\left[0, \frac{1}{2}\right]\right]$ |
| #9: | $\text{MPIV0(b)} = \left[\left[\frac{x}{x^2+4}, \frac{2}{x^2+4}\right]\right]$ |
| #10: | $\text{MPIV0(c)} = \left[\left[0, \frac{1}{x}\right]\right]$ |

Clearly, what is computed in *Derive* is the Moore–Penrose inverse of $\boldsymbol{c}$ in case $x \neq 0$. The special case $x = 0$ is ignored.

We now reconsider the matrices

$$\boldsymbol{A} = \begin{pmatrix} 1 & 0 \\ 2 & 0 \end{pmatrix}; \quad \boldsymbol{B} = \begin{pmatrix} x & 0 \\ 2 & 0 \end{pmatrix}; \quad \boldsymbol{C} = \begin{pmatrix} 1 & 0 \\ 2 & x \end{pmatrix}$$

from Sect. 3 and determine their rank. Clearly, $r(\boldsymbol{A}) = 1$, and $r(\boldsymbol{B}) = 1$ for any $x \in \mathbb{R}$. Since the value of $x$ is crucial as to whether the second column of $\boldsymbol{C}$ is a zero vector or not, we have

$$r(\boldsymbol{C}) = \begin{cases} 2 & \text{if } x \neq 0 \\ 1 & \text{if } x = 0 \end{cases}$$

Computing the rank of the three matrices in *Derive*

| | |
|---|---|
| #11: | RANK(A) = 1 |
| #12: | RANK(B) = 1 |
| #13: | RANK(C) = 2 |

we get $r(\boldsymbol{C}) = 2$, i.e. *Derive* again does not make a case differentiation. Apparently, the single $x$ value which turns the second column of $\boldsymbol{C}$ into a zero vector is neglected.

Finally we compute the Moore–Penrose inverse of the three matrices in *Derive* with the `MPI0` function, which is identical to the `MPI` function in Sect. 3 except that it calls the `MPIV0` function instead of the `MPIV` function in lines 3 and 11.

#14: $\mathrm{MPI0(A)} = \begin{bmatrix} ? & ? \\ ? & ? \end{bmatrix}$

#15: $\mathrm{MPI0(B)} = \begin{bmatrix} ? & ? \\ ? & ? \end{bmatrix}$

#16: $\mathrm{MPI0(C)} = \begin{bmatrix} 1 & 0 \\ -\frac{2}{x} & \frac{1}{x} \end{bmatrix}$

As we get $\boldsymbol{C}^+$ this time (more precisely the Moore–Penrose inverse of $\boldsymbol{C}$ if $x \neq 0$, i.e. the special case $x = 0$ is ignored), it turns out that the `MPI0` function is unable to compute the Moore–Penrose inverse of $\boldsymbol{A}$ and $\boldsymbol{B}$. Note that $\boldsymbol{C}$ is a nonsingular matrix for $x \neq 0$ such that $\boldsymbol{C}^+ = \boldsymbol{C}^{-1}$. Computing the inverse of $\boldsymbol{C}$ in *Derive* generates the same matrix as the `MPI0` function, i.e. *Derive* is consistent in terms of disregarding the special case $x = 0$.

#17: $C^{-1} = \begin{bmatrix} 1 & 0 \\ -\frac{2}{x} & \frac{1}{x} \end{bmatrix}$

The obvious reason for the inability of the `MPI0` function to compute the Moore–Penrose inverse of $\boldsymbol{A}$ and $\boldsymbol{B}$ is that it cannot handle the case where a column of the input matrix is a zero vector (like the second column of $\boldsymbol{A}$ or $\boldsymbol{B}$), or, more precisely, if the `MPIV0` function is called with a zero vector as a parameter. Note that if you divide 0 by 0 in *Derive*, "?" is displayed.

Considering the two alternatives for the computation of the Moore–Penrose inverse with *Derive* described in this paper, namely the `MPI` function (using the `MPIV` function) and the `MPI0` function (using the `MPIV0` function), we find that neither is always superior.

The `MPI` function works if the input matrix does not have any symbolic elements (even if there are zero vectors), but might fail if it cannot decide if a certain vector is a zero vector. On the other hand, the `MPI0` function can handle symbolic elements, but fails however, should a zero vector become the input of the `MPIV0` function during the computation of the Moore–Penrose inverse using Greville's method.

## References


[1] Ben-Israel, A., Greville, T.N.E.: Generalized Inverses: Theory and Applications, 2nd edn. Springer, New York (2003)
[2] Büning, H., Naeve, P., Trenkler, G., Waldmann, K.-H.: Mathematik für Ökonomen im Hauptstudium. Oldenbourg, München (2000)
[3] Greville, T.N.E.: Some Applications of the Pseudoinverse of a Matrix. SIAM Rev. **2**, 15–22 (1960)
[4] Moore, E.H.: On the Reciprocal of the General Algebraic Matrix (Abstract). B. Am. Math. Soc. **26**, 394–395 (1920)
[5] Penrose, R.: A Generalized Inverse for Matrices. Proc. Camb. Philos. Soc. **51**, 406–413 (1955)
[6] Rao, C.R.: A Note on a Generalized Inverse of a Matrix with Applications to Problems in Mathematical Statistics. J. Roy. Stat. Soc. B **24**, 152–158 (1962)
[7] Rao, C.R., Mitra, S.K.: Generalized Inverse of Matrices and Its Applications. Wiley, New York (1971)
[8] Schmidt, K.: Computing the Moore-Penrose Inverse of a Matrix with a Computer Algebra System. Internat. J. Math. Edu. Sci. Techn. **39**, 557–562 (2008)
[9] Schmidt, K., Trenkler, G.: Einführung in die Moderne Matrix-Algebra, 2nd edn. Springer, Berlin (2006)

# On Permutations of Matrix Products

Hans Joachim Werner and Ingram Olkin

**Abstract** It is well-known that trace$(AB) \geq 0$ for real-symmetric nonnegative definite matrices $A$ and $B$. However, trace$(ABC)$ can be positive, zero or negative, even when $C$ is real-symmetric nonnegative definite. The genesis of the present investigation is consideration of a product of square matrices $A = A_1A_2 \cdots A_n$. Permuting the factors of $A$ leads to a different matrix product. We are interested in conditions under which the spectrum remains invariant. The main results are for square matrices over an arbitrary algebraically closed commutative field. The special case of real-symmetric, possibly nonnegative definite, matrices is also considered.

## 1 Introduction

For a given product of not necessarily distinct factors it is natural to call two factors direct (or next-door) neighbors (with respect to this product) if in this product these two factors stand next to each other. For convenience, the first and the last factor in a product are also called direct neighbors of each other. The length of a product is defined as the number of its (not necessarily distinct) factors. So, in a product of length $\geq 3$, each of its factors has at least two (not necessarily distinct) neighbors. If all factors are distinct, each factor has exactly two direct neighbors. Moreover, if the product has length 3, then each factor is obviously a direct neighbor of the remaining two (not necessarily distinct) factors.

Now, let $p$ be a product of length $n$, and let $q$ be obtained from $p$ by a permutation of its $n$ (not necessarily distinct) factors. These two products are said to be DN-related to each other if each of the $n$ factors has in both products exactly the same direct neighbors. In which case, we write $p \sim_{\text{DN}} q$.

---

Hans Joachim Werner
Wirtschaftswissenschaftlicher Fachbereich, Statistische Abteilung, Universität Bonn, D-53113 Bonn, Germany
hjw.de@uni-bonn.de

In this note, we consider products of finite length $n$ of square matrices over an algebraically closed commutative field $\mathbb{F}$. Let $A_i$ $(i = 1, 2, \cdots, n)$ be $n$ matrix symbols for $m \times m$ matrices over $\mathbb{F}$, and let $A := A_1 A_2 \cdots A_n$ be the naturally ordered matrix product of these $n$ symbols $A_i$. As usual, let $\mathscr{S}_n$ denote the symmetric group on the natural numbers $1, 2, \cdots, n$, i.e., let $\mathscr{S}_n$ be the set of all the permutations of these symbols $1, 2, \cdots, n$. For each $\pi := (\pi_1, \pi_2, \cdots, \pi_n) \in \mathscr{S}_n$, let $A_\pi := \prod_{i=1}^{n} A_{\pi_i}$, and let $\mathscr{S}_A := \{A_\pi \mid \pi \in \mathscr{S}_n\}$, i.e., let $\mathscr{S}_A$ be the set of all products obtained from $A$ by permuting its factors. By varying $\pi \in \mathscr{S}_n$ we clearly obtain all possible $n!$ matrix products of length $n$ of $A_1, A_2, \cdots, A_n$. In other words, $\mathscr{S}_A$ can be considered as the group of all permutations on the $n$ matrix symbols $A_i$ $(i = 1, 2, \cdots, n)$. Because the relation $\sim_{\mathrm{DN}}$ is reflexive, symmetric and transitive, it is an equivalence relation, and so it is clear that any two equivalence classes are either equal or disjoint. Hence the collection of equivalence classes $\mathrm{DN}(B) := \{A_\pi \,|A_\pi \sim_{\mathrm{DN}} B\}$, $B \in \mathscr{S}_A$, forms a partition of $\mathscr{S}_A$. It is not difficult to see that there are exactly $(n-1)!/2$ disjoint equivalence classes for $n \geq 3$. For observe that the equivalence class $\mathrm{DN}(A_\pi)$ of

$$A_\pi = A_{\pi_1} A_{\pi_2} \cdots A_{\pi_n} \tag{1}$$

contains exactly all those products of the matrix symbols $A_i$ $(i = 1, 2, \cdots, n)$ that are obtained from (1) by cyclical and/or reverse re-orderings. Because there are $n$ cyclical and $n$ reverse re-orderings of the factors in $A_\pi$, each equivalence class therefore consists of $2n$ factor permutations of the $n$ matrix symbols, and so it is clear that we have, as claimed, $(n-1)!/2$ disjoint equivalence classes for $n \geq 3$.

In Sect. 3, we study some functions which are defined on $\mathscr{S}_A$, mainly the trace$(\cdot)$ and the spectrum$(\cdot)$. There, we particularly show that if $A_i$ $(i = 1, 2, \cdots, n)$ are all symmetric $m \times m$ matrices over $\mathbb{F}$, then these functions are constant on each equivalence class. Section 4 deals with the set of all symmetric nonnegative definite $m \times m$ matrices over the field $\mathbb{R}$ of real numbers. Section 2 contains some known results, from which our findings in the subsequent two chapters easily follow.

## 2 Prerequisite Results

For real or complex matrices the following results are well-known; cf. [3] or [5]. Because for matrices over an arbitrary algebraically closed commutative field $\mathbb{F}$ the according results can be established on identical lines the proofs are omitted.

**Fact 2.1.** [See, e.g., pp. 53–55 in [3].] *Let $A$ be a square matrix of order $m$ over the field $\mathbb{F}$. The characteristic polynomial of $A$, being defined as $c_A(\lambda) := \det(\lambda I_m - A)$, is a monic polynomial of degree $m$ with exactly $m$ (not necessarily distinct) roots $\lambda_1, \lambda_2, \cdots, \lambda_m \in \mathbb{F}$, called the eigenvalues of $A$. When writing the characteristic polynomial of $A$ as*

$$c_A(\lambda) = \lambda^m - c_1 \lambda^{m-1} + c_2 \lambda^{m-2} + \cdots + (-1)^m c_m,$$

*the following relationships hold between the coefficients $c_r$ $(r = 1,2,\cdots,m)$, the eigenvalues of A, the r-th compound $A^{(r)}$ and the principal minors of A:*

$$c_r = \text{trace}(A^{(r)}) = \sum \text{(all } r \times r \text{ principal minors)} \quad = \sum_{1\le i_1<i_2<\cdots i_r\le m} \lambda_{i_1}\lambda_{i_2}\cdots\lambda_{i_r}.$$

*Hence, in particular,*

$$c_1 = \text{trace}(A) = \sum_{i=1}^{m} \lambda_i \qquad \text{and} \qquad c_m = \det(A) = \prod_{i=1}^{n} \lambda_i.$$

For more details concerning compounds we refer, for instance, to the books by Aitken [1], Gröbner [2] or Marshall & Olkin [4].

**Fact 2.2.** [See, e.g., Exercise 6 on p. 56 in [3].] *Let A be a square matrix over $\mathbb{F}$. Then*

$$c_A(\lambda) = c_{A^t}(\lambda),$$

*where $A^t$ denotes the transpose of A. In other words, A and its transpose $A^t$ possess the same characteristic polynomial, and so these matrices have the same set of eigenvalues with corresponding algebraic multiplicities.*

**Fact 2.3.** [See, e.g., Exercise 7.1.19 on p. 503 in [5].] *Let A and B be square matrices of order m over the field $\mathbb{F}$. Then the matrices AB and BA have the same set of eigenvalues with corresponding algebraic multiplicities. Hence, in particular,*

$$\text{spectrum}(AB) = \text{spectrum}(BA).$$

For the sake of completeness as well as for easier reference, we also cite some well-known results for Hermitian matrices (over the field $\mathbb{C}$ of complex numbers).

**Fact 2.4.** [See, e.g., pp. 75–78 in [3].] *Let A be an Hermitian $m \times m$ matrix. Then all eigenvalues of A are real. Moreover, A is unitarily similar to the diagonal matrix $D =$ $\text{diag}(\lambda_1,\lambda_2,\cdots,\lambda_m)$ of its eigenvalues, i.e., there exists an $m \times m$ (unitary) matrix $U = (u_1,u_2,\cdots,u_m)$ such that*

$$UU^* = I_m \quad \text{and} \quad A = UDU^*$$

*or, equivalently,*

$$\sum_{i=1}^{m} u_i u_i^* = I_m \quad \text{and} \quad A = \sum_{i=1}^{m} \lambda_i u_i u_i^*,$$

*with $(\cdot)^*$ indicating as usual the conjugate transpose of $(\cdot)$. The pairs $(\lambda_i,u_i)$, $i = 1,2,\cdots,m$, are eigenpairs for A, i.e., $\lambda_i$ and $u_i$, satisfying $Au_i = \lambda_i u_i$, are eigenvalues and eigenvectors of A, respectively.*

For real-symmetric matrices the previous result allows the following version.

**Fact 2.5.** [See, e.g., pp. 75–78 in [3].] *Let $A$ be a real-symmetric $m \times m$ matrix. Then all eigenvalues of $A$ are real. Moreover, $A$ is orthogonally similar to the diagonal matrix $D = \operatorname{diag}(\lambda_1, \lambda_2, \cdots, \lambda_m)$ of its eigenvalues, i.e., there exists an $m \times m$ (orthogonal) real matrix $P = (p_1, p_2, \cdots, p_m)$ such that*

$$PP^t = I \quad \text{and} \quad A = PDP^t,$$

*or equivalently*

$$\sum_{i=1}^{m} p_i p_i^t = I_m \quad \text{and} \quad A = \sum_{i=1}^{m} \lambda_i p_i p_i^t.$$

*The pairs $(\lambda_i, p_i)$, $i = 1, 2, \cdots, m$, are eigenpairs for $A$, i.e., $\lambda_i$ and $p_i$, satisfying $Ap_i = \lambda_i p_i$, are eigenvalues and eigenvectors of $A$, respectively.*

Below we will also make use of the following two results.

**Fact 2.6.** [See, e.g., p. 559 in [5].] *Let $A$ be a real-symmetric nonnegative definite matrix. Then all its eigenvalues are nonnegative. If all its eigenvalues are positive, then $A$ is a positive definite matrix.*

**Fact 2.7.** [See, e.g., Exercise 7.2.16 in [5].] *Let $A$ and $B$ be diagonalizable matrices of the same order, say $m \times m$. Then $A$ and $B$ commute, i.e., $AB = BA$, if and only if $A$ and $B$ can be simultaneously diagonalized, i.e., if and only if*

$$A = XD_AX^{-1} \quad \text{and} \quad B = XD_BX^{-1}$$

*for some regular matrix $X = (x_1, x_2, \cdots, x_m)$ and some diagonal matrices $D_A = \operatorname{diag}(\lambda_1, \lambda_2, \cdots, \lambda_m)$ and $D_B = \operatorname{diag}(\mu_1, \mu_2, \cdots, \mu_m)$. For $i = 1, 2, \cdots, m$, the pairs $(\lambda_i, x_i)$ and $(\mu_i, x_i)$ are eigenpairs of $A$ and $B$, respectively.*

## 3 Main Results

In virtue of the Facts 2.1, 2.2, and 2.3 we now obtain the following.

**Theorem 3.1.** *For symmetric $m \times m$ matrices $A_1$, $A_2, \cdots, A_n$ over the field $\mathbb{F}$, let $A := \prod_{i=1}^{n} A_i$. Then we have*

$$c_A(\lambda) = c_{A_\pi}(\lambda),$$

*irrespective of $A_\pi \in \mathrm{DN}(A)$. Consequently,*

$$\begin{aligned} \operatorname{trace}(A^{(r)}) &= \sum (\text{all } r \times r \text{ principal minors of } A) \\ &= \sum (\text{all } r \times r \text{ principal minors of } A_\pi) \\ &= \operatorname{trace}(A_\pi^{(r)}), \end{aligned}$$

*irrespective of* $A_\pi \in \mathrm{DN}(A)$*, and so, in particular,*

$$\mathrm{trace}(A) = \mathrm{trace}(A_\pi)$$

*for all* $A_\pi$ *with* $A_\pi \sim_{\mathrm{DN}} A$.

*Proof.* Recall from Sect. 1 that $\mathrm{DN}(A)$ consists exactly of all those $2n$ matrix products that are obtainable from $A = A_1A_2\cdots A_n$ by cyclical and/or reverse re-orderings of the $n$ matrix factors in $A$. In virtue of Fact 2.1 and Fact 2.2, the claimed results now follow easily by means of Fact 2.3 and the fact that, for instance, $(A_1A_2\cdots A_n)^t = A_nA_{n-1}\cdots A_1$. Details are left to the reader. □

The previous theorem deserves some further emphasizing. For, it may be surprising that $\mathrm{trace}(A) = \mathrm{trace}(A_\pi)$, irrespective of $A_\pi \in \mathrm{DN}(A)$. Of course, $\det(A) = \det(A_\pi)$ does always hold and, needless to say, this is not surprising. But that the sum of the eigenvalues taken $r$ at a time are equal may again be surprising. Observe that trace is the sum of the eigenvalues taken 1 at a time, and that determinant is the sum of the eigenvalues taken $m$ at a time. Our previous theorem tells us that even the sum of all eigenvalues taken in between at a time are also equal, that is, $\mathrm{trace}(A^{(r)}) = \mathrm{trace}(A_\pi^{(r)})$.

Because for any matrix product $A$ of length 3, $\mathrm{DN}(A) = \mathscr{S}_A$, the following is an immediate consequence of the previous theorem observing that $(BC)^{(r)} = B^{(r)}C^{(r)}$.

**Corollary 3.2.** *Let* $A := A_1A_2A_3$*, with* $A_1$*,* $A_2$ *and* $A_3$ *being symmetric matrices of the same order* $m \times m$ *over the field* $\mathbb{F}$*. Then*

$$c_{A_1A_2A_3}(\lambda) = c_{A_{\pi_1}A_{\pi_2}A_{\pi_3}}(\lambda)$$

*for each permutation* $\pi = (\pi_1, \pi_2, \pi_3) \in \mathscr{S}_3$*. Hence, in particular,*

$$\mathrm{trace}(A_1^{(r)}A_2^{(r)}A_3^{(r)}) = \mathrm{trace}(A_{\pi_1}^{(r)}A_{\pi_2}^{(r)}A_{\pi_3}^{(r)}),$$

*irrespective of* $\pi = (\pi_1, \pi_2, \pi_3) \in \mathscr{S}_3$ *and* $r \in \mathbb{N}_m$*, where* $\mathbb{N}_m := \{r \in \mathbb{N} : 1 \le r \le m\}$.

If two of the three (not necessarily symmetric) square matrices $A_1$, $A_2$ and $A_3$ in the matrix product $A := A_1A_2A_3$ commute, then each matrix in $\mathscr{S}_A$ can obviously be obtained by a cyclical reordering of the factors of $A$ and/or by the commutation of the commuting factors, and so we obtain the following.

**Corollary 3.3.** *Let* $A := A_1A_2A_3$*, with* $A_1$*,* $A_2$ *and* $A_3$ *being such that at least two of the three* $m \times m$ *matrices commute. Then*

$$c_{A_1A_2A_3}(\lambda) = c_{A_{\pi_1}A_{\pi_2}A_{\pi_3}}(\lambda)$$

*and so*

$$\mathrm{trace}(A_1^{(r)}A_2^{(r)}A_3^{(r)}) = \mathrm{trace}(A_{\pi_1}^{(r)}A_{\pi_2}^{(r)}A_{\pi_3}^{(r)}),$$

*irrespective of* $\pi = (\pi_1, \pi_2, \pi_3) \in \mathscr{S}_3$ *and* $r \in \mathbb{N}_m$.

## 4 Special Case: Products of Length Three of Real-Symmetric Nonnegative Definite Matrices

We conclude this note by considering the special case of matrix products of length 3, say $A_1A_2A_3$, where all three factors are real-symmetric nonnegative definite matrices. Corollary 3.2 of the preceding section tells us that the matrices $A_1A_2A_3$ and $A_2A_1A_3$ have the same characteristic polynomial and so

$$\mathrm{trace}((A_1A_2 + A_2A_1)A_3) = 2\,\mathrm{trace}(A_1A_2A_3)$$

holds true. An interesting question is, whether for given such factors anything can be said about the signum of $\mathrm{trace}((A_1A_2 + A_2A_1)A_3)$.

So, let $A_1, A_2$ and $A_3$ be symmetric nonnegative definite $m \times m$ matrices over the field $\mathbb{R}$ of real numbers. Then, according to Fact 2.5, these matrices are orthogonally similar to some diagonal matrices and hence, for $i = 1, 2, 3$, $A_i$ can always be written as

$$A_i = \sum_{j=1}^{m} \lambda_{ij} x_{ij} x_{ij}^t,$$

where $(\lambda_{ij}, x_{ij})$ $(j = 1, 2, \cdots, m)$ are eigenpairs of $A_i$ and $\{x_{ij} \mid j = 1, 2, \cdots, m\}$ constitutes an orthonormal basis for $\mathbb{R}^m$. Then

$$\mathrm{trace}(A_1A_2) = \sum_{j=1}^{m}\sum_{k=1}^{m} \lambda_{1j}\lambda_{2k}(x_{1j}^t x_{2k})^2 \geq 0, \tag{2}$$

since, in view of Fact 2.6, all eigenvalues of a real-symmetric nonnegative definite matrix are nonnegative. One might be tempted to believe that this result can be extended to three factors. That this, however, is erroneous is illustrated by our next example.

**Example 4.1.** Consider the real-symmetric positive definite matrices

$$A_1 := \begin{pmatrix} 4 & -1.9 \\ -1.9 & 1 \end{pmatrix}, \quad A_2 := \begin{pmatrix} 1 & 0.9 \\ 0.9 & 1 \end{pmatrix} \quad \text{and} \quad A_3 := \begin{pmatrix} 1 & -1.4 \\ -1.4 & 2 \end{pmatrix}.$$

Then

$$A_1A_2A_3 = \begin{pmatrix} -0.09 & 0.194 \\ 0.006 & -0.02 \end{pmatrix} \quad \text{and} \quad A_2A_1A_3 = \begin{pmatrix} 3.69 & -5.206 \\ 2.694 & -3.8 \end{pmatrix},$$

and so trace$(A_1A_2A_3) = -0.11$ and trace$(A_2A_1A_3) = -0.11$. The traces are negative and coincide; the latter is in accordance with our findings in Section 2. The spectrum of both matrices as well as all other matrices from $\mathscr{S}_{A_1A_2A_3}$ is given by

$$\begin{aligned}\text{spectrum}(A_1A_2A_3) &= \{(\sqrt{61}-55)/1000, (-\sqrt{61}-55)/1000\}\\ &= \text{spectrum}(A_2A_1A_3).\end{aligned}$$

We conclude with emphasizing that therefore, without any further restrictive assumptions, nothing can be said about the signum of the trace of the product of three real-symmetric nonnegative definite matrices. The trace can be positive, negative, or even 0. If, however, the $m \times m$ matrices $A_1$, $A_2$, and $A_3$ are all real-symmetric nonnegative definite and, in addition, such that at least two of them commute, then, in view of Fact 2.6 and Fact 2.5, it is clear that the product of the commuting pair of matrices is itself a symmetric nonnegative definite matrix. Since in such a situation the product of $A_1A_2A_3$ can hence be considered as the product of two real-symmetric nonnegative definite matrices, it follows from the lines around (2) that the trace of $A_1A_2A_3$ is indeed also nonnegative. Needless to say, if $A_1$, $A_2$ and $A_3$ are real-symmetric positive definite and two of these matrices commute, then the trace of $A_1A_2A_3$ is necessarily positive.

## References


[1] Aitken, A.C.: Determinants and Matrices. Oliver & Boyd, Edinburgh-London (1952)

[2] Gröbner, W.: Matrizenrechnung. Bibliographisches Institut, Mannheim (1965)

[3] Lancaster, P.: Theory of Matrices. Academic, New York (1969)

[4] Marshall, A.W., Olkin, I.: Inequalities: Theory of Majorization and its Applications. Academic, New York (1979)

[5] Meyer, C.D.: Matrix Analysis and Applied Linear Algebra. SIAM, Philadelphia, USA (2000)

# Part VII
# Special Topics

*Models are to be used and not to believed.*

Henri Theil

# Some Comments on Fisher's $\alpha$ Index of Diversity and on the *Kazwini Cosmography*

Oskar Maria Baksalary, Ka Lok Chu, Simo Puntanen, and George P. H. Styan

**Abstract** Biodiversity, or biological diversity, is "the variety of life on our planet, measurable as the variety within species, between species, and the variety of ecosystems" [12, 41] and the most widely applied index of biodiversity is surely Fisher's $\alpha$, defined implicitly by $S = \alpha \log_e\{1 + (n/\alpha)\}$, where $n$ is the number of individuals and $S$ is the number of species. This index $\alpha$ was first proposed over 60 years ago by R. A. Fisher in a three-part joint paper with A. Steven Corbet and C. B. Williams [14]. We also present some comments on the diversity of the paintings by Johannes Vermeer (1632–1675) depicted on postage stamps updating our findings in [3]. The earliest study of biodiversity seems to be that reported in the *Kazwini Cosmography* c. 1283; this study involved 72 different kinds of *papillons* that were collected in what we believe was Baghdad in c. 900 AD. We also found some magic squares in the *Kazwini Cosmography*. Our list of references is annotated and contains hyperlinks to open-access and restricted-access files on the internet.

## 1 Introduction and Mise-en-scène

What is now widely known as Fisher's $\alpha$ Index[1] was first proposed over 60 years ago by R. A. Fisher in a three-part joint paper with A. S. Corbet & C. B. Williams [14] in which Fisher applied a logarithmic-series model to Corbet's data on Malayan

Oskar Maria Baksalary
Faculty of Physics, Adam Mickiewicz University, ul. Umultowska 85, PL 61-614 Poznań, Poland
baxx@amu.edu.pl

[1] Fisher's $\alpha$ is also known as Fisher's $\alpha$ log series (bio)diversity index or as Fisher's $\alpha$ index of (bio)diversity.

butterflies and to Williams's data on selected nocturnal *Lepidoptera*[2] caught in a light-trap at Rothamsted Experimental Station[3].

Fisher's $\alpha$ is defined implicitly by

$$S = \alpha \log_e \left(1 + \frac{n}{\alpha}\right),$$

where $n$ is the number of individuals and $S$ is the number of species. Fisher's $\alpha$ is always positive, and the larger its value, the more diverse is the underlying population. Moreover, $\alpha$ is common to all samples from a single population and so is a property of the population [51, p. 148].

In this paper we survey some of the underlying theory and look at some applications of Fisher's $\alpha$ to the study of the diversity within certain sets of butterflies, moths, and trees. In addition, we present some comments on diversity in paintings by Johannes Vermeer (1632–1675) depicted on postage stamps. We also comment on an early study of biodiversity and some magic squares reported in the *Kazwini Cosmography*[4] first published in the thirteenth century.

Biodiversity, or biological diversity, is the variety of life on our planet, measurable as the variety within species, between species, and the variety of ecosystems [12]. In her *Measuring Biological Diversity* [30, Cover 4], Anne Magurran points out that: The diversity of life on earth inspires fundamental ecological questions concerning the abundance of species and their distribution over space and time. The rapid loss of this diversity, primarily due to the impact of humanity, makes the need for effective ways of measuring biological diversity more important than ever. *We agree completely!*

Fisher's $\alpha$ was first proposed over 60 years ago in 1943 in a seminal paper [14] by Ronald Aylmer Fisher, later Sir Ronald Fisher (1890–1962), Alexander Steven Corbet (1896–1948), and Carrington Bonsor Williams (1889–1981). Fisher was a very well-known statistician, evolutionary biologist, and geneticist, while Corbet was a biochemist by training, who worked as a soil chemist and bacteriologist before becoming a lepidopterist[5]. Williams was "the first real quantitative empirical ecologist: a naturalist who was numerate" (Wigglesworth [48]), and the author of the wonderful book *Patterns in the Balance of Nature and Related Problems in Quantitative Ecology* [51], where [51, Plate I, page facing page 32] Williams shows a catch of $n$ = 219 individual *Lepidoptera* of $S$ = 42 different species captured in a light-trap at Rothamsted on the night of 23 July 1946; he finds that Fisher's $\alpha \simeq$ 15.44.

---

[2] *Lepidoptera* is a large order of insects, characterized by having 4 membranous wings covered with scales; it comprises the butterflies and moths [36].

[3] Rothamsted Experimental Station, located at Harpenden, Hertfordshire (just north of London), is one of the oldest agricultural research institutions in the world. It is now known as Rothamsted Research [49].

[4] *Book of the Marvels of Nature and the Singularities of Created Things,* by Zakariyyā' ibn Muḥammad ibn Maḥmūd Abū Yaḥyā al-Ḳazwīnī (c. 1203–1283).

[5] *The Butterflies of the Malay Peninsula* [10] with Henry Maurice Pendlebury (1893–1945) may well be Corbet's most important contribution to entomology.

In addition, Williams [51, Plate II, page facing page 33] shows $n = 65$ individual hawk moths[6] of $S = 19$ different species captured in a light-trap near Ibadan, Nigeria, in the single night 22 April 1953, and so Fisher's $\alpha \simeq 9.03$. From this we may conclude that certain 1946 Rothamsted *Lepidoptera* are quite a bit more diverse than certain 1953 Nigerian hawk moths.

We may also consider a set of $n = 75$ trees from $S = 30$ species planted on a certain property in northern Vermont[7]. We find Fisher's $\alpha \simeq 15.15$. These 75 trees may be categorized into 3 groups as follows:

1. 27 deciduous trees from 15 species: Fisher's $\alpha \simeq 13.99$,
2. 24 fruit trees from 8 species: Fisher's $\alpha \simeq 4.20$,
3. 24 evergreen trees from 7 species: Fisher's $\alpha \simeq 3.22$.

We conclude that the deciduous trees are by far the most diverse, while the fruit trees and evergreens have about the same diversity. It is interesting to see that the overall Fisher's $\alpha$ exceeds all three values of Fisher's $\alpha$ for the three component groups.

The paper by Fisher, Corbet & Williams [14] was entitled "The relation between the number of species and the number of individuals in a random sample of an animal population". In Part 3 entitled "A theoretical distribution for the apparent abundance of different species", Fisher introduces $\alpha$ index of diversity in a logarithmic-series model for Williams's data on Rothamsted *Lepidoptera* and for Corbet's data on Malayan butterflies, including Rajah Brooke's birdwings[8]; Sir James Brooke (1803–1868) was the first White Rajah of Sarawak[9].

## 2 Some Properties of Fisher's $\alpha$ Index of Diversity

In Fisher's logarithmic-series model, the expected number of species $S_k$ with $k$ individuals is

$$\mathrm{E}(S_k) = \frac{\alpha x^k}{k}, \qquad k = 1, 2, \ldots,$$

with two positive constants $\alpha$ and $x$, and so the expected number of species with just one individual is:

$$\mathrm{E}(S_1) = \alpha x.$$

---

6 A hawk moth is a moth of the family *Sphingidæ* or *Sphingina*; a sphinx-moth; so called for their manner of flight, which resembles the hovering and darting of a hawk." (OED [36]).

7 Between the towns of Franklin and Highgate near the Canadian border.

8 Birdwings are large, tropical papilionid butterflies native to mainland and archipelagic Southeast Asia and Australasia (with one Indian species) and are usually regarded as belonging to three genera: *Ornithoptera, Trogonoptera* and *Troides* [49].

9 Sarawak is one of the two Malaysian states on the island of Borneo [49].

By equating the total number of species $S$ to the sum of the $\mathrm{E}(S_k)$, Fisher showed that

$$S=\sum_{k=1}^{\infty}\mathrm{E}(S_k)=\sum_{k=1}^{\infty}\frac{\alpha x^k}{k}=-\alpha\log_e(1-x).$$

Moreover,

$$n=\sum_{k=1}^{\infty}kS_k=\sum_{k=1}^{\infty}k\frac{\alpha x^k}{k}=\sum_{k=1}^{\infty}\alpha x^k=\frac{\alpha x}{1-x}.$$

Eliminating $x$ yields

$$S=\alpha\log_e\left(1+\frac{n}{\alpha}\right),$$

which cannot readily be solved explicitly for $\alpha$. Since $\alpha>0$ and

$$x=\frac{n}{n+\alpha}=\frac{1}{1+\frac{\alpha}{n}},$$

we see that $x<1$ and that $x$ monotonically approaches 1 as $n$ increases with fixed $\alpha>0$. And so

$$\mathrm{E(S_1)}=\alpha\mathrm{x}<\alpha$$

and

$$\mathrm{E(S_1)}\rightarrow\alpha \quad \text{as} \quad n\rightarrow\infty.$$

Almost always in practice $x>0.9$ according to Magurran [30, p. 30], who also says that when $n/S>20$, then $x>0.99$. For Williams's Nigerian hawk moths we find that $n/S=65/19\simeq 3.42$ and $x=0.878$ and for his *Lepidoptera* $n/S=219/42\simeq 5.21$ and $x\simeq 0.93$. With 14 decimal places, we find that for $n/S>20.00000000000000$, then $x>0.989044487700025$.

Indeed

$$\frac{n}{S}=-\frac{x}{(1-x)\log_e(1-x)} \tag{1}$$

has a unique positive root $x$ for fixed $n/S>0$ and this root $x$ is monotonically increasing in $n/S$.

Plotted here is $x$ (vertical axis) vs. $n/S$ (horizontal axis), where $n$ is the number of individuals, $S$ is the number of species, and $x=\mathrm{E}(S_1)/\alpha$, where $S_1$ is the number of species with just 1 individual and $\alpha$ is Fisher's $\alpha$. Indeed Fisher found the solution for $\alpha$ to be "troublesome and indirect" [to compute] but gave a table [14, Table 9, p. 55] of $\log_{10}n/\alpha$ in terms of $\log_{10}n/S=0.40\,[0.01]\,3.59$. Rice & Demarais [37, Table 1, pp. 149–154 (1996)], give values of $\alpha$ in terms of $n-S=1\,[1]\,10\,[10]\,100\,[100]\,1000$ and $S=1\,[1]\,50\,[10]\,100\,[100]\,500,1000$. An explicit closed-form (but quite complicated) representation (involving 2 integrals) for $\alpha$ has been obtained quite recently by Kheyfits & Kheyfits [26, (2005)].

To compute Fisher's $\alpha$, we wrote a simple program using Newton's Method. Let

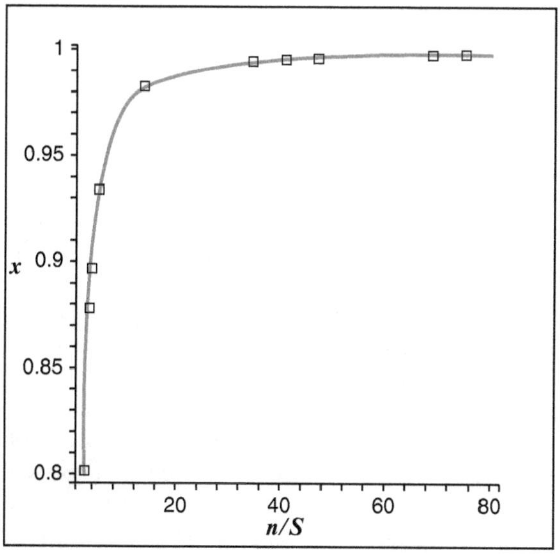


**Fig. 1** Plot of $x$ (vertical axis) vs. $n/S$ (horizontal axis) as defined by (1)

$$f(\alpha) = \alpha \log_e \left(1 + \frac{n}{\alpha}\right) - S.$$

Then the first derivative

$$f'(\alpha) = \log_e \left(1 + \frac{n}{\alpha}\right) - \frac{n}{n+\alpha},$$

and the $(i+1)$th iterate, with $i = 0, 1, \ldots,$

$$\alpha_{i+1} = \alpha_i - \frac{f(\alpha_i)}{f'(\alpha_i)} = \alpha_i - \frac{\alpha_i \log_e\left(1 + \frac{n}{\alpha_i}\right) - S}{\log_e\left(1 + \frac{n}{\alpha_i}\right) - \frac{n}{n+\alpha_i}}.$$

## 3 The Diversity of Vermeer Paintings Depicted on Postage Stamps

As noted in [3], Jerzy Baksalary (1944–2005) introduced the Dutch painter Johannes Vermeer[10] (1632–1675) to the first author after a visit[11] in 1986 to The Netherlands. As Jerzy later recalled, he was exploring the Rijksmuseum in Amsterdam marvelling

[10] Extensive information about Johannes Vermeer and his paintings is relatively given in the extremely competent and rich website [17].

[11] Jerzy Baksalary was then visiting Wageningen under a joint research agreement between the Department of Mathematics, Agricultural University at Wageningen (The Netherlands), and the Department of Mathematical and Statistical Methods, Academy of Agriculture in Poznań (Poland).

at the masterpieces there. At a certain moment he reached a corner where these four paintings by Vermeer were hanging:

1. *The little street* [P7],
2. *The milkmaid* [P9],
3. *Woman in blue reading a letter* [P14],
4. *The love letter* [P29].

Numbers in square brackets prefixed with the letter P refer to the list of 36 recognized Vermeer paintings, as given in Table 1 below.

Images of these four paintings and details from two of them are depicted on a sheetlet with six postage stamps issued by the Maldives[12] in celebration of the 200th anniversary of the Rijksmuseum (see Fig. 2 below).

These four paintings so delighted Jerzy that he spent the remaining time until the museum closed just looking at them (see Fig. 3 below). Later on, Jerzy decided that he would try to see all 36 "recognized" Vermeer paintings (see Table 1 below) with his own eyes. Unfortunately, Jerzy did not manage to achieve this task, with his untimely death in March 2005 at the age of 60 having seen only 20 of the 36 Vermeer paintings.

**Table 1** List of 36 recognized paintings attributed to Vermeer, according to Bailey [2] (in chronological order with current locations). Images of paintings identified with an asterisk have appeared (at least in part) on a postage stamp[13] and with a dagger in the selvage (only)

| |
|---|
| [P1] *Christ in the house of Martha and Mary* (1654/1655), National Gallery of Scotland, Edinburgh |
| [P2] *St Praxedis* (1655), Barbara Piasecka Johnson Collection, Princeton |
| [P3] *Diana and her companions* (1655/1656), Mauritshuis, The Hague |
| [P4]* *The procuress* (1656), Gemäldegalerie Alte Meister, Dresden |
| [P5]* *A girl asleep* (1657), Metropolitan Museum of Art, New York |
| [P6]* *Girl reading a letter at an open window* (1657), Gemäldegalerie Alte Meister, Dresden |
| [P7]* *The little street* (1657/1658), Rijksmuseum, Amsterdam |
| [P8]† *Officer and laughing girl* (1658), Frick Collection, New York |
| [P9]* *The milkmaid*[14] (1658/1660), Rijksmuseum, Amsterdam |
| [P10]* *The glass of wine* (1658/1660), Gemäldegalerie, Staatliche Museen zu Berlin |
| [P11] *The girl with two men* (1659/1660), Herzog Anton Ulrich-Museum, Braunschweig |
| [P12]* *View of Delft* (1660/1661), Mauritshuis, The Hague |
| [P13] *Girl interrupted at her music* (1660/1661), Frick Collection, New York |
| [P14]* *Woman in blue reading a letter* (1662/1664), Rijksmuseum, Amsterdam |
| [P15]* *The music lesson* (1662/1665), Royal Collection, London |
| [P16]* *Woman holding a balance* (1664), National Gallery of Art, Washington, DC |

[12] The Maldives (or Maldive Islands), officially the 'Republic of Maldives', is an island nation consisting of a group of atolls in the Indian Ocean. The Maldives are located south of India's Lakshadweep islands, about 700 km south–west of Sri Lanka. [49].

[13] For further information see Tables 2 & 3 below.

[14] Boyer [5] has pointed out that in France the Vermeer painting *The milkmaid* [P9] is known as *La laitière*, who is the subject of numerous advertisements by Nestlé, see, e.g., [11, 57]. For a stamp depicting *The milkmaid* [P9], see Figure 2 below, bottom row center.

[P17] *Woman with a pearl necklace* (1664), Gemäldegalerie, Staatliche Museen zu Berlin
[P18]* *Woman with a lute* (1664), Metropolitan Museum of Art, New York
[P19]* *Young woman with a jug* (1664/1665), Metropolitan Museum of Art, New York
[P20]* *The girl with a pearl earring* (1665), Mauritshuis, The Hague
[P21] *A lady writing* (1665), National Gallery of Art, Washington, DC
[P22] *Girl with a red hat* (1665), National Gallery of Art, Washington, DC
[P23]* *The concert* (1665/1666), Isabella Stewart Gardner Museum, Boston (stolen in March 1990)
[P24]* *The art of painting* (1666/1667), Kunsthistorisches Museum, Vienna
[P25]* *Head of a young woman* (1666/1667), Metropolitan Museum of Art, New York
[P26]][†] *Mistress and maid* (1667/1668), Frick Collection, New York
[P27]* *The astronomer* (1668), Musée du Louvre, Paris
[P28] *The geographer* (1668/69), Städel Museum, Frankfurt am Main
[P29]* *The love letter* (1669/1670), Rijksmuseum, Amsterdam
[P30]* *The lacemaker* (1669/1670), Musée du Louvre, Paris
[P31] *A young woman seated at the virginal* (1670), Art Gallery of Wynn, Las Vegas
[P32]* *Lady writing a letter with her maid* (1670), National Gallery of Ireland, Dublin
[P33] *Allegory of faith* (1671/1674), Metropolitan Museum of Art, New York
[P34]* *The guitar player* (1672), Kenwood House (Iveagh Bequest), London
[P35] *A lady standing at the virginal* (1673/1675), National Gallery, London
[P36]* *A lady seated at the virginal* (1673), National Gallery, London.

**Fig. 2** Vermeer paintings depicted on stamps from The Maldives

**Fig. 3** *The girl with a pearl earring* [P20] with Jerzy K. Baksalary at the Mauritshuis, The Hague, 25 September 2001 [Photo by Mirosława Baksalary]

Our list of 36 paintings in Table 1 differs from the list of 34 paintings given by Bailey [2], and from the list of 34 paintings in the English-language *Wikipedia* [49] article on Johannes Vermeer, and [5] from the list of 37 paintings in the French-language *Wikipédia* [50] article. Bailey [2] and the English-language *Wikipedia* [49] omit, as we do, *Girl with a flute* [National Gallery of Art, Washington, DC], but this is included by the French-language *Wikipédia*; Brown [6] says that *Girl with a flute* is by "circle of Vermeer".

Bailey [2] and the English-language *Wikipedia* [49] omit *A young woman seated at the virginal* [P31], which has been only relatively recently attributed to Vermeer. The English-language *Wikipedia* [49] also omits *St Praxedis* [P2], which Brown [6] says "is difficult to reconcile with the few known paintings by Vermeer".

Quite recently, Vermeer became widely known to the general public through the 1999 novel *Girl with a Pearl Earring* [8] by Tracy Chevalier and the 2003 movie [16] adapted by Olivia Hetreed from the novel. The movie starred Colin Firth as Vermeer and received three Oscar nominations. Three stamps depicting *The girl with a pearl earring* [P20] are shown in Fig. 4 below.

We have identified 83 postage stamps from 33 different "countries" that depict (all or part of) 23 of the 36 recognized paintings by Johannes Vermeer. Details are in Table 2 below. By "country" we mean a stamp-issuing region that issues or has issued its own postage stamps. Our findings here update those given in [3] with the

**Table 2** Vermeer stamps by country

| painting id | Painting title \ "country" | Aitutaki = Pacific Ocean TOTAL | Ajman | Fujeira | Kathiri --Seiyun (Aden) | Khor Fakkan | Manama -- Ajman | State of Oman | Ras al Khaima | Sharjah | Mutawakkilite Kingdom of Yemen | Yemen Arab Republic | Middle East TOTAL | Benin | Burundi | Comoros | Guinea | Ivory Coast | Liberia | Mali | Mozambique | Rwanda | Sierra Leone | Somaliland Republic | Spanish North Africa | Tanzania | Upper Volta | Africa TOTAL | France | German Democratic Republic | Netherlands | Europe TOTAL | Grenada | Guyana | Panama | United Nations | America TOTAL | Maldives = Asia TOTAL | GRAND TOTAL |
|---|---|---|---|---|---|---|---|---|---|---|---|---|---|---|---|---|---|---|---|---|---|---|---|---|---|---|---|---|---|---|---|---|---|---|---|---|---|---|
| | region | Pac | ME | ME | ME | ME | ME | ME | ME | ME | ME | ME | ME | Afr | Afr | Afr | Afr | Afr | Afr | Afr | Afr | Afr | Afr | Afr | Afr | Afr | Afr | Afr | Eur | Eur | Eur | Eur | Ame | Ame | Ame | Ame | Ame | Asia | |
| P1 | Christ in the house of Martha | | | | | | | | | | | | | | | | 1 | | | | | | | | | | | | | | | | | | | | | | 1 |
| P4 | The procuress | | | | | | | | | | | | | 1 | | | | | | | | | | | | | | 1 | | | | | | | | | | | 1 |
| P5 | A girl asleep | | | | | | | | | | | | | | | | | | | | | | 1 | | | | | 1 | | | | | | | | | | | 1 |
| P6 | Girl reading a letter at an | | | | | | | | 1 | | | | 1 | | 1 | | | | | | | | | | | | 1 | 2 | | 1 | | 1 | | | | | | | 4 |
| P7 | The little street | | | | | | | | | | | | | | | | | | | | | | | | | | | | | | | | | | | | | 2 | 2 |
| P9 | The milkmaid | | | | | | 2 | 1 | | | 1 | | 4 | 1 | | | | 1 | | | | 1 | | | | | | 3 | | | | | | | | | | 1 | 8 |
| P10 | The glass of wine | | | | 2 | | 2 | | | | | | 4 | | | | | | 1 | | | 1 | | | | | | 2 | | | | | | | | | | | 6 |
| P12 | View of Delft | | | | | | | | | | | | | | | | | | | | | | | | | | | | | | 1 | 1 | | | | | | | 1 |
| P14 | Woman in blue reading a letter | 2 | | | | | | | | | | | | | 1 | | | | | | | 1 | | | | | | 2 | | | 1 | 1 | | | | | | 1 | 6 |
| P15 | The music lesson | | | | | | | | | | | | | 1 | | | | | | | | | | | | | | 1 | | | | | | | | | | | 1 |
| P16 | Woman holding a balance | | | | | | 1 | | | | | | 1 | | | | 1 | | | | | | | | | | | | | | | | | | | | | | 2 |
| P18 | Woman with a lute | | | | | | | | | | | | | | | | | | | | | | 1 | | | | | 1 | | | | | | | | | | | 1 |
| P19 | Young woman with a jug | | | 1 | | | | | | | | | 1 | | | | | | | | 1 | 1 | | | | 1 | | 3 | | | | | | | | | | | 4 |
| P20 | The girl with a pearl earring | | | | 1 | | | | | | | | 1 | | | | | 1 | | | | | | | | | | 1 | 1 | | | 1 | | | | | | | 3 |
| P23 | The concert | | | | | | 2 | | | | | | 2 | | | | | | | | | | | | | | | | | | | | | | | | | | 2 |
| P24 | The art of painting | | 1 | | | | | | | | | | 1 | 1 | | | | | | | | 1 | | | | | | 2 | | | | | 1 | | | | 1 | | 4 |
| P25 | Head of a young woman | | | | | | | | | | | | | | | | | | | | | | | | 1 | 1 | | 2 | | | | | | | | | | | 2 |
| P27 | The astronomer | | | | | | | | | | | | | | | | | | | 1 | | | | | | | | 1 | | | | | | 1 | | | 1 | | 2 |
| P29 | The love letter | | | 2 | | | | | | | | | 2 | 1 | 1 | | | | | | | | | | | | | 2 | | | 2 | 2 | | | | | | 2 | 8 |
| P30 | The lacemaker | | 1 | 1 | | | | | | | | 1 | 3 | | 1 | 1 | | 1 | | | | | | | | | | 3 | 1 | | | 1 | | 1 | | | 1 | | 8 |
| P32 | Lady writing a letter with her | 2 | | | | | 1 | | | | | | 1 | | | | | | | | | | | | | | | | | | 1 | 1 | | | | 1 | 1 | | 5 |
| P34 | The guitar player | | | | 1 | 2 | | | | 2 | | | 5 | 1 | | | | 1 | | | | | | 1 | | | | 3 | | | | | | | | | | | 8 |
| P36 | A lady seated at the virginal | | 1 | | | | | | | | | | 1 | | | | | | | | | 1 | | | | | | 1 | | | | | | | 1 | | 1 | | 3 |
| | **# stamps** | **4** | 3 | 4 | 4 | 2 | 8 | 1 | 1 | 2 | 1 | 1 | **27** | **6** | 4 | 1 | 2 | 4 | 1 | 1 | 1 | 6 | 2 | 1 | 1 | 2 | 1 | **33** | 2 | 1 | 5 | **8** | 1 | 2 | 1 | 1 | **5** | **6** | **83** |
| | **# paintings** | 2 | | | | | | | | | | | 13 | | | | | | | | | | | | | | | 18 | | | | 7 | | | 1 | | 5 | 4 | 23 |

**Table 3** Fisher's $\alpha$ for selected datasets

| | | $n$ | $S$ | Fisher's alpha |
|---|---|---|---|---|
| Vermeer stamps | South Pacific | 4 | 2 | 1.59 |
| Vermont trees | evergreen | 24 | 7 | 3.32 |
| Vermont trees | fruit | 24 | 8 | 4.20 |
| Vermeer stamps | Asia | 6 | 4 | 5.24 |
| Williams | Nigerian hawk moths | 65 | 19 | 9.03 |
| Vermeer stamps | Middle East | 27 | 13 | 9.86 |
| Vermeer stamps | all countries | 83 | 23 | 10.53 |
| Vermont trees | deciduous | 27 | 15 | 13.90 |
| Williams | Rothamsted *Lepidoptera* | 219 | 42 | 15.44 |
| Vermeer stamps | Africa | 33 | 18 | 16.22 |
| Vermont trees | evergreen, fruit & deciduous | 75 | 30 | 18.53 |
| Vermeer stamps | Europe | 8 | 7 | 26.80 |
| Vermeer stamps | America | 5 | 5 | infinity |

recently-discovered sheetlet with six stamps from Benin[15] (see Fig. 5 below) issued in 2003 and two stamps from Guinea[16] (see Figs. 7 and 8 below) issued in 2007.

Included in the six stamps from Benin are the first stamps we have found to depict *The procuress* [P4] and *The music lesson* [P15], respectively, bottom right and top left. Details from several Vermeer paintings appear in the selvage[17]. In particular, on the right (center) and bottom (left) selvage appear details from the painting *Officer and laughing girl* [P8] (see Fig. 6 below), which we have not found depicted philatelically elsewhere.

The two stamps from Guinea appear in two sheetlets (see Figs. 7 and 8), each with one stamp. The stamp in the sheetlet shown in Fig. 7 depicts a detail from

[15] Benin, officially the Republic of Benin, is a country in west Africa known as Dahomey until 1975 [49].

[16] Guinea, officially Republic of Guinea (in French: République de Guinée), is a country in west Africa, known as French Guinea until 1958 [49].

[17] In philately the selvage is the margin of a pane of stamps that usually includes the plate number and other markings such as copyright notices.

**Fig. 4** Three stamps depicting *The girl with a pearl earring*: (*left*) Ivory Coast, c. 2001; (*center*) Kathiri State of Seiyun, c. 1967; (*right*) France 2008

the Vermeer painting *Woman holding a balance* [P16] and what is considered to be the only self-portrait of Vermeer [43, p. 9] taken from the upper left corner of *The procuress* [P4][18]. Depicted in the selvage are an enlarged detail from *The girl with a pearl earring* [P20] and the Vermeer painting *Mistress and maid* [P26], which we have (also) not found depicted philatelically elsewhere.

The second sheetlet (see Fig. 8), which was issued in celebration of "Painters of The Netherlands", includes images of six stamps depicting three paintings, two of each; the three paintings are by Van Gogh (1853–1890), Rembrandt (1606–1669), and Vermeer. The Vermeer painting is *Christ in the house of Martha and Mary* [P1], which we have not found depicted on any other postage stamp. This stamp also shows the same self-portrait of Vermeer (as shown on the stamp in Fig. 7). In the selvage upper right near the phrase "Vermeer (1632–1675)" is curiously not a portrait of Vermeer but in fact a self-portrait of Carel Fabritius (1622–1654), who is "generally supposed to be Vermeer's teacher" [43, p. 13]. Fabritius was a pupil of Rembrandt in the 1640s and then settled in Delft in 1650.

We compute Fisher's $\alpha$ index (Table 3 above) to assess the diversity of Vermeer paintings depicted on postage stamps. We find that the Vermeer paintings depicted on stamps from Europe are the most diverse ($n = 8, S = 5, \alpha = 26.80$), except for stamps from America for which we have infinite diversity with 5 stamps each depicting a different painting. The least diverse are the paintings depicted on stamps from the South Pacific ($n = 4, S = 2, \alpha = 1.59$) and Asia ($n = 6, S = 4, \alpha = 5.24$). Interestingly the paintings depicted on stamps from Africa ($n = 33, S = 18, \alpha = 16.22$) are just slightly more diverse than Williams's Rothamsted *Lepidoptera* ($n = 219, S = 42$, $\alpha = 15.44$).

[18] The detail from *The procuress* in the sheetlet from Benin cuts off this self-portrait and the portrait of another person.

**Fig. 5** Sheetlet from Benin including six postage stamps, issued in 2003

**Fig. 6** *Soldier and the laughing girl* [P8]

**Fig. 7** Sheetlet from Guinea including a single postage stamp, issued in 2007

**Fig. 8** Sheetlet from Guinea including six postage stamps, issued in 2007

## 4 A Biodiversity Study Reported in the *Kazwini Cosmography*

In his book, Williams [51, p. 17 (1974)] reports on what may have been the very first study of biodiversity. He observes that in *Chrestomathie arabe* [19] [45, 2nd ed., vol. 3, pp. 422–423 (1827)] by the French linguist and orientalist Antoine Isaac, baron Silvestre de Sacy (1758–1838), there is an extract (in French) from the thirteenth century Arabic cosmography[20] *Book of the Marvels of Nature and the Singularities of Created Things,* by Zakariyyā' ibn Muḥammad ibn Maḥmūd Abū Yaḥyā al-Ḳazwīnī (c. 1203–1283). We will refer to this book as the *Kazwini Cosmography* and to its author as *Kazwini*[21].

Our translation of the extract (in French) into English is:

The *Papillon*[22]. It is that little insect that flits around the torches incessantly and singes itself on their flames. *Khafif of Samarkand*, a friend of *Motadhed*, tells how finding himself one night in the company of this caliph and seeing a large number of *papillons* fluttering around the torches, the desire overtook him to gather them all together: in so doing they gathered a full measure called *macouc*; then in sorting them out, they counted 72 different kinds.

[19] *Chrestomathie = Chrestomathy = Anthology* = collection of selected passages or stories of an author. [33].

[20] A "cosmography" maps the general features of the universe; describes both heaven and Earth, but without encroaching on geography or astronomy. [49].

[21] Wüstenfeld [55] uses el-Cazwini, while for her comprehensive PhD dissertation on illustrations in this cosmography, Badiee [1] uses Qazwīnī; the National Library of Medicine [35] uses al-Qazwīnī. See also [18, 19, 20, 21, 22, 23, 24].

[22] In French *papillon* means butterfly and *papillon de nuit* is moth. Since not all moths are nocturnal and not all butterflies are diurnal and since we do not know what kind of insects were collected we will use *papillon*. Badiee [1, p. 219], citing Wüstenfeld [55, p. 443], mentions "moths and insects".

How diverse were these *papillons?* Clearly $S = 72$ but we need to know $n$ in order to compute Fisher's $\alpha$.

We pose the following questions:

(Q1) Who was *Kazwini*?

(Q2) Who was *Khafif of Samarkand*?

(Q3) Who was *Motadhed*?

(Q4) Where were these *papillons* collected?

(Q5) How big is a *macouc*?

(Q6) How many *papillons* in a litre?

We will suggest answers to the first 5 of these 6 questions.

In *The Encyclopedia of Islam* [13], Lewicki [28] says that: *Kazwini* was the greatest of Arabic cosmographers. "He was at the same time astronomer, geographer, geologist, mineralogist, botanist, zoologist and ethnographer. Like all his predecessors, *Kazwini* was a good compiler who neither produced a new fact nor created any new theory. Being, however, very learned and very cultivated at the same time, he succeeded in synthesizing all the facts known in his time about the sciences he studied. His principal merit lies in his having accomplished the raising of cosmography to a literary genre of an extremely high level."

From the National Library of Medicine website [35] we find that: the *Kazwini Cosmography* is the most well-known example of a genre of classical Islamic literature that was concerned with *mirabilia:* things which inspire wonder; miraculous events; wonderful, marvelous, astonishing, extraordinary things[23].

A cosmography was concerned with topics that challenged understanding. These could include aspects of God's creation that inspire awe, such as human anatomy or the variety of plants. The treatise covered all the wonders of the world, and the variety of the subject matter (humans and their anatomy, plants, animals, strange creatures at the edges of the inhabited world, constellations of stars, zodiacal signs, angels, and demons) provided great scope for the author.

We suggest that *Motadhed*'s friend *Khafif of Samarkand* refers to a man, probably a vizier, whose name is "Khafif" and who comes from Samarkand. In his book, Williams [51, p. 17] says "Khalif" rather than "Khafif", but we believe in "Khalif" there is a typo. In his French text in the *Chrestomathie arabe*, Silvestre de Sacy [45, p. 422 (1827)] uses "khalife" for "caliph": in contemporary French, "caliph" is "calife"[24].

This *Khafif* is, however, probably not the mystic and Sufi[25] from Iran: Mohammad Ibn Khafif also known as Sheikh-i Kabir. Born in Shiraz in 882, he was just 20 when *Motadhed* died (in 902).

[23] The genre was known as 'ajā'ib or 'jā'ib al-makhlūqāt literature, that is "wonders" or "wonders of creation".

[24] Many thanks to Christian Boyer for drawing our attention to this.

[25] One of a sect of Muslim ascetic mystics who in later times embraced pantheistic views. (OED [36]).

Samarkand[26] is a large city in Uzbekistan southwest of Tashkent. Dating from the 3rd or 4th millennium BC and destroyed by Genghis Khan[27] c. 1220, Samarkand was rebuilt as a centre of great splendour and opulence when it became (c. 1370) the capital of Tamerlane's empire. Tamerlane (1336–1405), or Emir Timur or Amir Temur or Timur the Lame, was a fourteenth century Turco–Mongol conqueror of much of western and central Asia, and founder of the Timurid Empire and Timurid dynasty (1370–1405) in central Asia, which survived until 1857 as the Mughal dynasty of India [49].

We believe that Motadhed refers to Abu'l-'Abbās Aḥmad ibn Ṭalḥa Bi'llāh Al-Mu'taḍid (c. 860–902) who was the Abbasid[28] caliph[29] of Baghdad from 892–902, and who in 898 was appointed the "governor of Transoxania"[30]. Born in the Persian town of Ḳazwīn[31] [28], *Kazwini* later moved to Baghdad. After Baghdad was taken by the Mongols in 1258, *Kazwini* retired from public life to devote himself entirely to scientific activities.

We suggest that our *papillons* were collected in Baghdad on a certain night sometime between 898 and 902 AD. Baghdad was once the center of Muslim civilization, and the home of many eminent scholars, artists, and poets. The period of its utmost glory is reflected in the *Thousand and One Nights*[32], in which many of the tales are set in Baghdad. The *Thousand and One Nights* tells the story of Queen Scheherazade, who must relate a series of stories to her malevolent husband, King Shahryar, to delay her execution.

In Silvestre de Sacy's *Chrestomathie arabe* [45, pp. 514–515], we find the following: Le *macouc* est la huitième partie du *kafiz* : c'est une mesure de capacité, usitée dans l'Iraq. The *makkūk* is one eighth of a *qafīz:* it is a measure of volume utilized in Iraq. We believe that *macouc* and *kafiz* are the French transliterations and *makkūk* and *qafīz* both the English and German transliterations from the Arabic of certain measures of capacity. In this paper, we will (now) use the (English and German) forms *makkūk* and *qafīz*.

---

[26] Samarcande in French, Samarqand in Uzbek, earlier Marakanda or Maracanda (in Greek).

[27] Genghis Khan (1162–1227) was a Mongol political and military leader who founded the Mongol Empire (1206–1368), the largest contiguous empire in world history. [49].

[28] Abbasid is the dynastic name generally given to the caliph of Baghdad, the second of the two great Muslim caliphates of the Arab Empire [25, 34, 49].

[29] In *The New American Cyclopædia* [39, p. 264] "caliph" is defined as the title of the successors of the prophet Mohammed (c. 571–632), also known as Muhammad, who established the religion of Islam and the Muslim community [49].

[30] In the *Encyclopædia Britannica*, it is noted that "Transoxania corresponds roughly to present-day Uzbekistan and parts of Turkmenistan and Kazakhstan".

[31] Ḳazwīn (also written as Casbin, Kasvin, Kazvin, or Qazvin), which is now in north-west Iran, was the location of a former capital of the Persian Empire. Destroyed by Genghis Khan in the thirteenth century, it is also where the famous coup d'état was launched that led to the rise of the first Pahlavi dynasty in 1921.

[32] The *Thousand and One Nights* was also known as *The Book of The Thousand Nights and One Night* [7], *The Book of a Thousand Nights and a Night, 1001 Arabian Nights, The Nightly Entertainments*, or *The Nights*.

In *Islamische Masse und Gewichte* by Walther Hinz [15, p. 48] in German and in its English translation by Marcinkowski [31, p. 71], we find that in the course of the 10th century, two *qafīz* had emerged in Iraq. The larger *qafīz* measure in Baghdad and in Kūfah[33] contained 8 *makkūk*[34] ... which we calculate to be 60 litres on the average. The smaller *qafīz* measure, which had been current in Baṣrah[35] and Wāṣiṭ[36] amounted to 4 *makkūk* ... and was thus calculated, on the average, at 30 litres. And so we conclude that in the 10th century in Iraq a *makkūk* was equivalent to

$$\frac{60}{8} = \frac{30}{4} = 7\tfrac{1}{2} \text{ litres}$$

not only in Baghdad, and Kūfah but also in Baṣrah and Wāṣiṭ.

However, "the Syrian *makkūk* was of a completely different size. During the 12th century in Aleppo[37] ... one *makkūk* ... amounted to about 19 litres [15, p. 44], [31, p. 65]. Since we believe our *papillons* were collected in Baghdad, we conclude that the volume of our *makkūk* was $7\frac{1}{2}$ litres[38].

The question remains: How many *papillons* in a litre?

## 5 Magic squares in *Kazwini's Cosmography*

We found online open-access digitized copies of 4 Persian translations of *Kazwini*'s *Cosmography* in the National Library of Medicine [35]. Folio 310a of copy #MS P 3 has two magic squares depicted, one is $3 \times 3$, see Fig. 9 and Fig. 10 below, and the other $5 \times 5$; the complete folio 310a is shown in Fig. 6 above. A third magic square, which is $4 \times 4$, is partially visible, possibly from the verso of folio #310a. A *magic square* of order $n$ is an arrangement of $n^2$ numbers, usually distinct integers, in a square, such that the $n$ numbers in all rows, all columns, and both diagonals sum to the same number—the *magic constant* or *magic sum*.

The $3 \times 3$ magic square in the upper left part of folio #310a and a translation[39] are shown in Fig. 10.

[33] Kūfah is a medieval city in Iraq, about 145 km south of Baghdad, that was a centre of Arab culture and learning from the 8th to the 10th century.

[34] 1 *qafīz* = 8 *makkūk* or equivalently, as stated in endnote (107), 1 *makkūk* = 1/8 *qafīz*.

[35] Baṣrah, also spelled Basra, is the second largest city of Iraq and is located about 55 km from the Persian Gulf and 545 km from Baghdad.

[36] Wāṣiṭ is one of the governorates of Iraq in the east of the country. Its name comes from the Arabic word meaning "middle", as it lies along the Tigris about midway between Baghdad and the Persian Gulf.

[37] Aleppo, or Halab in Arabic, is a city and province in northern Syria. It is the second largest city in Syria after Damascus. It is one of the oldest cities in the region, known in antiquity as Khalpe, to the Greeks as Beroea, and to the Turks as Halep *Wikipedia.*

[38] Williams [51, p. 17] mentions that a *makkūk* might contain between $\frac{1}{2}$ and 1 litre (we think this may be a bit low to find 72 species of *papillons*).

[39] Many thanks to Aisha Iftekhar for providing us with this translation and to Amir Memartoluie and Neda Zare Neyney for their help with other translations from *Kazwini*'s Arabic and Persian texts.

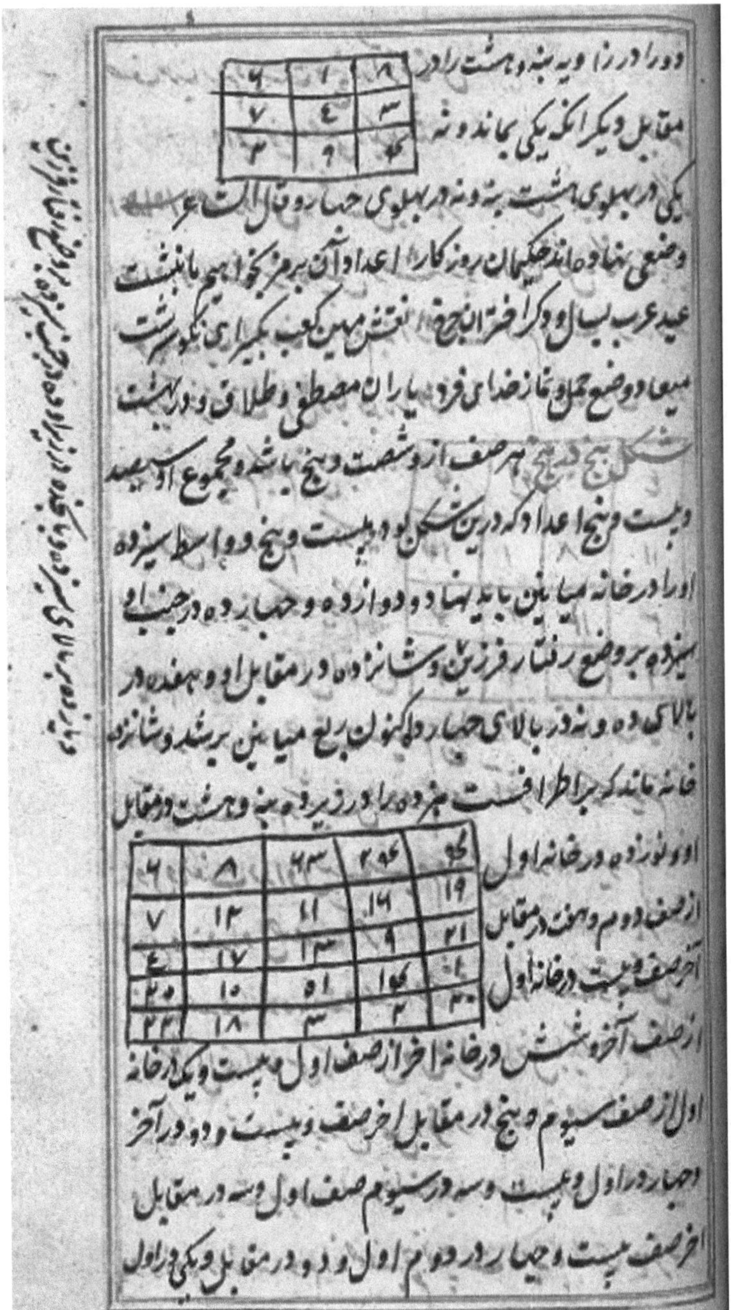

**Fig. 9** Folio #310a in the Persian translation #MS P 3 [35] of the *Kazwini Cosmography*

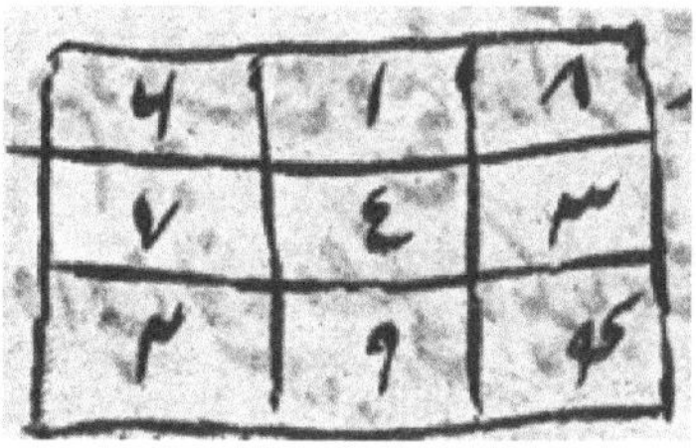

| | | |
|---|---|---|
| 6 | 1 | 8 |
| 7 | 5 | 3 |
| 2 | 9 | 4 |

**Fig. 10** The 3 × 3 magic square in the upper left part of folio #310a [35]

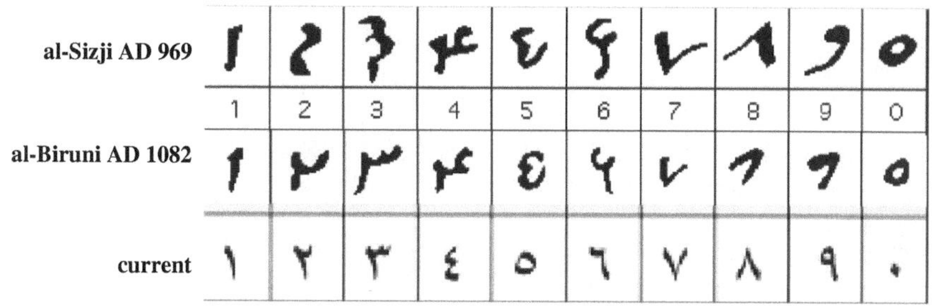


**Fig. 11** Three sets of Hindu–Arabic numerals

The glyphs[40] used here, which we will call Hindu–Arabic numerals[41], and in the other two magic squares in folio #310a, seem to be the same as those used in a treatise by the well-known Persian mathematician Abu Arrayhan Muhammad ibn Ahmad al-Biruni (973–1048) copied in 1082. al-Biruni was also a "physicist, encyclopedist, philosopher, astronomer, astrologer, traveller, historian, pharmacist, and teacher" [49].

The Hindu–Arabic numerals in the top row of the 3 × 3 magic square seem to come from a work of Abu Said Ahmad ibn Muhammad ibn Abd al-Jalil al-Sijzi (c. 945–1020), an Islamic astronomer and mathematician, who wrote on the geometry of spheres. The numerals changed their form somewhat 100 years later. In the middle row in Fig. 11 are the numerals as they appear in a 1082 copy of one of the astronomical texts of Abu Arrayhan Muhammad ibn Ahmad al-Biruni (973–1048).

Between 969 and 1082 the biggest change in the numerals was that the 2 and the 3 have been rotated through 90°. This came about since scribes wrote on a scroll which they wound from right to left across their bodies as they sat cross-legged. The scribes therefore, instead of writing from right to left (as Arabic is written), wrote in lines from top to bottom. The script was rotated when the scroll was read and the characters were then in the correct orientation [29].

[40] A glyph is the shape given in a particular typeface to a specific grapheme or symbol [49].

[41] Hindu–Arabic numerals, also known as Arabic numerals, Indian numerals, and Hindu numerals, form the basis of the European number systems which are now widely used [49].

**Acknowledgement** This paper is an expanded version of the paper [47] by Styan & Puntanen presented at the International Biometric Conference (IBC-2006): Montréal (Québec), Canada: 17 July 2006 (see also [46]), and Sect. 3 of this paper is an updated (and shortened) version of the paper by Baksalary & Styan [3]. Many thanks to Marlen Stamp Company of Great Neck, New York, for leading us to the two sheetlets from Guinea. We also are grateful to Christian Boyer, Laure Boyer, R. William Farebrother, Aisha Iftekhar, Amir Memartoluie, Joanna Modławska, Neda Zare Neyney, Evelyn Matheson Styan, Götz Trenkler, Denys J. Voaden, Jane Thomas Zaring, and Philip B. Zaring for their help. Oskar Maria Baksalary would like to express his sincere thanks to the Alexander von Humboldt Foundation for its financial support. This research was supported in part by the Natural Sciences and Engineering Research Council of Canada.

## References

[1] Julie Anne Oeming Badiee (1978). *An Islamic Cosmography: The Illustrations of the Sarre Qazwīnī.* Ph. D. dissertation[42], University of Michigan, Ann Arbor, xvii + 466 pp.

[2] Martin Bailey (1995). *Vermeer.* Phaidon, London.

[3] Oskar Maria Baksalary & George P. H. Styan (2008). Some comments on the diversity of Vermeer paintings depicted on postage stamps. *Discussiones Mathematicae—Probability and Statistics,* 28, 65–83.

[4] BiStamp.com (2008). Stamps & paper money from ex-USSR states. Online open-access: web archive (http://www.bistamp.com/inter/BiStamp.nsf/Title!OpenPage).

[5] Christian Boyer (2008). Personal communication, 14 February 2008.

[6] Christopher Brown (1996). Washington and The Hague: Vermeer (exhibition review). *The Burlington Magazine*, 138 (1117), 281–283. Full-text pdf on JSTOR (http://www.jstor.org/stable/pdfplus/887007.pdf).

[7] Sir Richard Francis Burton (2001). *The Arabian Nights: Tales from a Thousand and One Nights.* Translated, with a preface and notes, by Sir Richard Francis Burton (1821–1890) (http://en.wikipedia.org/wiki/Richard_Francis_Burton); introduction by A. S. Byatt. Modern Library, New York, 2001, xxxvi + 872 pp., ISBN 0-375-75675-2. [Reprint edition (2004), 1104 pp., ISBN 0-812-97214-7.]

[8] Tracy Chevalier (1999). *Girl with a Pearl Earring*. Dutton, New York.

[9] A.-L. de Chézy (1827). Extraits du livre des *Merveilles de la nature et des singularités des choses créés* par Mohammed Kazwini, Fils de Mohammed. Traduits par Antoine Leonard de Chézy (1773–1832) (http://en.wikipedia.org). In *Chrestomathie arabe* [45, pp. 387–516]. [Italicization in title copied from Contents page vii (vij [*sic*]).]

[42] From the Preface [1, page v]: "The purpose of this dissertation will be to examine in depth an illustrated manuscript of Zakariyya' ibn Muhammad ibn Mahmud al-Qazwinl's cosmography *The Wonders of Creation and Oddities of Existence = 'Aja'ib al-Makhluqat wa Ghara'ib al-Mawjudat.* This manuscript was acquired in the early years of this century by Friedrich Sarre (1865–1945) in Algiers, and became known through major exhibitions in Munich and Berlin in 1910 and London in 1931."

[10] A. Steven Corbet & H. M. Pendlebury [Henry Maurice Pendlebury (1893–1945)] (1934). *The Butterflies of the Malay Peninsula.* [First published in 1934; 2nd edition revised by A. Steven Corbet and edited by N. D. Riley, pub. Oliver & Boyd, Edinburgh, 1956; 3rd edition, revised and enlarged by Lt. Col. J. N. Eliot, 1978; 4th edition revised by Lt. Col. J. N. Eliot with plates by Bernard d'Abrera, pub. Malayan Nature Society, Kuala Lumpur, 1992.]

[11] Daily*motion:* Upload Videos (2008). LaLaitière-Ravaillac. Online open-access: video (http://www.dailymotion.com/related/1292171/video/xse64_lalaitiereravaillac_ads).

[12] Dandelion watch (2008). Glossary. Online open-access: article (http://www.naturewatch.ca/english/plantwatch/dandelion/glossary.html). Website developed in partnership with Environment Canada.

[13] *The Encyclopaedia of Islam* (1960/2008). New edition prepared by a number of leading orientalists; H. A. R. Gibb *et al.*, editorial committee. Brill Academic Publishers, Leiden, 12 vols., ISBN 90-04-09239-0, 90-04-12803-4. Online edition edited by P. Bearman, Th. Bianquis, C.E. Bosworth, E. van Donzel & W.P. Heinrichs, Brill Online, 2008. {http://www.encislam.brill.nl}{Online restricted access:} Brill Online. [Includes "al-Mu'taḍid Bi'llāh, Abu'l-'Abbās Aḥmad b. Ṭalḥa" by H. Kennedy [25], "al-Mu'taḍid Bi'llāh, Abū Ámr 'Abbād b. Muḥammad b. 'Abbād." by E. Lévi-Provençal [27], and "al-Qazwīnī, Zakariyya' ibn Muḥammad" by L. Richter-Bernburg [38].]

[14] R. A. Fisher [Sir Ronald Aylmer Fisher (1890–1962). (http://www-groups.dcs.st-and.ac.uk/history/Biographies/Fisher.html)], A. S. Corbet [Alexander Steven Corbet (1897–1948). (http://www.nhm.ac.uk/research-curation/collections-library/collections-management/collections-navigator/transform.jsp?rec=/ead-recs/nhm/uls-a340905.xml)] & C. B. Williams [Carrington Bonsor Williams (1889–1981) (http://www.jstor.org)] (1943). The relation between the number of species and the number of individuals in a random sample of an animal population. *Journal of Animal Ecology*, 12 (1), 42–58. Full-text pdf on JSTOR (http://www.jstor.org/stable/pdfplus/1411.pdf).

[15] Walther Hinz (1955). *Islamische Maße und Gewichte: Umgerechnet ins metrische System,* Handbuch der Orientalistik herausgegeben von Bertold Spuler, Ergänzungsband 1, Heft 1, E. J. Brill, Leiden, [v] + 66 pp. [Reprinted 1970 & translated into English as Marcinkowski [31].]

[16] The Internet Movie Database (2008). *Girl with a Pearl Earring,* 2003, movie: United Kingdom/Luxembourg. Screenplay adapted by Olivia Hetreed from the novel [8] by Tracy Chevalier and starring Colin Firth[43] as Vermeer. Online open-access: photos and reviews (http://imdb.com/title/tt0335119/).

[43] A set of nine stamps depicting Colin Firth was issued by the "Amurska Republic" in 2002, a year before the movie *Girl with a Pearl Earring* was released (in 2003). According to *Wikipedia* [49], "The Amur Oblast is a federal subject [administrative subdivision] of Russia, situated about 8,000 km east of Moscow on the banks of the Amur and Zeya Rivers; it shares its border with China in the south." Stamps from the "Amurska Republic" are "Local issues without official status" [4]. Wood [52, pp. 2, 95] says "Amur [*sic*] is a province in Siberia" and that stamps were issued (already) in 1920; see also [42, pp. 108–109].

[17] Jonathan Janson (2007). *The Complete Vermeer Catalogue*. Online open-access: website (http://www.essentialvermeer.com/), Essential Vermeer Resources, updated December 7, 2007.

[18] Kazwini (1973). *'Ajā'ib al-makhlūqāt wa-gharāib al-mawjūdāt* [*Book of the Marvels of Nature and the Singularities of Created Things*], by Zakariyyā' ibn Muḥammad ibn Maḥmūd Abū Yaḥyā al-Ḳazwīnī (*c*. 1203–1283) (http://en.wikipedia.org/wiki/Zakariya_al-Qazwini), In Arabic, edited translation by Faruq Sa'd. pub. Dar al-Afaq al-Jadidah, Bayrut [Beirut, Lebanon], 526 pp.

[19] Kazwini (1996). *'Ajā'ib al-makhlūqāt wa-gharāib al-mawjūdāt* [*Book of the Marvels of Nature and the Singularities of Created Things*], by Zakariyyā' ibn Muḥammad ibn Maḥmūd Abū Yaḥyā al-Ḳazwīnī (*c*. 1203–1283). In Persian, with citations in Arabic, edited translation by Ja'far Mudarris Sadiqi. Bazkhvani-i mutun 6, pub. Nashr-i Markaz, Tihran [Tehran, Iran], 37 + 567 pp.

[20] Kazwini (2001). Miniatury arabskie [Arabic miniatures]: Al-Kazwini *Kosmografia* Irak 1280 r. Illustration from *'Ajā'ib al-makhlūqāt wa-gharāib al-mawjūdāt* [*Book of the Marvels of Nature and the Singularities of Created Things*], by Zakariyyā' ibn Muḥammad ibn Maḥmūd Abū Yaḥyā al-Ḳazwīnī (*c*. 1203–1283). Online open-access: Krakow Pedagogical University, Krakow, Poland: © Stanisław Skórka, updated 25 August 2001 (http://www.wsp.krakow.pl/whk/galeriar/alkazwini3.html).

[21] Kazwini (2006). Illustrations from *'Ajā'ib al-makhlūqāt wa-gharāib al-mawjūdāt* [*Book of the Marvels of Nature and the Singularities of Created Things*], by Zakariyyā' ibn Muḥammad ibn Maḥmūd Abū Yaḥyā al-Ḳazwīnī (*c*. 1203–1283), MS P 1 [44, P 1, p. 330]. Online open-access: National Library of Medicine [35], Bethesda, updated: 23 February 2006 (http://www.nlm.nih.gov/hmd/arabic/natural_hist3.html). [The manuscript consists of 335 leaves.]

[22] Kazwini (2006). Illustrations from *'Ajā'ib al-makhlūqāt wa-gharāib al-mawjūdāt* [*Book of the Marvels of Nature and the Singularities of Created Things*], by Zakariyyā' ibn Muḥammad ibn Maḥmūd Abū Yaḥyā al-Ḳazwīnī (*c*. 1203–1283), MS P 2 [44, P 2, p. 330]. Online open-access: National Library of Medicine [35], Bethesda, updated: 23 February 2006 (http://www.nlm.nih.gov/hmd/arabic/natural_hist4.html). [Manuscript dated 1788–1789.]

[23] Kazwini (2006). Illustrations from *'Ajā'ib al-makhlūqāt wa-gharāib al-mawjūdāt* [*Book of the Marvels of Nature and the Singularities of Created Things*], by Zakariyyā' ibn Muḥammad ibn Maḥmūd Abū Yaḥyā al-Ḳazwīnī (*c*. 1203–1283), MS P 3 [44, P 3, p. 330]. Online open-access: National Library of Medicine [35], Bethesda, updated: 23 February 2006 (http://www.nlm.nih.gov/hmd/arabic/natural_hist5.html). [Folio 310a on image page 8.]

[24] Kazwini (2006). Illustrations from *'Ajā'ib al-makhlūqāt wa-gharāib al-mawjūdāt* [*Book of the Marvels of Nature and the Singularities of Created Things*], by Zakariyyā' ibn Muḥammad ibn Maḥmūd Abū Yaḥyā al-Ḳazwīnī (*c.* 1203–1283)], MS P 29 [44, P 29, p. 330]. Online open-access: National Library of Medicine [35], Bethesda, updated: 23 February 2006 (http://www.nlm.nih.gov/hmd/arabic/natural_hist6.html). [Manuscript signed by the scribe, Sayyid Ḥusayn Yazdī. Undated, but probably completed by 1546.]

[25] H. Kennedy (1960/2008). al-Mu'taḍid Bi'llāh, Abu'l-'Abbās Aḥmad b. Ṭalḥa. In *The Encyclopaedia of Islam* [13, vol. 7, p. 760, col. 1] & online restricted-access: Brill OnLine (http://www.encislam.brill.nl/subscriber/uid = 1417/entry?entry = islam_SIM-5640).

[26] Alexander Kheyfits & Dmitry Kheyfits (2005). Fisher's $\alpha$ revisited: closed-form representation and computation. *Journal of Interdisciplinary Mathematics,* 8 (2), 215–225.

[27] E. Lévi-Provençal (1960/2008). al-Mu'taḍid Bi'llāh, Abū Ámr 'Abbād b. Muḥammad b. 'Abbād. %u'l-'Abbās Aḥmad b. Ṭalḥa. In *The Encyclopaedia of Islam* [13, vol. 7, p. 761, column 1] and {http://www.encislam.brill.nl/subscriber/uid=1417/entry?entry=islam_SIM-5641 online restricted-access:} Brill OnLine.

[28] T. Lewicki (1960/2008). al-Ḳazwīnī, Zakariyyā' b. Muḥammad b. Maḥmūd Abū Yaḥyā. In *The Encyclopaedia of Islam* [13, vol. 4, pp. 865–867] and online restricted access: Brill OnLine (http://www.encislam.brill.nl/subscriber/uid = 1417/entry?entry = islam_SIM-4093).

[29] *The MacTutor History of Mathematics Archive* (2008). Online open-access: web archive created by John J. O'Connor and Edmund F. Robertson, School of Mathematics and Statistics, University of St Andrews, Scotland, UK (http://www-groups.dcs.st-and.ac.uk/ history/BiogIndex.html).

[30] Anne E. Magurran (2004). *Measuring Biological Diversity.* Blackwell Science, Oxford.

[31] M. Ismail Marcinkowski (2003). *Measures and Weights in the Islamic World: An English Translation of Walther Hinz's Handbook "Islamische Maße und Gewichte"* (Foreword by Professor C. E. Bosworth, F.B.A.) International Institute of Islamic Thought and Civilization (ISTAC), International Islamic University Malaysia (IIUM), Kuala Lumpur xxii + 98 pp., ISBN 983-9379-27-5. [Translation into English of the 1970 reprint of Hinz [15, (1955)].]

[32] Julie Scott Meisami & Paul Starkey, eds. (1998). *Encyclopedia of Arabic Literature.* Routledge, London, 1998, 2 vols., xvii + 857 pp., ISBN 0-415-06808-8 (set).

[33] *Merriam-Webster's Collegiate Dictionary*, 11th edition (2003). Merriam-Webster, Springfield, Massachusetts. Available as paper copy, on CD-ROM, and restricted-access online (https://member.collegiate.com/subscribe.php).

[34] Sir William Muir (2001). *The Caliphate: Its Rise, Decline, and Fall.* Adamant Media Corporation, 626 pp., ISBN 1-40219327-0.

[35] National Library of Medicine (2006). Islamic medical manuscripts. Online open-access: Catalogue: Natural History (http://www.nlm.nih.gov/hmd/arabic/natural_hist2.html) and Glossary of terms: National Library of Medicine, Bethesda (http://www.nlm.nih.gov/hmd/arabic/glossary.html).

[36] *Oxford English Dictionary: The Definitive Record of the English Language* (2002), edited by John Simpson & Edmund Weiner. 20 volume set in 5 boxes, 22000 pp. & online restricted-access: Oxford University Press (http://www.oed.com/).

[37] C.G. Rice & S. Demarais (1996). A table of values for Fisher's $\alpha$ log series diversity index. *Texas Journal of Science,* 48, 147–158. [See also online open-access table: USACERL, Champaign, Illinois (http://nhsbig.inhs.uiuc.edu/general_stats/alpha.lst).]

[38] L. Richter-Bernburg (1998). al-Qazwīnī, Zakariyya' ibn Muḥammad (C.600-82/C.1203-83). In *Encyclopedia of Arabic Literature* [32, vol. 2, pp. 637–638].

[39] George Ripley & Charles A. Dana, eds. (1858). *The New American Cyclopædia: A Popular Dictionary of General Knowledge, Volume IV: Brownson–Chartres*, D. Appleton, New York, 16 vols., 1858–1863. [vol. IV pub. 1858. Revised as [40]. Online open-access: Google Books (http://books.google.com/books?).

[40] George Ripley & Charles A. Dana, eds. (1881–1883). *The American Cyclopædia: A Popular Dictionary of General Knowledge*, D. Appleton, New York, 17 vols. [vol. 17 = Index by T.J. Conant. Revised version of [39] and apparently not digitized. Earlier version [39] has "New" in title!].

[41] Michael L. Rosenzweig (1995). *Species Diversity in Space and Time.* Cambridge University Press. [Reprinted (with corrections) 1996–2002.]

[42] Stuart Rossiter & John Flower (1986). *The Stamp Atlas: A Unique Assembly of Geography, Social and Political History, and Postal Information.* W.H. Smith, London.

[43] Norbert Schneider (2000). *Vermeer 1632–1675: Veiled Emotions.* Benedikt Taschen Verlag GmbH, Köln.

[44] Dorothy M. Schullian & Francis E. Sommer (1950). *A Catalogue of Incunabula and Manuscripts in the Army Medical Library.* Schuman, New York, xiii + 361 pp.

[45] Baron Silvestre de Sacy (1827). *Chrestomathie arabe, ou Extraits de divers écrivains arabes, tant en prose qu'en vers, avec une traduction française et des notes, A l'usage des Élèves de l'École royale et spéciale des Langues orientales vivantes*, Tome III, seconde édition, corrigée et augmentée [by http://en.wikipedia.org/wiki/Silvestre_de_SacyAntoine Isaac, baron Silvestre de Sacy (1758–1838)], imprimé par autorisation du roi, á l'Imprimerie royale [Paris]. [See "Extraits du livre des *Merveilles de la nature et des singularités des choses créés* par Mohammed Kazwini, Fils de Mohammed" by A.-L. de Chézy [9] on pp. 387–516.]

[46] George P.H. Styan (2006). Some notes on an early Islamic study of biodiversity and on some early Islamic magic squares. Unpublished manuscript, 33 pp.

[47] G.P.H. Styan & S. Puntanen (2006). Fisher's alpha index of biodiversity: 1943–2005. Talk presented (by G. P. H. Styan) at the XXIII$^{rd}$ International Biometric Conference (IBC-2006): Montréal (Québec), Canada: 17 July 2006 [abstract #453.pdf in IBC2006-CD].

[48] Sir Vincent Wigglesworth [Sir Vincent Brian Wigglesworth (1899–1994) (http://en.wikipedia.org/wiki/Sir_Vincent_Brian_Wigglesworth)] (1982). Carrington Bonsor Williams: 7 October 1889–12 July 1981. *Biographical Memoirs of Fellows of the Royal Society*, 28, 666–684. (1982). Full-text pdf on JSTOR (http://www.jstor.org/stable/pdfplus/769914.pdf).

[49] Wikimedia Foundation (2008). *Wikipedia, The Free Encyclopedia* Open-access web archive (in English) (http://en.wikipedia.org).

[50] Wikimedia Foundation (2008). *Wikipédia, l'encyclopédie libre.* Open-access web archive (in French) (http://fr.wikipedia.org).

[51] C.B. Williams (1964). *Patterns in the Balance of Nature and Related Problems in Quantitative Ecology.* Academic Press, London. [Biography of Carrington Bonsor Williams (1889–1981) full-text pdf on JSTOR [48] (http://www.jstor.org/stable/pdfplus/769914.pdf)].

[52] Kenneth A. Wood (1983/1985). *Where in the World? An Atlas for Stamp Collectors.* Van Dahl Publications, Albany, Oregon. [1st printing © 1983, 2nd printing 1985].

[53] Ferdinand Wüstenfeld (1848). *Zakarija Ben Muhammed Ben Mahmud el-Cazwini's Kosmographie, Erster Theil: Die Wunder der Schöpfung.* Aus den Handschriften der Bibliotheken zu Berlin, Gotha, Dresden und Hamburg. In German, edited translation of Part 1 of *'Ajā'ib al-makhlūqāt wa-gharāib al-mawjūdāt* [*Book of the Marvels of Nature and the Singularities of Created Things*], by Zakariyyā' ibn Muḥammad ibn Maḥmūd Abū Yaḥyā al-Ḳazwīnī (*c.* 1203–1283)], Verlag der Dieterichschen Buchhandlung, Göttingen. [Part 2 published in 1849 [54], reprinted in 1967 [55].]

[54] Ferdinand Wüstenfeld (1849). *Zakarija Ben Muhammed Ben Mahmud el-Cazwini's Kosmographie, Zweiter Theil: Die Denkmäler der Länder.* Aus den Handschriften des Hn. Dr. Lee und der Bibliotheken zu Berlin, Gotha, und Leyden. In German, edited translation of Part 2 of *'Ajā'ib al-makhlūqāt wa-gharāib al-mawjūdāt* [*Book of the Marvels of Nature and the Singularities of Created Things*], by Zakariyyā' ibn Muḥammad ibn Maḥmūd Abū Yaḥyā al-Ḳazwīnī (*c.* 1203–1283), Verlag der Dieterichschen Buchhandlung, Göttingen. [Part 1 published in 1848 [53], reprinted in 1967 [55].]

[55] Ferdinand Wüstenfeld (1967). *Zakarija Ben Muhammed Ben Mahmud el-Cazwini's Kosmographie, Erster Theil: Die Wunder der Schöpfung & Zweiter Theil: Die Denkmäler der Länder.* In German, edited translation of *'Ajā'ib al-makhlūqāt wa-gharāib al-mawjūdāt* [*Book of the Marvels of Nature and the Singularities of Created Things*], by Zakariyyā' ibn Muḥammad ibn Maḥmūd Abū Yaḥyā al-Ḳazwīnī (*c.* 1203–1283), Dr. Martin Sändig oHG,

Wiesbaden. ["Genehmigter Neudruck der Ausgabe von 1848–1849 des Verlages der Dieterichschen Bunchhandlung (Titel-Nummer 1941)." Reprint edition of [53, 54] published in 1848/1849, two volumes in one].

[56] Richard A. Young & Thomas J. Glover (1996). *Measure for Measure*. Sequoia Publishing, Littleton, Colorado. [7th printing: May 2006].

[57] YouTube: Broadcast Yourself (2008). Groupies pour La Laitière. Online video (http://fr.youtube.com/watch?v=rcyW3bHgQbE). [The audio for this video is the song Twist and Shout (http://en.wikipedia.org/wiki/Twist_and_Shout) written by Phil Medley and Bert Russell. It was originally recorded by the Topnotes and then covered by The Isley Brothers and later by The Beatles (http://en.wikipedia.org/wiki/The_Beatles), with John Lennon (1940–1980) (http://en.wikipedia.org/wiki/John_Lennon) on the lead vocals, and originally released on their first album *Please Please Me (http://en.wikipedia.org/wiki/Please_Please_Me).]*

# Ultimatum Games and Fuzzy Information

Philip Sander and Peter Stahlecker

**Abstract** We consider the proposer's decision process in an ultimatum game where his uncertainty with respect to the responder's preferences and the associated acceptance threshold is modeled by a fuzzy set. Employing a three-step defuzzification strategy we determine the proposer's best possible claim which depends on his beliefs and his attitude towards risk. Furthermore, we derive an explicit solution for a specific class of fuzzy sets. From a more abstract point of view we analyze a game in which one player has a non-continuous objective function and where the uncertain point of discontinuity is determined by the other player's strategy.

## 1 Introduction

Classical game-theoretical predictions regarding the ultimatum game are inconsistent with the results of numerous economic experiments.[1] These empirical observations show that proposers usually claim between 50 and 60 per cent of the amount to share and virtually never exceed 80 per cent, while responders are far from accepting every positive offer.[2]

Based on these results and similar experiments with respect to comparable games a couple of authors have developed models which take into account the possibility

Peter Stahlecker
Universität Hamburg, Institut für Statistik und Ökonometrie, Von-Melle-Park 5, D-20146 Hamburg, Germany
peter.stahlecker@uni-hamburg.de

[1] See, e.g., [15] who were the first to consider the ultimatum game as well as Thaler [28], Camerer and Thaler [7], Güth [14], Roth [24], Camerer [6] and Oosterbeek et al. [22] who summarize many empirical results.

[2] These and more 'stylized facts' can be found in Sobel [26], p. 398 or Falk and Fischbacher [10], p. 303. Fehr and Schmidt [11], p. 827 outline the results of ten studies and detect, based on a total of 875 observations, that actually 71% of the offers amount to a share of 40–50% of the total sum.

of players having *social preferences*.[3] According to those theories the responders' rejections of proposals offering a positive amount to them show that they prefer to forgo a certain amount of money by refusing a claim perceived as being 'unfair' or 'unkind'. Furthermore, many experiments indicate that proposers seem to be in a position to anticipate the responders' *acceptance thresholds* and correspondingly align their claims.

This paper focuses on the question how a proposer incorporates his uncertainty regarding the responder's acceptance threshold to derive his optimal decision. For this purpose we do not apply a stochastic model framework, but simply assume that the proposer has some *vague beliefs* about the responder's acceptance threshold. This vagueness is represented by a *fuzzy set* and its corresponding membership function which reflects the proposer's subjective beliefs with respect to the responder's acceptance threshold. We apply an economically sensible defuzzification strategy to capture the proposer's degree of pessimism (or optimism respectively) and we derive an explicit (crisp) solution to his utility maximization problem for a specific case.

In the following Sect. 2 we present the model which depicts the proposer's decision problem under uncertainty. The related defuzzification strategy is presented and the existence of an optimal solution is proved in Sect. 3. In Sect. 4 we explicitly characterize the proposer's optimal behavior for an example. Concluding remarks and possibilities to generalize the model are presented in Sect. 5.

## 2 The Model

An ultimatum game describes a situation in which two players bargain about the split of a fixed amount $s \in \mathbb{R}, s > 0$, of money that is paid to them only if they come to a settlement. The bargaining procedure specifies that at first the proposer (player 1) announces a certain claim $p \in \Omega$, with $\Omega = [0,s]$ being his strategy set.[4] In the second step the responder (player 2) accepts or refuses this claim depending on his personal acceptance threshold $a \in \Omega$. If $p \leq a$ holds the claim does not exceed the acceptance threshold and, therefore, is accepted by player 2. If the claim is higher than his acceptance threshold ($p > a$) the responder rejects it and both players obtain a payoff of 0.

In the following we analyze the proposer's decision process in this setting. To keep the exposition simple we assume that the proposer is acting as a pure income maximizer.[5] Then, his utility function can be represented by $u : \Omega \times \Omega \to \Omega$ with

[3] See Bolton [4], Rabin [23], Levine [21], Fehr and Schmidt [11], Bolton and Ockenfels [5], Charness and Rabin [8], Dufwenberg and Kirchsteiger [9] and Falk und Fischbacher [10]. Sobel [26] provides a comprehensive survey about models which allow for social or interdependent preferences.

[4] In general, $\Omega = [0,s]$ with $s$ denoting the sum of money to divide can be considered. For $\Omega = [0,1]$ we may also interpret $p \in \Omega$ as the share of the total amount which is demanded by the proposer..

[5] This assumption is backed by, e.g., Roth et al. [25] who assert that positive offers in an ultimatum game mostly occur due to the proposer's *fear of rejection*.

$$u(p,a) = \begin{cases} p, & \text{if } p \leq a, \\ 0, & \text{if } p > a \end{cases} \tag{1}$$

and, thus, merely depends on his claim $p$ and the responder's acceptance threshold $a$. After the proposer has made a claim, his utility may only amount to $u = 0$ (due to rejection for $p > a$) or $u = p$ (due to acceptance for $p \leq a$).

Whether the responder accepts or rejects a certain claim is determined by his acceptance threshold $a$ which the proposer does not know. In a stochastic setting the proposer would regard $a$ as a random variable with a perfectly known probability distribution. In contrast to that we suppose in this paper that the proposer only has some vague ideas about $a$ which can be represented by a fuzzy set $\mathscr{B} = \{(a, m(a)) \,|\, a \in \Omega\}$, where $m : \Omega \to [0,1]$ denotes the membership function.[6] Here, for every $a \in \Omega$ the value $m(a)$ is the degree of membership to which $a$ belongs to $\mathscr{B}$, where $m(a) = 0$ reflects nonmembership (e.g. an acceptance threshold considered as impossible) and $m(a) = 1$ refers to most plausible acceptance thresholds.[7] Observe that degrees of membership must not be treated like probabilities. In particular, the membership values do not have to add up to one.[8]

Since the set of acceptance thresholds is fuzzy the proposer cannot directly derive his utility maximizing decision. In order to obtain a crisp optimal claim $p$ the multi-valuedness in $a$ has to be eliminated by suitable modifications of the objective function. To achieve this aim we present an appropriate defuzzification strategy which incorporates the proposer's attitude towards deviations of the acceptance threshold from its most plausible values. In doing so we obtain a crisp objective function.

## 3 A Defuzzification Strategy

As an essential part of the chosen defuzzification strategy[9] we firstly introduce the so-called $\alpha$-cuts of the fuzzy set $\mathscr{B}$ which are defined by

$$\mathscr{B}_\alpha = \{a \in \Omega \,|\, m(a) \geq \alpha\}, \tag{2}$$

for all $\alpha \in (0,1]$. According to this definition, the crisp set $\mathscr{B}_\alpha$ comprises all acceptance thresholds $a$ having a degree of membership $m(a) \geq \alpha$. For $\alpha = 0$ we define

$$\mathscr{B}_0 := \mathrm{cl}\{a \in \Omega \,|\, m(a) > 0\} \tag{3}$$

as the support of the fuzzy set $\mathscr{B}$ where $\mathrm{cl}\{\cdot\}$ denotes the closure of $\{\cdot\}$. To further characterize the $\alpha$-cuts we assume that the membership function $m : \Omega \to [0,1]$ is

[6] For a comprehensive introduction to the theory of fuzzy sets the reader is referred to [3] and [29].

[7] A crisp set $\mathscr{B} \subset [0,1]$ is included as a special case when $m$ is a mapping into the set $\{0,1\}$. Then $m$ would reduce to the usual characteristic function.

[8] Klir and Wierman [20] extensively discuss the relationship between fuzzy set theory and stochastic theory.

[9] To this point see [27, 1, 16, 17, 2]

continuous, quasi-concave, and onto. Hence, every $\alpha$-cut is a closed interval in $\Omega$, i.e., there exist functions $\underline{b}, \overline{b} : [0,1] \to \Omega$ with $\underline{b} \leq \overline{b}$ so that $\mathscr{B}_\alpha = [\underline{b}(\alpha), \overline{b}(\alpha)]$ holds for all $\alpha \in [0,1]$.[10] Since $\mathscr{B}_{\alpha_1} \subseteq \mathscr{B}_{\alpha_2}$ for all $\alpha_1, \alpha_2 \in [0,1]$ with $\alpha_1 > \alpha_2$, it follows that $\underline{b}(\cdot)$ ($\overline{b}(\cdot)$) is monotonically increasing (decreasing). For the sake of simplicity we assume that (at least) $\underline{b}$ is strictly monotone.[11]

We now present our three-step defuzzification strategy used to derive the proposer's optimal claim. In the *first step* of this procedure we assume that for a given claim $p$ and a given $\alpha$-cut $\mathscr{B}_\alpha$ the proposer usually will only be interested in those acceptance thresholds which lead to his lowest and highest utility. Then for every $\alpha \in [0,1]$ we obtain[12]

$$\min_{a \in \mathscr{B}_\alpha} u(p,a) = \begin{cases} p, \text{ if } p \leq \underline{b}(\alpha), \\ 0, \text{ if } p > \underline{b}(\alpha), \end{cases} \tag{4}$$

$$\max_{a \in \mathscr{B}_\alpha} u(p,a) = \begin{cases} p, \text{ if } p \leq \overline{b}(\alpha), \\ 0, \text{ if } p > \overline{b}(\alpha). \end{cases} \tag{5}$$

In the *second step* of the defuzzification strategy we assume that the proposer's objective function represents a weighted average of a pessimistic and optimistic attitude regarding the responder's acceptance threshold, i.e., to some extent the proposer keeps in mind the worst case as well as the best case scenario when deriving his optimal decision. According to the well-known Hurwicz principle and based on (4) and (5) we define a weighted average of the worst case and the best case for every $\alpha$-cut as $u_q : \Omega \times [0,1] \to \Omega$ with

$$\begin{aligned} u_q(p,\alpha) &= q \cdot \min_{a \in \mathscr{B}_\alpha} u(p,a) + (1-q) \cdot \max_{a \in \mathscr{B}_\alpha} u(p,a) \\ &= \begin{cases} p, & \text{if } p \leq \underline{b}(\alpha), \\ (1-q)p, & \text{if } \underline{b}(\alpha) < p \leq \overline{b}(\alpha), \\ 0 & \text{if } \overline{b}(\alpha) < p, \end{cases} \end{aligned} \tag{6}$$

for all $(p,\alpha) \in \Omega \times [0,1]$, where $q \in [0,1]$ denotes a given weighting parameter which represents the proposer's degree of pessimism (increasing in $q$).

The weighted mean $u_q(p,\alpha)$ of the worst case and the best case in equation (6) still depends on the $\alpha$-cut. Hence, the *third step* of the defuzzification strategy aims at eliminating this dependency by aggregating over all $\alpha$-cuts. In order to achieve this goal we assume a given continuous function $w : [0,1] \to \mathbb{R}_+$ and define the proposer's new objective function $Z_q : \Omega \to \Omega$ by

[10] $m$ is quasi-concave, iff for all $\alpha \in [0,1]$ $\mathscr{B}_\alpha$ is a convex set. Therefore every $\alpha$-cut is an interval in $\Omega$ and, due to the continuity of the membership function, these intervals are closed. Observe that $\underline{b}(0)$ and $\overline{b}(0)$ are the boundary points of the closed set $\mathscr{B}_0$.

[11] $\underline{b}$ is strictly monotone, if the reasonable assumption $m(0) = 0$ holds.

[12] Suppose $p \leq \underline{b}(\alpha)$. Then we have for all $a \in \mathscr{B}_\alpha = [\underline{b}(\alpha), \overline{b}(\alpha)] : p \leq a$, $u(p,a) = p$, and thus trivially $\min_{a \in \mathscr{B}_\alpha} u(p,a) = p$. Assume $p > \underline{b}(\alpha)$. Then there exists $a \in \mathscr{B}_\alpha$ with $p > a$, $u(p,a) = 0$ and therefore $\min_{a \in \mathscr{B}_\alpha} u(p,a) = 0$. The proof of (5) can be given analogically.

$$Z_q(p) = \int_0^1 w(\alpha) \cdot u_q(p,\alpha) d\alpha \tag{7}$$

for all $p \in \Omega$.

To see that (7) is well-defined and to obtain an explicit formula of the objective function which corresponds to (6) we consider the sets

$$\mathscr{A}_1(p) = \{\alpha \in [0,1] \,|\, p \le \underline{b}(\alpha)\}, \tag{8}$$
$$\mathscr{A}_2(p) = \{\alpha \in [0,1] \,|\, \underline{b}(\alpha) < p \le \overline{b}(\alpha)\}, \tag{9}$$
$$\mathscr{A}_3(p) = \{\alpha \in [0,1] \,|\, \overline{b}(\alpha) < p\} \tag{10}$$

as well as the associated indicator functions

$$\mathbf{1}_{\mathscr{A}_i(p)}(\alpha) = \begin{cases} 1, \text{ if } \alpha \in \mathscr{A}_i(p), \\ 0, \text{ if } \alpha \notin \mathscr{A}_i(p), \end{cases} \tag{11}$$

for any given $p \in \Omega$ and $i = 1,2,3$. Then using (6) we can restate $Z_q$ as

$$\begin{aligned} Z_q(p) &= \int_0^1 w(\alpha) \cdot u_q(p,\alpha) d\alpha \\ &= \int_0^1 w(\alpha) \left(p\mathbf{1}_{\mathscr{A}_1(p)}(\alpha) + (1-q)p\mathbf{1}_{\mathscr{A}_2(p)}(\alpha) + 0 \cdot \mathbf{1}_{\mathscr{A}_3(p)}(\alpha)\right) d\alpha \\ &= p\int_0^1 \mathbf{1}_{\mathscr{A}_1(p)}(\alpha) w(\alpha) d\alpha + (1-q)p\int_0^1 \mathbf{1}_{\mathscr{A}_2(p)}(\alpha) w(\alpha) d\alpha. \end{aligned} \tag{12}$$

Now observe that we have $(m(p),1] \subseteq \mathscr{A}_1(p) \subseteq [m(p),1]$ for any $p \in \Omega$ with $p \le \underline{b}(1)$, $\mathscr{A}_1(p) = \{\}$ for any $p \in \Omega$ with $p > \underline{b}(1)$, and $[0,m(p)) \subseteq \mathscr{A}_2(p) \subseteq [0,m(p)]$ for any $p \in \Omega$.[13] Therefore, we obtain in case of $p \le \underline{b}(1)$

$$Z_q(p) = p\int_{m(p)}^1 w(\alpha) d\alpha + (1-q)p\int_0^{m(p)} w(\alpha) d\alpha \tag{13}$$

and in case of $p > \underline{b}(1)$

$$Z_q(p) = (1-q)p\int_0^{m(p)} w(\alpha) d\alpha. \tag{14}$$

[13] Here we use the assumption that $\underline{b}$ is strictly monotone.

By combining (13) and (14) and defining

$$W(x) := \int_0^x w(\alpha)d\alpha \tag{15}$$

we finally arrive at the proposer's new objective function $Z_q(p)$ with

$$Z_q(p) = \begin{cases} p\int_0^1 w(\alpha)d\alpha - qp\int_0^{m(p)} w(\alpha)d\alpha, & \text{if } p \le \underline{b}(1), \\ (1-q)p\int_0^{m(p)} w(\alpha)d\alpha, & \text{if } p > \underline{b}(1), \end{cases}$$
$$= \begin{cases} pW(1) - qpW(m(p)), & \text{if } p \le \underline{b}(1), \\ (1-q)pW(m(p)), & \text{if } p > \underline{b}(1) \end{cases} \tag{16}$$

for any $p \in \Omega$. Then we can state the following result:

**Proposition 1.** There exists at least one maximal point $p^* \in \Omega$ of $Z_q$.

*Proof.* By assumption $m : \Omega \to [0,1]$ is continuous. Therefore, the function $W \circ m : \Omega \to \mathbb{R}$ with $W \circ m(p) := W(m(p)) = \int_0^{m(p)} w(\alpha)d\alpha$ is also continuous. Hence, $Z_q$ is continuous for all $p \in \Omega \setminus \{\underline{b}(1)\}$. Since we have $Z_q(\underline{b}(1)) = (1-q)\underline{b}(1)W(1) = \lim_{p\downarrow \underline{b}(1)} Z_q(p)$, $Z_q(p)$ is continuous for all $p \in \Omega$. Because $\Omega$ is a nonempty and compact set the existence of a maximal point $p^* \in \Omega$ of $Z_q$ follows from a well-known theorem of Weierstrass. □

*Remark 1.* Suppose we start from the unbounded strategy set $\Omega \subseteq \mathbb{R}$. Then it is important that the support $\mathscr{B}_0$ of the fuzzy set $\mathscr{B}$ is closed and bounded, i.e. $\mathscr{B}_0 = [\underline{b}(0), \overline{b}(0)], \underline{b}(0), \overline{b}(0) \in \mathbb{R}_+$ with $\underline{b}(0) < \overline{b}(0)$. In that case we obtain from (16)

$$Z_q(p) \le \underline{b}(0)W(1) = Z_q(\underline{b}(0))$$

for any $p \notin \mathscr{B}_0$ and therefore

$$\max_{p\in\Omega} Z_q(p) = \max_{p\in\mathscr{B}_0} Z_q(p).$$

Because we then may restrict our considerations to the nonempty and compact set $\mathscr{B}_0$, proposition 1 obviously remains valid and allows for the modeling of situations in which the size of the amount to share is not known with certainty.

In the following we will not discuss the general case in full detail but prefer to derive the proposer's optimal decision for an intuitive example.

## 4 A Particular Model

We assume that the membership function $m : \Omega \to [0,1]$ is defined by

$$m(a) = \begin{cases} 0, & \text{if } a \le a_0, \\ \left(\frac{a-a_0}{a_1-a_0}\right)^{\rho}, & \text{if } a_0 < a \le a_1, \\ 1, & \text{if } a_1 < a \le a_2, \\ \left(\frac{a_3-a}{a_3-a_2}\right)^{\delta}, & \text{if } a_2 < a \le a_3, \\ 0, & \text{if } a_3 < a, \end{cases} \tag{17}$$

where $a_i \in \Omega, i = 0, \ldots, 3$ with $0 \le a_0 < a_1 \le a_2 < a_3$ and $\delta, \rho \in \mathbb{R}$ with $\delta > 0, \rho > 0$ are given parameters. The membership function (17) is continuous, quasi-concave and onto. An example is depicted in Fig. 1 for $\Omega = [0,s]$ and parameter values of $a_0 = 0$, $a_1 = 0.7s$, $a_2 = 0.8s$, $a_3 = s$, $\rho = 3$ and $\delta = 0.5$.

Observe that we have

$$\mathscr{B}_\alpha = \left[\underline{b}(\alpha), \overline{b}(\alpha)\right]$$

with

$$\underline{b}(\alpha) = a_0 + (a_1 - a_0)\,\alpha^{1/\rho} \quad \text{and} \quad \overline{b}(\alpha) = a_3 - (a_3 - a_2)\,\alpha^{1/\delta}$$

for any $\alpha \in [0,1]$. Subsequently, the function $w : [0,1] \to \mathbb{R}_+$ is supposed to be given by

$$w(\alpha) = (1+\gamma)\alpha^{\gamma} \tag{18}$$

with $\gamma > -1$ so that by (15)

$$W(x) = x^{1+\gamma} \tag{19}$$

for all $x \in [0,1]$ and $W(m(p)) = m(p)^{1+\gamma}$ for all $p \in \Omega$.

By remark 1, we may restrict the search for the maximum of $Z_q$ to the support $\mathscr{B}_0 = [a_0, a_3]$ of the fuzzy set $\mathscr{B}$.

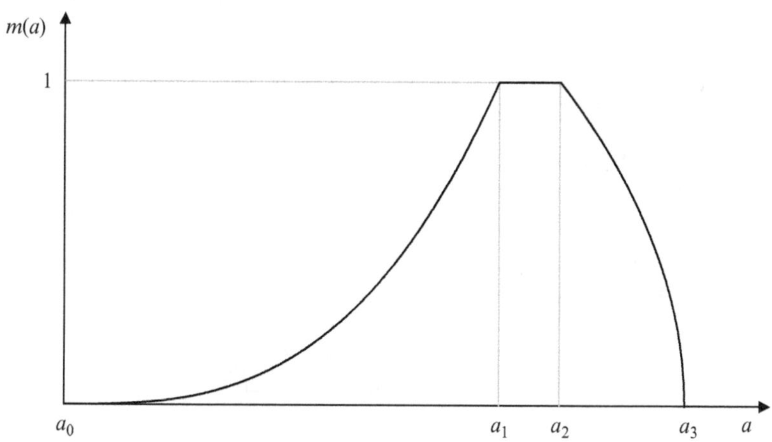


**Fig. 1** A specific membership function

Since we have $m(p) = 1$ for any $p \in [a_1, a_2]$ and therefore

$$Z_q(p) = (1-q)p \leq (1-q)a_2 = Z_q(a_2),$$

we only have to analyze the left branch of $m$ over $\mathscr{B}_0$, i.e., $a_0 \leq p \leq a_1$, and the right branch of $m$ over $\mathscr{B}_0$, i.e., $a_2 \leq p \leq a_3$. Then by (17) the relevant part of $W(m(p))$ reduces to

$$W(m(p)) = \begin{cases} \left(\frac{p-a_0}{a_1-a_0}\right)^{\rho(1+\gamma)}, \text{ if } a_0 \leq p \leq a_1, \\ \left(\frac{a_3-p}{a_3-a_2}\right)^{\delta(1+\gamma)}, \text{ if } a_2 \leq p \leq a_3. \end{cases} \tag{20}$$

To simplify the notation a bit we define for all $\lambda \in [0,1]$

$$p_l(\lambda) = \lambda a_1 + (1-\lambda)a_0 = a_0 + \lambda \Delta_l \tag{21}$$

with

$$\Delta_l = a_1 - a_0 > 0 \tag{22}$$

and

$$p_r(\lambda) = \lambda a_2 + (1-\lambda)a_3 = a_3 - \lambda \Delta_r \tag{23}$$

with

$$\Delta_r = a_3 - a_2 > 0, \tag{24}$$

implying $p_l([0,1]) = [a_0, a_1]$ and $p_r([0,1]) = [a_2, a_3]$. Furthermore, we set

$$\eta_l = \rho(1+\gamma), \tag{25}$$

$$\eta_r = \delta(1+\gamma), \tag{26}$$

where $\eta_l, \eta_r > 0$ holds due to $\rho, \delta > 0$ and $\gamma > -1$. From (20) in combination with (21), (23), (25), and (26) we obtain

$$W(m(p)) = \begin{cases} \lambda^{\eta_l}, \text{ if } p = p_l(\lambda), \\ \lambda^{\eta_r}, \text{ if } p = p_r(\lambda), \end{cases} \tag{27}$$

and

$$Z_q(p) = \begin{cases} (a_0 + \lambda \Delta_l)(1 - q\lambda^{\eta_l}), \text{ if } p = p_l(\lambda), \\ (1-q)(a_3 - \lambda \Delta_r)\lambda^{\eta_r}, \text{ if } p = p_r(\lambda). \end{cases} \tag{28}$$

We now consider the problem of maximizing the proposer's objective $Z_q(p)$ for the case $p = p_l(\lambda)$ and the case $p = p_r(\lambda)$, respectively, with $\lambda$ satisfying the constraint $0 \leq \lambda \leq 1$. Comparing $\max_{\lambda \in [0,1]} Z_q(p_l(\lambda))$ with $\max_{\lambda \in [0,1]} Z_q(p_r(\lambda))$, we get the complete solution of the proposer's optimization problem.

We know from proposition 1 and remark 1 that a solution always exists. Unfortunately, however, it is possible, that there is no explicit solution for the case $a_0 > 0$. To obtain an explicit solution we therefore suppose that $a_0 = 0$. Then by (21) and (22) it follows $p_l(\lambda) = \lambda a_1$, $\Delta_l = a_1$, and $\Delta_l/(\Delta_l + a_1\eta_l) = 1/(1+\eta_l)$. For this constellation we derive the proposer's optimal claim $p^*$.

**Proposition 2.** Suppose that $a_0 = 0$ and $q \in [0,1]$.
(i) There exists exactly one $p_l^*$ maximizing $Z_q$ on $[0,a_1]$. It is given by

$$p_l^* = \begin{cases} a_1, & \text{if } q < \frac{1}{1+\eta_l}, \\ \lambda_l^* a_1, & \text{if } q \geq \frac{1}{1+\eta_l}, \end{cases}$$

with

$$\lambda_l^* = \left( \frac{1}{q(1+\eta_l)} \right)^{\frac{1}{\eta_l}}. \tag{29}$$

(ii) Let $q = 1$. Then any $p \in [a_2,a_3]$ yields $Z_q(p) = 0$ which is the maximum of $Z_q(p)$ on $[a_2,a_3]$.

Let $q < 1$. Then there exists exactly one $p_r^*$ maximizing $Z_q$ on $[a_2,a_3]$. It is given by

$$p_r^* = \begin{cases} a_2, & \text{if } a_3 < a_2(1+\eta_r), \\ a_3 - \lambda_r^* \Delta_r = \frac{a_3}{1+\eta_r}, & \text{if } a_3 \geq a_2(1+\eta_r), \end{cases}$$

with

$$\lambda_r^* = \frac{a_3 \eta_r}{\Delta_r(1+\eta_r)}. \tag{30}$$

*Proof.* The proof of proposition 2 is given in the appendix.

Observe that $p_l^*$ depends on the pessimism parameter $q$, while $p_r^*$ does not. Using the results of proposition 2 in connection with (28) we obtain

$$Z_q(p_l^*) = \begin{cases} (1-q)a_1, & \text{if } q < \frac{1}{1+\eta_l}, \\ \frac{a_1 \eta_l}{1+\eta_l} \left( \frac{1}{q(1+\eta_l)} \right)^{\frac{1}{\eta_l}}, & \text{if } q \geq \frac{1}{1+\eta_l} \end{cases} \tag{31}$$

and

$$Z_q(p_r^*) = \begin{cases} (1-q)a_2, & \text{if } a_3 < a_2(1+\eta_r), \\ (1-q) \left( \frac{a_3}{1+\eta_r} \right) \left( \frac{a_3 \eta_r}{\Delta_r(1+\eta_r)} \right)^{\eta_r}, & \text{if } a_3 \geq a_2(1+\eta_r), \end{cases} \tag{32}$$

where we have set $p_l^* = p_l(\lambda_l^*)$ and $p_r^* = p_r(\lambda_r^*)$.

Because the global maximum of the objective function is given by

$$\max_{p \in \Omega} Z_q(p) = \max[Z_q(p_l^*), Z_q(p_r^*)].$$

we have to distinguish between four different cases. (Here we use Proposition 2 in connection with (31) and (32).)[14]

**Case 1:** $q \geq \frac{1}{1+\eta_l}$ and $a_3 \geq a_2(1+\eta_r)$:

[14] In case 3 we have $Z_q(p_l^*) \leq Z_q(p_r^*)$ with equality if $a_3 = a_2(1+\eta_r)$ and $a_1 = a_2$.

$$p^* = \begin{cases} a_1 \left(\frac{1}{q(1+\eta_l)}\right)^{\frac{1}{\eta_l}}, & \text{if } \frac{a_1\eta_l}{1+\eta_l}\left(\frac{1}{q(1+\eta_l)}\right)^{\frac{1}{\eta_l}} > (1-q)\left(\frac{a_3}{1+\eta_r}\right)\left(\frac{a_3\eta_r}{\Delta_r(1+\eta_r)}\right)^{\eta_r}, \\ a_1 \left(\frac{1}{q(1+\eta_l)}\right)^{\frac{1}{\eta_l}} \text{ or } \frac{a_3}{1+\eta_r}, & \text{if } \frac{a_1\eta_l}{1+\eta_l}\left(\frac{1}{q(1+\eta_l)}\right)^{\frac{1}{\eta_l}} = (1-q)\left(\frac{a_3}{1+\eta_r}\right)\left(\frac{a_3\eta_r}{\Delta_r(1+\eta_r)}\right)^{\eta_r}, \\ \frac{a_3}{1+\eta_r}, & \text{else.} \end{cases} \tag{33}$$

**Case 2:** $q \geq \frac{1}{1+\eta_l}$ and $a_3 < a_2(1+\eta_r)$:

$$p^* = \begin{cases} a_1 \left(\frac{1}{q(1+\eta_l)}\right)^{\frac{1}{\eta_l}}, & \text{if } \frac{a_1\eta_l}{1+\eta_l}\left(\frac{1}{q(1+\eta_l)}\right)^{\frac{1}{\eta_l}} > (1-q)a_2, \\ a_1 \left(\frac{1}{q(1+\eta_l)}\right)^{\frac{1}{\eta_l}} \text{ or } a_2, & \text{if } \frac{a_1\eta_l}{1+\eta_l}\left(\frac{1}{q(1+\eta_l)}\right)^{\frac{1}{\eta_l}} = (1-q)a_2, \\ a_2, & \text{else.} \end{cases} \tag{34}$$

**Case 3:** $q < \frac{1}{1+\eta_l}$ and $a_3 \geq a_2(1+\eta_r)$:

$$p^* = \frac{a_3}{1+\eta_r}. \tag{35}$$

**Case 4:** $q < \frac{1}{1+\eta_l}$ and $a_3 < a_2(1+\eta_r)$:

$$p^* = a_2. \tag{36}$$

To further explain our results we turn to a numerical example, which is based on the parameter values underlying Fig. 1 (with $s = 1$).

**Numerical example:** Let the membership function be defined by $a_0 = 0$, $a_1 = 0.7$, $a_2 = 0.8$, $a_3 = 1$, $\rho = 3$ and $\delta = 0.5$. Furthermore, let $\gamma = 1$ so that $w(\alpha) = 2\alpha$ and $W(x) = x^2$ is chosen. This results in $\Delta_l = 0.7$ and $\Delta_r = 0.2$. From (25) and (26) we obtain $\eta_l = 6$ and $\eta_r = 1$.

Firstly, due to

$$a_3 = 1 < 1.6 = a_2(1+\eta_r)$$

we find that only cases 2 and 4 have to be considered. Furthermore, based on cases 2 and 4 (and (34) and(36)) we obtain that for all values of the pessimism parameter $q \in [0,1]$ with $q \geq \frac{1}{7}$ and

$$39.331q(1-q)^6 < 1$$

the proposer's optimal claim is given by

$$p^*(q) = 0.7\left(\frac{1}{7q}\right)^{\frac{1}{6}}.$$

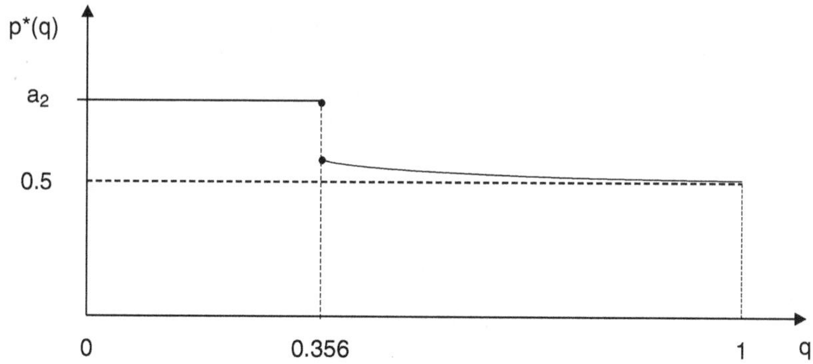


**Fig. 2** Optimal claim $p^*(q)$

All lower pessimism parameters lead to the optimal claim

$$p^* = a_2 = 0.8.$$

For our example the critical value of the pessimism parameter $q$ where the regime switches is given by

$$q_c \cong 0.356.$$

The relationship between the optimal claim $p^*$ and the pessimism parameter $q$ is illustrated in Fig. 2. Optimistic proposers with $q \leq 0.35$ claim $p^* = a_2 = 0.8$. For higher degrees of pessimism the regime switches leading to optimal claims between $p^*(0.36) = 0.60$ and $p^*(1) = 0.51$.[15] Hence, the optimal claim of a purely pessimistic proposer ($q = 1$) is quite close to half of the pie. If we furthermore assume that the degree of pessimism is equally distributed in the population then according to our example we obtain that about $1/3$ of the people would claim 80% and about $2/3$ of the people would claim between 50 and 60% of the pie. This outcome is rather close to the results of the studies summarized by Fehr and Schmidt [11] (see footnote 2).

## 5 Concluding Remarks

We have analyzed the proposer's decision process in an ultimatum game, when he does not know the responder's acceptance threshold and, thus, is uncertain with respect to his optimal claim. The responder's possible rejection of positive offers may, e.g., be caused by his aversion to unequal payoffs or by the fact that he perceives certain offers as unkind.

---

[15] Observe that $p^*(q_c)$ consists of two optimal $p^*$-values, i.e. $p^*(q)$ is not a function for $q = q_c$.

The proposer's uncertainty with respect to the responder's acceptance threshold is modeled by means of a fuzzy set. We derive the proposer's optimal claim by choosing an economically reasonable defuzzification strategy.

Proceeding in this way we show by means of a numerical example with plausible parameter constellation that the most frequent behavior patterns observed in economic experiments on ultimatum games can be explained by our decision model.

Furthermore, the presented model is accessible to empirical testing. In order to conduct such a test one would have to elicit the parameters of the proposer's membership function and his degree of pessimism before or after he discloses his actual claim.[16] In a subsequent step it could be tested whether the claim the model predicts for the elicited parameter values adequately approximates the actual decision of the respective proposer.

It may be seen as a disadvantage of our model that the proposer while aiming at pure income maximization expects the responder to behave in a different way. However, the results of numerous experiments on ultimatum games show that the proposer, in contrast to the responder, is not interested in his counterpart's behavior and payoff. Hence, the proposer's potentially positive offers may only be attributed to his fear of rejection, i.e., we exclude generosity on part of the proposer. This can be regarded as a restriction if one takes into account that the stylized facts (see the introduction) are virtually exclusively based on experiments at western universities and only partially prevail if behavior in other cultural environments is considered.[17] Moreover, generosity of the first (moving) player is often observed in experiments regarding the dictator game where the second player has to accept any distribution chosen by the first player.[18] The usual explanation of this phenomenon is that the first player possesses an aversion to (excessive) *advantageous* inequality and, hence, maximizes his utility by leaving a share of the amount to the second player. These aspects, which would permit to differentiate which part of the proposer's claim is due to fear of rejection and which part is based on generosity, should be taken into consideration for a generalization of our model.[19]

---

[16] Fischbacher and Gächter [12] propose an approach to identify the players' preferences and beliefs in the context of experiments concerning the provision of a public good.

[17] Henrich et al. [18] conduct economic experiments in fifteen small-scale societies in developing countries. They on one hand observe for some societies that offers frequently exceed half of the amount and that, on the other hand, many of these 'hyperfair' offers are rejected. Such kind of behavior or preferences is neither captured by the five stylized facts nor by the underlying assumptions of this paper.

[18] See, e.g., [28, 13, 19].

[19] With respect to the dictator game the cited studies find that both aspects affect the first player's behavior. The average payoff of the second player in the dictator game is positive, but lower than in the ultimatum game. To put it in the words of Sobel [26], p. 399: 'Intuition suggests that at least some of the behavior in the ultimatum game comes from generosity and not fear of rejection.'

## Appendix

*Proof of proposition 2.* (i) Suppose that $p = p_l(\lambda)$. Then by (21), (22), (28) and $a_0 = 0$ we obtain $p_l(\lambda) = a_1\lambda$,

$$Z_q(p_l(\lambda)) = a_1\lambda(1-q\lambda^{\eta_l}) = a_1\lambda - qa_1\lambda^{\eta_l+1}, \tag{37}$$

and

$$Z_q'(p_l(\lambda)) = a_1 - qa_1(\eta_l+1)\lambda^{\eta_l}.$$

First assume that $q = 0$. Then we have $p_l^* = a_1$.

Now suppose that $q > 0$. Then the function given by (37) is strictly concave on $[0,1]$.

Assume that

$$q < \frac{1}{1+\eta_l}.$$

Then we have for all $\lambda \in [0,1]$

$$Z_q'(p_l(\lambda)) \geq a_1 - a_1 q(1+\eta_l) > a_1 - a_1 = 0,$$

i.e., $Z_q(p_l(\cdot))$ is monotonically increasing on $[0,1]$. Thus, $p_l(1) = a_1$ is the maximal point.

Now let

$$q \geq \frac{1}{1+\eta_l}$$

be valid. Because $Z_q(p_l(\cdot))$ is strictly concave on $[0,1]$ and we have $q(1+\eta_l) \geq 1$ it follows from $Z_q'(p_l(\lambda)) = 0$, that the unique maximal point $p_l^* = p(\lambda_l^*)$ of $Z_q$ on $[0,a_1]$ is given by the unique solution $\lambda_l^* \in [0,1]$ of the equation

$$\lambda^{\eta_l} - \frac{1}{q(1+\eta_l)} = 0.$$

(ii) Let now $p = p_r(\lambda) = a_3 - \lambda\Delta_r$ so that by (28)

$$Z_q(p_r(\lambda)) = (1-q)(a_3-\lambda\Delta_r)\lambda^{\eta_r} \geq 0 = Z_q(p_r(0))$$

for any $\lambda \in [0,1]$.

For $q = 1$ we have $Z_q(p_r(\lambda)) = 0$ for any $\lambda \in [0,1]$, i.e., for any $p \in [a_2,a_3]$.

Suppose that $q < 1$. Moreover, let

$$\theta := \frac{a_3\eta_r}{\Delta_r(1+\eta_r)} > 0.$$

Then we get for any $\lambda > 0$

$$Z_q'(p_r(\lambda)) = (1-q)\lambda^{\eta_r-1}[a_3\eta_r - \Delta_r(1+\eta_r)\lambda],$$

where

$$Z_q'(p_r(\lambda)) \begin{cases} > 0, \text{ if } 0 < \lambda < \theta, \\ = 0, \text{ if } \lambda = \theta, \\ < 0, \text{ if } \lambda > \theta, \end{cases} \tag{38}$$

i.e., $Z_q(p_r(\cdot))$ is strictly increasing on $(0, \theta)$ (decreasing on $(\theta, \infty)$).

Now observe that the condition $a_3 < a_2(1+\eta_r)$ ($a_3 \geq a_2(1+\eta_r)$) is equivalent to $\theta > 1$ ($\theta \leq 1$).

Suppose that $\theta > 1$. Then by (38) $Z_q(p_r(\cdot))$ is strictly increasing on $(0,1]$. Therefore $\lambda_r^* = 1$ and $p_r^*(1) = a_3 - 1 \cdot \Delta_r = a_2$ is the unique maximal point of $Z_q$ on $[a_2, a_3]$.

Assume that $\theta \leq 1$. Then by (38) $\lambda_r^* = \theta \in [0,1]$ and $p_r^*(\lambda_r^*) = a_3 - \lambda_r^* \Delta_r = \frac{a_3}{1+\eta_r}$ is the unique maximal point of $Z_q$ on $[a_2, a_3]$. □

## References


[1] Arnold, B.F., Größl, I., Stahlecker, P.: Competitive supply behaviour when price information is fuzzy. J. Econ. **72**, 45–66 (2000)

[2] Arnold, B.F., Hauenschild, N., Stahlecker, P.: Monopolistic price setting under fuzzy information. Eur. J. Oper. Res. **154**, 787–803 (2004)

[3] Bandemer, H., Gottwald, S.: Fuzzy sets, fuzzy logic, fuzzy methods with applications. Wiley, New York (1995)

[4] Bolton, G.E.: A comparative model of bargaining: Theory and evidence. Am. Econ. Rev. **81**, 1096–1136 (1991)

[5] Bolton, G.E., Ockenfels, A.: A Theory of equity, reciprocity and competition. Am. Econ. Rev. **90**, 166–194 (2000)

[6] Camerer, C.: Behavioral game theory: Experiments in strategic interaction. Princeton University Press, Princeton (2003)

[7] Camerer, C., Thaler, R.H.: Ultimatums, dictators, manners. J. Econ. Persp. **9**, 209–219 (1995)

[8] Charness, G., Rabin, M.: Understanding social preferences with simple tests. Q. J. Econ. **117**, 817–869 (2002)

[9] Dufwenberg, M., Kirchsteiger, G.: A theory of sequential reciprocity. Games Econ. Behav. **47**, 268–298 (2004)

[10] Falk, A., Fischbacher, U.: A theory of reciprocity. Games Econ. Behav. **54**, 293–315 (2006)

[11] Fehr, E., Schmidt, K.M.: A theory of fairness, competition and co-operation. Quart. J. Econ. **114**, 817–868 (1999)

[12] Fischbacher, U., Gächter, S.: Heterogeneous social preferences and the dynamics of free riding in public goods. IZA Discussion paper **2011** (2006)

[13] Forsythe, R., Horowitz, J.L., Savin, N.E., Sefton, M.: Fairness in simple bargaining experiments. Games Econ. Behav. **6**, 347–369 (1994)

[14] Güth, W.: On ultimatum bargaining experiments – a personal view. J. Econ. Behav. Organ. **27**, 329–344 (1995)

[15] Güth, W., Schmittberger, R., Schwarze, B.: An experimental analysis of ultimatum bargaining. J. Econ. Behav. Organ. **3**, 367–388 (1982)
[16] Hauenschild, N., Stahlecker, P.: Precautionary saving and fuzzy information. Econ. Lett. **70**, 107–114 (2001)
[17] Hauenschild, N., Stahlecker, P.: Nash-Equilibria in a heterogeneous oligopoly with fuzzy information. Rev. Econ. Design **8**, 165–184 (2003)
[18] Henrich, J., Boyd, R., Bowles, S., Camerer, C., Fehr, E., Gintis, H. (eds.) Foundations of human sociality. Oxford University Press, Oxford (2004)
[19] Hoffman, E., McCabe, K., Smith, V.L.: Social distance and other-regarding behavior in Dictator Games. Am. Econ. Rev. **86**, 653–660 (1996)
[20] Klir, G.J., Wierman, M.J.: Uncertainty-based information: Elements of generalized information theory. Physica, Heidelberg (1999)
[21] Levine, D.K.: Modeling altruism and spitefulness in experiments. Rev. Econ. Dyn. **1**, 593–622 (1998)
[22] Oosterbeek, H., Sloof, R., Van De Kuilen, G.: Cultural differences in ultimatum game experiments: evidence from a Meta-analysis. Exper. Econ. **7**, 171–188 (2004)
[23] Rabin, M.: Incorporating fairness into game theory and economics. Am. Econ. Rev. **83**, 1281–1302 (1993)
[24] Roth, A.E.: Bargaining Experiments. In: Kagel, J.H., Roth, A.E. (eds.) Handbook of experimental economics, pp. 253–348. Princeton University Press, Princeton (1995),
[25] Roth, A.E., Prasnikar, V., Okuno-Fujiwara, M., Zamir, S.: Bargaining and market behavior in Jerusalem, Ljubljana, Pittsburgh, and Tokyo: An experimental study. Am. Econ. Rev. **81**, 1068–1095 (1991)
[26] Sobel, J.: Interdependent preferences and reciprocity. J. Econ. Lit. **43**, 392–436 (2005)
[27] Stahlecker, P., Größl, I., Arnold, B.F.: Monopolistic competition and supply behaviour under fuzzy price information. Homo Oecon. **15**, 561–579 (1999)
[28] Thaler, R.H.: Anomalies: the ultimatum game. J. Econ. Persp. **2**, 195–206 (1988)
[29] Zimmermann, H.-J.: Fuzzy set theory – and its applications (4th edn.). Kluwer, Dordrecht (2001)

# Are Bernstein's Examples on Independent Events Paradoxical?

Czesław Stępniak and Tomasz Owsiany

**Abstract** Bernstein gave two examples showing that a collection of pairwise independent random events need not to be jointly independent. These examples were numbered by Stoyanov among the most fascinating counterexamples in probability. Considering the minimal sample size for existing $n$ independent and pairwise independent but jointly dependent random events we reveal the fact that the second situation is more often. In consequence it is rather a rule than a paradox.

## 1 Introduction

In 1946 ([1], p. 47) Bernstein gave two examples showing that a collection of pairwise independent events need not to be jointly independent. In both examples, the sample space has four outcomes, all equally likely. These examples were numbered by Stoyanov ([4], Sect. 3.1) among the most fascinating counterexamples in probability. Some more refined versions of this phenomenon were studied by Wang, Stoyanov and Shao ([5]).

General case of three pairwise independent events has been considered by Derriennic and Kłopotowski [2] and Stępniak [3]. It was noted in [2] and proved in [3], that the Bernstein's examples are optimal in the sense that there is no smaller examples of this type, or others of this size. This arises two questions concerning the minimal sample size for existing $k$ independent events and for existing $k$ events, each $k-1$ of them is independent.

It is shown here that the first answer is $2^k$, while the second one is $2^{k-1}$. It means that the joint independence is more exceptional than $(k-1)$-*independence* of $k$ events. In this context the Bernstein type examples are not paradoxical.

Czesław Stępniak
Institute of Mathematics, University of Rzeszów, Al. Rejtana 16 A, 35-959 Rzeszów, Poland
cees@univ.rzeszow.pl

By the way a necessary and sufficient condition for independence of a finite number of events is derived. This condition is the main tool in our consideration.

## 2 Independence of Events

Let us recall that $k$ probability events $A_1, A_2, ..., A_k$ in a probability space $(\Omega, \mathcal{F}, P)$ are *jointly independent*, if

$$P(A_{i_1} \cap A_{i_2} \cap ... \cap A_{i_r}) = P(A_{i_1})P(A_{i_2})...P(A_{i_r}) \tag{1}$$

for every subset $\{i_1, i_2, ..., i_r\}$ of $\{1, 2, ..., k\}$. Obviously the joint independence implies independence of any subset of the set $A_1, A_2, ..., A_k$. Without loss of generality we may (and do) assume that none of our events is degenerate (i.e. its probability is zero or one), since its presence does not affect the overall or a subset independence. For any $A \in \mathcal{F}$ we introduce the symbols $A^0 = A$ and $A^1 = \Omega \setminus A$.

**Lemma 1.** *If $A_1, ..., A_k$ are independent events then $A_1^{j_1}, ..., A_k^{j_k}$ are also independent for any $j_i \in \{0, 1\}$.*

*Proof.* First note that the collection $A_1^{j_1}, ..., A_k^{j_k}$ may be obtained from $A_1, ..., A_k$ in a finite number of steps, each replacing a single set $A_{i_0}$ by its complement $A_{i_0}^1$ . Thus we only need to show that any such replacement preserves independence.

If $i_0 \notin \{i_1, ..., i_r\}$ then (1) remains unchanged. Otherwise, it takes the form

$$P(A_{i_0}^1 \cap [\bigcap_{i \neq i_0} A_{j_i}]) = P(A_{i_0}^1) \prod_{i \neq i_0} P(A_{j_i})$$

To verify this condition, we note that the events $A = A_{i_0}$ and $B = \bigcap_{i \neq i_0} A_{j_i}$ are independent. This takes the independence of $A^1$ and $B$, and, in consequence, implies the assertion of the Lemma.

It is clear that the condition $P(\bigcap_{i=1}^k A_i) = \prod_{i=1}^k P(A_i)$, similarly as pairwise independence, does not imply the independence of $A_1, ..., A_k$. The following theorem will be a key tool in the further consideration.

**Theorem 1.** *Random events $A_1, ..., A_k$ are independent, if and only if,*

$$P(\bigcap_{i=1}^{k} A^{j_i}) = \prod_{i=1}^{k} P(A_i^{j_i}) \tag{2}$$

for any $j_1, ..., j_k \in \{0, 1\}$.

*Proof.* Implication (1) $\Longrightarrow$ (2) follows directly from Lemma 1. Now for fixed but arbitrary $A_{i_1}, ..., A_{i_r}$ denote the remaining events by $B_1, ..., B_{k-r}$ and consider the intersections of type $B_{j_1,\ldots,j_{k-r}} = \bigcap_{i=1}^{k-r} B_i^{j_i}$. We observe that the sets $B_{j_1,\ldots,j_{k-r}}$ and $B_{j'_1,\ldots,j'_{k-r}}$ are disjoint for $(j_1,\ldots,j_{k-r}) \neq (j'_1,\ldots,j'_{k-r})$ and $\bigcup_{\{(j_1,\ldots,j_{k-r}):j_i=0,1\}} B_{j_1,\ldots,j_{k-r}} = \Omega$. Thus, by (2),

$$\begin{aligned}
P(A_{i_1}\cap\ldots\cap A_{i_r}) &= P(A_{i_1}\cap\ldots\cap A_{i_r}\cap[\bigcup_{\{(j_1,\ldots,j_{k-r}):j_i=0,1\}} B_{j_1,\ldots,j_{k-r}}])\\
&= \sum_{\{(j_1,\ldots,j_{k-r}):j_i=0,1\}} P(A_{i_1})\cdots P(A_{i_r})\prod_{i=1}^{k-r} P(B_i^{j_i})\\
&= P(A_{i_1})\cdots P(A_{i_r}) \sum_{\{(j_1,\ldots,j_{k-r}):j_i=0,1\}} P(\bigcap_{i=1}^{k-r} B_i^{j_i})\\
&= P(A_{i_1})\cdots P(A_{i_r}) P(\bigcup_{\{(j_1,\ldots,j_{k-r}):j_i=0,1\}} \bigcap_{i=1}^{k-r} B_i^{j_i})\\
&= P(A_{i_1})\cdots P(A_{i_r})
\end{aligned}$$

completing the implication (2) $\Longrightarrow$ (1) and, in consequence, the proof of the theorem.

Now we are ready to prove the main result in this note.

## 3 Joint Independence and Subset Independence

**Theorem 2.** *Let $A_1\ldots,A_k$ be nondegenerate independent events in a probability space $(\Omega,\mathscr{F},P)$ with finite $\Omega = \{\omega_1,\ldots,\omega_n\}$. Then $n \geq 2^k$.*

*Proof.* Consider the family $\mathscr{A} = \{A_{j_1 j_2 \ldots j_k} = A_1^{j_1}\cap\ldots\cap A_k^{j_k}\colon\ j_i \in \{0,1\}\}$ of $2^k$ random events generated by $A_1,\ldots,A_k$. It is clear that all these events are disjoint. On the other hand, by Theorem 1, each of them is not empty. This implies the desired result.

One can ask whether the condition $n \geq 2^k$ is also sufficient for existing $k$ independent events. The answer is included in the following lemma.

**Lemma 2.** *For any probability model $(\Omega,\mathscr{F},P)$ with equally likely sample space $\Omega$ of size $n = 2^k$ there exist $k$ nondegenerate independent events.*

*Proof.* It will be convenient to write the sample space in the form $\Omega = \{\omega_0,\ldots,\omega_{n-1}\}$ and to identify each $\omega_i$ with the binary representation $(i_1,\ldots,i_k)$ of the integer $i$. Defining $A_j = \{(i_1,\ldots,i_k) : i_j = 1\}$ we satisfy the desired condition.

Now we shall prove the following lemma.

**Lemma 3.** *For any probability model $(\Omega, \mathscr{F}, P)$ with equally likely sample space $\Omega$ of size $n = 2^k$ there exist $k+1$ nondegenerate events such that each $k$ of them are independent.*

*Proof.* As in the proof of Lemma 2, we define $A_j = \{(i_1, ..., i_k) : i_j = 1\}$ for $j = 1, ..., k$ and, moreover, $A_{k+1} = \{(i_1, ..., i_k) : \sum_{j+1}^{k} i_j \text{ is even}\}$. By Theorem 1, we only need to verify that for any $i = 1, ..., k$, $P(A_{k+1} \cap [\bigcap_{r \neq i} A_r^{j_r}]) = \frac{1}{2^k}$. Really,

$$P(A_{k+1} \cap [\bigcap_{r \neq i} A_r^{j_r}]) = P(\bigcap_{r \neq i} A_r^{j_r}) P(A_{k+1} / A_r^{j_r}) = \frac{1}{2^k},$$

implying the desired result.

Let us end this note by the following conclusion.

**Conclusion 3** *Condition for existing k independent events is more restrictive than one for existing k dependent events, each $k-1$ of them are independent. In thus, an example with k independent events is more exceptional than an example of Bernstein type.*

## References


[1] Bernstein, S.N.: Theory of Probability (4th edn., in Russian). Gostechizdat, Moscow-Leningrad (1946)

[2] Derriennic, Y., Kłopotowski, A.: Bernstein's example of three pairwise independent random variables. Sankhya A **62**, 318–330 (2000)

[3] Stępniak, C.: Bernstein's examples on independent events. College Math. J. **38**, 140–142 (2007)

[4] Stoyanov, J.: Counterexamples in Probability. Wiley, New York (1987)

[5] Wang, Y.H., Stoyanov, J., Shao, Q.-M.: On independence and dependence properties of a set of random events. Amer. Statist. **47**, 112–115 (1993)

# A Classroom Example to Demonstrate Statistical Concepts

Dietrich Trenkler

**Abstract** A hands-on example to alleviate the transition process from probability models to statistical inference is proposed which can be used in a statistics lecture. It is very easy to grasp and has the benefit that many skills learned so far can be applied. Concepts like parameter estimation, confidence intervals and testing are addressed in this context. It can serve as reference when more abstract concepts such as unbiasedness, mean square error, pivot quantities, confidence level or p-values are treated later on in the course.

## 1 Introduction

Being involved in teaching statistics in an economics department, I have often experienced that students have difficulties in the transition process from probability models to statistical inference. Dealing with topics such as the binomial or normal distribution is mastered by and large, but things turn for the worse when it comes to implementing them by, for example, estimating relevant parameters. So I always look for hands-on examples to pave the way for a better understanding. Furthermore, I was motivated by some advice from Professor Herbert Büning, a teacher of mine and one of Götz' friends: *A lecture has failed if the students did not laugh once.* Gimmicks like Bortkiewicz' famous horse-kick data or anthropometric data from a football championship or the *Tour de France* are very welcome in this respect.

*Gauß' Schmaus*
Osnabrücker Studentenfutter

Dietrich Trenkler
Fachbereich Wirtschaftswissenschaften, Universität Osnabrück, D-49069 Osnabrück, Germany
Dietrich,Trenkler@Uni-Osnabrück.de

Another example is the well-known birthday problem, see e.g. [2, page 46] or [3, page 67]. It forms the background of the following bet in the classroom: *I claim that there are at least two persons in the first row in this room having a common birthday.* Of course, I insure myself in advance by choosing at least 50 persons or so. Much to the students' surprise, I usually win in this way. And I show my gratitude to those two persons who turn out to have a common birthday by handing them bags of gorp equipped with a sticker which is aptly translated as *Gauss' feast – Osnabrück student fodder*".

The birthday problem is somewhat similar to another well-known technique, namely the capture-recapture method. Like the former, it is also very easy to grasp. Furthermore, it turns out to be well-suited for illustrating topics related to point or interval estimation and testing. These are dealt with in the next sections.

## 2 The Statistical Model

Suppose there are $N$ persons in a (class) room and that $N$ is too large to count them all. The objective is to make inferences about $N$. Something like the capture-recapture method comes to mind, but how are we to perform the capture stage? The idea is to choose one or several criteria whose distribution is known or for which some reasonable assumptions can be made. For instance, it seems appropriate to assume that the probability of a person having a birthday this or last week is 2/52=1/26. Another criterion is left-handedness. A small perusal on the internet gives hints that the probability of a person being left-handed is about 0.1.

Let $X$ be the number of persons in the room having an attribute whose fraction $p$ is approximately known. Then it is obvious that $X$ follows a binomial distribution and hence

$$\mathrm{E}[X] = Np \quad \text{and} \quad \mathrm{Var}[X] = Np(1-p)\,. \tag{1}$$

In this way it turns out that the quantity of interest is part of a model the students are already familiar with. The main difference is that now $N$ is unknown, thus turning a probability model into a statistical one.

## 3 Point Estimation

We first tackle the problem of finding surrogate values for $N$ in the statistical model, namely estimates. We would like to use an estimator $\hat{N}$, i.e. a random variable which is related to $N$ in some way. To this end, we exploit an idea known from the capture-recapture technique: Setting $\hat{N} := X/p$, it follows from (1)

$$\mathrm{E}[\hat{N}] = N \quad \text{and} \quad \mathrm{Var}[\hat{N}] = N\frac{1-p}{p}\,. \tag{2}$$

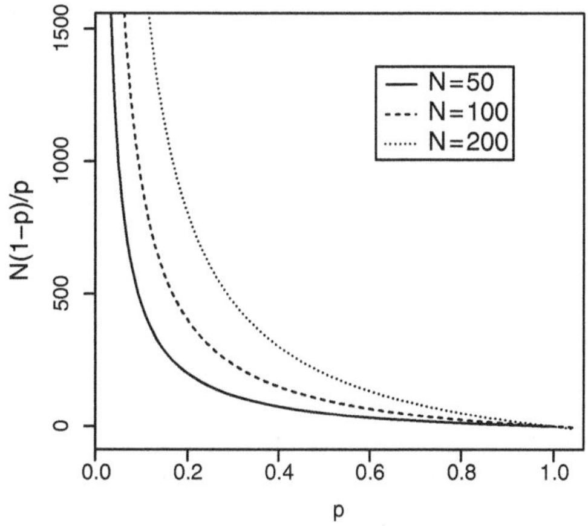


**Fig. 1** $\mathrm{Var}[\hat{N}]$ for $N = 50$, 100 and 200

Equation (2) may be used to illustrate the theoretical concepts of unbiasedness and mean squared error. In fact, $\mathrm{E}[\hat{N}] = N$ means that $\hat{N}$ does not go astray. On the other hand, there are differences in accuracy depending on how $p$ is chosen, as measured by the mean square error which coincides with $\mathrm{Var}[\hat{N}]$ in this case, see Fig. 1.

Very small values of $p$ may lead to unsatisfactory estimates. For instance, choosing $p = 1/365$ (probability of a person having a birthday on this very day) renders $P(\hat{N} = 0) = 0.9973^N$, which is 0.87, 0.76 and 0.58 for $N = 50$, 100 and 200. Furthermore, the next largest value that $\hat{N}$ can attain is $1/p = 365$, which means that values of $N \approx 180$ can dramatically be under- or overestimated. On the other hand, very large values of $p$ may cause a large number of persons to be counted (take for instance the number of people less than 50 years old: Then presumably $X = N - 1$!). So there is a trade-off between accuracy and effort, which is a typical feature of point estimators.

In the following we consider the criteria B: *Number $X_B$ of people having a birthday this or last week* and L: *Number $X_L$ of persons being left-handed.* Setting

$$\hat{N}_\mathrm{B} := \frac{X_\mathrm{B}}{2/52} = 26X_\mathrm{B} \quad \text{and} \quad \hat{N}_\mathrm{L} := \frac{X_\mathrm{L}}{1/10} = 10X_\mathrm{L}\,,$$

we have

$$\mathrm{Var}[N_\mathrm{B}] = 25N \quad \text{and} \quad \mathrm{Var}[N_\mathrm{L}] = 9N\,. \tag{3}$$

Although both estimators are unbiased $\hat{N}_\mathrm{L}$ is the preferred one because it has a smaller variance. Figure 2 illustrates this by means of theoretical quantile-boxplots of $\hat{N}_\mathrm{B}$ and $\hat{N}_\mathrm{L}$ for various values of $N$, see D. Trenkler [4]. In a nutshell, a theoretical quantile-boxplot displays the lower and upper quartiles and the median by a box and the extreme quantiles (1%, 5%, 95%, 99%) by points. The 10%- and the 90%-

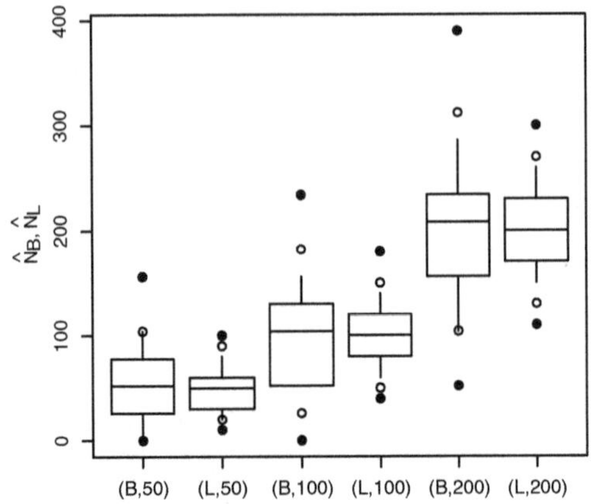


**Fig. 2** Theoretical quantile-boxplots of $\hat{N}_B$ and $\hat{N}_L$ for $N = 50$, 100 and 200

quantiles comprise the whiskers of the plot. In this way, one can get a rough impression of a distribution's characteristics such as location, variability and skewness. For the example at hand, the boxes are shifted upwards as $N$ is increasing due to the unbiasedness of the estimators and the boxplots of $\hat{N}_L$ are more compressed than those of $\hat{N}_B$, due to the smaller variances.

Having established the theoretical properties of both estimators, I finally asked those students fulfilling one of these criteria to raise their hands. There were 3 B-students (needless to say that all of them were given a bag of gorp ...) and 6 L-students leading, to $\hat{N}_B = 26 \times 3 = 78$ and $\hat{N}_L = 10 \times 6 = 60$. A discussion followed by asking: *What do these results mean? Do you think they are contradictory?*

The next step was to ask if we can do better than $\hat{N}_B$ or $\hat{N}_L$ by using some kind of combination. For instance, it is straightforward to see that the arithmetic mean $\hat{N}_M := (\hat{N}_B + \hat{N}_L)/2$ is also an unbiased estimator, and assuming independence, it follows that

$$\mathrm{Var}[N_M] = 8.5N\,,$$

which is only a minor improvement compared to (3).

To persue this approach more systematically, let us consider an estimator of the form

$$\hat{N}_0 := \alpha \hat{N}_B + \beta \hat{N}_L$$

and try to find $\alpha$ and $\beta$ such that $\mathrm{E}[\hat{N}_0] = N$ and $\mathrm{Var}[\hat{N}_0]$ attains a minimum. The first property reveals that $\hat{N}_0$ is of the form

$$\hat{N}_0 := \alpha \hat{N}_B + (1 - \alpha)\hat{N}_L\,.$$

Minimizing

$$\mathrm{Var}[\hat{N}_0] = N\left(\alpha^2 \frac{1-p_\mathrm{B}}{p_\mathrm{B}} + (1-\alpha)^2 \frac{1-p_\mathrm{L}}{p_\mathrm{L}}\right)$$

with respect to $\alpha$ leads to

$$\hat{N}_0 = \frac{p_\mathrm{B}(1-p_\mathrm{L})X_\mathrm{B} + p_\mathrm{L}(1-p_\mathrm{B})X_\mathrm{L}}{p_\mathrm{B}(1-p_\mathrm{L}) + p_\mathrm{L}(1-p_\mathrm{B})}$$

with

$$\mathrm{Var}[\hat{N}_0] = \frac{(1-p_\mathrm{B})(1-p_\mathrm{L})N}{p_\mathrm{B}(1-p_\mathrm{L}) + p_\mathrm{L}(1-p_\mathrm{B})} .$$

Setting $p_\mathrm{B} = 1/26$ and $p_\mathrm{L} = 1/10$, one gets

$$\hat{N}_0 = 0.26\hat{N}_\mathrm{B} + 0.74\hat{N}_\mathrm{L} = 0.26 \times 78 + 0.74 \times 60 = 64.68$$

with $\mathrm{Var}[\hat{N}_0] = 6.62$. Compared to $\hat{N}_\mathrm{M}$, this is an intuitive result saying that we should pay more attention to that estimator with the smaller variance.

A fifth estimator is considered. We can assume that criterion B and criterion L are independent. Thus, the probability for a person having property B or L is

$$p_{\mathrm{B}\cup\mathrm{L}} := p_\mathrm{B} + p_\mathrm{L} - p_\mathrm{B} \times p_\mathrm{L} = 1/26 + 1/10 - 1/260 = 0.1346$$

leading to the estimator $\hat{N}_{\mathrm{B}\cup\mathrm{L}} := 7.43 X_{\mathrm{B}\cup\mathrm{L}}$ with

$$\mathrm{Var}[\hat{N}_{\mathrm{B}\cup\mathrm{L}}] = 6.43N .$$

Comparing variances, there are only minor differences between $\hat{N}_{\mathrm{B}\cup\mathrm{L}}$ and $\hat{N}_0$.

Table 1 summarizes the results of this section so far.

Finally, by asking: *Having observed 9 B- or L-persons, what does this tell us about N?*, one can touch on maximum likelihood estimation. Since $\hat{N}_{\mathrm{B}\cup\mathrm{L}} = X_{\mathrm{B}\cup\mathrm{L}}/p_{\mathrm{B}\cup\mathrm{L}}$ and $X_{\mathrm{B}\cup\mathrm{L}}$ follows a binomial distribution, we have

$$P\left(\hat{N}_{\mathrm{B}\cup\mathrm{L}} = \frac{x}{p}\right) = P(X_{\mathrm{B}\cup\mathrm{L}} = x) = \binom{N}{x} p_{\mathrm{B}\cup\mathrm{L}}^x (1-p_{\mathrm{B}\cup\mathrm{L}})^{N-x} , x = 0, 1, \ldots, N . \quad (4)$$

This equation, together with (2), shows that the distribution of $\hat{N}_{\mathrm{B}\cup\mathrm{L}}$ heavily depends on $p_{\mathrm{B}\cup\mathrm{L}}$ and $N$, and that the magnitude of the unknown $N$ will have an impact on it. In fact, from (4) we can compute the probability that $(\hat{N}_{\mathrm{B}\cup\mathrm{L}} = 9)$ will happen, namely

**Table 1** Results for five estimators of $N$ with $\hat{X}_\mathrm{B} = 3$ and $\hat{X}_\mathrm{L} = 6$

| Estimator | Variance | Estimate |
|---|---|---|
| $\hat{N}_\mathrm{B} = 26X_\mathrm{B}$ | $25N$ | $26 \times 3 = 78$ |
| $\hat{N}_\mathrm{L} = 10X_\mathrm{L}$ | $9N$ | $10 \times 6 = 60$ |
| $\hat{N}_\mathrm{M} = 0.5\hat{N}_\mathrm{B} + 0.5\hat{N}_\mathrm{L}$ | $8.5N$ | $0.5 \times 78 + 0.5 \times 60 = 69$ |
| $\hat{N}_0 = 0.26\hat{N}_\mathrm{B} + 0.74\hat{N}_\mathrm{L}$ | $6.62N$ | $0.26 \times 78 + 0.74 \times 60 = 64.68$ |
| $\hat{N}_{\mathrm{B}\cup\mathrm{L}} = 7.43X_{\mathrm{B}\cup\mathrm{L}}$ | $6.43N$ | $7.43 \times 9 = 66.87$ |

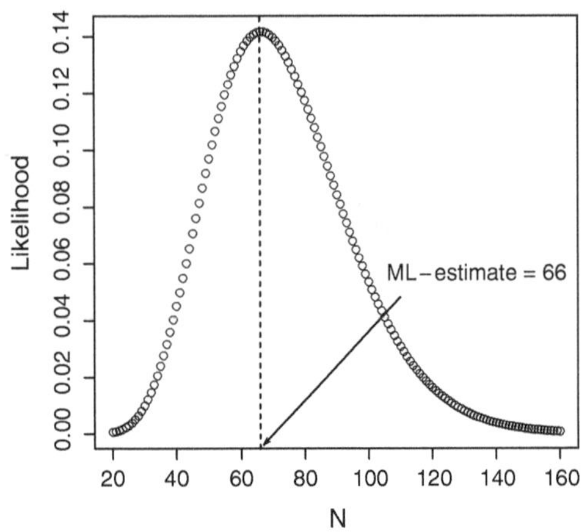


**Fig. 3** Likelihoodfunction for $N$

$$P(\hat{N}_{\mathrm{B}\cup\mathrm{L}} = 9) = \binom{N}{9} 0.1346^9 (1 - 0.1346)^{N-9} .$$

But *we have* observed ($\hat{N}_{\mathrm{B}\cup\mathrm{L}} = 9$), so we can preclude some values of $N$ which are not likely. Thus, it is suggested to have a look at the likelihood function as in Fig. 3.

The maximum of this function is located at $\hat{N}_{\mathrm{ML}} = 66$ lying in the interval $[9/0.1346 - 1, 9/0.1346] = [65.86, 66.86]$. In general it can be shown that $\hat{N}_{\mathrm{ML}}$ is located at the integer(s) in the interval $[X_{\mathrm{B}\cup\mathrm{L}}/p_{\mathrm{B}\cup\mathrm{L}} - 1, X_{\mathrm{B}\cup\mathrm{L}}/p_{\mathrm{B}\cup\mathrm{L}}]$, which is pretty close to our ad hoc estimator $\hat{N}_{\mathrm{B}\cup\mathrm{L}} = X_{\mathrm{B}\cup\mathrm{L}}/p_{\mathrm{B}\cup\mathrm{L}}$.

## 4 Confidence Intervals

To develop confidence intervals, I am very fond of the elegant concept of (approximate) pivotal quantities, see Mood et al. [1]. The example considered here is well-suited to give a first insight since at this point students have gained knowledge about the binomial distribution and the normal distribution. They also have heard about the connection between these two. Thus, I can make use of the following:

- $X_{\mathrm{L}}$ follows a binomial distribution,
- $X_{\mathrm{L}}$ follows an approximate normal distribution,
- $\hat{N}_{\mathrm{L}} = X_{\mathrm{L}}/p_{\mathrm{L}}$ is a linear transformation of $X_{\mathrm{L}}$,
- The standardization of $\hat{N}_{\mathrm{L}}$,

$$Z_{\mathrm{L}} := \frac{\hat{N}_{\mathrm{L}} - N}{\sqrt{9N}}$$

follows an approximate standard normal distribution $\mathcal{N}(0,1)$.

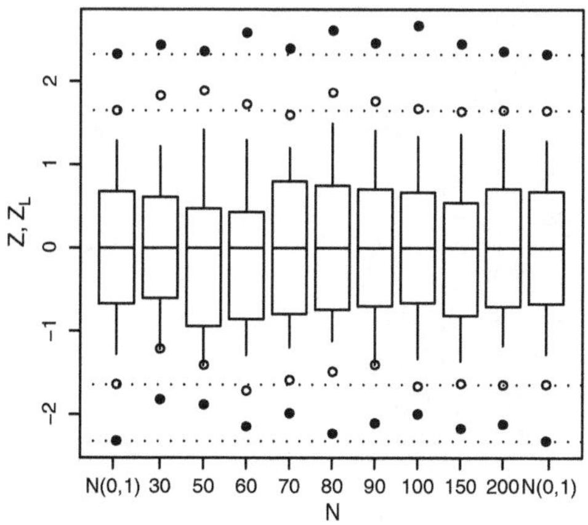


**Fig. 4** Theoretical quantile-boxplots of $\hat{N}_\mathrm{L}$ for various values of $N$. Furthermore the theoretical quantile-boxplot of a standard normal distribution is added to the left and to the right

The correspondence between the exact distribution of $Z_\mathrm{L}$ and the standard normal distribution is illustrated by theoretical quantile-boxplots in Fig. 4.

As can be seen from the plots, the approximations improve with increasing $N$. This also applies to the extreme quantiles, so that we can write:

$$P\left(-z \le \frac{\hat{N}_\mathrm{L} - N}{\sqrt{9N}} \le +z\right) \approx 1 - \alpha$$

for sufficiently large $N$. Here, $z = z_{1-\alpha}$ is the appropriate quantile of the standard normal distribution. Similar considerations apply to $\hat{N}_0$, because it is a linear combination of independent random variables following approximate normal distributions. Thus,

$$P\left(-z \le \frac{\hat{N}_0 - N}{\sqrt{6.62N}} \le +z\right) \approx 1 - \alpha\,.$$

Especially for $z = 2$, we have $P(\psi_1(N) \le 4) \approx 0.95$ and $P(\psi_2(N) \le 4) \approx 0.95$ where

$$\psi_1(N) := \frac{(\hat{N}_\mathrm{L} - N)^2}{9N} \quad \text{and} \quad \psi_2(N) := \frac{(\hat{N}_0 - N)^2}{6.62N}\,.$$

Figure 5 shows the functions $\psi_1$ and $\psi_2$.

The set of all values $N$ such that $\psi_j(N) \le 4$ is an approximate 95 percent confidence interval for $N$. We can find explicit bounds by solving the quadratic equations $\psi_j(N) = 4$ for $N$ yielding

$$\mathrm{CI}_\mathrm{L}(N) = 18 + \hat{N}_\mathrm{L} \mp \sqrt{324 + 36\hat{N}_\mathrm{L}}$$

for $\hat{N}_\mathrm{L}$ and

$$\mathrm{CI}_0(N) = 13.24 + \hat{N}_0 \mp \sqrt{175.3 + 26.48\hat{N}_0}$$

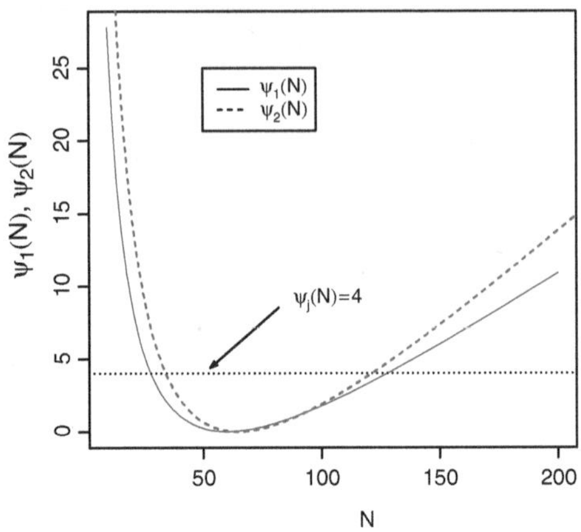


**Fig. 5** Obtaining approximate 95% confidence intervals for $N$. The values of $N$ for which $\psi_j(N) = 4$ define approximate 95%-confidence intervals of $N$

for $\hat{N}_0$. Inserting the values from Table 1 delivers $\mathrm{CI_L}(N) = [28.16, 127.84]$ and $\mathrm{CI_0}(N) = [34.43, 121.41]$.

Which is the preferred interval? Of course, we use that interval with the shorter length, which is $\mathrm{CI_0}(N)$ here. But does this happen in general?

At this point one can mention that a confidence interval's length plays a similar role as the mean square error for estimators, but establishing optimal properties may become a challenging problem. For instance, by solving a quadratic equation, the confidence limits based on an estimator $\hat{N}$ associated with $p$ is given by

$$\mathrm{CI}_p(N) = \frac{2\hat{N}p + (1-p)z^2 \mp \sqrt{(1-p)z^2[4\hat{N}p + (1-p)z^2]}}{2p}$$

Its length is

$$L(p) = \frac{\sqrt{(1-p)z^2[4\hat{N}p + (1-p)z^2]}}{p},$$

which is a monotonically decreasing function in $p$, as is readily found by differentiation. Hence, $\mathrm{CI}_{p_1}(N)$ will be broader than $\mathrm{CI}_{p_2}(N)$ for $p_1 < p_2$. Furthermore, $L(p) \to \infty$ for $p \to 0$ and $L(p) \to 0$ for $p \to 1$, which corresponds to our reasoning in the context of estimating $N$.

# 5 Tests

Someone claims that the number of persons in this room is larger than 150. Given $\hat{N}_0 = 64.68$, how can we test this hypothesis? We have argued that $\hat{N}_0$ follows an approximate normal distribution $\mathcal{N}(N, 6.52N)$. Figure 6 shows theoretical quantile-

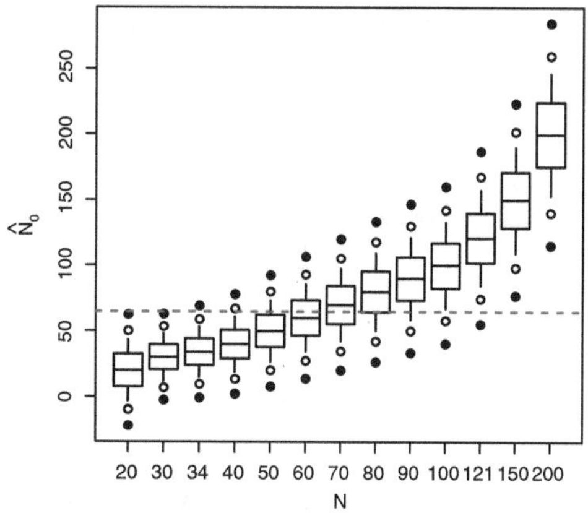


**Fig. 6** Theoretical quantile-boxplots of $\mathcal{N}(N, 6.62N)$ distributions

boxplots of $\mathcal{N}(N, 6.52N)$ distributions for various values of $N$. Furthermore, the location of $\hat{N}_0 = 64.68$ is displayed by a dashed line.

This line divides the set of $N$ values into two parts: Those which are in accordance with the observed value of $\hat{N}_0$, and those which are not. For instance, it is not very likely that $N = 150$ persons are present, because the corresponding boxplot lies well below the line. Similarly, $N = 20$ is not likely. In fact, the transition from acceptance to rejection is blurred and here is the point where the concepts of significance levels and p-values can be addressed. Furthermore, confidence intervals can be explained as the set of all plausible values of $N$. The theoretical quantile boxplots corresponding to $N = 34$ and $N = 121$ bracket the 95% confidence interval for $N$.

## 6 Conclusions

Students very much dislike abstract teaching. So explaining concepts using problems which are easy to grasp are quite welcome. Estimating the number of persons in a room is one such possibility. It is a proposal to alleviate the transition from probability models to statistical inference. Topics like independence of events, binomial distribution, normal distribution, expectation and variance are exploited to pave the way for several goals such as estimation, construction of confidence intervals and tests. At the same time, the derivations are relatively simple and should meet the students' capabilities acquired so far.

## References

[1] Mood, A.M., Graybill, F.A., Boes, D.C.: Introduction to the Theory of Statistics, 3rd edn. McGraw-Hill, Singapore (1974)

[2] Parzen, E.: Modern Probability Theory and Its Applications. Wiley, New York (1960)
[3] Székely, G.J.: Paradoxa, Klassische und neue Überraschungen aus Warscheinlichkeitsrechnung und Mathematischer Statistik. Harry Deutsch, Frankfurt am Main (1990)
[4] Trenkler, D.: Quantile-boxplots. Commun. Stat. Simul. Comput. **31**, 1–12 (2002)

## Epilogue

My brother Götz has not only been an excellent researcher as witnessed by the other authors of this Festschrift but also a terrific teacher. It is due to him that I came to love mathematics in general and statistics in particular. I owe him a lot.

*All the best, big brother.*

# Selected Publications of Götz Trenkler

## Monographs

1. Nichtparametrische Statistische Methoden (with H. Büning). De Gruyter, Berlin (1978; 2nd edn. 1994)
2. Biased Estimators in the Linear Regression Model. Mathematical Systems in Economics Vol. 58. Oelgeschlager, Gunn & Hain, Meisenheim and Cambridge, MA. (1981)
3. Data Analysis and Statistical Inference (with S. Schach). Festschrift in Honour of Friedhelm Eicker. Verlag Josef Eul, Bergisch Gladbach (1992)
4. Lexikon der populären Irrtümer (with W. Krämer). Eichborn Verlag, Frankfurt (1996)
5. Mathematik für Ökonomen im Hauptstudium (with H. Büning, P. Naeve and K.-H. Waldmann). Oldenbourg, München (2000)
6. Das neue Lexikon der populären Irrtümer (with W. Krämer and D. Krämer). Eichborn Verlag, Frankfurt (2000)
7. Mathematical Statistics with Applications in Biometry (with J. Kunert). A Festschrift in Honour of Siegfried Schach. Verlag Josef Eul, Bergisch Gladbach (2001)
8. Einführung in die Moderne Matrix-Algebra (with K. Schmidt). Springer, Berlin (1998; 2nd edn. 2006)

## Research Papers

1. On a New Method of Testing Statistical Hypotheses. Annals of the Institute of Statistical Mathematics A **28**, 371–384 (1976)
2. An Iteration Estimator for the Linear Model. COMPSTAT 1978, pp. 126–131. Physica, Wien (1978)
3. A New Class of Biased Estimators in the Linear Model. Allgemeines Statistisches Archiv **63**, 76–77 (1979)

4. Generalized Mean Squared Error Comparisons of Biased Regression Estimators. Communications in Statistics A **9**, 1247–1259 (1980)
5. Experimental Coronary Air Embolism. Assessment of Time Course of Myocardial Ischemia and the Protective Effect of Cardiopulmonary Bypass (with T. Stegmann et al.). The Thoracic and Cardiovascular Surgeon **28**, 141–149 (1980)
6. A Comparison of the Least-Squares-Estimator with Biased Alternatives (with D. Trenkler). Operations Research Proceedings 1980, pp. 218–227. Springer, Berlin (1981)
7. On a Generalized Iteration Estimator. In: Büning, H., Naeve, P. (eds.) Computational Statistics, pp. 315–335. De Gruyter, Berlin (1981)
8. Total Anomalous Pulmonary Venous Connection: Surgical Treatment in 35 Infants (with T. Stegmann et al.). The Thoracic and Cardio-vascular Surgeon **29**, 299–302 (1981)
9. Estimable Functions and Reduction of Mean Square Error (with D. Trenkler). In: Henn, R. et al. (eds.) Methods of Operations Research 44, pp. 225–234. Oelgeschlager, Gunn & Hain, Königstein and Cambridge, MA. (1981)
10. Partitions, Sufficiency and Undominated Families of Probability Measures. Annals of the Institute of Statistical Mathematics **34**, 151–160 (1982)
11. Therapeutic Measures in Experimental Coronary Air Embolism (with T. Stegmann, W. Daniel and H.G. Borst). Langenbecks Archiv für Chirurgie **357**, 213–214 (1982)
12. Mallows' $C_p$ and Optimal Ridge Constants. In: Henn, R. et al. (eds.) Methods of Operations Research 46, pp. 157–166. Oelgeschlager, Gunn & Hain, Königstein and Cambridge, MA. (1983)
13. Biased Linear Estimators and Estimable Functions. Scandinavian Journal of Statistics **10**, 53–55 (1983)
14. On the Generalized Variance of Biased Estimators in the Linear Model. Allgemeines Statistisches Archiv **67**, 199–201 (1983)
15. A Note on Superiority Comparisons of Homogeneous Linear Estimators (with D. Trenkler). Communications in Statistics A **12**, 799–808 (1983)
16. Minimum Mean Square Error Ridge Estimation (with D. Trenkler). Sankhya A **46**, 94–101 (1984)
17. On the Euclidean Distance between Biased Estimators (with D. Trenkler). Communications in Statistics A **13**, 273–284 (1984)
18. A Simulation Study Comparing Some Biased Estimators in the Linear Model (with D. Trenkler). Computational Statistics Quarterly **1**, 45–60 (1984)
19. Some Further Remarks on Multicollinearity and the Minimax Conditions of the Bock Stein-like Estimator. Econometrica **52**, 1067–1069 (1984)
20. On the Performance of Biased Estimators in the Linear Regression Model with Correlated Errors. Journal of Econometrics **25**, 179–190 (1984)
21. On Heterogeneous Versions of the Best Linear and the Ridge Estimator (with P. Stahlecker). Proceedings of the First International Tampere Seminar on Linear Models (1983), 301–322 (1985)

22. Mean Square Error Matrix Comparisons of Estimators in Linear Regression. Communications in Statistics A **14**, 2495–2509 (1985)
23. Updating the Ridge Estimator when Additional Explanatory Variables or Observations are Available (with B. Schipp and D. Trenkler). Computational Statistics Quarterly **2**, 135–141 (1985)
24. Does Weak Consistency of an Estimator Imply Asymptotic Unbiasedness? Statistica Neerlandica **39**, 241–242 (1985)
25. Linear Constraints and the Efficiency of Combined Forecasts (with E.P. Liski). Journal of Forecasting **5**, 197–202 (1986)
26. Mean Square Error Matrix Comparisons between Mixed Estimators (with E. Freund). Statistica **46**, 493–501 (1986)
27. Mean Square Error Matrix Comparisons among Restricted Least Squares Estimators. Sankhya A **49**, 96–104 (1987)
28. Partial Minimax Estimation in Regression Analysis (with F. Hering and P. Stahlecker). Statistica Neerlandica **41**, 111–128 (1987)
29. Quasi Minimax Estimation in the Linear Regression Model (with P. Stahlecker). Statistics **18**, 219–226 (1987)
30. Minimax Estimation with Additional Linear Restrictions: A Simulation Study (with B. Schipp and P. Stahlecker). Communications in Statistics B **17**, 393–406 (1987)
31. Iterative Improvements of a Partial Minimax Estimator in Regression Analysis (with F. Hering, B. Schipp and P. Stahlecker). Proceedings of the Second International Tampere Conference in Statistics, pp. 679–690 (1987)
32. Some Remarks on a Ridge-Type-Estimator and Good Prior Means. Communications in Statistics A **17**, 4251–4256 (1988)
33. Full and Partial Minimax Estimation in Regression Analysis with Additional Linear Constraints (with P. Stahlecker). Linear Algebra and its Applications **111**, 279–292 (1988)
34. Mean Square Error Matrix Superiority of the Mixed Regression Estimator under Misspecification (with P. Wijekoon). Statistica **49**, 65–71 (1989)
35. Mean Square Error Matrix Superiority of Estimators under Linear Restrictions and Misspecification (with P. Wijekoon). Economics Letters **30**, 141–149 (1989)
36. Mean Square Error Matrix Improvements, Mean Square Error Matrix Improvements and Admissibility of Linear Estimators (with J.K. Baksalary and E.P. Liski). Journal of Statistical Planning and Inference **23**, 313–325 (1989)
37. A Curious Result on Least Squares Estimation in Linear Regression. International Journal of Mathematical Education in Science and Technology **20**, 475–476 (1989)
38. A Note on Generalized Ridge Estimators (with J.K. Baksalary and P. Pordzik). Communications in Statistics A **19**, 2871–2877 (1990)
39. Mean Square Error Matrix Comparisons of Optimal and Classical Predictors and Estimators in Linear Regression (with H. Toutenburg). Computational Statistics and Data Analysis **10**, 297–305 (1990)

40. Testing the Stability of Regression Coefficients using Recursive Residuals (with P. Michels). Australian Journal of Statistics **32**, 293–312 (1990)
41. Mean Squared Error Matrix Comparisons between Biased Estimators – An Overview of Recent Results (with H. Toutenburg). Statistical Papers **31**, 165–179 (1990)
42. Matrix Mean Square Error Comparisons Based on a Certain Covariance Structure (with G. Ihorst). Communications in Statistics B **19**, 1035–1043 (1990)
43. Linear and Ellipsoidal Restrictions in Linear Regression (with P. Stahlecker). Statistics **22**, 163–176 (1991)
44. Mean Square Error Matrix Comparisons between Biased Restricted Least Squares Estimators. Sankhya A **53**, 309–319 (1991)
45. Covariance Adjustment in Biased Estimation (with J.K. Baksalary). Computational Statistics & Data Analysis **12**, 221–230 (1991)
46. Nonnegative and Positive Definiteness of Matrices Modified by two Matrices of Rank one (with J.K. Baksalary). Linear Algebra and its Applications **151**, 169–184 (1991)
47. Computation of Expectation and Variance for the Binomial Distribution (with H. Knautz, in German). Praxis der Mathematik **33**, 24 (1991)
48. Dürer, Moore-Penrose, Drazin ... (with H. Knautz, in German). Praxis der Mathematik **33**, 183–184 (1991)
49. A Matrix Equation (with B. Schipp). Econometric Theory **7**, 422–423 (1991)
50. On the MSE-Matrix Comparisons of Vector-Valued Estimators. Statistica Neerlandica **45**, 343 (1991)
51. Proxy Variables and Mean Square Error Dominance in Linear Regression (with H. Toutenburg). Journal of Quantitative Economics **8**, 443–442 (1992)
52. Pre-test Estimation in the Linear Regression Model with Competing Linear Constraints (with H. Hessenius). Statistica **52**, 17–31 (1992)
53. Some Further Results on Hermitian Matrix Inequalities (with B. Schipp and J.K. Baksalary). Linear Algebra and its Applications **160**, 119–129 (1992)
54. Optimal Estimation Methods under Weakened Linear Restrictions in Regression (with E.P. Liski and H. Toutenburg). Computational Statistics & Data Analysis **14**, 527–536 (1992)
55. Pretest Procedures and Forecasting in the Regression Model under Restrictions (with H. Toutenburg). Journal of Statistical Planning and Inference **30**, 249–256 (1992)
56. Nonlinear Unbiased Estimation in Linear Models (with S. Gnot, H. Knautz and R. Zmyslony). Statistics **23**, 5–16 (1992)
57. A Note on the Expected Value of a Moore-Penrose Inverse (with H. Knautz). International Journal of Mathematical Education in Science and Technology **23**, 155 (1992)
58. On Convexity of Certain Classes of Estimators. In: Trenkler, G., Schach, S. (eds.) Data Analysis and Statistical Inference, Festschrift in Honour of Friedhelm Eicker, pp. 83–90. Verlag Josef Eul, Bergisch Gladbach (1992)
59. Dropping Variables Versus Use of Proxy Variables in Linear Regression (with P. Stahlecker). Journal of Statistical Planning and Inference **50**, 65–75 (1993)

60. Minimum Mean Square Error Estimation in Linear Regression (with E.P. Liski and H. Toutenburg). Journal of Statistical Planning and Inference **37**, 203–214 (1993)
61. Some Further Results on the Use of Proxy Variables in Prediction (with P. Stahlecker). The Review of Economics and Statistics **75**, 707–711 (1993)
62. MSE-Improvement of the Least Squares Estimator by Dropping Variables (with E.P. Liski). Metrika **40**, 263–269 (1993)
63. A Note on Comparing Stochastically Restricted Estimators in a Regression Model. Biometrical Journal **35**, 125–128 (1993)
64. Minimax Estimation in Linear Regression with Singular Covariance Structure and Convex Polyhedral Constraints (with P. Stahlecker). Journal of Statistical Planning and Inference **36**, 185–196 (1993)
65. Leverage and Cochrane-Orcutt Estimation in Linear Regression (with D. Stemann). Communications in Statistics A **22**, 1315–1333 (1993)
66. MSE-Matrix Superiority of the Mixed over the Least Squares Estimator in the Presence of Outliers (with G. Ihorst). Communications in Statistics A **22**, 1865–1877 (1993)
67. On the Correlation between $\bar{X}$ and $S^2$ (with H. Knautz). Statistics & Probability Letters **16**, 235–237 (1993)
68. A Note on the Correlation between $S^2$ and the Least Squares Estimator in the Linear Regression Model (with H. Knautz). Statistical Papers **34**, 237–246 (1993). Corrigendum: Statistical Papers **35**, 42 (1994)
69. Characterizations of Oblique and Orthogonal Projectors. In: Calinski, T., Kala, R. (eds.) Proceedings of the International Conference on Linear Statistical Inference, LINSTAT 93, pp. 255–270. Kluwer, Dordrecht (1994)
70. Using Nonnegative Minimum Biased Quadratic Estimation for Variable Selection in the Linear Regression Model (with S. Gnot and H. Knautz). In: Calinski, T., Kala, R. (eds.) Proceedings of the International Conference on Linear Statistical Inference, LINSTAT 93, pp. 65–71. Kluwer, Dordrecht (1994)
71. Admissible Nonnegative Invariant Quadratic Estimation in Linear Models with Two Variance Components (with S. Gnot and D. Stemann). In: Calinski, T., Kala, R. (eds.) Proceedings of the International Conference on Linear Statistical Inference, LINSTAT 93, pp. 129–137. Kluwer, Dordrecht (1994)
72. Singular Magic Squares. International Journal of Mathematical Education in Science and Technology **25**, 595–597 (1994)
73. Pre-test Estimation in the Linear Regression Model with Competing Restrictions (with P. Pordzik). Linear Algebra and its Applications **210**, 123–137 (1994)
74. On the Information Loss of the Cochrane-Orcutt-Estimation Procedure (with D. Stemann). Journal of Quantitative Economics **10**, 227–234 (1994)
75. On the Moore-Penrose Inverse of a Completely Symmetric Matrix. Journal of Statistical Computation and Simulation **49**, 230–231 (1994)
76. A Matrix Formulation on How Deviant an Observation Can Be. The American Statistician **48**, 60–61 (1994)

77. Assessing Coverage-Probabilities for Approximate Minimax Estimators with Respect to Interval Restrictions (with B. Schipp and P. Stahlecker). Journal of Statistical Computation and Simulation **50**, 59–74 (1994)
78. Improved Estimation by Weak Covariance Adjustment Technique (with G. Ihorst). Discussiones Mathematicae **15**, 189–201 (1995)
79. Mean Square Error Matrix Superiority of Empirical Bayes Estimators under Misspecification. Test **4**, 187–205 (1995)
80. Moore-Penrose Inverse of a Matrix Product with Normal Matrix. Econometric Theory **11**, 653–654 (1995)
81. Some Bounds for Bias and Variance of $S^2$ under Dependence (with H. Knautz). Scandinavian Journal of Statistics **22**, 121–128 (1995)
82. Nonnegative Minimum Biased Quadratic Estimation in the Linear Regression Model (with S. Gnot and R. Zmyslony). Journal of Multivariate Analysis **54**, 113–125 (1995)
83. A New Characterization of Orthogonal Projectors (with S.-O. Troschke). Elemente der Mathematik **50**, 171 (1995)
84. The Common Mean, Non-Negative Definite Matrices and APL (with D. Trenkler). Vector **12**, 107–112 (1995)
85. An Objective Stability Criterion for Selecting the Biasing Parameter from the Ridge Trace (with D. Trenkler). Industrial Mathematics **45**, 93–104 (1995)
86. On the Singularity of the Sample Covariance Matrix. Journal of Statistical Computation and Simulation **52**, 172–173 (1995)
87. Estimation of Parameters in the Linear Regression Model with Equicorrelated Errors (with S. Gnot and D. Stemann). International Conference on Statistical Methods and Statistical Computing for Quality and Productivity Improvement. Proceedings Volume, pp. 624–631 (1995)
88. Pre-Test Estimation in the Linear Regression Model under Stochastic Restrictions (with P. Wijekoon). Ceylon Journal of Science: Physical Sciences **2**, 57–64 (1995)
89. The Bayes Estimator in a Misspecified Linear Regression Model. Test **5**, 113–123 (1996)
90. Nonnegative Quadratic Estimation of the Mean Squared Errors of Minimax Estimators in the Linear Regression Model (with S. Gnot). Acta Applicandae Mathematicae **43**, 71–80 (1996)
91. Minimax Adjustment Technique in a Parameter Restricted Linear Model (with P. Stahlecker and H. Knautz). Acta Applicandae Mathematicae **43**, 139–144 (1996)
92. Hypothesis Testing Using Affine Linear Estimators (with P. Stahlecker and H. Knautz). Acta Applicandae Mathematicae **43**, 153–158 (1996)
93. Prediction and the Choice Between Two Restricted Regression Models (with J. Groß). Journal of Quantitative Economics **12**, 125–131 (1996)
94. On the Least Squared Distance Between Affine Subspaces (with J. Groß). Linear Algebra and its Applications **237/238**, 269–276 (1996)
95. Records Tests for Trend in Location (with J. Diersen). Statistics **28**, 1–12 (1996)

96. A General Investigation of the Mean Square Error Matrix Superiority in Linear Regression (with G. Ihorst). Statistica **56**, 15–25 (1996)
97. MSE Comparisons of Restricted Least Squares Estimators in Linear Regression Model – Revisited (with P. Pordzik). Sankhya B **58**, 352–359 (1996)
98. Wozu braucht man Ökonometrie? Prisma. Wochenmagazin zur Zeitung **40**, 44 (1996)
99. On the Equality of OLSE and BLUE in a Partitioned Linear Model (with S. Puntanen and J. Groß). In: Manly, B.J.F. (ed.) The Proceedings of the A.C. Aitken Centenary Conference, pp. 143–152. University of Otago (1996)
100. When do Linear Transforms of Ordinary Least Squares and Gauss-Markov Estimator coincide? (with J. Groß) Sankhya A **59**, 175–178 (1997)
101. On Matrices Whose Moore-Penrose-Inverse is a Scalar Multiple of its Transpose (with S.-O. Troschke). The Mathematical Gazette **81**, 470–472 (1997)
102. Generalized and Hypergeneralized Projectors (with E. Liski and J. Groß). Linear Algebra and its Applications **264**, 463–474 (1997)
103. On the Equality of Usual and Amemiya' s Partially Generalized Least Squares Estimator (with J. Groß). Communications in Statistics – Theory and Methods **26**, 2075–2086 (1997)
104. Restrictions and Projections in Linear Regression (with J. Groß). International Journal of Mathematical Education in Science and Technology **28**, 465–468 (1997)
105. Mean Square Error Matrix Properties of Bayes Estimation for Incorrect Prior Information under Misspecification (with L. Wei and S. Tamaschke). Journal of the Italian Statistical Society **6**, 273–284 (1997)
106. Four Square Roots of the Vector Cross Product. The Mathematical Gazette **82**, 100–102 (1998)
107. On the Product of Oblique Projectors (with J. Groß). Linear and Multilinear Algebra **41**, 247–260, (1998)
108. On the Equality of Linear Statistics in General Gauss-Markov Model (with J. Groß). In: Mukherjee, S.P., Basu, S.K., Sinha, B.K. (eds.) Frontiers in Probability and Statistics, pp. 189–194. Narosa, New Delhi (1998)
109. Another Look at Cubes and Inverses of Magic Squares. The Mathematical Gazette **82**, 288–289 (1998)
110. Vector Equations and Their Solutions. International Journal of Mathematical Education in Science and Technology **29**, 455–459 (1998)
111. Necessary and Sufficient Conditions for Superiority of Misspecified Restricted Least Squares Regression Estimator (with E. Liski and J. Groß). Journal of Statistical Planning and Inference **71**, 109–116 (1998)
112. Using First Differences as a Device against Multicollinearity (with H. Toutenburg). In: Galata, R., Küchenhoff, H. (eds.) Econometrics in Theory and Practice, pp. 131–135. Physica, Heidelberg (1998)
113. Two Results on the Efficiency of the Almon Lag Technique. Journal of Quantitative Economics **14**, 17–22 (1998)
114. On the Exchangeability of Transformations and the Arithmetic Mean. Journal of Statistical Computation and Simulation **61**, 305–307 (1998)

115. On the Eigenvalues of $3 \times 3$ Magic Squares. International Journal of Mathematical Education in Science and Technology **30**, 307–308 (1999)
116. The Vector Cross Product in $C^3$ (with J. Groß and S.-O. Troschke). International Journal of Mathematical Education in Science and Technology **30**, 549–555 (1999)
117. On Semi-Orthogonality and a Special Class of Matrices (with S.-O. Troschke and J. Groß). Linear Algebra and its Applications **289**, 169–182 (1999)
118. On Maximum Likelihood Estimators in Certain Multivariate Normal Mixed Models with Two Variance Components (with S. Gnot, D. Stemann and A. Urbánska-Motyka). Tatra Mountains Mathematical Publications **17**, 1–9 (1999)
119. A Remark on Prediction Problems in Regression Analysis. Journal of Statistical Research **33**, 67–70 (1999)
120. Nonsingularity of the Difference of Two Oblique Projectors (with J. Groß). SIAM Journal on Matrix Analysis and Applications **21**, 390–395 (1999)
121. On Properties of $3 \times 3$ Semi-Magic Squares (with S.-O. Troschke and J. Groß). International Journal of Mathematical Education in Science and Technology **30**, 861–865 (1999)
122. Die einfache dritte Aufgabe. Elemente der Mathematik **54**, 38 (1999)
123. Some Further Results on the Efficiency of the Cochrane-Orcutt Estimator (with D. Stemann). Journal of Statistical Planning and Inference **88**, 205–214 (2000)
124. Quaternions. Further Contributions to a Matrix Oriented Approach (with S.-O. Troschke and J. Groß). Linear Algebra and its Applications **326**, 205–213 (2000)
125. On a Generalization of the Covariance Matrix of the Multinomial Distribution. In: Heijmans, R.D.H., Pollock, D.S.G., Satorra, A. (eds.) A Festschrift for Heinz Neudecker, pp. 67–73. Kluwer, Dordrecht (2000)
126. The Equality Between Linear Transforms of Ordinary Least Squares and Best Linear Unbiased Estimator (with H.J. Werner and J. Groß). Sankhya A **63**, 118–127 (2001)
127. The Moore-Penrose Inverse of a Semi-magic Square is Semi-magic (with K. Schmidt). International Journal of Mathematical Education in Science and Technology **32**, 624–629 (2001)
128. The Vector Cross Product from an Algebraic Point of View. Discussiones Mathematicae, General Algebra and Applications **21**, 67–82 (2001)
129. A Comparison of the Ridge and the Iteration Estimator. In: Friedmann, R., Knüppel, L., Lütkepohl, H. (eds.) Econometric Studies. A Festschrift in Honour of Joachim Frohn, pp. 73–87. Lit Verlag, München (2001)
130. On the Efficiency of the Cochrane-Orcutt Estimator in the Serially correlated error components regression model for panel data (with S.H. Song). Communications in Statistics – Simulation and Computation **30**, 195–207 (2001)
131. Weighted records tests for splitted series of observation (with J. Diersen). In: Kunert, J., Trenkler, G. (eds.) Mathematical statistics with applications in

biometry. A Festschrift in Honour of Siegfried Schach, pp. 163–178. Verlag Josef Eul, Bergisch-Gladbach, Köln (2001)

132. Magic Squares, Melancholy and the Moore-Penrose inverse (with D. Trenkler). Image **27**, 3–10 (2001)
133. On efficiently estimable parametric functionals in the general linear model with nuisance parameters (with P. Pordzik). In: Ullah, A., Wang, A.T.K., Chaturvedi, A. (eds.) Handbook of Applied Econometrics and Statistical Inference, pp. 45–52. Dekker, New York (2002)
134. Maximum likelihood estimation in mixed normal models with two variance components (with S. Gnot, D. Stemann and A. Urbǻnska-Motyka). Statistics **36**, 283–302 (2002)
135. The Moore-Penrose inverse and the vector product. International Journal of Mathematical Education in Science and Technology **33**, 431–436 (2002)
136. Third and fourth moment matrices of vec($X'$) in multivariate analysis (with H. Neudecker). Linear Algebra and its Applications **354**, 223–229 (2002)
137. Complementing remarks on a matrix extension of a Cauchy-Schwarz inequality (with J. Groß). Journal of Statistical Computation and Simulation **72**, 26–29 (2002)
138. On the square root of $aa' + bb'$ (with D. Trenkler). The College Mathematics Journal **34**, 39–41 (2003)
139. A revisitation of formulae for the Moore-Penrose inverse of modified matrices (with J.K. Baksalary and O.M. Baksalary). Linear Algebra and its Applications **372**, 207–224 (2003)
140. Estimation of the Cross-product of Two Mean Vectors (with H. Neudecker and R. Zmyslony). International Journal of Mathematical Education in Science and Technology **34**, 928–935 (2003)
141. Matrices Which Take a Given Vector into a Given Vector – Revisited. The American Mathematical Monthly **111**, 50–52 (2004)
142. An Extension of Lagrange's Identity to Matrices. International Journal of Mathematical Education in Science and Technology **35**, 245–249 (2004)
143. Most-perfect Pandiagonal Magic Squares and Their Moore-Penrose Inverse. International Journal of Mathematical Education Science and Technology **35**, 697–701 (2004)
144. A Multivariate Version of Samuelson's Inequality (with S. Puntanen). Linear Algebra and its Applications **410**, 143–149 (2005)
145. On Generalized Quadratic Matrices (with R.W. Farebrother). Linear Algebra and its Applications **410**, 244–253 (2005)
146. Multivariate Data, the Arithmetic Mean and Exchangeability of Transformation (with E. Frauendorf and H. Neudecker). Linear Algebra and its Applications **410**, 87–95 (2005)
147. Estimation of the Kronecker and Inner Product of Two Mean Vectors in Multivariate Analysis (with H. Neudecker). Discussiones Mathematicae, Probability and Statistics **25**, 207–215 (2005)

148. The Sherman-Morrison Formula and Eigenvalues of a Special Bordered Matrix (with D. Trenkler). Acta Mathematica Universitatis Comenianae **74**, 255–258 (2005)
149. On Oblique and Orthogonal Projectors. In: Brown, P., Liu, S., Sharma, D. (eds.) Contributions to Probability and Statistics: Applications and Challenges. Proceedings of the International Statistics Workshop, pp. 178–191. World Scientific, Singapore (2006)
150. On the Approximate Variance of a Nonlinear Function of Random Variables. In: Brown, P., Liu, S., Sharma, D. (eds.) Contributions to Probability and Statistics: Applications and Challenges. Proceedings of the International Statistics Workshop, pp. 172–177. World Scientific, Singapore (2006)
151. Haramard, Khatri-Rao, Kronecker and Other Matrix Products (with S. Liu). International Journal of Information and Systems Science **1**, 1–18 (2007)
152. On the Product of Rotations (with D. Trenkler). International Journal of Mathematical Education in Science and Technology **39**, 94–104 (2007)
153. Image Philatelic Corner reprise (with G.P.H. Styan). Image **38**, 9–12 (2007)
154. On Singular $3 \times 3$ Semi-diagonal Latin Squares (with D. Trenkler). The Mathematical Gazette **91**, 126–128 (2007)
155. A Philatelic Excursion with Jeff Hunter in Probability and Matrix Theory (with G.P.H. Styan). Journal of Applied Mathematics and Decision Sciences, Volume 2007, Article ID 13749, 1–10 (2007)
156. On the Product of Rotations (with D. Trenkler). International Journal of Mathematical Education in Science and Technology **39**, 94–104 (2008)
157. A Remark on Proofs of the Vector Triple Product law. International Journal of Mathematical Education in Science and Technology **39**, 423 (2008)
158. Characterizations of EP, Normal, and Hermitian Matrices (with O.M. Baksalary). Linear and Multilinear Algebra **56**, 299–304 (2008)
159. Hadamard, Khatri-Rao, Kronecker and Other Matrix Products (with S. Liu). International Journal of Information and Systems Science **4**, 160–177 (2008)

Götz Trenkler is also an enthusiastic teacher. This is well documented by his numerous contributions to the Problems and Solutions Section of the following journals:

- American Mathematics Monthly
- Econometric Theory
- Elemente der Mathematik
- Image
- Mathematic Magazine
- Statistical Papers
- Statistica Neerlandica
- The College Mathematics Journal
- The Mathematical Gazette.

FSC
www.fsc.org
MIX
Papier | Fördert gute Waldnutzung
FSC® C083411

Kolibri 360 GmbH
Ferdinand-Jühlke-Straße 7
99095 Erfurt, Deutschland
produktsicherheit@kolibri360.de